Probiotics as Antifungals

M. Y. Sreenivasa • Deepthi B V
Charith Raj Adkar Purushothama
Premila N Achar
Editors

Probiotics as Antifungals

Application in Plant, Animal, Human Health

Springer

Editors
M. Y. Sreenivasa
Department of Studies in Microbiology
University of Mysore
Mysuru, Karnataka, India

Deepthi B V
Department of Studies in Microbiology
University of Mysore
Mysuru, Karnataka, India

Charith Raj Adkar Purushothama
Department of Biochemistry and
Functional Genomics
Université de Sherbrooke
Sherbrooke, QC, Canada

Premila N Achar
Department of Molecular and Cellular Biology
Kennesaw State University
Kennesaw, GA, USA

ISBN 978-981-92-0001-6 ISBN 978-981-92-0002-3 (eBook)
https://doi.org/10.1007/978-981-92-0002-3

This Springer imprint is published by the registered company Springer Nature Singapore Pte Ltd.
The registered company address is: 152 Beach Road, #21-01/04 Gateway East, Singapore 189721, Singapore

Introduction

Fungal diseases and mycotoxin contamination pose a significant threat to human health, animal welfare, food safety, agricultural productivity, and environmental sustainability. Pathogenic fungi such as *Fusarium*, *Aspergillus*, *Alternaria*, *Penicillium*, *Botrytis*, *Colletotrichum*, and a wide range of other molds are responsible for diseases including smuts, wilts, rots, mildews, and anthracnose, thereby severely impacting agricultural productivity and livestock health. Many of these fungi produce toxic secondary metabolites known as mycotoxins such as aflatoxins, fumonisins, trichothecenes, ochratoxins, and deoxynivalenol that are carcinogenic, neurotoxic, nephrotoxic, and teratogenic, causing serious health disorders in animals and humans. Furthermore, fungal pathogens such as *Candida*, *Cryptococcus*, *Fusarium*, *Aspergillus*, *Mucor*, and other dermatophytes are capable of causing a wide spectrum of infections in humans, ranging from superficial and cutaneous diseases to invasive and life-threatening systemic mycoses, particularly in immunocompromised individuals.

The management strategies such as physical and chemical methods used to control fungal infections and treat fungal colonization and intoxication in hosts have become increasingly ineffective. Further, the widespread emergence of resistance to antifungal drugs and chemical fungicides, coupled with drug-associated toxicity and ecological concerns, has underscored the urgent need for alternative, biologically based, environmentally sustainable, and effective antifungal strategies. In this context, probiotics and other beneficial microorganisms have emerged as promising biological agents with multifaceted antifungal capabilities.

Probiotics have long been associated with gut health and immune modulation in host organisms, and over the past decade, have been increasingly recognized for their ability to antagonize fungal pathogens through diverse mechanisms. These include competitive exclusion of pathogens, production of antifungal bioactive metabolites, disruption of fungal biofilms, modulation of host immune responses, and detoxification of mycotoxins. In plant systems, probiotics also exhibit indirect antifungal activity by triggering plant defense mechanisms such as induced systemic resistance, enhanced antioxidant responses, and modulation of phytohormone signaling pathways.

Mycoses and mycotoxin-related illnesses in livestock, poultry, aquaculture, and other animals result in poor feed efficiency, feed refusal, retardation in growth and performance, reproductive toxicity, immunosuppression, oxidative stress, genotoxic and carcinogenic effects, organ damage, and increased mortality. Probiotic interventions play a critical role in alleviating these effects through multiple mechanisms, including competitive exclusion, production of antifungal compounds such as short-chain fatty acids, organic acids, and bacteriocins, immunomodulation resulting in the enhancement of gut health, and improvement of growth performance, feed conversion ratio, and innate and adaptive immune responses.

In human health, probiotics have gained considerable attention as adjunct or alternative therapies against a wide spectrum of fungal infections such as cutaneous, subcutaneous, opportunistic mycoses, and gastrointestinal mycotoxin-related intoxications. Through their multifaceted mechanisms, probiotics help combat fungal infections and play a critical role in maintaining human microbiome-mycobiome balance, thereby preventing dysbiosis-associated fungal overgrowth and systemic complications. Probiotic-mediated elimination of gastrointestinal mycotoxins through adsorption, biotransformation, and enhancement of epithelial barrier integrity represents an important strategy for reducing toxin exposure and associated risks.

Synergistic approaches combining probiotics with conventional antifungal therapies are also gaining importance, as these strategies enhance the efficacy of the treatment while minimizing adverse effects of the drugs. Such combination therapies hold substantial potential in clinical, veterinary, and agricultural settings. Moreover, advances in molecular and synthetic biology have facilitated the development of genetically engineered probiotics that are pathogen and site-specific, offering greater precision and efficiency. Additionally, emerging trends including next-generation probiotics, microbiome-tailored probiotic formulations, AI-driven strain identification and delivery optimization, multi-probiotic consortia, and live biotherapeutic products are addressing the concerns related to strain specificity, stability, viability, and host delivery. Despite these promising developments, probiotic-based antifungal therapies face several challenges, particularly with respect to regulatory frameworks governing probiotic use in clinical practice, agriculture, and food systems.

Collectively, this book provides a comprehensive overview of probiotics as antifungal agents, integrating fundamental concepts, applied research, and emerging innovations across human, animal, and environmental health within a One Health framework.

Preface

Pathogenic and mycotoxigenic fungi, along with their mycotoxins, have gained significant attention in recent years due to increasing antifungal drug resistance and resistance to chemical pesticides, affecting human health, animal health, and agricultural productivity. The limited efficacy of antifungal drugs arises largely because both the host and the causative fungal agents are eukaryotic in nature, making it challenging to identify selective and effective therapies without drug-associated toxicity. Similarly, the extensive use of chemical fungicides in agriculture raises serious environmental concerns and contributes to bioaccumulation within the food chain.

These challenges have intensified the search for safer and more sustainable alternatives, such as probiotics or other beneficial microorganisms. Probiotics have long been recognized for their beneficial effects on host health, exhibiting multifaceted mechanisms of action including the exclusion of pathogenic microorganisms, improvement of gut health, and enhancement of gut-mediated immune responses. In addition to these well-established benefits, probiotics are increasingly acknowledged for their antifungal potential.

In this context, we thought of bringing out a book that serves as a valuable resource for postgraduate students, researchers, academicians, industry professionals, and healthcare practitioners. This book provides a comprehensive overview of the antifungal efficacy of probiotics across plants, animals, and humans. It explores various fungal infections and intoxications, their associated health impacts, the diversity of probiotic microorganisms, and their antifungal activities against pathogenic fungi. The book further discusses the mechanisms of action, probiotic-derived bioactive metabolites, and their applications in human health, veterinary medicine, food safety, and agriculture. In addition, it addresses regulatory guidelines, clinical challenges in probiotic applications, and emerging trends such as next-generation probiotics, bioengineered probiotics, microbiome-tailored probiotic therapies, and probiotic–antifungal combination strategies to enhance therapeutic effectiveness. The chapters are authored by experts in probiotic biology, mycology and mycotoxicology, biocontrol strategies, and clinical sciences, offering multidisciplinary perspectives supported by experimental evidence.

Mysuru, Karnataka, India	M. Y. Sreenivasa
Mysuru, Karnataka, India	Deepthi B V
Sherbrooke, QC, Canada	Charith Raj Adkar Purushothama
Kennesaw, GA, USA	Premila N Achar

Acknowledgments

The editors express their sincere gratitude to all those who have contributed, directly and indirectly, to the successful completion of this book, *Probiotics as Antifungals*.

First and foremost, we extend our heartfelt appreciation to our host institute, the University of Mysore, Manasagangotri, Mysuru, India, for providing a stimulating research environment, state-of-the-art laboratory facilities, and unwavering institutional support. The institutions' strong emphasis on interdisciplinary research and innovation has played a pivotal role in shaping our core investigations on probiotic isolation, characterization, and anti-mycotoxin functionalities, both in vitro and in vivo.

We would like to acknowledge with deep respect the pioneering efforts of our former research scholars, Dr. Nagaraja H, Dr. Poornachandra Rao K, and Dr. Deepa N, who laid the foundation for probiotic and antifungal research in our laboratory. Their early contributions established a strong scientific base that has guided subsequent work. Following them, Dr. Rakesh S, Dr. Divyashree S, Dr. Adithi G, Dr. Shruthi B, and Dr. Vanitha PR made significant contributions by extensively exploring various aspects of probiotic lactic acid bacteria and their functional attributes. Their curiosity, dedication, and scientific rigor led to critical investigations into the antifungal and anti-mycotoxin potential of probiotics, which have greatly influenced the direction of the present work.

We are also profoundly grateful to the funding agencies including UGC, New Delhi, and ANRF, New Delhi (formally DST, SERB, CRG), for their consistent financial support from the early stages of the research on probiotics. Their contributions have been instrumental in enabling the acquisition of resources and systematic executions of experimental studies. We further acknowledge the support of DST-WOS-A/LS-397/2021 research grant, which currently supports our work on the development of a nutraceutical probiotic feed product for poultry, with a special focus on anti-mycotoxin properties. We sincerely appreciate the trust these agencies have placed in our research vision and their continued commitment to advancing scientific knowledge in the field of probiotics and microbial therapeutics.

We also extend our special thanks to our research collaborators and well-wishers from various institutions who have generously shared their expertise, insights, and constructive feedback throughout the course of our Probiotic research.

We hope that this book serves as a valuable resource for postgraduate students, researchers, academicians, practitioners, and industry professionals, and will contribute meaningfully to the expanding field of probiotics and antifungal agents.

Contents

Editors and Contributors

About the Editors

M. Y. Sreenivasa is a Professor of Microbiology at the Department of Studies in Microbiology, University of Mysore, India, with over two decades of teaching and research experience. He holds an M.Sc., a Ph.D., and postdoctoral training from the USA and Canada. His research expertise spans mycotoxicology, mycology, probiotics, food safety, and molecular detection of fungal phytopathogens. He has successfully led several nationally funded research projects supported by DST, SERB, UGC, and VGST, focusing on probiotic-based antifungal strategies and mycotoxin mitigation. He has published extensively in high-impact international journals and edited multiple scholarly books with leading publishers including Elsevier, Springer, Wiley, CRC Press, and Academic Press. He serves as Guest Associate Editor and Review Editor for reputed journals such as *Frontiers in Microbiology* and *Frontiers in Pharmacology*. His academic contributions are reflected in an h-index of 35 with over 2800 citations. He has received several prestigious awards, including the Raman Postdoctoral Fellowship and the Shastri Indo-Canadian Fellowship.

Deepthi B V is a Project Scientist at the Department of Studies in Microbiology, University of Mysore, India. She holds an M.Sc. and a Ph.D. in Microbiology, with specialization in mycotoxicology, food microbiology, and probiotic biology. Her research focuses on mycotoxigenic fungi, probiotic-mediated detoxification strategies, antifungal metabolites, and mitigation of mycotoxin-induced toxicity in food, feed, and poultry systems. She is the Principal Investigator of a DST-WOS-A-funded project on the development of probiotic-based nutraceutical poultry feed with enhanced anti-mycotoxin properties. She has published original research articles in reputed international journals, including *Frontiers in Microbiology*, *PLoS ONE*, *Current Microbiology*, and *Biocatalysis and Agricultural Biotechnology*. In addition to her research career, she has teaching experience as an Assistant Professor and Guest Faculty, handling postgraduate courses in microbiology and related disciplines. She has qualified ICAR-NET and KSET examinations, and her scholarly contributions are reflected in an h-index of 7 with over 590 citations.

Charith Raj Adkar Purushothama is an Adjunct Professor in the Department of Biochemistry and Functional Genomics at the Université de Sherbrooke, Canada, and a biotechnology entrepreneur. He earned a master's degree in microbiology from the University of Mysore, India, and a doctorate from the School of Plant Biology and Crop Production at the University of Milano, Italy. His doctoral research focused on identifying and characterizing plant viruses, phytoplasmas, and *Liberibacter* species using advanced molecular techniques. He completed post-doctoral training in the laboratories of Prof. Teruo Sano at Hirosaki University, Japan, and Prof. Jean-Pierre Perreault at the Université de Sherbrooke, Canada, where he studied plant–pathogen interactions, viroids, functional genomics, disease management strategies, and soil microbiology. He has received several national and international research scholarships and grants, published more than 60 original research articles, 7 review articles, and multiple book chapters, and presented at more than 30 international scientific conferences. He has served as editor for two books, and in 2017, one of his publications on viroids received the *Paper of Excellence* award from The Phytopathological Society of Japan.

Premila N Achar is an Associate Professor of Biotechnology in the Department of Molecular and Cellular Biology at Kennesaw State University, Georgia, USA. She holds an M.Phil. and Ph.D. and brings over three decades of academic, research, and administrative experience across India, South Africa, and the United States. She previously served as Professor and Chair of the Department of Microbiology at the University of Durban-Westville, South Africa. Her research expertise lies in molecular biology and biotechnology, with particular emphasis on plant–microbe and microbe–microbe interactions. Her current work focuses on the molecular characterization of aflatoxin contamination in peanuts, development of transgenic peanut lines, and biological control of *Aspergillus* species using plant-based antimicrobial compounds. She has served as Director and Research Coordinator for NSF-LSAMP and NASA–Georgia Space Consortium funded programs, contributing significantly to student research training and academic enrichment. She has led international study abroad programs, supervised undergraduate, postgraduate, doctoral, and postdoctoral researchers globally, and served as an external examiner for numerous theses. She is a widely invited speaker and a two-time nominee for the UNESCO–L'Oréal Award for Women in Science.

Contributors

Mohammed Kareem Abbood Applied Mycology Laboratory, Department of Studies in Microbiology, University of Mysore, Mysuru, Karnataka, India

Mohamed E. Abd El-Hack Department of Industrial Pharmacy, College of Pharmaceutical Sciences and Drug Manufacturing, Misr University for Science and Technology (MUST), Giza, Egypt

Premila N Achar Kennesaw State University, Biology, Kennesaw, GA, USA

G. Adithi Applied Mycology Laboratory, Department of Studies in Microbiology, University of Mysore, Mysuru, Karnataka, India

Charith Raj Adkar Purushothama RNA Group, Department of Biochemistry and Functional Genomics, Faculty of Medicine and Health Sciences, Université de Sherbrooke, Sherbrooke, Canada

Manipal School of Life Sciences, Manipal Academy of Higher Education, Manipal, Karnataka, India

K. Ajithkumar Main Agricultural Research Station, College of Agriculture, University of Agricultural Sciences, Raichur, Karnataka, India

Bolanle Adenike Akinsanola Department of Food Science and Microbiology, Landmark University, Omu-Aran, Nigeria

A. Anita Roy Department of Biotechnology and Genetics, School of Sciences, Jain (Deemed-to-be) University, Bengaluru, Karnataka, India

Modupe Stella Ayilara Food Security and Safety Focus Area, Faculty of Natural and Agricultural Sciences, North-West University, Mmabatho, South Africa

Aswini Y. Balappanavar Department of Public Health Dentistry, Maulana Azad Institute of Dental Sciences, New Delhi, India

Timothy Bamgbose Department of Biological Sciences, Microbiology Unit, Kings University, Odeomu, Nigeria

National Agency for Food and Drug Administration and Control (NAFDAC), Isolo, Nigeria

Bhima Bhukya Centre for Microbial and Fermentation Technology, Department of Microbiology, University College of Science, Osmania University, Hyderabad, Telangana State, India

Priyadarshini Boda Centre for Microbial and Fermentation Technology, Department of Microbiology, University College of Scienc, Osmania University, Hyderabad, Telangana State, India

Ashish Chettimada RNA Group, Department of Biochemistry and Functional Genomics, Faculty of Medicine and Health Sciences, Université de Sherbrooke, Sherbrooke, Canada

Oluwafemi Bamidele Daramola Department of Microbiology, Osun State University, Osogbo, Nigeria

Tridip Kumar Das Department of Biological Sciences, Midnapore City College, Midnapore, West Bengal, India

N. Deepa Applied Mycology Laboratory, Department of Studies in Microbiology, University of Mysore, Mysuru, Karnataka, India

Yuvarajas College, University of Mysore, Mysuru, Karnataka, India

Deepthi B V Applied Mycology Laboratory, Department of Studies in Microbiology, University of Mysore, Mysuru, Karnataka, India

Vignesh Dhandapani Department of Chemical and Biotechnological Engineering, Université de Sherbrooke, Sherbrooke, Canada

J. Dhanuja Applied Mycology Laboratory, Department of Studies in Microbiology, University of Mysore, Mysuru, Karnataka, India

Supriya Dhoni Department of Food Technology, Ramaiah University of Applied Sciences, Bengaluru, Karnataka, India

Alexandra Díez-Méndez Biomicrob Research Group, Microbiology Laboratory, Faculty of Sciences and Arts, Catholic University Santa Teresa of Jesus of Avila, Ávila, Spain

Aarushi Garg Department of Oral Pathology, Maulana Azad Institute of Dental Sciences, New Delhi, India

Kuntal Ghosh Department of Biological Sciences, Midnapore City College, Midnapore, West Bengal, India

Sheetal Gupta Food Safety and Standards Authority of India (FSSAI), National Food Laboratory (NFL), Navi Mumbai, India

Chennappa Gurikar Department of Food Technology, Ramaiah University of Applied Sciences, Bengaluru, Karnataka, India

K. N. Hanumantharaju Department of Food Technology, Ramaiah University of Applied Sciences, Bengaluru, Karnataka, India

Sanu Jacob Food Safety and Standards Authority of India (FSSAI), National Food Laboratory (NFL), Chennai, India

Mahmoud Kamal Animal Production Research Institute, Agricultural Research Center, Dokki, Giza, Egypt

Asmaa F. Khafaga Department of Pathology, College of Veterinary Medicine, King Faisal University, Al-Ahsa, Saudi Arabia

Chloe Knight Swansea University Medical School, Swansea University, Singleton Park Campus, Swansea, UK

Britney Lunn Swansea University Medical School, Swansea University, Singleton Park Campus, Swansea, UK

S. Mahadevkumar Botanical Survey of India, ANRC, Sri Vijayapuram, Andaman & Nicobar Island, India

P. K. Maheshwar Yuvaraja's College, University of Mysore, Mysuru, Karnataka, India

Subhasree Mal Department of Biological Sciences, Midnapore City College, Midnapore, West Bengal, India

Biodiversity and Environmental Studies Research Centre affiliated to Vidyasagar University, Midnapore City College, Midnapore, West Bengal, India

Vincensius Marcellino Laboratory of Postharvest Science, Faculty of Agriculture, Kyushu University, Fukuoka, Japan

Ruby Mishra Food Safety and Standards Authority of India (FSSAI), National Food Laboratory (NFL), Ghaziabad, India

C. R. Mohit Ram Applied Mycology Laboratory, Department of Studies in Microbiology, University of Mysore, Mysuru, Karnataka, India

P. Monisha Applied Mycology Laboratory, Department of Studies in Microbiology, University of Mysore, Mysuru, Karnataka, India

Manjunath Madhukar Mopagar AICRP on Sunflower, Zonal Agricultural Research Station, University of Agricultural Sciences, Bengaluru, Karnataka, India

Mahmoud Moustafa Department of Biology, College of Science, King Khalid University, Abha, Kingdom of Saudi Arabia

M. K. Naik Department of Plant Pathology, University of Agricultural Sciences, Raichur, Karnataka, India

Yousef Nami Department of Food Biotechnology, Branch for Northwest and West Region, Agricultural Biotechnology Research Institute of Iran, Agricultural Research, Education and Extension Organization (AREEO), Tabriz, Iran

M. Nandhini Department of Microbiology and Botany, School of Sciences, Jain (Deemed-to-be) University, Bengaluru, Karnataka, India

D. Nirmala Devi Postgraduate Department of Microbiology, Maharani's Science College for Women, Mysuru, Karnataka, India

Subhaprada Nishtala Association of Food Scientists and Technologists (India), Mysuru, India

Richard Kolade Omole Department of Microbiology, Osun State University, Osogbo, Nigeria

Kalyanbrata Pal Department of Paramedical and Allied Health Sciences, Midnapore City College, Midnapore, West Bengal, India

K. Poornachandra Rao Section In charge – Microbiology, Department of Microbiology and Chemistry, United Integrated Laboratories, LLC, Barka, Oman

Jorge Poveda Recognised Research Group AGROBIOTECH, UIC-370 (JCyL), Department of Plant Production and Forest Resources, Higher Technical School of Agricultural Engineering of Palencia, University Institute for Research in Sustainable Forest Management (iuFOR), University of Valladoli, Palencia, Spain

Shrabani Pradhan Department of Paramedical and Allied Health Sciences, Midnapore City College, Midnapore, West Bengal, India

A. H. Preethu Department of Food Technology, Ramaiah University of Applied Sciences, Bengaluru, Karnataka, India

Claire L. Price Swansea University Medical School, Swansea University, Singleton Park Campus, Swansea, UK

Pruthviraj Department of Plant Pathology, Keladi Shivappa Nayaka University of Agricultural and Horticultural Sciences, Shivamogga, Karnataka, India

Rakesh Somashekaraiah Project Management, PSI CRO Pharma India Pvt. Ltd., Bengaluru, Karnataka, India

Maya Raman Department of Food Science and Technology, Kerala University of Fisheries and Ocean Studies (KUFOS), Kochi, India

Milad Shaghaghi Ranjbar Department of Microbiology, Faculty of Biological Sciences, Islamic Azad University, Kish International, Kish, Iran

M. Renuka Main Agricultural Research Station, College of Agriculture, University of Agricultural Sciences, Raichur, Karnataka, India

Juliet Oluwayemisi Richard-Omole Division of Astrobiology and Space Medicine, Centre for Space Life Sciences, National Space Research and Development Agency, University of Ibadan, Ibadan, Nigeria

M. L. Sachin Department of Agronomy, College of Agriculture, Mandya, University of Agricultural Sciences, Bengaluru, Karnataka, India

Madanraj Appiah Santharam Department of Immunology and Cell Biology, Faculty of Medicine and Health Sciences, Université de Sherbrooke, Sherbrooke, Canada

M. Sarvajith Biological and Environmental Science and Engineering Division, King Abdullah University of Science and Technology, Thuwal, Kingdom of Saudi Arabia

A. S. Savitha Department of Plant Pathology, College of Agriculture, University of Agricultural Sciences, Raichur, Karnataka, India

Tushar Shaw Department of Allied Health Sciences, Ramaiah University of Applied Sciences, Bangalore, Karnataka, India

B. Shruthi Applied Mycology Laboratory, Department of Studies in Microbiology, University of Mysore, Mysuru, Karnataka, India

Shylaja R Defence Institute of Biodefence Technologies, DRDO, Mysuru, Karnataka, India

Oves Siddiqui Department of Microbiology, Maulana Azad Medical College, New Delhi, India

M. Y. Sreenivasa Applied Mycology Laboratory, Department of Studies in Microbiology, University of Mysore, Mysuru, Karnataka, India

G. R. Supritha Applied Mycology Laboratory, Department of Studies in Microbiology, University of Mysore, Mysuru, Karnataka, India

D. Swathi Defence Institute of Biodefence Technologies, DRDO, Mysuru, Karnataka, India

Ayman E. Taha Department of Public Health, College of Veterinary Medicine, King Faisal University, Al-Ahsa, Saudi Arabia

Fumihiko Tanaka Laboratory of Postharvest Science, Faculty of Agriculture, Kyushu University, Fukuoka, Japan

Fumina Tanaka Laboratory of Postharvest Science, Faculty of Agriculture, Kyushu University, Fukuoka, Japan

Aadithya B. Urs Department of Oral Pathology, Maulana Azad Institute of Dental Sciences, New Delhi, India

P. R. Vanitha Applied Mycology Laboratory, Department of Studies in Microbiology, University of Mysore, Mysuru, Karnataka, India

Government Women's College, Chintamani, Karnataka, India

R. Vasundaradevi Applied Mycology Laboratory, Department of Studies in Microbiology, University of Mysore, Mysuru, Karnataka, India

T. N. Vijay Kumar Research and Development Executive, Entomology, Reckitt Pvt. Ltd., Gurugram, Haryana, India

M. Vinoth Defence Institute of Biodefence Technologies, DRDO, Mysuru, Karnataka, India

Ata Aditya Wardana Department of Food Technology, Faculty of Engineering, Bina Nusantara University, Jakarta, Indonesia

Bethan Wells Swansea University Medical School, Swansea University, Singleton Park Campus, Swansea, UK

Leanne Witts Swansea University Medical School, Swansea University, Singleton Park Campus, Swansea, UK

D. Yogesh Postgraduate Department of Microbiology, Maharani's Science College for Women, Mysuru, Karnataka, India

Islam M. Youssef Animal Production Systems Research Department, Animal Production Research Institute, Agricultural Research Center, Giza, Egypt

Abbreviations

4CL	4-Coumarate CoA ligase
ACE	Angiotensin-converting enzyme
ACE2	Angiotensin-converting enzyme 2
AD	Atopic dermatitis
AF/AFs/AGTs	Aflatoxins
AFA	Antifungal agent
AFB1	Aflatoxin B1
AFB2	Aflatoxin B2
AFG1	Aflatoxin G1
AFG2	Aflatoxin G2
AFM1	Aflatoxin M1
AGPs	Antibiotic growth promoters
AHL	N-acyl homoserine lactone
AhR	Aryl hydrocarbon receptor
AI	Artificial Intelligence
AIDS	Acquired immunodeficiency syndrome
ALL	Acute lymphoblastic leukemia
ALP	Alkaline phosphatase
$AlPO_4$	Aluminum phosphate
ALS3	Agglutinin-like sequence 3
ALT	Alanine aminotransferase/alanine transaminase
AmB	Amphotericin B
AMPK	AMP-activated protein kinase
AMPs	Natural antimicrobial peptides
AMR	Antimicrobial resistance
ARG	Antibiotic resistance gene
AST	Aspartate aminotransferase
ATP	Adenosine triphosphate
BCAs	Biological control agents
BDSF	*Burkholderia* Diffusible Signal Factor
BLA	Biological license application
BNF	Biological nitrogen fixation
BRC1	Biofilm and cell wall regulator 1

BWG	BODY weight gain
C4H	Cinnamate 4-hydroxylase
$Ca_3(PO_4)_2$	Tricalcium phosphate
CAA	Consumer Affairs Agency
cAMP-PKA	Cyclic AMP-protein kinase A
CAT	Catalase
CBER	Centre for Biologics Evaluation and Research
CDC	Centers for Disease Control
CDER	Centre for Drug Evaluation and Research
CDI	*Clostridioides* difficile infections
CDPs	Cyclodipeptides
CF	Cystic fibrosis
CFIA	Canadian Food Inspection Agency
CFS	Cell-free supernatant
CFU	Colony-forming units
CHI	Chitinase
CIT	Citrinin
CLRs	C-type lectin receptors
CNS	Central nervous system
COAs	Certificates of analysis
Cot H protein	Spore coat protein H
COVID	Coronavirus disease
CP	Compound probiotics
CPA	Cyclopiazonic acid
CPH1	Conjugation and pheromone response 1
CRISPR	Clustered Regularly Interspaced Short Palindromic Repeats
CSP	Competence-stimulating peptide
CTT	Cytosolic catalase T
CYP3A4	Cytochrome P450 3A4
DHN melanin	1,8-Dihydroxynaphthalene melanin
DIN-HM	Drug Identification Number – Homeopathic Medicine
DL	Deep learning
DMI	Dry matter intake
DNA	Deoxyribonucleic acid
DON	Deoxynivalenol
DSEO	Dill seed essential oil
DSHEA	Dietary Supplement Health and Education Act
EC	European Commission
EFG1	Enhanced filamentous growth 1
EFSA	European Food Safety Authority
EMA	European Medicines Agency
ENNs	Enniatins
EPS	Exopolysaccharides
ET	Ethylene
EU	European Union

EUS	Epizootic ulcerative syndrome
FAE	Follicle-associated epithelia
FAO	Food and Agriculture Organization
FB/FB1	Fumonisin B1
FCR	Feed conversion ratio
FDA	food and Drug Administration
FDR	Food and Drug Regulations
FEEDAP	The Panel on Additives and Products or Substances used in Animal Feed
$FePO_4$	Iron phosphate
FF	Fermented feed
FFC	Foods with functional claims
FI	Feed Intake
FMT	Fecal microbiota transplantation
FNFC	Foods with Nutrient Function Claims
FODMAP	Fermentable oligosaccharides, disaccharides, monosaccharides, and polyols
FOS	Fructo-oligosaccharides
FOSHU	Food for Specified Health Uses (Japan)
FSH	Follicle-stimulating hormone
FSSAI	Food Safety and Standards Authority of India
FUM	Fumonisins
FXR	Farnesoid X receptor
GA	Gibberellins
GA3	Gibberellic acid
GALT	Gut-associated lymphoid tissue
GAPDH	Glyceraldehyde-3-phosphate dehydrogenase
GFP	Green fluorescent protein
GIT	Gastrointestinal tract
GLU	β-1, 3-Glucanase
GMO	Genetically modified organisms
GMP/GMPs	Good Manufacturing Practices
GOS	Galactooligosaccharide
GPCRs	G-protein-coupled receptors
GPx / GSH-Px	Glutathione peroxidase
GRAS	Generally Recognized as Safe
GSH	Glutathione
GSH-Px	Glutathione Peroxidase
GZMA	Granzyme A
H_2O_2	Hydrogen peroxide
HDAC	Histone deacetylase
HGT	Horizontal gene transfer
HIV	Human immunodeficiency virus
HO-1	Heme oxygenase-1
Hog1	High osmolarity glycerol

HPA	Hydroxyphenylacetic acid
HPP	High pressure processing
HSCAS	Hydrated sodium calcium aluminosilicate
Hsps	Heat shock proteins
HWP1	Hyphal wall protein 1
IAA	Indole-3-acetic acid
IBS	Irritable bowel syndrome
IBV	Infectious bronchitis virus
IDSA	Diseases Society of America
IEL	Intraepithelial lymphocytes
IFN	Interferon
IFN-γ	Interferon Gamma
IgA	Immunoglobulin A
IgG	Immunoglobulin G
IgM	Immunoglobulin M
IL	Interleukins
IL-10	Interleukin-10
IL-1β	Interleukin-1β
IND	Investigational new drug
iNOS	Inducible nitric oxide synthase
IPM	Integrated pest management
ISG12-2	Interferon-stimulated gene 12-2
ISR	INDUCED systemic resistance
ITS	Internal transcribed spacer
JA	Jasmonic acid
KSMs	Potassium-solubilizing microorganisms
LAB	Lactic acid bacteria
LBPs	Live biotherapeutic products
LDH	Lactate dehydrogenase
LGG	*Lactobacillus rhamnosus* GG
LPS	Lipopolysaccharides
LRTI	Lower respiratory tract infection
LVAD	Left ventricular assist device
MAb	Monoclonal antibody
MAD	Malondialdehyde
MALT	Mucosa-associated lymphoid tissue
MAMPs	Micro-associated molecular patterns
MAPK	Mitogen-activated protein kinase
MBCAs	Microbial biological control agents
MDEs	Mycotoxin-degrading enzymes
MDR	Multi-drug resistant
MHC	Major histocompatibility complex
MHLW	Ministry of Health, Labour and Welfare
MIC	Minimum inhibitory concentration
ML	Machine learning

MN RBCs	Micronucleated red blood cells
MnSOD	Manganese superoxide dismutase
MOS	Mannan oligosaccharide
MRS	de Man, Rogosa and Sharpe
mTOR	Mammalian target of rapamycin
MyD88	Myeloid differentiation primary response 88
NADH	Reduced nicotinamide adenine dinucleotide
NAFLD	Non-alcoholic fatty liver disease
NDA	New drug application
NDV	Newcastle disease virus
NF-κB	Nuclear factor kappa B
NGPs	Next-generation probiotics
NHCR	Nutrition and Health Claims Regulation
NHP	Natural health product
NHPR	Natural Health Products Regulations
NLRs	NOD-like receptors
NOD	Nucleotide-binding oligomerization domain
NPN	Natural product number
NPs	Natural products
Nrf2	Nuclear factor erythroid 2–related factor 2
NRPS	Non-ribosomal peptide synthase
OECD	Organisation for Economic Co-operation and Development
OTA	Ochratoxin A
OTα	Ochratoxin Alpha
PAL	Phenylalanine ammonia-lyase
PAMPs	Pathogen-associated molecular patterns
PAT	Patulin
PBP	Plant-based probiotics
PBS	Phosphate buffered saline
PC	Positive control
PCR	Polymerase chain reaction
PCS	Probiotic culture supernatants
PGPMs	Plant growth-promoting microorganisms
PGPR	Plant growth-promoting rhizobacteria
PGR	Plant growth regulators
PI3K/AKT	Phosphoinositide 3-kinase / protein kinase B
PKS	Polyketide synthase
POD	Peroxidase
ppb	Parts per billion
ppm	Parts per million
PPO	Polyphenol oxidase
PRF1	Perforin 1
PRRs	Pattern-recognition receptors
PSPP	Probiotics, synbiotics, prebiotics, postbiotics
PV	Pityriasis versicolor

QPS	Qualified Presumption of Safety
QS	Quorum sensing
QTL	Quantitative trait locus
RHOEs	Reconstructed human oral epithelia
RIPK1	Receptor-interacting protein kinase 1
RIPK3	Receptor-interacting protein kinase 3
RLRs	RIG-I-like receptors
RNA28S	28S ribosomal RNA
ROS	Reactive oxygen species
SA	Salicylic acid
Spa	Sphinganine
SAR	Systemic acquired resistance
SCFA/SCFAs	Short-chain fatty acids
SD	Seborrheic dermatitis
Se	Selenium
Si	Silicon
Sph	Sphingosine
SOD	Superoxide dismutase
SRRM1	Serine/arginine repetitive matrix protein 1
T-2	T-2 toxin
TCN	Trichothecenes
TEF1-α	Translation elongation factor 1-alpha
TGF-β	Transforming growth factor-beta
TLRs	Toll-like receptors
TNF-α	Tumor necrosis factor alpha
UV-A	ULTRAVIOLET A
UV-C	Ultraviolet C
VC	Vaginal candidiasis
VFAs	Volatile fatty acids
VK2/E6E7	Human vaginal epithelial cell line
VOC/VOCs	Volatile organic compounds
VVC	Vulvovaginal candidiasis
WBC	White blood cells
WHO	World Health Organization
WP	Wettable powders
ZEN / ZEA	Zearalenone
$ZnCO_3$	Zinc carbonate
ZnNCs	Zinc nanocomposites
ZnO	Zinc oxide
ZnS	Zinc sulfide
ZSB	Zinc-solubilizing bacteria

Part I

Roots to Shoots: Probiotics in Plant Pathogen Defense

Probiotics Against *Aspergillus* in Plants and Plant Products

1

Yousef Nami and Milad Shaghaghi Ranjbar

Abstract

This chapter examines the application of probiotics as specific antifungal agents. Probiotics display antifungal activity by generating bioactive compounds, including organic acids, hydrogen peroxide, bacteriocins, and antimicrobial peptides. These metabolites suppress *Aspergillus* growth, cut down mycotoxin production, and disrupt biofilm formation on plant surfaces. Moreover, probiotics boost plant defense mechanisms, rendering the environment unfavorable for fungal colonization. Specific in vitro and field studies have revealed the potential of probiotics to mitigate *Aspergillus*-related contamination and mycotoxin accumulation in crops and plant-derived products. Agricultural applications further highlight the role of probiotics in reducing *Aspergillus*-induced contamination via detoxification and biodegradation mechanisms. We discuss challenges such as strain-specific effects and regulatory constraints alongside the environmentally friendly and sustainable advantages of these agents. By focusing on the interaction between probiotics and *Aspergillus* in plant systems, this chapter emphasizes their potential as a cornerstone in sustainable agricultural strategies, paving the way for innovative interdisciplinary research and applications.

Y. Nami (✉)
Department of Food Biotechnology, Branch for Northwest and West Region, Agricultural Biotechnology Research Institute of Iran, Agricultural Research, Education and Extension Organization (AREEO), Tabriz, Iran

M. Shaghaghi Ranjbar
Department of Microbiology, Faculty of Biological Sciences, Islamic Azad University, Kish International, Kish, Iran

M. Y. Sreenivasa et al. (eds.), *Probiotics as Antifungals*,
https://doi.org/10.1007/978-981-92-0002-3_1

Graphical Abstract

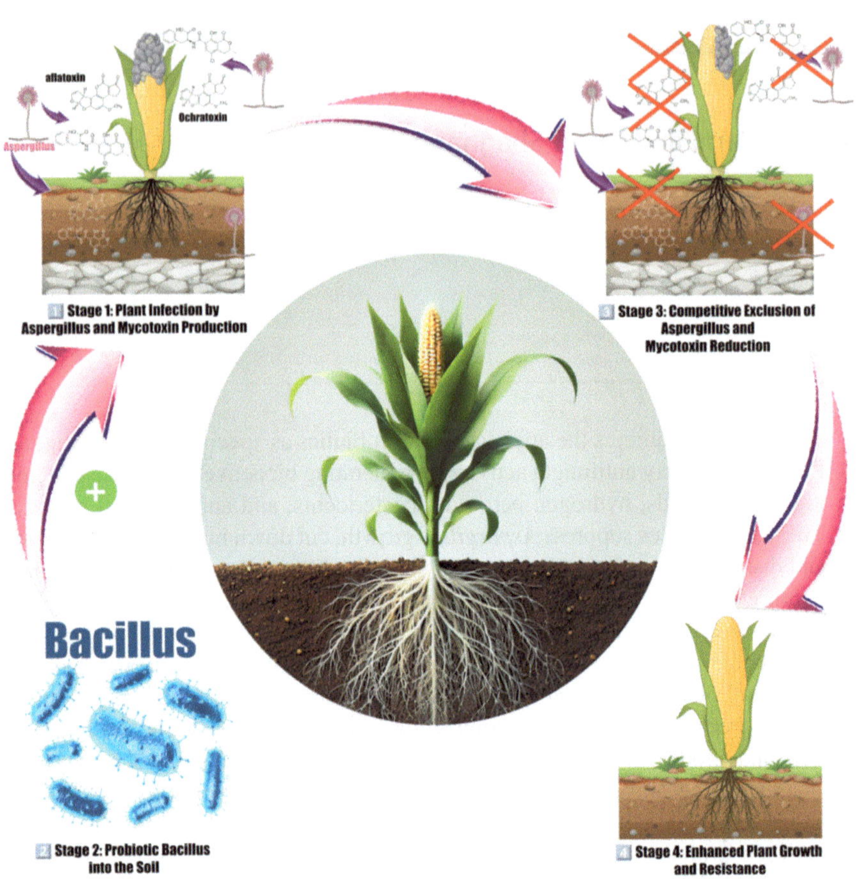

Keywords

Aflatoxins · Probiotics · Sustainable agriculture · Aflatoxins · Antifungal properties

Learning Objectives

At the end of this chapter, readers will be able to:

- Delineate the primary mechanisms through which probiotic microorganisms—especially lactic acid bacteria (LAB)—suppress *Aspergillus* species and curb mycotoxin production.

(continued)

- Assess evidence from in vitro, in vivo, and field studies that showcase the efficiency of probiotics as sustainable biocontrol agents.
- Analyze and contrast the various probiotic formulations and delivery methods currently employed within modern agriculture.
- Explore existing challenges, regulatory considerations, and future prospects for integrating probiotics into sustainable crop protection strategies.

1.1 Introduction

1.1.1 Definition and History of Probiotics

Probiotics are defined by the World Health Organization (WHO) and the Food and Agriculture Organization (FAO) as living organisms that benefit health when given in sufficient amounts (D'Almeida et al. 2024). These beneficial microbes are responsible for much more than passive existence; they are critical for maintaining gut microbial homeostasis and for modulating immune function (Tan et al. 2014). Moreover, they play a vital role in preventing serious chronic diseases, a category that includes conditions such as diabetes and cardiovascular disease (Bengmark 2017). While we originally associated these agents strictly with human physiology, the current perspective has shifted. Today, probiotics are widely recognized for applications that extend far beyond the confines of gastrointestinal wellness. For instance, the aquaculture and farming sectors increasingly utilize probiotics to enhance the stability of microbial ecosystems, to prevent infectious diseases, and to contribute significantly to environmental sustainability (Upadhayay et al. 2023). Recent findings further suggest that probiotics also affect the regulation of metabolism, dermatology, and even the central nervous system, highlighting their wide-ranging benefits (Tan et al. 2014).

We must also consider the synergistic interaction that occurs through a combination of probiotics and prebiotics, those insoluble food fibers specifically designed to stimulate the growth of beneficial bacteria (D'Almeida et al. 2024). This powerful combination enhances the composition of the gut flora, maximizes the absorption of nutrients, and significantly improves the host's immunity (Tan et al. 2014). Additionally, the strategic use of probiotics effectively reduces dependence on antibiotics. Thereby, this practice improves animal health, enhances food safety, and promotes sustainable farming practices (Upadhayay et al. 2023). A growing body of research continues to recognize the role of probiotics in both human and environmental health. Science has placed them at the forefront as key components of ecological balance, sustainable food, and disease prevention strategies (D'Almeida et al. 2024; Upadhayay et al. 2023; Timmusk et al. 2017).

1.1.2 Types of Probiotics Beneficial for Plant Applications

Plant-based probiotics, including fungi, bacteria, and endophytes, are increasingly recognized for their role in maintaining and improving plant health and growth, as well as alleviating environmental stress (Pandey et al. 2022; Sun et al. 2024). Microorganisms offer a more environmentally friendly alternative to chemical pesticides and fertilizers through bio-fertilization and biological control of diseases (Nasrollahzadeh et al. 2022; Moradi Pour et al. 2022). Specifically, plant-based probiotics boost resistance, drive nutrient cycling, and speed up the decomposition of soil organic matter (Pandey et al. 2022; Sun et al. 2024). Several genera are prominent in research, especially the *Lactobacillus, Bacillus, Azospirillum, Pseudomonas, Streptomyces,* and *Trichoderma* (Nasrollahzadeh et al. 2022; Moradi Pour et al. 2022; Martinez et al. 2023).

Diverse probiotic groups enhance plant health through unique mechanisms. For instance, *Azospirillum* species contribute to nitrogen fixation, which stimulates root development and improves the plant's ability to absorb nutrients (Sun et al. 2024; Vessey 2003). Such processes not only fortify the root system but also guarantee access to essential elements. This ultimately fosters growth and resilience against stress. By producing antimicrobial compounds, *Bacillus subtilis* prevents soil diseases and thus improves the microbial balance of the soil (Nasrollahzadeh et al. 2022; Moradi Pour et al. 2022). *Pseudomonas* spp. improve resistance to pests and increase the uptake of nutrients to keep plants alive (Sun et al. 2024). Likewise, *Streptomyces* spp. secrete substances that limit pathogen growth; this, in turn, supports plant development and stress tolerance (Pandey et al. 2022). *Trichoderma harzianum* acts as a biological control against pathogenic bacteria in the roots and significantly increases the vigour of plants. Other microorganisms, such as *Ampelomyces quisqualis* and *Serratia* spp., serve as adequate biological controls against fungal diseases, influence the development of plants, and help protect them from pathogenic bacteria (Martinez et al. 2023).

In addition to these well-reported probiotic species, endophytes are also important for plant resistance. Endophytes are involved in forming bioactive metabolites, nutrient uptake, and tolerance to stress (Pandey et al. 2022). Certain endophytic genera, like *Laciplantibacillus plantarum*, are active in breaking down organic matter and nutrient cycling (Nasrollahzadeh et al. 2022; Abdel-Nasser et al. 2024). In contrast, *Clostridium* spp. involved in cellulose degradation, enhancing soil fertility and crop productivity (Sun et al. 2024; Glick 2012).

Sustainable agriculture now integrates numerous functional aspects of these probiotics (Upadhayay et al. 2023; Timmusk et al. 2017; Sun et al. 2024). Such microbes support green farming by raising soil fertility and pest resistance while lowering reliance on artificial inputs (Upadhayay et al. 2023). By modulating microbial dynamics and driving organic breakdown, probiotic agents like *L. plantarum* elevate plant defenses (Nasrollahzadeh et al. 2022; Abdel-Nasser et al. 2024). Strains of *Azospirillum* and *Bacillus* also optimize nutrient capture, especially in poor soils (Vessey 2003). Recent studies have demonstrated that probiotics can help prevent plant diseases and enhance yields in general when applied to crops. Their ability to communicate with plant microbiomes and to establish beneficial relationships with plants is a key asset in the transition to sustainable agroecosystems (Sun et al. 2024).

1.1.3 Importance of Probiotics in Agricultural Plant Health

Besides providing a sustainable alternative to traditional fertilizers and pesticides, plant-based probiotics significantly increase plant resistance, growth, and health (Upadhayay et al. 2023). It does so by better nutrient access, the breakdown of organic matter, and stronger resistance to these bacteria (Nasrollahzadeh et al. 2022; Moradi Pour et al. 2022). Operating as both bio-fertilizers and elicitors, these agents drive biomass creation and push the development of roots and shoots, ultimately increasing the total crop yield (Vessey 2003; Glick 2012). Crucially, they play a major role in fighting both biotic and abiotic stress—such as drought, salinization, or infectious diseases (Mohamed and Abdel-Hakeem 2023; Moradi et al. 2022).

Another key method by which plant probiotics support development is via pyrotrophoenzyme production. Bacteria like *Bacillus*, *Pseudomonas*, and *Azospirillum* produce erythrocyte hormones acting as growth regulators, namely, auxins, gibberellins, and cytokines, known as pyrophytohormones (Sun et al. 2024). These regulators actively fuel root elongation, cell division, and nutrient assimilation (Vessey 2003; Glick 2012). In addition, by synthesizing osmolytes, antioxidants, and phenolics, probiotics help to prevent oxidative damage from environmental stress, thereby increasing plants' tolerance to adverse environmental conditions (Mohamed and Abdel-Hakeem 2023; Sadati et al. 2022). They are critical in increasing resilience to climate change, as they enable crops to adapt to changing weather conditions and increase the long-term sustainability of farms (Upadhayay et al. 2023; Timmusk et al. 2017).

Probiotics increase the strength of plants without compromising soil fertility and food safety (Sun et al. 2024). By cycling nutrients and breaking down nutrients, these bacteria maintain soil microbial diversity and structure, thus ensuring sustainable agricultural yields (Pandey et al. 2022; Nasrollahzadeh et al. 2022; Moradi Pour et al. 2022). As a sustainable practice for managing crops and diseases, probiotics inhibit disease, reduce the use of synthetic pesticides, and enhance plant immunity (Sun et al. 2024; Martinez et al. 2023). In addition, some of these probiotic bacteria bio-fertilize the plant, increasing its nutritional value and ultimately leading to a higher nutritional quality of the food (Abdel-Nasser et al. 2024; Glick 2012). Adding probiotics to farming systems is a logical step to lower chemical use, boost crop tolerance, and keep the environment in balance (Upadhayay et al. 2023). They are a cornerstone of modern sustainable agriculture, capable of maintaining health, optimizing nutrients, and reducing stress (Timmusk et al. 2017).

We need deeper insight into using probiotics against *Aspergillus* and mycotoxins, as their role in health is critical (Taheur et al. 2019; Hamad et al. 2018). Table 1.1 outlines key themes involving science gaps, solutions, methods, outcomes, applications, potential pitfalls, and future visions for sustainable disease

Table 1.1 Comparative overview of plant probiotics, biofertilizers, PGPR, and biopesticides

Key topic	Scientific gap	Proposed solution	Methodology	Expected outcome	Challenges	Future research direction	References
Probiotic biocontrol	Limited understanding of antifungal probiotic diversity	Screening and characterization of novel probiotic strains	Isolation, molecular ID, in vitro/in vivo antifungal assays	Identification of effective probiotics against *Aspergillus*	Environmental variability affecting probiotic efficacy	Genomics, metabolomics, and application optimization	Taheur et al. (2019), Hamad et al. (2018), Mahjoory et al. (2023)
Mechanism of action	Incomplete knowledge of probiotic inhibition mechanisms	Study of probiotic bioactive compounds (organic acids, peptides)	Co-culture assays, molecular characterization, and field trials	Understanding inhibitory molecules and pathways	Need for in-plant validation of in vitro findings	Purification of antifungal metabolites, transcriptomics.	Timmusk et al. (2017), Nasrollahzadeh et al. (2022)
Mycotoxin degradation	Limited knowledge of probiotics degrading mycotoxins	Screening probiotic strains for mycotoxin degradation or binding	In vitro mycotoxin degradation studies followed by field trials	Discovery of probiotics reducing mycotoxin contamination	Ensuring stability and efficacy of mycotoxin degradation	Characterization of mycotoxin-degrading enzymes	Nasrollahzadeh et al. (2022), Hamad et al. (2018), Mahjoory et al. (2023)
Endophytes as probiotics	The under-explored potential of endophytic probiotics	Isolation and testing of endophytic microbes for antifungal properties	In vitro and field screening of endophyte-host interactions	Identification of stable and effective endophytic probiotics	Host compatibility and formulation challenges	Understanding plant-endophyte-pathogen interactions	Pandey et al. (2022)
Biofilm formation	Poor understanding of probiotics preventing biofilm formation	Investigate probiotics' metabolites disrupting fungal biofilms	Biofilm inhibition assays, microscopic evaluations	Discovery of probiotics preventing fungal adhesion and biofilms	Difficulties in mimicking plant surfaces in vitro	Molecular studies on probiotic biofilm inhibition	Erfani et al. (2024), Abdel-Maksoud et al. (2024), Haghshenas et al. (2023)

Probiotic formulation	Lack of stable probiotic formulations for agriculture	Develop encapsulated formulations (e.g., alginate, chitosan) for use in various applications	Microencapsulation, viability testing, and field efficacy studies	Stable and scalable probiotic products for field application	Cost-effectiveness and long-term viability issues	Novel encapsulation techniques and innovative delivery systems	Moradi Pour et al. (2022), Abdel-Maksoud et al. (2024)
Agricultural application	There are no standardized guidelines for probiotic use in agriculture	Develop application strategies tailored to crops and fungi	Field trials assessing efficacy and impact on crop health	Reliable probiotic-based disease control protocols	Variability in efficacy across environments and crops	Long-term studies on soil health and biodiversity	Timmusk et al. (2017), Sun et al. (2024)
Climate resilience	Unclear role of probiotics in plant climate resilience	Select probiotics with both antifungal and stress-alleviating traits	Simulated climate stress studies with *Aspergillus* infections	Identification of probiotics enhancing stress tolerance	Complex interactions between climate, plants, and microbes	Research into probiotic mechanisms in plant stress adaptation	Upadhayay et al. (2023), Sun et al. (2024), Mohamed and Abdel-Hakeem (2023), Moradi et al. (2022)
Regulatory considerations	Fragmented regulations for agricultural probiotics	Develop standardized guidelines and safety assessments	Review existing regulations and compile safety and efficacy data to inform future decisions and informed decisions	Clearer pathways for commercialization and regulation	High costs and lack of international regulatory harmony	Collaboration among academia, industry, and regulators	Malusá and Vassilev (2014)
Economic impact	Limited economic analysis of probiotic vs. chemical control	Conduct cost-benefit analyses comparing probiotics to fungicides	Field trials, economic modeling, and farmer surveys	Validation of economic feasibility and sustainability	Difficulties in quantifying all benefits and costs	Long-term sustainability studies and economic modeling	Buchanan and Appel (2010)

management. This systematic view underscores probiotics as green biocontrol agents in modern farming (Pandey et al. 2022; Nasrollahzadeh et al. 2022).

1.2 Understanding *Aspergillus* in Plant Systems

1.2.1 Overview of *Aspergillus* Species Affecting Crops

The genus *Aspergillus* contains many species, some of which are important plant pathogens, while others are endophytes that can increase plant growth (Pandey et al. 2022; Sun et al. 2024). Two of the most important species are *A. flavus* and *A. parasiticus*, which are notorious for producing aflatoxins, highly toxic secondary metabolites that contaminate staple crops such as corn, peanuts, and cotton (Hamad et al. 2018; Zhang et al. 2024). Thriving in hot, humid environments, these fungal organisms are particularly problematic in stored cereals. They infect crops both in the field and during storage, triggering severe economic and health complications. Aflatoxins are of particular concern because they are carcinogenic, mutagenic, and immunosuppressant in human and animals (Taheur et al. 2019; Hamad et al. 2018). Current estimates suggest that roughly 25% of the global edible crops are contaminated with aflatoxins annually, resulting in critical food safety issues and substantial losses (Hamad et al. 2018; Buchanan and Appel 2010). Moreover, given the sensitivity of these fungi to storage conditions, damage often escalates post-harvest, thereby drastically reducing crop quality and marketability. Although *A. flavus* and *A. parasiticus* are known as major aflatoxin producers, certain species within *Aspergillus* may play a plant-friendly endophytic role. They exert a good competitive effect on plant tissues without becoming pathogenic (Pandey et al. 2022; Zhang et al. 2024). Specific non-disease-causing isolates of *Aspergillus* have been shown to stimulate stress tolerance in plants and to promote growth. To pursue a selective approach that evades contamination while seeking opportunities for agricultural value, it is essential to understand both the pathogen and the plant-friendly endophyte (Pandey et al. 2022; Sun et al. 2024).

Contamination by *Aspergillus* directly threatens agricultural production and food safety; therefore, control measures must be highly effective (Taheur et al. 2019; Hamad et al. 2018; Bagheri-Gavkosh et al. 2009). Control achieved by adding non-aflatoxigenic strains of *A. flavus* has proven very effective in reducing aflatoxin levels in crops (Hamad et al. 2018; Zhang et al. 2024). When combined with improved storage methods and the development of resistant varieties, this can effectively ensure an economical means of preventing *Aspergillus* from affecting agricultural production (Buchanan and Appel 2010).

1.2.2 Epidemiology of *Aspergillus* Contamination in Plant Products

Mycotoxin contamination, especially by organisms like *A. flavus*, is a global phenomenon that directly impacts food security and economic viability (Taheur et al.

2019; Hamad et al. 2018). Current estimates suggest that mycotoxins contaminate roughly 25% of all edible crops worldwide; consequently, toxins such as aflatoxins and ochratoxins exert a significant impact on both humans and animals (Hamad et al. 2018; Buchanan and Appel 2010). The environment, in particular temperature and humidity, remains a critical factor in the growth of *Aspergillus* (Bagheri-Gavkosh et al. 2009). Suboptimal storage conditions, including insufficient drying and excessive moisture, favor the growth of *Aspergillus* and heighten the risk of contamination (Taheur et al. 2019; Hamad et al. 2018). Among the *Aspergillus* species, *A. flavus* stands out as the major producer of aflatoxins and is most often found in cereals such as maize and peanuts (Hamad et al. 2018; Zhang et al. 2024). Typically, contamination begins in the field and persists in storage after harvest; therefore, both stages demand close monitoring. Appropriate field management practices to minimize fungal infections include crop rotation and the use of resistant varieties (Sun et al. 2024). Similarly, post-harvest processes like drying and controlled storage conditions should be applied to minimize contamination risks. This problem is most acute in tropical and subtropical areas where temperature and relative humidity are most conducive to the formation of aflatoxins by fungi (Bagheri-Gavkosh et al. 2009). Novel approaches to biocontrol, such as the introduction of *A. flavus* isolates other than Aflatoxigenic species, have effectively controlled contamination with aflatoxin (Hamad et al. 2018; Zhang et al. 2024). Integrated management practices, spanning both pre- and post-harvest phases, form the basis for effectively controlling *Aspergillus* contamination and ensuring food safety (Timmusk et al. 2017).

1.2.3 Mycotoxins Produced by *Aspergillus*

Mycotoxins from the *Aspergillus* genus are toxic metabolites that seriously threaten plant health, food safety, and human health (Taheur et al. 2019). Mycotoxins are secondary metabolites of mycobacteria and may cause widespread damage before and after harvesting, including the contamination of food products during processing (Bagheri-Gavkosh et al. 2009). The presence of these toxic substances results in substantial economic losses, including reduced product quality, decreased livestock productivity, and reduced foreign trade due to higher food safety standards (Hamad et al. 2018; Buchanan and Appel 2010).

1.2.3.1 Aflatoxin

The most toxic and extensively studied mycotoxins are aflatoxins produced primarily by *A. flavus* and *A. parasiticus* (Hamad et al. 2018; Zhang et al. 2024). Aflatoxins are highly toxic and carcinogenic; they have been reported near agricultural products such as maize, peanuts, cottonseed, and tree nuts (Taheur et al. 2019; Bagheri-Gavkosh et al. 2009). *A. flavi* represents a group of species capable of producing aflatoxins. These toxins are particularly interesting because they are resistant, meaning they can withstand changes in many food processing steps, such as cooking and drying, making them very difficult to destroy (Hamza et al. 2022). Toxins exert their carcinogenic effect by covalently attaching to DNA, inducing a genetic mutation that leads to hepatocellular carcinoma in animals and humans (Tan et al. 2014). Due to the high health risk, aflatoxin contamination of foodstuffs is a priority for regulatory action and food safety measures (Hamad et al. 2018; Buchanan and Appel 2010).

1.2.3.2 Ochratoxin

Another significant class of mycotoxins derived from *Aspergillus* species specifically from *A. niger*, *A. ochraceus*, and *A. carbonarius*, alongside *Penicillium verrucosum*, is Ochratoxin A (Taheur et al. 2019). Taxonomically, the species responsible for generating Ochratoxin fall into two primary clusters: *Circumdati* (e.g., *A. ochraceus*) and *Nigri* (e.g., *A. niger* and *A. carbonarius*). Contamination by OTA permeates agricultural and food networks; it appears across diverse plant and animal commodities (Taheur et al. 2019; Erfani et al. 2024). The toxicological profile of OTA includes teratogenic (causing congenital disabilities), embryotoxic (toxic to the developing embryo), carcinogenic (causing cancer), hepatotoxic (toxic to the liver), immunotoxic (toxic to the immune system), and nephrotoxic (toxic to the kidneys) (Tan et al. 2014). Due to these multiple toxicities, OTA remains a public health priority, especially in regions with intensive agriculture yet inadequate storage infrastructure (Buchanan and Appel 2010).

Mycotoxins (mainly aflatoxins and ochratoxins) are mainly caused by *Aspergillus* species, with serious consequences for food safety and public health (Taheur et al. 2019). While trichothecenes, fumonisins, and zearalenones are also vital mycotoxins, they are more frequently linked to *Fusarium* rather than *Aspergillus*. To mitigate health risks driven by mycotoxins, understanding their properties is essential; this knowledge enables us to refine our monitoring and containment strategies (Hamad et al. 2018; Buchanan and Appel 2010).

1.2.4 Impact of *Aspergillus* on Plant Yield and Food Safety

Crop infection by *Aspergillus* gravely diminishes farm output and food quality through yield loss and the production of toxic mycotoxins (Taheur et al. 2019). Ochratoxins, which are nephrotoxic and immunosuppressive, are produced by *A. niger* and *A. ochraceus*, whereas *Aspergillus flavus* synthesizes aflatoxins, which are potent hepatotoxic carcinogens. The three most significantly affected large crops are cottonseed (*Gossypium hirsutum*), peanuts (*Arachis hypogaea*), and maize (*Zea mays*) (Taheur et al. 2019; Bagheri-Gavkosh et al. 2009). In situations where post-harvest facilities are inadequate, these infections cause substantial economic losses. High humidity and poor storage conditions create favorable environments for fungal growth, leading to ongoing contamination as mycotoxins remain stable throughout food processing (Hamad et al. 2018). In addition to chemical and biological control, strategies such as using competitive, non-toxic strains of *Aspergillus*, enzymatic hydrolysis, and physical interventions like irradiation and heat treatment are employed (Zhang et al. 2024; Hamza et al. 2022). The prevention of *Aspergillus*-induced contamination and ensuring food security necessitate an integrated approach that includes biotech-nological innovations, enhanced storage facilities, and preventive agronomic practices (Timmusk et al. 2017; Malusá and Vassilev 2014). While probiotics, biopesticides, biofertilizers, and PGPR are often used synonymously, they refer to separate but sometimes overlapping functions of microorganisms (Pandey et al. 2022). We define probiotics in plant systems as beneficial microorganisms that, applied in sufficient quantities, can improve the health or resistance of plants through the manipulation of their microbiome. This differs from the previous definition of biopesticides, which are organisms that express pesticidal activity, and biofertilizers, which increase nutrient availability. The definition of PGPR is broad, encompassing a diverse array of growth-promoting bacteria present in the rhizosphere (Sun et al. 2024). Some of these genera and/or species may serve in multiple capacities, either synergistic or otherwise, depending on changes in context; however, the benefits of host-mediated health highlighted by the term probiotic are akin to those proposed in medical microbiology (Vessey 2003; Glick 2012; Malusá and Vassilev 2014).

1.3 Antifungal Properties of Probiotics Against *Aspergillus* in Plants

1.3.1 Mechanisms of Antifungal Action

1.3.1.1 Competitive Exclusion and Antimicrobial Metabolite Production by Probiotics

Two closely related mechanisms drive the antifungal action of LAB, especially the genus *Lactiplantibacillus* towards *Aspergillus* spp.: competitive exclusion and the generation of antimicrobial metabolites. While *L. plantarum* is important in the guts of humans and other animals (and in the food we ferment), *Lactiplantibacillus* species are frequently recovered from plant-associated settings. These include the rhizosphere and phyllosphere of various crops, such as maize, cabbage, and grapes. Thus, *L. plantarum* represents an organism type with potentially dual ecological relevance and biocontrol capability (Mahjoory et al. 2023).

LAB grows rapidly, colonizing the plant surface and occupying ecological niches while simultaneously preventing the establishment of phytopathogens via competition for nutrients and space (Sadati et al. 2022). Colonization of an ecological niche occurs through the adherence, invasion, and proliferation of teleomorph mycelium, which establishes a protective chemical barrier against microbial antagonism. Simultaneously, LAB species produce bioactive molecules such as lactic acid, phenylacetic acid, and bacteriocins that lead to a reduced pH of the surrounding environmental condition and inhibit fungal physiology. This biochemical antagonism inhibits the vegetative growth of *Aspergillus* spp. and halts their mycotoxin biosynthesis (including harmful mycotoxins like aflatoxins and ochratoxins) (Hamad et al. 2018).

Even with the potential benefits of these properties, LAB is usually transient in plant habitation due to several factors. These include limited structures for adhesion to plant surfaces, competition with indigenous microorganisms, and environmental stressors (e.g., desiccation and UV light) (Erfani et al. 2024). To mitigate this drawback and improve efficacy at the farm level, several alternatives were explored, such as microencapsulation, protective carrier systems, or co-application with appropriate PGPR that are more tolerant to environmental stressors; hereby, the co-application enhances colonization stability and biocontrol ability (Moradi Pour et al. 2022).

The biocontrol product *Aflasafe*, utilizing native non-aflatoxigenic strains of *A. flavus*, represents one of the rarest examples of competitive exclusion for mycotoxins. This product is being used in crops such as maize and groundnuts across several African countries, including Nigeria, Kenya, Senegal, and Ghana, to replace toxigenic strains with non-toxigenic strains, thereby reducing aflatoxin contamination on farms. Although infrequently, there have been reported instances of the *A. flavus* strains recovering the aflatoxin-inducing ability of the toxigenic strains due to horizontal gene transfer (HGT). Events like this underscore the importance of continuous genetic monitoring of microbial biocontrol agents, accompanied by legislative oversight, to ensure long-term biosafety (Ab EL-Fadeel et al. 2022).

1.3.1.2 Detoxification and Biodegradation of Mycotoxins

Biological systems, particularly those employing LAB, play a crucial role in mycotoxin detoxification through enzymatic degradation and microbial biosorption, specifically involving toxins like OTA and aflatoxins. Figure 1.1 illustrates a visual representation of the probiotic mycotoxin degradation pathways, including key microbial strains, enzymes, and metabolic products (Hamad et al. 2018).

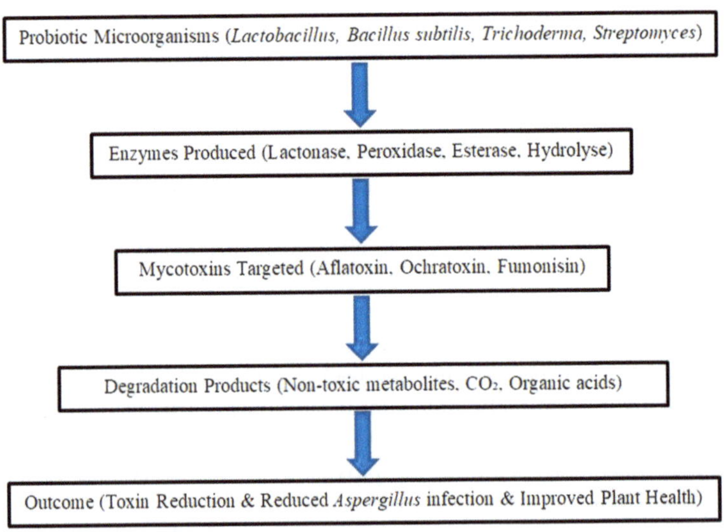

Fig. 1.1 Simplified probiotic-mediated detoxification of mycotoxins and plant health improvement

Probiotics, such as *Lactobacillus*, *Bacillus subtilis*, and *Trichoderma*, have been shown to interact with and degrade mycotoxins, including Aflatoxin and Ochratoxin. These probiotics reduce harmful mycotoxins, facilitating the conversion of toxic compounds into non-toxic metabolites such as CO_2 and Organic Acids. The diagram also highlights the role of enzymes such as lactonase and peroxidase in breaking down mycotoxins, thereby promoting plant health by preventing fungal infections caused by *Aspergillus* and *Penicillium*. While biosorption entails the adherence of mycotoxins to the bacterial cell surface, thereby diminishing their bioavailability, enzymatic degradation involves the action of specialized enzymes, such as lactonase and peroxidase, to break down toxic substances. Also, by releasing bacteriocins and lactic acid, which are inhibitory to the biosynthesis of mycotoxin, LAB directly inhibits the growth of *Aspergillus flavus* and *Aspergillus parasiticus*. Toxin production at the cellular level is inhibited during this detoxification process mediated by LAB, thereby improving the safety of the food through an eco-friendly alternative to mechanical and chemical decontamination methods (Mahjoory et al. 2023).

1.3.2 In Vitro Studies on Probiotics Against *Aspergillus* in Plant Systems

In vitro studies illustrate the efficacy of probiotics, particularly *Lactobacillus* species, as biocontrol agents against *Aspergillus* (Taheur et al. 2019). Probiotic Culture Supernatants (PCS) of *Bifidobacterium bifidum*, *Lactobacillus acidophilus*, and *Lactobacillus plantarum* inhibit *Aspergillus flavus* and *Aspergillus parasiticus*, inhibiting up to 77% of aflatoxin B1 (AFB1) production and inhibiting Aflatoxin G1 and G2 synthesis completely (Hamad et al. 2018). *Limosilactobacillus fermentum* strains are highly effective antifungals that are resistant to mimicked gastrointestinal stresses, indicating the double potential for plant biocontrol and human gut well-being (Adithi et al. 2022; Mahjoory et al. 2023) (Fig. 1.2). The findings suggest the potential of probiotics in reducing fungal infestations and mycotoxin biosynthesis, but further field applications need to be assessed under fluctuating environmental conditions to optimize their use (Ab EL-Fadeel et al. 2022).

1.3.3 In Vivo Studies and Field Applications

In vivo and in-field studies suggest the biocontrol potential of *Lactobacillus* species, particularly against *Aspergillus* species (Abdel-Nasser et al. 2024). PCS of *Bifidobacterium bifidum*, *Lactobacillus acidophilus*, and *Laciplantibacillus plantarum* suppress aflatoxin B1 production by as much as 77% (Hamad et al. 2018), and *Limosilactobacillus fermentum* strongly suppresses the growth of *Aspergillus* under in vitro gastrointestinal conditions (Mahjoory et al. 2023). Field trials witness their efficacy in preventing mycotoxin contamination and inducing plant resistance to fungal infection, thereby facilitating the production of a safe and sustainable food crop (Hamza et al. 2022). Research has to be conducted to modulate their activity under varying climatic conditions (Martinez et al. 2023).

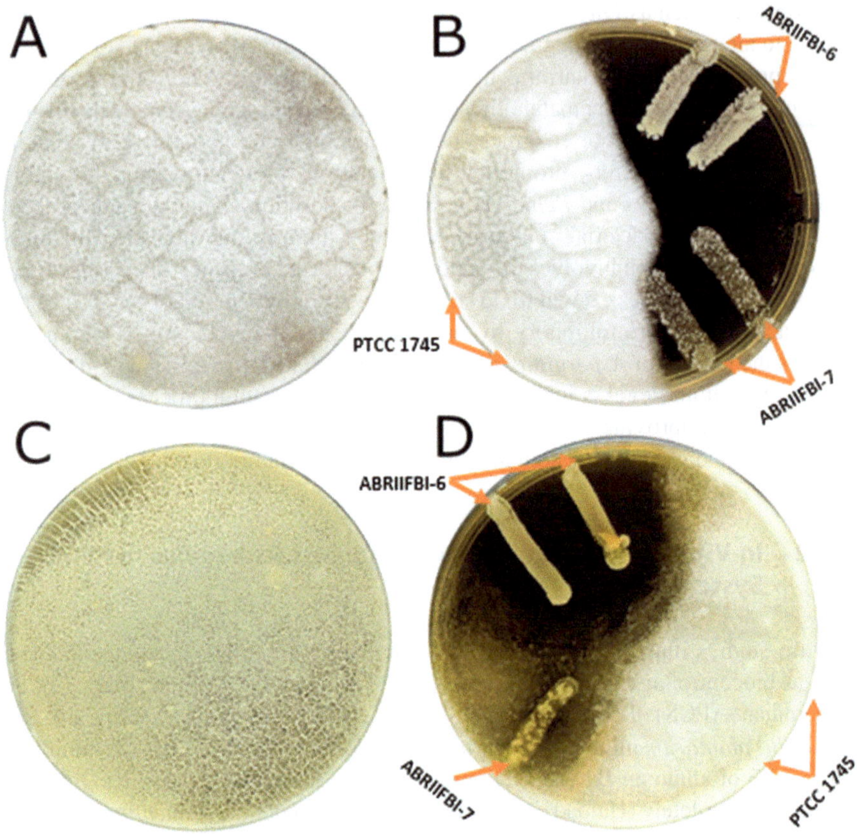

Fig. 1.2 Antifungal activity of potential probiotic strains against *Aspergillus flavus* and *Aspergillus niger* (Mahjoory et al. 2023). (**a**) *A. flavus* covered the entire plate in the absence of probiotic. (**b**) The potential probiotics ABRIIFBI-6 and ABRIIFBI-7 were able to inhibit the growth of *A. flavus*, whereas PTCC 1745, used as a control, was unable to do so. (**c**) *A. niger* covered the entire plate in the absence of probiotics. (**d**) The potential probiotic ABRIIFBI-6 was able to inhibit the growth of *A. niger* similarly to *A. flavus*, but ABRIIFBI-7 only slightly inhibited the growth of *A. niger*

1.4 Probiotic Formulations and Delivery Methods for Agricultural Use

1.4.1 Types of Probiotic Formulations for Crops

In enhancing crop productivity, resistance, and sustainability, probiotics and plant growth-promoting microorganisms (PGPMs) play a pivotal role. They work by promoting maximum nutrient uptake, antagonizing pathogens, and enhancing plant immunity (Sun et al. 2024). Ideal formulations must ensure microbial viability, stability, and effective delivery under field conditions (Abdel-Maksoud et al. 2024).

Mass-scale formulations encompass liquid, powder, granule, encapsulated, pellet, and seed coating technology. Consider liquid formulations, such as those of *Pseudomonas fluorescens*; while effective for rapid colonization, they remain sensitive to cold temperatures. Conversely, powder and granule formulations based on peat or biochar maintain controlled release via the soil. Encapsulation products such as nanoencapsulation and microencapsulation protect against stress to microbes and provide long-term activity (Moradi Pour et al. 2022). Synthetic biology technologies and increasingly sophisticated biopolymer coatings are driving the development of probiotic delivery, multispecies consortia, and engineered strains toward enhanced biofertilization and biocontrol (Upadhayay et al. 2023). Such operations ensure the adoption of green technology for agriculture in the face of upcoming agricultural challenges.

1.4.2 Delivery Methods for Probiotic Application to Plants

Plant probiotics enhance plant health, tolerance, and sustainability through effective delivery mechanisms that ensure optimal colonization and viability (Pandey et al. 2022). Key methods include seed treatment, foliar spraying, and soil planting. Treating seeds induces early symbiosis, leading to stronger seedling development, enhanced root elongation, and increased disease resistance (Sun et al. 2024). Advanced polymeric coatings and nanoencapsulation improve microbial stability and controlled release (Abdel-Maksoud et al. 2024). The foliar application safeguards aerial plant parts, promoting systemic resistance and stress tolerance. Although environmental factors may influence microbial persistence, these effects can be mitigated through the use of nanoencapsulation and biopolymer stabilizers (Erfani et al. 2024). Soil application targets the rhizosphere, improving nutrient solubilization, nitrogen fixation, and organic matter decomposition through drenching, granular formulations, and irrigation systems (Moradi et al. 2022). Furthermore, integrating nanotech-nology enhances probiotic delivery, maintaining microbial viability and allowing the controlled release of bioactive components (Upadhayay et al. 2023). Recognizing an ideal delivery system considering crop type, environment, and probiotic characteristics maximizes benefits, thereby improving sustainable agriculture and reducing dependence on chemical inputs.

1.4.3 Stability and Bioavailability Issues

Stability and bioavailability are crucial for accurately stimulating plant growth promotion, tolerance, and disease prevention by plant probiotics (Pandey et al. 2022). However, microbial survival is highly sensitive to environmental stressors such as UV radiation, temperature, and water stress (Abdel-Maksoud et al. 2024). To overcome these challenges, microencapsulation and nanoencapsulation are two novel stabilization techniques developed to provide controlled release and enhanced microbial viability (Moradi Pour et al. 2022). Among the biodegradable carriers offering greater protection and adherence to roots are alginate, chitosan, and starch polymers (Hamza et al. 2022). Integrating responsive biopolymer coatings and nanotech-nology is expected to enhance the sustainability and bioavailability of probiotics while reducing chemical usage in agriculture (Upadhayay et al. 2023).

1.5 Case Studies and Practical Applications

1.5.1 Successful Probiotic Interventions for *Aspergillus* Control in Crops

Probiotics employed in agriculture have demonstrated significant success in reducing mycotoxin accumulation and *Aspergillus* contamination (Hamad et al. 2018). Species of *Lactobacillus,* notably *Lacticaseibacillus casei* and *Laciplantibacillus plantarum*, display potent antifungal activity. Specifically, *L. casei* markedly cuts aflatoxin levels in contaminated grains and nuts. Simultaneously, *Limosilactobacillus fermentum* when paired with silymarin improves biocontrol against *A. flavus*, while cell-free supernatants of *L. plantarum* alter toxin production pathways (Sadati et al. 2022).

Although not harmful, bacterial probiotics such as *Saccharomyces cerevisiae* also effectively inhibit mycotoxins in maize, particularly fumonisin B1 (Hamza et al. 2022). In parallel, *A. flavus* strains compete with toxigenic strains to suppress aflatoxin production (Taheur et al. 2019). Collectively, these treatments highlight the potential of probiotics in biocontrol integration. Future research is now focusing on optimizing formulation, delivery, and probiotic consortia for broader agricultural applications (Martinez et al. 2023).

1.5.2 Effects on Mycotoxin Reduction

Probiotics, particularly among *Lactobacillus* species, have emerged as effective biocontrol agents for lowering mycotoxin contamination in food crops. Strains such as *L. plantarum* and *Lactobacillus kefir* bind and eliminate OTA, thereby reducing its bioavailability (Mahjoory et al. 2023). Antifungal agents derived from plants, like thyme oil, also demonstrate synergistic activity with probiotics, thus inhibiting aflatoxin production in *Aspergillus parasiticus*. Among yeast-based probiotics, *Saccharomyces cerevisiae* further decreases mycotoxin levels (Hamza et al. 2022; Shruthi et al. 2025). While non-toxigenic strains of *Aspergillus* replace their toxigenic counterparts to lower aflatoxin levels (Taheur et al. 2019), other microorganisms such as *Bacillus subtilis* and *Trichoderma viride* inhibit fungal mycotoxin production (Martinez et al. 2023). Ultimately, the combination of non-toxigenic fungi, multi-strain probiotic consortia, and botanical antifungals constitutes a sustainable approach. This strategy controls mycotoxin levels, enhances food safety, and reduces the application of chemical fungicides (Upadhayay et al. 2023).

1.6 Future Directions and Research Opportunities

1.6.1 Emerging Trends in Probiotic-Based Plant Protection

Currently, LAB and plant growth-promoting rhizobacteria stand at the forefront of probiotic applications regarding plant protection, emerging as sustainable substitutes for conventional pesticides (Pandey et al. 2022). Specifically, *L. plantarum* and *L. kefiri* represent two strains that effectively reduce the production of mycotoxins, especially aflatoxins derived from *Aspergillus* species. Probiotics have become more stable and bioavailable due to advancements in nanotechnology, such as nanoencapsulation, which makes them suitable for foliar spraying, soil application, and seed treatment (Moradi Pour et al. 2022). Furthermore, the synergistic effects of multi-strain microbial consortia paired with plant-derived bioactive compounds boost pathogen resistance and stress tolerance (Upadhayay et al. 2023). Looking ahead, future research will integrate synthetic biology, metagenomics, and precision agriculture to maximize probiotic applications, ensuring improved crop resilience, reduced dependence on chemicals, long-term soil health, and environmental sustainability (Timmusk et al. 2017).

1.6.2 Discovery of Novel Probiotic Strains with Antifungal Properties

Uncovering novel probiotic strains possessing antifungal properties is crucial for guaranteeing long-term plant protection (Hamad et al. 2018). A significant lactic acid bacterium, *L. plantarum*, employs bioactive metabolites such as organic acids and bacteriocins to inhibit fungal development and mycotoxin formation (Mahjoory et al. 2023). Moreover, by producing secondary metabolites that mimic those of their hosts, endophytic fungi contribute to plant defense, illustrating a co-evolutionary adaptation (Zhang et al. 2024). To optimize probiotic strains for agriculture and food safety applications, future inquiries should explore multispecies microbial consortia while employing genomics, metabolomics, and bioinformatics (Timmusk et al. 2017).

1.6.3 Sustainable Agriculture and the Role of Probiotics

Using plant probiotics in agricultural management constitutes a more environmentally friendly alternative to synthetic agrochemicals, as it enhances plant growth, nutrient absorption, and stress resilience (Pandey et al. 2022). Additional probiotic microorganisms, including endophytes and plant growth-promoting rhizobacteria, play a vital role in maintaining ecological balance and soil vitality (Haghshenas et al. 2023). Endophytic bacteria promote biocontrol, disease resistance, and stress tolerance by increasing crop hardness. Research should be targeted toward discovering the most suitable microbial consortia for specific environmental conditions to attain maximum agricultural productivity. Seed treatment, foliar sprays, and soil nutrient efficacy are increased due to nanotechnology's higher probiotic viability and controlled release properties (Moradi Pour et al. 2022). Finally, biological nitrogen fixers and plant bioactive compounds like *Azotobacter* species improve mycotoxin control and fertility of the soil (Sun et al. 2024).

1.7 Challenges and Limitations

1.7.1 Variability in Probiotic Efficacy

The effectiveness of probiotics in protecting crops against mycotoxin contamination depends on various factors, including microbial strain specificity, environmental conditions, microbial–plant interactions, and regulatory distinctions. Specific strains of *Lactobacillus* act as potent suppressors of *Aspergillus*, whereas others prove ineffective owing to variations in antimicrobial production and biofilm formation (Mahjoory et al. 2023). Fundamentally, LAB contain typical genetic types that define their antifungal activity. Furthermore, environmental drivers such as pH, temperature, and microbial competition heavily influence probiotic colonization (Erfani et al. 2024). Additionally, the absence of standardized global regulations creates a significant barrier to confirming efficacy (Malusá and Vassilev 2014). Addressing these obstacles requires guided screening via bioinformatics and the careful selection of strains, followed by formulation optimization tailored for application in specific agro-environments (Timmusk et al. 2017).

1.7.2 Regulatory Considerations for Agricultural Probiotics

On a global scale, the regulatory landscape for agricultural probiotics remains disjointed, characterized by varying criteria, approval procedures, and definitions applied across different regions. Unlike human probiotics, plant-based probiotics (PBPs) operate without a unified regulatory framework. This void obstructs standardization, stalls market growth, and weakens customer trust. While the EU mandates comprehensive safety assessments that can delay permits, PBPs are regulated in the United States under more general classifications, such as biopesticides or biofertilizers. In sharp contrast, the United States governs PBPs under broader umbrellas, often labeling them as biopesticides or biofertilizers. Likewise, in Australia and Brazil, pre-market evaluations are stringent. Science-based regulatory systems should define specific legal categories for agricultural probiotics and standardize safety assessments, including toxicological and environmental evaluations to enhance global acceptance. Establishing global standards for microbiological quality control, claims, and labeling is essential. Ultimately, harmonized regulations will enhance industry confidence, accelerate commercialization, and facilitate the sustainable integration of probiotics into agriculture (Malusá and Vassilev 2014).

1.8 Conclusion

1.8.1 Summary of Key Findings

This study presents alternative perspectives, particularly regarding plant-based probiotics, which are crucial in sustainable agriculture and food safety. Notably, *Lactobacillus kefir* FR7, an important contributor to LAB, has shown antifungal activity against *Aspergillus* species, reducing mycotoxin contamination and providing an eco-friendly biocontrol agent (Erfani et al. 2024). Endophytic microorganisms enhance plant stress resistance, nutrient uptake, and strength. Meanwhile, essential oils provide a natural means of food preservation. Strain specificity, environmental conditions, and regulatory discrepancies remain pressing concerns. Global standardization is a critical stepping stone to consumer trust and safety of use (Malusá and Vassilev 2014). Ultimately, integrating microbiome engineering, nanotechnology, and precision agriculture will enhance the scalability and efficiency of probiotic-based biocontrol (Timmusk et al. 2017).

1.8.2 Implications for Sustainable Plant Health Strategies

Plant probiotics, such as lactic acid bacteria (LAB) and endophytes, play a crucial role in promoting sustainable agriculture by enhancing vigor and resistance and increasing plant productivity. Beyond growth, these microbes enhance stress resistance in the plant, augment nutrient efficiency, and induce resistance to plant diseases, offering a biological alternative to chemical use (Erfani et al. 2024).

PGPR enhances resource utilization and reduces reliance on synthetic agrochemicals while promoting soil health concurrently (Pandey et al. 2022). Integrating nanotechnology with plant probiotics significantly enhances microbial stability, targeted delivery, and efficiency, thereby reducing the environmental impact of agriculture while optimizing yield and quality (Timmusk et al. 2017).

Endophytes and bioprotective cultures serve as effective biocontrol measures against fungal pathogens, enhancing plant defense mechanisms and promoting the production of secondary metabolites. Essential oils from medicinal plants inhibit fungal infestation and maintain post-harvest quality, while toxigenic fungal isolates mitigate mycotoxin accumulation, thus improving food safety. However, it is essential to recognize that even non-aflatoxigenic strains, such as those used in *Aflasafe*, may regain the ability to produce toxins through horizontal gene transfer under specific field conditions. This necessitates continued genetic monitoring and regulatory oversight to ensure long-term biosafety of such biocontrol strategies (Martinez et al. 2023). Future research should concentrate on enhancing microbial utilization, establishing global regulations, and utilizing synthetic biology, precision agriculture, and computational methods for microbial selection. These technologies will facilitate the global adoption of sustainable plant health initiatives, reducing chemical usage and promoting environmentally friendly agriculture, thereby enhancing global food security (Timmusk et al. 2017).

1.9 Chapter Summary

This chapter has delivered an all-encompassing review regarding the function of probiotics as antifungal agents combating *Aspergillus* species within plants and plant-derived products. Specifically, probiotic microorganisms—notably lactic acid bacteria—display robust antifungal activity. They achieve this through mechanisms like competitive exclusion, the production of antimicrobial metabolites, and the enzymatic detoxification of mycotoxins. Research spanning both in vitro and in vivo contexts has underscored their capacity to stifle *Aspergillus* growth while simultaneously reducing contamination from aflatoxin and ochratoxin. We explored probiotic-based formulations and novel delivery methods, including encapsulation and nanotechnology, as effective strategies. These approaches bolster microbial stability and bioavailability within agricultural systems. Practical case studies showcased the use of probiotics in crop protection. Meanwhile, emerging research highlighted the critical role of multi-strain microbial consortia and synthetic biology in molding the future of probiotic-based plant health strategies. Notwithstanding regulatory and environmental obstacles, probiotics stand as a promising cornerstone of sustainable agriculture. They offer eco-friendly alternatives to chemical fungicides and contribute significantly to global food safety and security.

References

Ab EL-Fadeel RH et al (2022) Biological control of some types of mycotoxins by probiotic bacteria and yeast strains. Egypt J Dairy Sci:88–96. https://doi.org/10.21608/ejds.2022.171199.1008

Abdel-Maksoud G et al (2024) Biosynthesis of titanium dioxide nanoparticles using probiotic bacterial strain *Lactobacillus rhamnosus* and evaluation of their biocompatibility and antifungal activity. Biomass Convers Biorefinery 14(19):23961–23983

Abdel-Nasser M et al (2024) Evaluating the efficacy of probiotic bacterial strain *Lactobacillus plantarum* for inhibition of fungal strains associated with historical manuscript deterioration: an experimental study. Fungal Biol 128(6):1992–2006

Adithi G, Rakesh S, Divyashree S, Shruthi B, Sreenivasa MY (2022) Assessment of probiotic and antifungal activity of *Lactiplantibacillus plantarum* MYSAGT3 isolated from locally available herbal juice against mycotoxigenic *Aspergillus* species. Food Biosci 50:102118. https://doi.org/10.1016/j.fbio.2022.102118

Bagheri-Gavkosh S et al (2009) Inhibitory effects of Ephedra major host on *Aspergillus parasiticus* growth and aflatoxin production. Mycopathologia 168:249–255

Bengmark S (2017) Inflammation and microbiota and gut reconditioning. In: Inflammation: from molecular and cellular mechanisms to the clinic, pp 1609–1660

Buchanan RL, Appel B (2010) Combining analysis tools and mathematical modeling to enhance and harmonize food safety and food defense regulatory requirements. Int J Food Microbiol 139: S48–S56

D'Almeida AP et al (2024) Plant-based probiotic foods: current state and future trends. Food Sci Biotechnol 33(15):3401–3422

Erfani A et al (2024) Specific species of probiotic bacteria as bio-preservative cultures for control of fungal contamination and spoilage in dairy products. Int Dairy J 151:105863

Glick BR (2012) Plant growth-promoting bacteria: mechanisms and applications. Scientifica 2012: 963401

Haghshenas B et al (2023) Probiotic properties and antimicrobial evaluation of silymarin-enriched *Lactobacillus* bacteria isolated from traditional curd. Sci Rep 13(1):10916

Hamad GM et al (2018) Combination probiotic supernatants reduce growth and aflatoxin production by *Aspergillus* spp. in food contamination. Am J Food Sci Technol 6:57–67

Hamza MF et al (2022) Functionalization of magnetic chitosan microparticles for high-performance removal of chromate from aqueous solutions and tannery effluent. Chem Eng J 428:131775

Mahjoory Y et al (2023) Antifungal activity of potential probiotic *Limosilactobacillus fermentum* strains and their role against toxigenic aflatoxin-producing aspergilli. Sci Rep 13(1):388

Malusá E, Vassilev N (2014) A contribution to set a legal framework for biofertilizers. Appl Microbiol Biotechnol 98:6599–6607

Martinez Y et al (2023) Biotechnological development of *Trichoderma*-based formulations for biological control. Appl Microbiol Biotechnol 107(18):5595–5612

Mohamed NG, Abdel-Hakeem MA (2023) Chitosan nanoparticles enhance drought tolerance in tomatoes (*Solanum lycopersicum* L.) via gene expression modulation. Plant Gene 34:100406

Moradi Pour M, Saberi Riseh R, Skorik YA (2022) Sodium alginate–gelatin nanoformulations for encapsulation of *Bacillus velezensis* and their use for biological control of *Pistachio* gummosis. Materials 15(6):2114

Moradi P et al (2022) Silicon nanoparticles and methyl jasmonate improve physiological response and increase expression of stress-related genes in strawberry cv. Paros under salinity stress. Silicon 14(16):10559–10569

Nasrollahzadeh A et al (2022) Antifungal preservation of food by lactic acid bacteria. Foods 11(3): 395

Pandey SS et al (2022) Plant probiotics–endophytes pivotal to plant health. Microbiol Res 263: 127148

Sadati SYR et al (2022) Zinc oxide nanoparticles enhance drought tolerance in wheat via physio-biochemical changes and stress gene expression. Iran J Biotechnol 20(1):e3027

Shruthi B, Deepthi BV, Adithi G, Vanitha PR, Sreenivasa MY (2025) *Saccharomyces cerevisiae* MYSY1 as a promising probiotic isolate with antifungal properties against foodborne *Aspergillus* species. Biocatal Agric Biotechnol 67:103749. https://doi.org/10.1016/j.bcab.2025.103649

Sun W, Shahrajabian MH, Soleymani A (2024) The roles of plant-growth-promoting rhizobacteria (PGPR)-based biostimulants for agricultural production systems. Plants 13(5):613

Taheur FB et al (2019) Use of lactic acid bacteria for the inhibition of *Aspergillus flavus* and *Aspergillus carbonarius* growth and mycotoxin production. Toxicon 166:15–23

Tan J et al (2014) The role of short-chain fatty acids in health and disease. Adv Immunol 121:91–119

Timmusk S et al (2017) Perspectives and challenges of microbial application for crop improvement. Front Plant Sci 8:49

Upadhayay VK et al (2023) Synergistic impact of nanomaterials and plant probiotics in agriculture: a tale of two-way strategy for long-term sustainability. Front Microbiol 14:1133968

Vessey JK (2003) Plant growth promoting rhizobacteria as biofertilizers. Plant Soil 255:571–586

Zhang Q et al (2024) Genomic insights into *Aspergillus tamarii* TPD11: enhancing polyphyllin production and uncovering potential therapeutic applications. BMC Genomics 25(1):9

Tackling *Fusarium*: Probiotics in Action

2

N. Deepa, G. Adithi, Rakesh Somashekaraiah, P. K. Maheshwar, and M. Y. Sreenivasa

Abstract

Fusarium is considered as notorious for its role in devastating agricultural losses due to its mycotoxin production, which poses significant challenges in the field of global food security, animal, and human health. Even after application of physical, traditional control methods like chemical fungicides and biological methods, there has been the development of resistant strains, which often leads to environmental concerns. In recent periods, probiotics have emerged as a promising, eco-friendly alternative for combating *Fusarium* and its species. The study explores the application of probiotics involving bacteria and fungi in vindicating *Fusarium* infection in plants by suppressing mycotoxin production and by enhancing plant and soil health. Antagonism is said to be the mechanism involved in the production of antifungal metabolites, induction of systemic resistance, and competition for niches and nutrients in host plants. Probiotics are said to contribute to soil microbiome balance and foster resilience against *Fusarium* outbreaks. Numerous trials have been in practise for the reduction in *Fusarium* incidence and

N. Deepa
Applied Mycology Laboratory, Department of Studies in Microbiology, University of Mysore, Mysuru, Karnataka, India

Yuvarajas College, University of Mysore, Mysuru, Karnataka, India

G. Adithi · M. Y. Sreenivasa (✉)
Applied Mycology Laboratory, Department of Studies in Microbiology, University of Mysore, Mysuru, Karnataka, India
e-mail: mys@microbiology.uni-mysore.ac.in

R. Somashekaraiah
Project Management, PSI CRO Pharma India Pvt. Ltd., Bengaluru, Karnataka, India

P. K. Maheshwar
Yuvarajas College, University of Mysore, Mysuru, Karnataka, India

M. Y. Sreenivasa et al. (eds.), *Probiotics as Antifungals*,
https://doi.org/10.1007/978-981-92-0002-3_2

its associated mycotoxins in crops by the potential field-level implementation. Here we discuss the latest advances in the application of probiotics for *Fusarium* management, their limitations, and the steps to employ for the large-scale adoption towards global agriculture by reducing synthetic chemicals.

Graphical Abstract

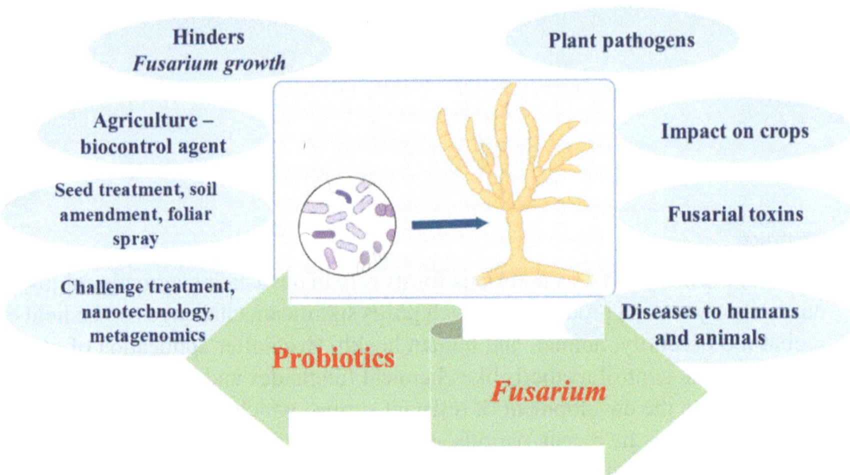

Keywords

Agriculture · Disease · Infection · Mycotoxins · Pathogen

Learning Objectives

- To understand the threat posed by *Fusarium* to the agricultural sector and associated environmental challenges.
- To explore probiotics as a sustainable solution by elucidating their role and mechanisms of action in agriculture.
- To analyze the challenges and future directions of probiotic applications for *Fusarium* management, with emphasis on strain specificity, environmental influences, and scalability.
- To promote sustainable agricultural practices by evaluating the environmental and economic benefits of substituting chemical fungicides with probiotic-based approaches strategies for long-term crop health.

2.1 Introduction

The topic of the study is a fascinating exploration of how beneficial microorganisms like probiotics are employed to combat *Fusarium*, a pathogenic fungus that is known for causing significant agricultural losses and health concerns. *Fusarium*, being a filamentous fungus worldwide, comprises abundant species, among them some are saprophytes and some are pathogens (Dean et al. 2012). *Fusarium* species exhibit significant genetic and ecological diversity, enabling them to infect a broad range of hosts and survive in varied environments. Some important species of *Fusarium* widely studied for the mycotoxins, such as *Fusarium oxysporum, F. graminearum, F. solani, F. verticillioides,* and *F. proliferatum* (Ma et al. 2013). These species of *Fusarium* are commonly found in plant debris and soil, and they are found to affect plants, animals, and humans (Sreenivasa et al. 2011). Certain species produce specific mycotoxins, causing a dual threat to agriculture and health (Compant et al. 2005). Some species are highly resistant to traditional fungicides, which are an alternative approach in their management by physical, chemical, and biological. Most physical and chemical methods to reduce the growth of *Fusarium* and its toxin concentrations in food or feed have shown to be rather ineffective or to be difficult to implement into the production process. Hence, the biological control method is the best alternative for the management of fungi and its mycotoxins (Deepa et al. 2021).

In recent days, lactic acid bacteria (LAB) and some strains of *Saccharomyces cerevisiae* var. *bourardii* have immense potential as tools in tackling the problem of mycotoxins in cereal-based foods (Fig. 2.1) (Adithi et al. 2025; Shruthi et al. 2024, 2025; Rakesh et al. 2021; Deepthi et al. 2017). Lactic acid bacteria, which are a commensal of the human gut, have been applied as a safe biocontrol agent to minimize the growth and production of fumonisin-producing *F. verticillioides. Lactobacillus rhamnous* and *S. cerevisiae* inhibited *F. moniliforme* growth in vitro and eliminated FB1 from the bodies of mature rats in vivo by orally administering the doses of biocontrol agents (Al-masri et al. 2011). *Saccharomyces cerevisiae* as a biocontrol agent controls the *F. moniliforme* growth as well as FB1 production in seeds (Fareid 2011). *Pediococcus pentosaceus* has GRAS (Generally Regarded as Safe) status, has been widely used in the fermentation of foods, and could be suitable as a biocontrol organism to improve the quality of ensilage. Antifungal activity in *P. pentosaceus* culture supernatant, which is pH-dependent, was observed at the end phase of growth, inhibiting *F. verticillioides* and *F. proliferatum* growth in vitro. Antifungal metabolites of the biocontrol agent were proven to be heat stable and resistant to proteolytic enzymes.

The recent trend of biological control using probiotics is widely practiced. Probiotics are live microorganisms that, when applied in adequate amounts, confer health benefits to the host, such as human and animal health (Raaijmakers 2009). These beneficial microbes also enhance water, soil, and overall health of ecosystem. Probiotics not only maintain a healthy gut but also play a critical role in maintaining balance and improving resilience in ecosystems by supporting beneficial microbial communities and suppressing the harmful ones (Ongena and Jacques 2008).

Fig. 2.1 Probiotics against *Fusarium* species

Probiotics in recent days have been found to be a versatile tool in ecosystem management, promoting health and resilience across natural and agricultural systems. They also offer an eco-friendly approach as an alternative to chemical fertilizers, aligning with sustainable development goals through improving productivity by reducing environmental degradation and fostering biodiversity (Pieterse et al. 2014).

Applications of these probiotics are in the field of agriculture, wastewater treatment, aquaculture, and environmental remediation. These probiotics enhance soil health through nutrient cycling, soil structure improvement, and reduction in chemical inputs to the field. Species of *Bacillus* and *Pseudomonas* facilitate decomposition of organic matter along with nutrient mineralization (Harman et al. 2004). Similarly, *Rhizobium* and *Azospirillum* species promote nitrogen fixation and phosphate solubilization by accessing nutrients for plants. Certain probiotics produce polysaccharides and biofilms which bind to soil particles by improving its structure, aeration, and water retention. They enhance natural nutrient availability by reducing the need for synthetic fertilizers and its associated environmental impacts. Probiotics act as a biocontrol agent in plants by suppressing diseases through antagonism and competition mechanisms by producing antifungal, antibacterial compounds, and outcompeting pathogens for nutrients and its colonization sites, respectively (Raaijmakers 2009). Species like *Trichoderma* combat soilborne pathogen *Fusarium* and *Bacillus subtilis* protect crops by producing antibiotics and enhancing its root health. This is how probiotics trigger plants' natural immune responses, enabling them to resist infections better by causing the induction of systemic resistance (Compant et al. 2005).

2.2 Overview of *Fusarium* Species and Their Impact

The impact of *Fusarium* mainly leads to loss in the agricultural sector through reduction in crop yield by causing infections in plants like wilt, rot, blight, and also contamination with mycotoxins (Table 2.1), reducing the market value for crops, posing safety concerns, leading to quality degradation of crops (Leslie and Summerell 2006). Certain economic consequences will be faced by causing direct losses from reduced harvests; trade restrictions will be applied for mycotoxin-contaminated exports, increasing costs for disease management, including for resistant crop varieties and fungicides (Ploetz 2006).

Consumption of mycotoxin-contaminated food leads to acute and chronic illness, like gastrointestinal distress, cancer, and immune suppression in humans, whereas in animals it leads to reduced productivity, mortality in livestock, and certain reproductive issues (Goswami and Kistler 2004; Sreenivasa et al. 2006; Dass et al. 2007). The persistent spores of *Fusarium* contribute to long-term soil contamination, causing fields unsuitable for susceptible crops, and the overuse of chemical fungicides causes environmental pollution and resistance development. To overcome the impact of *Fusarium,* it requires integrated strategies like the development of resistant crop varieties, advanced monitoring and diagnostic tools for early detection and management (Desjardins 2006).

Table 2.1 *Fusarium* species affecting crops

Fusarium pathogen	Disease caused	Crop affected	Toxin produced
Fusarium oxysporum	*Fusarium wilt* vascular wilts	Tomatoes, bananas (Panama disease), and cucurbits.	Trichothecene, Beauvericin
Fusarium graminearum	Head blight	Wheat, barley, and maize	Deoxynivalenol
F. equiseti	Root and stem rot	Cereal crops	Trichothecene
F. culmorum	Root rot, head blight	Wheat, barley	DON, zearalenone
Fusarium solani	Root and stem rots	Like beans, potatoes, and peas	Moderate toxins
Fusarium verticillioides	Stalk and ear rot	Maize	Fumonisins
Fusarium proliferatum	Dry rots, leaf spots, root rot	Variety of crops, including onions, garlic, and cereals like maize, barley, wheat, feed	Fumonisins and beauvericin

2.2.1 Plant Pathogens and Toxigenic Effects

Numerous *Fusarium* species, which are plant pathogens, not only reduce the growth and yield of the plant, causing a disease, but also produce their secondary metabolite called toxins that affect humans and animals that feed on such cereal-based food and feed (Sreenivasa et al. 2008; Summerell 2010). *Fusarium oxysporum* proved to be a notorious pathogen causing vascular wilts in tomatoes, pigeon pea, chickpea, chilli, banana, and cotton with the names *Fusarium* wilt, Panama disease, cotton wilt, respectively (Thangavelu and Mustaffa 2010). This pathogen invades the plant's vascular system, blocking water and nutrient transport, leading to symptoms like wilting, yellowing of leaves, stunted growth, and finally death of the plant, and *Fusarium graminearum* affects wheat and barley crops; this pathogen produces Deoxynivalenol and Zearalenone toxins that cause nausea and vomiting and lead to reproductive disorders and immune suppression (Rini and Sulochana 2007).

The leading pathogens *Fusarium verticillioides* and *Fusarium proliferatum* cause *Fusarium* root rot in beans and peas, *Fusarium* ear rot in maize, *Fusarium* stalk rot in maize, and *Fusarium* fruit rot among various crops. This pathogen attacks the plant tissues, causing decay and necrosis, and often enters through wounds and stressed plant systems too (Pertot 2017). These two species of *Fusarium* produce a toxin, Fumonisin B1, found in maize, which leads to esophageal cancer, liver damage, and neural tube defects when consuming the contaminated food and feed (Marasas et al. 2004). Trichothecene, Beauvericin, and Enniatins are emerging mycotoxins with potential antimicrobial and cytotoxic effects produced by several *Fusarium* species (Goswami and Kistler 2004).

2.2.2 Economic and Ecological Significance of *Fusarium*

Many factors involves and corelate to economic significance of *Fusarium* which includes direct losses, increased management costs, market and trade impact, global food security, Reduction in agricultural yield due to diseases caused by fungal pathogens like *Fusarium* wilt, head blight, root rot leads to direct losses where post-harvest losses from mycotoxin contamination render crops as unsafe for consumption and sale (Placinta et al. 1999; Deepa et al. 2016). Due to these diseases, it increases management costs by investing in fungicides, resistant crop varieties, crop rotation practices, storage, and to mitigate mycotoxin contamination (Chakraborty and Newton 2011). Due to contaminated produce, it faces loss of market access by reducing competitiveness and also countries face stringent export regulations through mycotoxin contamination limits (Munkvold 2003). These *Fusarium* species, as they affect staple crops like wheat, maize, and bananas, pose a threat to food supply chains, especially in developing countries, as global security is at a crossroads. Ecological significance of *Fusarium* faces soil microbial dynamics, persistence in the environment, biodiversity impact, and climate change interactions (Pertot 2017).

Species of *Fusarium* compete with beneficial microbes in soil, dominating niches due to its ability to produce antimicrobial compounds, alternating soil microbiomes, reducing soil health and crop productivity over time (McMullen et al. 2012). Some *Fusarium* species produce hardy chlamydospores and mycotoxins that persist in soil and plant debris for years, making fields unsuitable for susceptible crops without remediation effects (Munkvold 2003). Due to the wide usage of chemical fungicides to control *Fusarium,* it indirectly affects non-target species and disrupts the ecological balance, causing biodiversity impact. Due to unfavorable climatic conditions like change in temperature, humidity levels, carbon dioxide concentrations influence *Fusarium* growth, spread, and its toxin production, in which warmer climates and increased rainfall create favorable conditions for *Fusarium* epidemics (Schulz and Boyle 2005).

2.2.3 Impact on Crops

Fusarium species inflict significant damage to various crops, leading to both quantitative and qualitative impacts.

- *Maize* is infected by *F. verticillioides* and *F. proliferatum*, causing *Fusarium* ear rot and stalk rot. The stalk rot infection causes damage by weakening the plants and causing them to fall over, which complicates harvesting, whereas ear rot damages kernels, resulting in poor grain quality (Lamichhane and Venturi 2015). These infections in maize reduce the yield at 10–30% in areas with high disease pressure. The pathogen contaminates the grains with its mycotoxin Fumonisin (FUM) in food and feed, leading to financial loss. Consumption of such infected food or feed leads to cancer, liver damage, and neural tube defects as it mainly affects the liver, lungs, and kidneys (Berendsen et al. 2012).
- *Wheat, barley, sorghum, other cereals, and oil seeds* get infected from pathogen *Fusarium graminearum*, *F. culmorum*, and *F. sporotrichioides*, causing *Fusarium* head blight with production of mycotoxin deoxynivalenol (DON), Zearalenone (ZEN), T2, and HT2 toxins. Infection in wheat with the pathogen and toxin causes shriveled grains, reducing the grain weight and yield in the market ranging 10–70% during severe epidemics (Backhouse 2014). These affected grains may be rejected or downgraded, leading to economic losses for farmers. Consumption of such infected food or feed leads to nausea, vomiting, immune suppression, estrogenic effects, skin irritation, and reproductive issues (Chakraborty and Newton 2011).
- *Banana, tomato, beans, and peas* are infected due to pathogen *F. oxysporum* and *F. solani*, causing *Fusarium* wilt in banana and tomato, whereas root rot in beans and peas. The infection causes yellowing of leaves, wilting, and eventually plant death. Sometimes, entire plantations can be wiped out in regions where the pathogen is endemic, leading to a severe threat to livelihoods and global supply chains (Pertot 2017).

2.2.4 Secondary Effects by Toxin Contaminants in Food and Feed Chain

Fusarium causes both direct damage to crop yield and indirect damage through toxin contamination in food chains. These impacts pose significant challenges to global agriculture, human health, and economic stability. Chronic exposure to *Fusarium* mycotoxins leads to cancer, immune deficiencies, and endocrine disruption in humans (Pestka 2010). Many reports have been recorded as mycotoxicosis in regions where contaminated grains were consumed. Animal health reports and mycotoxin-contaminated feed consumption reduced livestock productivity and fertility with symptoms pertaining to feed refusal, organ damage, weight loss, and death in severe cases (Pinton and Oswald 2014). Certain economic losses will be faced by imposing strict regulations on permissible mycotoxin levels, which leads to crop rejection and also reduces the market access by disposal of contaminated feed and food. Majorly in the regions depending on the staple crops like wheat, banana, and maize threaten food supplies by yield and quality loss by food security agencies (Alassane-Kpembi et al. 2017). Trade restrictions will be imposed on certain countries as they need to face barriers to export contaminated crops, affecting international trade and the economy. Farmers' misconceptions about the usage of chemical fungicides that harm non-target organisms, reducing soil biodiversity, need to be stopped (Kowalska and Habrowska-Górczyńska 2020). To overcome these infections and contamination, the only way to be followed is managing *Fusarium* infection, including resistant crop varieties, biological controls like probiotics, and stringent monitoring of mycotoxins, which are crucial to mitigate *Fusarium* adverse effects.

2.2.5 Limitations of Conventional Methods as Control Measures

Conventional control measures are useful but face significant limitations in combating *Fusarium* species. Chemical fungicides are hampered by resistance, environmental impact, and incomplete efficacy, and the other is breeding resistant crops, even though it is a slow and complex process.

Chemical Fungicides Chemical fungicides have been a primary tool for managing *Fusarium* species infections in crops. It has certain limitations like efficacy challenges, development of resistance, health and environmental concerns, and economic burden. Many fungicides are ineffective against mycotoxin production by *Fusarium* even after fungal growth is inhibited (Minervini and Dell'Aquila 2008). These chemical fungicides often fail to provide complete protection against *Fusarium*, particularly for soilborne pathogens like *F. oxysporum*. Some *Fusarium* species can survive as chlamydospores or colonize plant debris, reducing fungicide efficacy. Certain fungicides cause resistance on repeated and widespread use, leading to the selection of resistant *Fusarium* strains, making them less effective over time. Resistance to one fungicide often leads to resistance to others within the same class, which is also a limiting factor (Marasas et al. 2004). Fungicides also cause a limiting factor for health concerns and environmental safety, as the excessive fungicide use contaminates soil and water systems, disrupting microbial communities and harming non-target organisms. Fungicides can pose risks to human and animal health, causing carcinogenicity and endocrine disruption (Voss et al. 2007). Due to this, certain countries have banned some fungicides with strict regulations on limiting their availability and effectiveness in *Fusarium* management. These fungicides also cause economic burden for farmers due to frequent application of fungicides, leading to an increase in production cost, and also during severe outbreaks with high disease pressure scenarios, these fungicides fail to prevent significant losses (Warth et al. 2013).

Breeding-Resistant Crops Developing crop varieties resistant to *Fusarium* is an essential strategy, with certain limitations; as the technology is a time-consuming process, it shows complexity of resistance, yield penalties, due to limited genetic resources (Alberts et al. 2016). *Fusarium* exhibits high genetic diversity and adaptability, often outpacing breeding efforts with the emergence of new virulent strains. Conventional breeding methods require several years, as long breeding cycles to identify and cross desirable traits, test progeny, and release-resistant varieties (Paul et al. 2008). Resistance may vary depending on environmental conditions like temperature, humidity, and soil type. Resistance to *Fusarium* often involves multiple genes, complicating the selection process, as resistance to *F. graminearum* for *Fusarium* head blight in wheat involves multiple quantitative trait loci (QTLs) (Yin 2009). Breeding for disease resistance can sometimes result in reduced yield or quality, making resistant varieties less attractive to farmers, as these resistance traits may not fully eliminate mycotoxin contamination (Van der Fels-Klerx 2016). Some *Fusarium* species, such as *F. oxysporum*, have host-specific pathotypes requiring unique resistance genes for each crop (Buerstmayr 2009). The Cavendish variety of banana requires genetic uniformity, making breeding for resistance challenging due to the limited availability of resistant genes in germplasm collections, hindering progress (Miedaner and Juroszek 2021).

2.3 Probiotics in Combating *Fusarium* Pathogen

Fusarium, posing significant threats to agricultural productivity and ecosystem health, causes devastating diseases in plants and contaminates food and food-based products with harmful mycotoxins (Zhao 2014). Probiotics, when applied to soil or plants, offer a promising eco-friendly solution to manage pathogens. Probiotics combat fungal pathogens by many methods, like competitive exclusion, production of antifungal compounds, induction of systemic resistance in plants, disruption of pathogen communication, enhancement of plant growth and health (Syed-Ab-Rahman 2018). Probiotics also play their major role in aquaculture by regulating water quality, which breaks down organic waste and reduces ammonia and nitrate levels through nitrifying and denitrifying processes and suppressing the growth of algal and fungal pathogens in water. These gut microbes restore the ecological balance, leading to environmental remediation by degrading toxic pollutants including hydrocarbons, pesticides, heavy metals, and bioremediation with the help of *Pseudomonas* and *Mycobacterium,* and also restore ecological function and biodiversity in degraded environments by replenishing beneficial microbes (Deepa et al. 2022; Adithi et al. 2024).

Probiotics help the ecosystem by mitigating climate change effects, promoting carbon sequestration in soil through microbial activity, and reducing greenhouse methane gas emissions from agriculture, which alters microbial communities in manure and rumen. These probiotics are serving as a promoter in sustainable agriculture, becoming an integral to Integrated Pest management and organic farming systems by reducing reliance on chemical inputs by fostering sustainable and eco-friendly practices (Deepa and Sreenivasa 2019).

2.3.1 Probiotics Against *Fusarium Species*

Lactobacillus paracasei isolated from sourdough inhibited *F. proliferatum* and *F. graminearum* in liquid media via production of organic acids and low pH effects, while fungal growth decreased, mycotoxin (trichothecene) production increased (Rakesh et al. 2019; Divyashree et al. 2024). *Lactobacillus plantarum* and *Lactobacillus rhamnosus* screened LAB for antifungal effects and found that *L. rhamnosus* VT1's cell-free supernatant (CFS) inhibited *F. graminearum* growth by 75–84% under optimized incubation (Banu et al. 2021). *Fusarium graminearum* and its DON degradation were reported to *L. plantarum* MYS6 and were also shown to act against fumonisin-producing *F. proliferatum* in poultry feed, reducing both fungal colonization and toxin accumulation (Deepthi et al. 2016; Rakesh et al. 2021). In recent studies, Adithi et al. (2025) proved *Lactobacillus plantarum* isolated from vellapam for detoxification of fumonisin produced by *F. verticillioides*. *Lactiplantibacillus* derived from tropical fruits is reported to minimize the growth of *Fusarium oxysporum* (Vasundaradevi et al. 2024). *Lactobacillus* spp. like *L. plantarum, L. rhamnosus*, and *L. paracasei* show promising antifungal activity against *Fusarium*, along with mycotoxin binding/degradation potential (Table 2.2).

2.3.2 Mechanisms of Probiotics to Hinder *Fusarium* Growth

In competitive exclusion, probiotic microbes colonize the plant surface leaves and roots that occupy niches which will be taken over by *Fusarium* species. *Bacillus subtilis* and *Pseudomonas fluorescens* colonize plant roots, which form a barrier against *F. oxysporum* (Schisler et al. 2011). These probiotics produce antibiotics, enzymes, and secondary metabolites like lipopolypeptides, volatile organic compounds, and enzymes that inhibit or kill fungal pathogens. Iturin and Surfacin from *Bacillus subtilis* disrupt fungal cell membranes, glotoxin from *Trichoderma* targets fungal metabolic pathways, and enzymes like chitinase and glucanases break

Table 2.2 Probiotics against *Fusarium species*

Probiotic	Fusarium species	Mode of action	References
Lactobacillus plantarum	*F. verticillioides*	Fumonisin detoxification in animal feed and cereals	Deepthi et al. (2017), Adithi et al. (2025)
Lactobacillus rhamnosus	*F. proliferatum*	Reduces the bioavailability of fumonisins and moniliformin both in in vitro and in animal models	El-Nezami (2007)
Lactobacillus plantarum MYS6	*F. proliferatum*	Fumonisin degradation	Deepthi et al. (2016)
Lactobacillus plantarum	*F. equiseti*	Potential for postharvest grain	Abdel-Rahman (2016)
Pediococcus acidilactici	*F. equiseti*	Feed detoxification	Abdel-Rahman (2016)
Lactobacillus fermentum	*F. equiseti*	Reduces growth of pathogen	Adithi et al. (2024)
Lactobacillus plantarum	*F. solani*	Degrade fusaric acid and phytotoxicity	Gowda (2018)
Lactobacillus rhamnosus	*F. graminearum*	Degrading DON Degrading ZEN Feed detoxification	Gratz et al. (2005), Rakesh et al. (2022)
Lactobacillus plantarum			
Pediococcus pentosaceus			
Lactobacillus plantarum	*F. oxysporum*	Reduces fusaric acid toxicity	Banu et al. (2021), Vasundaradevi et al. (2024)
Lactobacillus paracasei	*F. proliferatum, F. graminearum*	Inhibition of the fungus	
Oter LAB species L. crispatus, L. acidophilus, L. fermentum	*Fusarium* spp.	Multi-toxin binding/ detoxification	El-Nezami (2007)

down fungal cell walls, rendering them ineffective (Deepa and Sreenivasa 2017). The plant immune system will be triggered or stimulated by probiotic microbes by induced systemic resistance (ISR) or systemic acquired resistance (SAR). *Pseudomonas fluorescens* induces ISR among maize plants, protecting against *Fusarium* and other fungal species (Banu et al. 2021). *Fusarium* interaction with the host can be suppressed by quorum quenching, coordinating infection processes where probiotics interfere during signaling pathways, which disrupts fungal communication and virulence. Probiotics improve the plant vigor, which makes crops less susceptible to *Fusarium* infections by nutrient uptake and solubilization, hormonal stimulation like auxin production, and promoting the root and shoot growth, drought, and salinity conditions with stress alleviation predispose plants to fungal infections (Zhao 2014).

2.4 Probiotics in Agriculture

Probiotics in present day, as a biocontrol agent, have been increasingly used in agriculture to manage *Fusarium* pathogens. Application of probiotics in agriculture reduces the reliance on chemical fungicides by promoting more sustainable agricultural practices (Fig. 2.2).

2.4.1 Probiotic Control Strategies in Seed, Soil, and Foliar Applications

Seed treatment with probiotic microbes helps to establish beneficial microbial populations in soil along with the plant's life cycle that promotes against *Fusarium* pathogens (Larkin and Fravel 2002). *Bacillus subtilis* was applied as a probiotic against *Fusarium graminearum,* causing *Fusarium* head blight in wheat seeds, by allowing the coating of the seeds with probiotic culture (Schisler et al. 2006). The treatment established a beneficial microbial community on the seeds, enhancing seedling vigor by reducing initial colonization by *Fusarium* and DON contamination, and also yield was observed to be higher in treated crops compared with untreated control (Harman 2006) (Table 2.3).

Soil amendment with probiotics promotes beneficial microbial communities in the rhizosphere, which enhances plant defenses by inhibiting *Fusarium* growth (Vinale et al. 2008). *Trichoderma harzianum* is applied to banana plantations against *Fusarium oxysporum,* causing Panama disease. Soil was treated with *Trichoderma* either through direct inoculation or by using organic compost amendment (Liu

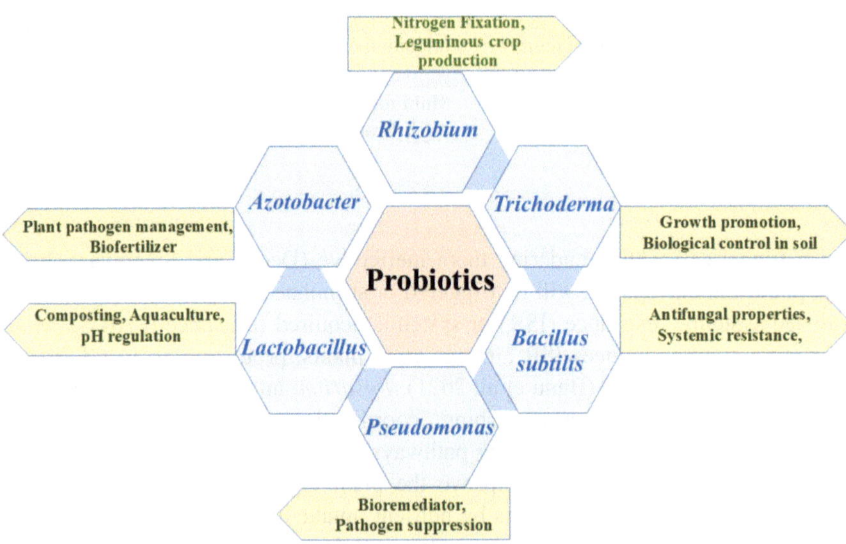

Fig. 2.2 Probiotics in Agriculture

Table 2.3 Biocontrol strategies against *Fusarium* pathogens

Biocontrol strategy	Target stage	Benefits	References
Seed treatment	Germination	Root colonization, early protection	Spadaro and Gullino (2015)
Soil amendment	Rhizosphere	Suppresses soilborne *Fusarium*, improves soil health	Abd-Elsalam and Prasanna (2020)
Foliar spray	Flowering/heading	Reduces foliar infection, mycotoxins	Pantazi et al. (2019)
Challenge treatment (seed+soil+foliar)	Multi-stage biocontrol application	Holistic protection across growth stages	Zhang (2020a, b)
Nanotechnology	Nano-fungicides, nano-carriers, toxin sensors	Targeted action, real-time monitoring	Rai (2016)
AI-based precision farming	Drones, sensors, AI prediction models	Early detection and site-specific control	Mahlein (2016)
Farmer education campaigns	Capacity building, awareness raising	Behavioral change, better food safety practices	Okoth (2016)
16S rRNA / ITS amplicon sequencing	Community profiling (bacteria/fungi)	Early protection	Abd-Elsalam and Prasanna (2020)
Shotgun metagenomics	Functional gene analysis, taxonomic resolution	Reduces mycotoxins	Abd-Elsalam and Prasanna (2020)
Metatranscriptomics	Active gene expression during pathogen stress	Improves soil health	Zhang (2020a, b)
Network analysis	Microbe–microbe interaction modeling	Pathogen suppression	Zhang (2020a, b)

2020). The probiotic treatment reduced disease incidence and plant mortality, promoting healthier banana growth and better yields (Bonanomi 2010) (Table 2.3).

Foliar application of probiotic solution allows for direct interaction with plant surfaces, offering protection against airborne or foliar *Fusarium* infections. *Pseudomonas fluorescens* was used as a biocontrol agent against *Fusarium oxysporum,* causing *Fusarium* wilt in tomato (Khan et al. 2004). Foliar spray of *Pseudomonas fluorescens,* which acts as live bacteria forming a protective biofilm, was applied to the tomato plant surface through Induced Systemic Resistance (ISR) (Zhang 2020a, b). Treated plants showed reduced wilt symptoms with increased root and shoot growth, with improved plant health and productivity (Abdel-Rahman 2016) (Table 2.3).

2.4.2 Integrated Probiotic Strategies for Sustainable Agriculture

Challenge inoculation treatment involves a combination of seed + soil + foliar applications to maximize the protective effect against *Fusarium.* A combination of *Bacillus subtilis* as seed treatment, *Trichoderma harzianum* in soil amendments, and

Pseudomonas fluorescens as foliar spray was applied against *Fusarium verticillioides* to control ear rot disease in maize (Palazzini et al. 2016). The integrated treatment significantly lowered the levels of *Fusarium* infection in maize ears, reducing fumonisin contamination. Challenge inoculation treatment of probiotics provided superior protection against *Fusarium* compared to individual treatments (Spadaro and Gullino 2015). Probiotics these days offer a sustainable and versatile approach to managing *Fusarium* diseases in the agriculture sector. Initially, from seed treatment to soil amendments and foliar spray technique each and every individual method promises to reduce the load of pathogen by improving plant health and enhancing the yield of it (Zhang 2020a, b). Some studies explain the integration of probiotic treatments with traditional agricultural practices that effectively control *Fusarium* by maintaining economic and environmental sustainability (Table 2.3).

Present trend nanotechnology integrates with probiotics as an emerging technology to enhance the effectiveness of probiotics in agriculture by improving stability, delivery, and activity (Rai 2016). The nano encapsulation technique can be practiced by encapsulating probiotics in nanomaterials that protect them from certain environmental stress like UV radiation, high temperature, and desiccation. Encapsulation of probiotics improves the controlled release of probiotics overtime making sure that they remain effective for longer periods (Abd-Elsalam and Prasanna 2020). Nanoparticles are designed in such a way that they target specific areas of the plant, like leaves or roots, allowing for application. This method improves the penetration of probiotics inside the plant tissues by enhancing their ability to colonize in plant leaves or roots by interacting more effectively with *Fusarium* pathogens (Narayanan and Park 2022) (Table 2.3).

Artificial intelligence and *machine learning* is the future for agriculture, along with management of plant diseases. This technology might enhance probiotics' effectiveness by enabling precise and data-driven approaches. AI-driven tools like automated tractors and drones can be used to apply an optimal load of probiotics in exact locations, minimizing waste and maximizing efficacy, where they are needed based on data collected, such as plant health, pathogen level, and soil conditions in the field, using sensors as disease management being the key to the technology (Mahlein 2016). Machine learning algorithms help to predict outbreaks of *Fusarium* pathogen infection based on soil conditions, historical data, and weather patterns, which allows for timely targeted probiotic application that prevents the disease, which is a major issue in the agriculture sector (Pantazi et al. 2019) (Table 2.3).

Awareness campaigns in recent days to educate farmers are required to adopt the widespread use of probiotics. These initiatives might help the farmers to understand the benefits and application of probiotics to adopt as practice (FAO 2022). Farmers may be trained through conducting workshops, demonstrations, and training programs that help them to learn about the science behind the application of probiotics by integrating it in the farming system, showcasing probiotics' benefits against *Fusarium* diseases (Gozzard 2021). Agricultural extension services can help farmers by incorporating probiotics into integrated pest management strategies (IPM) that provide support and guidance to farmers. Governments can also collaborate with agricultural organizations to educate farmers on the use of probiotics by tax breaks, subsidies, and certification programs for sustainable farming practices, which reduces economic barriers to adoption of probiotic-based solutions against *Fusarium* pathogens (Chilaka et al. 2017). Advanced integration techniques in biotechnology through AI, nanotechnology, metagenomics, and genetic engineering hold immense potential to enhance probiotics' effectiveness in agriculture (Okoth 2016).

Metagenomic Insights into Probiotic–Plant–Fungus Interactions Metagenomics is the study of genetic material from environmental samples, including interactions between plants and pathogens like *Fusarium* (Trivedi et al. 2020). Metagenomics approach helps to study the microbial communities by identifying beneficial microbes like probiotics that interact with the entire microbiome of soil, plant tissues, roots, and suppress the *Fusarium* pathogen contributing to plant resistance (Mendes et al. 2011). Certain probiotics are present that induce systemic resistance in plants and with other microbes that enhance their antifungal activity by improving disease resistance and plant health (Chaudhry 2020).

2.5 Advantages Over Conventional Strategies for *Fusarium* Control

Probiotics in agriculture have several advantages over conventional methods, such as the use of resistant crop breeding and chemical fungicides. Benefits of probiotics in agriculture include cost-effectiveness, sustainability, and environmental safety (Ghosh et al. 2022) (Table 2.4).

Sustainability Probiotics contribute more to sustainable agricultural practices, which enhances soil health by reducing chemical dependencies and promoting biodiversity. Probiotics improve the soil microbiome through enhancing nutrient cycling, overall soil fertility, and organic matter decomposition. The healthy soil supports crop production, reducing its need for pesticides and synthetic fertilizers (Mahlein 2016). On application of probiotics as a natural bio-control agent, farmers can reduce the use of chemical fungicides that are harmful to soil organisms, human health, and non-target species. Probiotics need not lead to the development of resistance in pathogens that are sustainable for managing *Fusarium* infections. Probiotics, as beneficial microbes, interact with the environment and plant through increasing microbial diversity in the rhizosphere (Bonanomi et al. 2018).

Cost-effectiveness The initial cost for developing and applying probiotics seems to be high, but their long-term benefits make them cost-effective in comparison to traditional methods. Probiotics reduce the need for costly chemical fungicide applications that require multiple treatments throughout the season by preventing disease outbreaks through increasing crop yields (Deepthi et al. 2017a; Berendsen

Table 2.4 Traditional and modern approaches of biocontrol strategies

Aspect	Traditional approaches	Modern approaches
Target specificity	Broad-spectrum fungicides	Targeted biocontrol and probiotics
Mycotoxin control	Often ineffective	Active degradation or binding by microbes and nanoparticles
Detection and timing	Post-infection symptom-based	AI-driven early detection
Soil and ecosystem impact	Depletes microbial diversity	Enhances soil health and resilience
Resistance risk	High (especially with triazoles)	Low risk due to multifactorial mechanisms
Cost and accessibility	Expensive, chemical-based	Low-cost, sustainable practices
Scientific insight	Limited to visible symptoms	Microbiome and gene-level insights (metagenomics)
Farmer impact	High input cost, variable success	Sustainable, adaptable, knowledge-based approaches

et al. 2018; Deepthi et al. 2022). Probiotic application increases plant growth and yield, leading to economic returns. Healthy crops seem to be more resistant to disease, improving overall farm productivity. Probiotics promote environmental sustainability and reduce contamination of water, soil. Long-term costs related to environmental damage like water pollution and soil degradation will be minimized (Xiong 2021).

Environmental Safety Probiotics proved to be an alternative to chemical fungicides in terms of environmental impact. Probiotics specifically target *Fusarium* and other harmful pathogens without harming beneficial microbes such as earthworms, soil microbes, and pollinators (Trivedi et al. 2020). Probiotics, unlike chemical fungicides, do not leave harmful chemical residues on crops and make them safer for human consumption by reducing contamination risk in food chains. These probiotics do not cause risk to soil and water systems by contaminating groundwater as chemical fungicides do by disrupting aquatic ecosystems (Lucas et al. 2015).

2.5.1 Integration with Other Management Practices

Probiotics alone serve as a magical biocontrol agent, but when combined with an Integrated Pest Management strategy, they focus on combining multiple control methods for managing pests and its diseases by minimizing the environmental impact (Sharma 2016). Probiotics can be applied in conjunction with the crop rotation process that reduces *Fusarium* infection buildup in soil. Probiotics also contribute to soil health, which enhances plant resistance to pathogens by improving crop resilience. Probiotics can be used as an alternative to chemical fungicide applications that limit the use of chemicals to control disease. A combination of probiotics with other biocontrol agents, like natural predators, enhances overall disease management along with crop protection (Pantazi et al. 2019). Probiotics also trigger systemic resistance complemented by other IPM strategies, along with resistant varieties and a physical barrier. It can also be applied for integration of probiotics with monitoring systems to track *Fusarium* pathogen levels detection through microbial shifts in soil or plant tissues for early intervention and targeted management (Mahlein 2016).

2.6 Future Perspectives

- Advanced applications in probiotic research are required by strain optimization and genomic insights through continuous screening and engineering of probiotic strains with enhanced antifungal activity, compatibility with diverse crop systems, resilience to environmental stressors, leveraging genomics and metagenomics that can suppress *Fusarium* and boost plant health.
- Integration of probiotics with agricultural practices by formulation development in large-scale agriculture and development of multi-functional products combining probiotics with biofertilizers or biostimulants for crop protection and growth enhancement.
- Precision agriculture and data integration through microbiome monitoring and tailored applications with predictive tools and models.
- Streamlining the approval and commercialization of probiotic-based solutions by framing certain policies and regulations in recurring sustainability incentives for agricultural practices.
- Educating farmers and creating public awareness by engaging with the community to promote awareness among consumers about the role of probiotics as safe and healthier.
- Strengthening research networks by international collaborations to share data, resources, and best practice in probiotic research and addressing climate challenges to explore how probiotics mitigate the effects of climate on Fusarium prevalence and crop health.

2.7 Chapter Summary

Fusarium and its species are a substantial threat to global agriculture, associated with crop disease and producing harmful mycotoxins. Probiotics have emerged as an eco-friendly resolution to manage *Fusarium*. Probiotics, with its multiple mechanisms, combat *Fusarium*, also contribute to soil health, and promote the growth of the plant, preventing pathogen infection. Many studies, as discussed in the above sections, have shown promising results with reduced *Fusarium* infections increasing crop yields and lowering mycotoxin levels. Future efforts need to be focused on optimizing probiotic strains by improving delivery methods, integrating probiotic strains into farming practices by addressing scalability challenges. So that we can move further towards more sustainable agricultural practices from the persistent threat of *Fusarium*.

Conflict of Interest The authors confirm that they have no conflict of interest.

Author Contribution Conceptualization and writing DN; Supervision, review, and editing MYS and PKM; re-editing, construction of figures and tables AG and RS.

Study Questions

- Discuss the advantages of foliar sprays using probiotic formulations over fungicides.
- How can nanotechnology be integrated with probiotics for improved *Fusarium* control?
- Design an integrated pest management plan that includes seed treatment, soil inoculation, and foliar spraying against *Fusarium* in maize.
- How can precision agriculture technologies support the timing of probiotic applications?
- Critically assess the potential for using metagenomics to customize probiotic formulations for different soil types.

References

Abdel-Rahman MA (2016) Detoxification of fungal toxins by probiotics: a review. Toxins 8(5):157

Abd-Elsalam KA, Prasanna R (2020) Nanotechnology for sustainable agriculture and crop protection. Environ Nanotechnol Monit Manage 14:100282

Adithi G, Divyashree S, Shruthi B et al (2024) Evaluation of *Limosilactobacillus fermentum* MYSAGAM1 isolated from herbal Amla juice as probiotic candidate with antifungal characteristics against *Fusarium equiseti*. Food Biosci 58:103843

Adithi G, Rakesh S, Shruthi B et al (2025) Biological decontamination of mycotoxigenic *Fusarium verticillioides* using *Lactiplantibacillus plantarum* MYSN128 isolated from traditional vellappam: implications for food safety and global mycotoxin management. Curr Microbiol 82:404

Alassane-Kpembi I, Puel O, Pinton P et al (2017) Co-exposure to low doses of the food contaminants deoxynivalenol and nivalenol has a synergistic inflammatory effect on intestinal explants. Arch Toxicol 91(7):2677–2687

Alberts JF, Van Zyl WH, Gelderblom WCA (2016) Biologically based methods for control of fumonisin prosucing *Fusarium* species and reduction of the fumonisins. Front Microbiol 7:548

Al-Masri SA, El-Safty SMS, Nada SA, Amra HA (2011) *Saccharomyces cerevisiae* and probiotic bacteria potentially inhibit fumonisin B1 production in vitro and in vivo. J Am Sci 7(1):198–205

Backhouse D (2014) Global distribution of Fusarium graminearum and chemotype variation. Plant Pathol 63(1):20–31

Banu JG et al (2021) Biodegradation of fusaric acid by probiotics and its role in alleviating wilt symptoms. J Appl Microbiol 131(1):38–47

Berendsen RL, Pieterse CM, Bakker PA (2012) The rhizosphere microbiome and plant health. Trends Plant Sci 17(8):478–486

Berendsen RL, Vismans G, Yu K, Song Y, de Jonge R, Burgman WP, Burmølle M, Herschend J, Bakker PAHM, Pieterse CMJ (2018) Disease-induced assemblage of a plant-beneficial bacterial consortium. ISME J 12:1496

Bonanomi G (2010) Soil suppressiveness to fungal pathogens: a review of mechanisms and applications. Soil Biol Biochem 42(7):1187–1197

Bonanomi G, Lorito M, Vinale F, Woo SL (2018) Organic amendments, beneficial microbes, and soil microbiota: toward a unified framework for disease suppression. Annu Rev Phytopathol 25(56):1–20

Buerstmayr H (2009) Molecular mapping of QTLs for Fusarium head blight resistance in wheat. Euphytica 170(1–2):131–143

Chakraborty S, Newton AC (2011) Climate change, plant diseases and food security: an overview. Plant Pathol 60(1):2–14

Chaudhry V (2020) Comparative metagenomic analysis reveals microbial signatures of root-zone microbiome in wheat treated with biocontrol agents. Sci Rep 10(1):1–13

Chilaka CA, De Boevre M, Atanda OO et al (2017) The status of Fusarium mycotoxins in sub-Saharan Africa: a review of emerging trends and post-harvest mitigation strategies towards food control. Toxins 9(1):19

Compant S, Duffy B, Nowak et al (2005) Use of plant growth-promoting bacteria for biocontrol of plant diseases: principles, mechanisms of action, and future prospects. Appl Environ Microbiol 71(9):4951–4959

Dass RS, Sreenivasa MY, Janardhana GR (2007) High incidence of *Fusarium verticillioides* in animal and poultry feed mixtures produced in Karnataka, India. Plant Pathol J 6(2):174–178

Dean R, Kan JAL, Pretorius ZA et al (2012) The top 10 fungal pathogens in molecular plant pathology. Mol Plant Pathol 13(4):414–430

Deepa N, Sreenivasa MY (2019) Biocontrol strategies for effective management of phytopathogenic fungi associated with cereals. In: Singh JS, Singh DP (eds) New and future developments in microbial biotechnology and bioengineering. Elsevier, Amsterdam, pp 177–189. https://doi.org/10.1016/B978-0-444-64191-5.00013-4. ISBN 9780444641915

Deepa N, Charith Raj AP, Sreenivasa MY (2016) Nested PCR method for the early detection of fumonisin-producing *Fusarium verticillioides* in pure cultures, cereal samples and plant parts. Food Biotechnol 30(1):18–29

Deepa N, Sreenivasa MY (2017) Fumonisins: a review on its global occurrence, epidemiology, toxicity and detection. J Vet Med Res 4(6):1093. ISSN: 2378-931

Deepa N, Achar P, Sreenivasa MY (2021) Current perspectives of biocontrol agents for management of Fusarium verticillioides and its fumonisin in cereals – a review. J Fungi 7:776. https://doi.org/10.3390/jof7090776

Deepa N, Chennappa G, Deepthi BV et al (2022) Antifungal potential of *Azotobacter* species and its metabolites against *Fusarium verticillioides* and biodegradation of fumonisin. J Appl Microbiol 133(4):2430–2444

Deepthi BV, Rao KP, Chennapa G et al (2016) Antifungal attributes of *Lactobacillus plantarum* MYS6 against Fumonisin producing *Fusarium proliferatum* associated with poultry feeds. PLoS One 11(6):e0155122

Deepthi BV, Rakesh S, Poornachandra Rao K et al (2017) *Lactobacillus plantarum* MYS6 ameliorates fumonisin B1-induced hepatorenal damage in broilers. Front Microbiol 8:2317

Deepthi BV, Gnanaprakash AP, Sreenivasa MY (2017a) Effect of γ-irradiation on fumonisin-producing *Fusarium* associated with animal and poultry feed mixtures. 3 Biotech 7:57. https://doi.org/10.1007/s13205-017-0693-x

Deepthi BV, Deepa N, Vanitha PR, Sreenivasa MY (2022) Stress responses on the growth and mycotoxin biosynthesis of *Fusarium proliferatum* associated with stored poultry feeds. Appl Food Res 2(1):100091. https://doi.org/10.1016/j.afres.2022.100091

Desjardins AE (2006) *Fusarium* mycotoxins: chemistry, genetics, and biology. APS Press

Divyashree S, Ramith R, Sreenivasa MY (2024) Evaluation of new candidate probiotic *Lactobacillus* strains isolated from a traditional fermented food- multigrain-millet dosa batter. Food Biosci 57:103450. https://doi.org/10.1016/j.fbio.2023.103450. IF 5.240 (Elsevier)
El-Nezami H (2007) Ability of dairy strains of lactic acid bacteria to bind aflatoxin M1 and fumonisin B1 in a food model. Food Chem Toxicol 45(9):1636–1642
FAO. (2022). Empowering farmers to manage mycotoxin risks. https://www.fao.org/3/cc0125en/cc0125en.pdf
Fareid MA (2011) Biocontrol of *Fusarium* moulds and Fumonisin B1 production. World Rural Obs 3(3):53–61
Ghosh S et al (2022) Biopesticides as a sustainable alternative to chemical pesticides. Microbiol Res 263:127058
Goswami S, Kistler HC (2004) Heading for disaster: *Fusarium graminearum* on cereal crops. Mol Plant Pathol 5(6):515–525
Gowda NKS (2018) Role of probiotics in decontaminating mycotoxins in feed. Toxins 10(3):124
Gozzard E (2021) Precision agriculture tools for fungal pathogen monitoring. Trends Biotechnol 39(5):445–457
Gratz SW, Mykkanen H, El-Nezami HS (2005) Probiotic bacteria and their detoxification of DON and other trichothecenes. Toxins 7(11):4755–4773
Harman GE (2006) Overview of mechanisms and uses of *Trichoderma* spp. Phytopathology 96(2): 190–194
Harman GE, Howell CR, Viterbo A et al (2004) Trichoderma species—opportunistic, avirulent plant symbionts. Nat Rev Microbiol 2(1):43–56
Khan NI et al (2004) Application of bacterial antagonists to reduce head blight and mycotoxins in wheat. Plant Dis 88(11):1298–1302
Kowalska K, Habrowska-Górczyńska DE (2020) Zearalenone as an endocrine disruptor in humans. Environ Toxicol Pharmacol 78:103389
Lamichhane JR, Venturi V (2015) Synergisms between microbial pathogens in plant disease complexes: a growing trend. Front Plant Sci 6:385
Larkin RP, Fravel DR (2002) Efficacy of various biocontrol organisms in suppressing *Fusarium* wilt of tomato. Biol Control 23(3):240–248
Leslie JF, Summerell BA (2006) The *Fusarium* laboratory manual. Blackwell Publishing
Liu J (2020) Seed coating with *Bacillus subtilis* suppresses *Fusarium* wilt in tomato. Biol Control 142:104169
Lucas et al (2015) The role of *Plasmophora viticola* effectors in suppressing host immunity. Plant Pathol 64(4):932–947
Ma LJ, Geiser DM, Proctor RH et al (2013) *Fusarium* pathogenomics. Ann Rev Microbiol 67(1): 399–416
Mahlein A-K (2016) Plant disease detection by imaging sensors—status and challenges. Annu Rev Phytopathol 54:669–689
Marasas WFO, Riley RT, Hendricks KA (2004) Fumonisins disrupt sphingolipid metabolism, folate transport, and neural tube development in embryo models. Environ Health Perspect 112(5):565–570
McMullen M, Bergstrom G, De Wolf E et al (2012) A unified effort to fight an enemy of wheat and barley: *Fusarium* head blight. Plant Dis 96(12):1712–1728
Mendes R, Kruijt M, De Bruijn I et al (2011) Deciphering the rhizosphere microbiome for disease-suppressive bacteria. Science 332(6033):1097–1100
Miedaner T, Juroszek P (2021) Climate change will influence disease resistance breeding in wheat in Europe. Theor Appl Genet 134:1771–1785
Minervini F, Dell'Aquila ME (2008) Zearalenone and reproductive function in farm animals. Int J Mol Sci 9(12):2570–2584
Munkvold GP (2003) Cultural and genetic approaches to managing mycotoxins in maize. Annu Rev Phytopathol 41:99–116

Narayanan KB, Park HH (2022) Multifunctional silver nanoparticles inhibit *Fusarium* spp. and DON biosynthesis. J Hazard Mater 423:127164

Okoth S (2016) Awareness and mitigation of mycotoxins in Africa: the role of agricultural extension and farmer education. Curr Opin Food Sci 11:38–43

Ongena M, Jacques P (2008) *Bacillus* lipopeptides: versatile weapons for plant disease biocontrol. Trends Microbiol 16(3):115–125

Palazzini JM et al (2016) Field application of bacillus reduces *Fusarium* head blight and DON accumulation. Biol Control 100:54–60

Pantazi XE et al (2019) Detection of *Fusarium* head blight in wheat using UAV and AI-based spectral imaging. Comput Electron Agric 160:75–83

Paul PA, Lipps PE, Hershman DE et al (2008) Efficacy of triazole-based fungicides for *Fusarium* head blight and deoxynivalenol control in wheat: a multivariate meta-analysis. Phytopathology 98(9):999–1011

Pertot I (2017) Integrated pest and disease management in greenhouse crops. Plant Dis 101(4):1–13

Pestka JJ (2010) Deoxynivalenol: mechanisms of action, human exposure, and toxicological relevance. Arch Toxicol 84(9):663–679

Pieterse CM, Zamioudis C, Berendsen RL et al (2014) Induced systemic resistance by beneficial microbes. Annu Rev Phytopathol 52(1):347–375

Pinton P, Oswald IP (2014) Effect of deoxynivalenol and other type B trichothecenes on the intestine: a review. Toxins 6(5):1615–1643

Placinta CM, D'Mello JPF, Macdonald AMC (1999) A review of worldwide contamination of cereal grains and animal feed with *Fusarium* mycotoxins. Anim Feed Sci Technol 78(1–2): 21–37

Ploetz RC (2006) Fusarium wilt of banana is caused by several pathogens referred to as *Fusarium oxysporum* f. sp. *cubense*. Phytopathology 96(6):653–656

Raaijmakers JM (2009) Microbial antagonists and the mechanisms of disease suppression in the rhizosphere. Annu Rev Phytopathol 47:153–181

Rai M (2016) Nano-based control of mycotoxin contamination in food and feed. Crit Rev Food Sci Nutr 56(11):1793–1807

Rakesh S, Shruthi S, Deepthi BV (2019) Probiotic properties of lactic acid bacteria isolated from Neera: a naturally-fermenting coconut palm nectar. Front Microbiol 10:1382

Rakesh S, Walid M, Adithi G (2021) Probiotic and antifungal attributes of *Levilactobacillus brevis* MYSN105, isolated from an Indian traditional fermented food Pozha. Front Microbiol 12: 696267

Rakesh S, Walid M, Hammami R, Sreenivasa MY (2022) *Levilactobacillus brevis* MYSN105, isolated from an Indian traditional fermented food. The assembled genome was deposited in GenBank under bioproject accession number PRJNA679667

Rini CR, Sulochana KK (2007) Management of Fusarium wilt in tomato using biological control agents. J Trop Agric 45(1–2):21–28

Schisler DA et al (2006) Biological control of Fusarium head blight using microbes from wheat anthers. Phytopathology 96(5):429–437

Schisler DA et al (2011) Biological control of *Fusarium head blight* with beneficial microbes. Plant Dis 95(4):528–535

Schulz B, Boyle C (2005) The endophytic continuum. Mycol Res 109(6):661–686

Sharma P (2016) Integration of *Trichoderma* with fungicides for Fusarium wilt suppression in chickpea. Biol Agric Hortic 32(3):188–200

Shruthi B, Adithi G, Divyashree S et al (2024) Probiotic and functional attributes of yeasts isolated from different traditional fermented foods and products. Probiotics Antimicrob Proteins 17(6): 4080–4098. https://doi.org/10.1007/s12602-024-10342-z

Shruthi B, Deepthi BV, Adithi G, Vanitha PR, Sreenivasa MY (2025) *Saccharomyces cerevisiae* MYSY1 as a promising probiotic isolate with antifungal properties against foodborne *Aspergillus* species. Biocatal Agric Biotechnol 67:103749. https://doi.org/10.1016/j.bcab.2025.103649

Spadaro D, Gullino ML (2015) Improving the efficacy of biocontrol agents against Fusarium using combined applications. Crop Prot 67:117–127

Sreenivasa MY, Dass RS, Charith Raj AP, Janardhana GR (2006) Molecular detection of fumonisin-producing Fusarium species of freshly harvested maize kernels using polymerase chain reaction (PCR). Taiwania 51(4):251–257

Sreenivasa MY, Dass RS, Charith Raj AP, Janardhana GR (2011) Mycological evaluation of Maize grains produced in Karnataka (India) for the post harvest fungal contamination. World Appl Sci J 13(4):688–692

Sreenivasa MY, Dass RS, Charith Raj AP, Janardhana GR (2008) PCR-based detection of genus *Fusarium* and fumonisin-producing isolates from freshly harvested sorghum grains grown in Karnataka, India. J Food Saf 28:236–237

Summerell BA (2010) *Fusarium*: Paul E. Nelson Memorial Symposium. APS Press

Syed-Ab-Rahman SF (2018) The role of microbial endophytes in suppressing *Fusarium* pathogens. Biol Control 116:26–39

Thangavelu R, Mustaffa MM (2010) Role of biological agents in the management of Fusarium wilt of banana. Indian J Plant Prot 38(1):18–25

Trivedi P, Leach JE, Tringe SG et al (2020) Plant-microbiome interactions: from community assembly to plant health. Nat Rev Microbiol 18(11):607–621

Van der Fels-Klerx HJ (2016) Mycotoxins in a changing world. Food Addit Contam Part A 33(10): 1606–1619

Vasundaradevi R, Sarvajith M, Divyashre S, Deepa N (2024) Antifungal and probiotic synergy of *Lactiplantibacillus* derived from tropical fruits: efficacy against *Fusarium oxysporum*. Sci Rep 15:2144

Vinale F, Sivasithamparam K, Ghisalberti EL et al (2008) Trichoderma–plant–pathogen interactions. Soil Biol Biochem 40(1):1–10

Voss KA, Smith GW, Haschek WM (2007) Fumonisins: toxicokinetics, mechanism of action and toxicity. Anim Feed Sci Technol 137(3–4):299–325

Warth B, Sulyok M, Berthiller F et al (2013) New insights into the human metabolism of the *Fusarium* mycotoxins deoxynivalenol and zearalenone. Toxicol Lett 220(1):88–94

Xiong C (2021) Bioinoculants enhance microbial network complexity and *Fusarium* wilt suppression. Microbiome 9(1):1–15

Yin Y (2009) Genetic variation and population structure of *Fusarium graminearum* from wheat in China. Fungal Genet Biol 46(12):936–943

Zhang S (2020a) Biocontrol of Fusarium head blight using *Bacillus velezensis*. Front Microbiol 11: 584779

Zhang S (2020b) Integration of *Bacillus* in seed and foliar treatment for Fusarium head blight control. Front Microbiol 11:584779

Zhao X (2014) Biological control of *Fusarium graminearum* by *Bacillus subtilis* strain SG6 and its effect on mycotoxin accumulation. Biol Control 72:39–48

Further Reading

Trivedi P, Leach JE, Tringe SG, Sa T, Singh BK (2020) Plant-microbiome interactions: from community assembly to plant health. Nat Rev. Microbiol 18(11):607–621

Xiong C (2021) Bioinoculants enhance microbial network complexity and *Fusarium* wilt suppression. Microbiome 9(1):1–15

Abd-Elsalam KA, Prasanna R (2020) Nanotechnology for sustainable agriculture and crop protection. Environ Nanotechnol Monit Manage 14:100282

Harnessing Probiotics Interactions to Mitigate *Alternaria* Pathogenicity

3

M. Sarvajith, R. Vasundaradevi, and M. Y. Sreenivasa

Abstract

Alternaria spp. are major phytopathogens causing severe crop diseases and significant economic losses worldwide. In addition to damaging plant health, *Alternaria* produces mycotoxins that affect food safety and public health. *Bacillus, Lactobacillus, Pseudomonas* spp., among others, have been suggested as eco-friendly antifungal agents for managing *Alternaria* infections. This chapter discusses antagonistic mechanisms of probiotic microorganisms by the production of antifungal metabolites, competitive interactions, etc. Additionally, host plant resistance is also detailed with examples from lab to field trials. Finally, strategies for post-harvest application formulations and challenges for stability and scalability are examined.

M. Sarvajith (✉)
Biological and Environmental Science and Engineering Division, King Abdullah University of Science and Technology, Thuwal, Kingdom of Saudi Arabia
e-mail: Sarvajith.manjunath@kaust.edu.sa

R. Vasundaradevi · M. Y. Sreenivasa
Applied Mycology Laboratory, Department of Studies in Microbiology, University of Mysore, Mysuru, India

M. Y. Sreenivasa et al. (eds.), *Probiotics as Antifungals*,
https://doi.org/10.1007/978-981-92-0002-3_3

Graphical Abstract

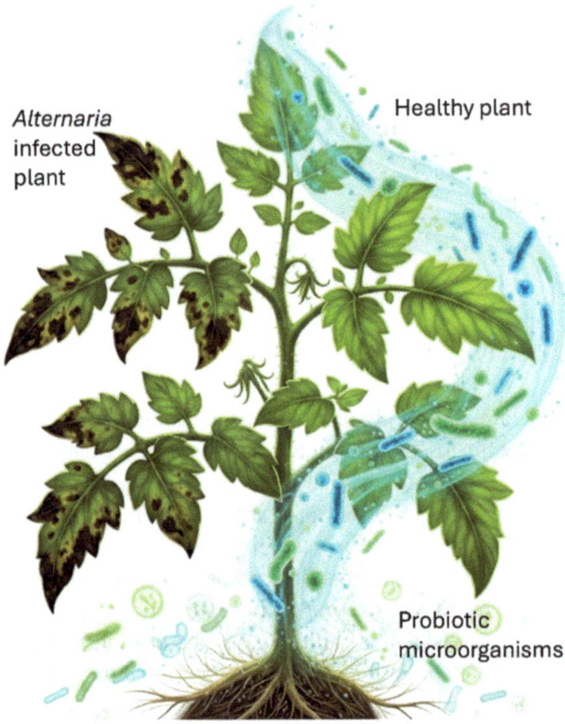

Keywords

Antifungal activity · Plant immunity · Sustainable agriculture · Biofilms · Post-harvest control

Learning Objectives

By the end of this chapter, readers will be able to:

- Understand the biology, taxonomy, and pathogenic mechanisms of *Alternaria* species as phytopathogens.
- Explain the mechanisms by which probiotic microorganisms suppress *Alternaria* infections.
- Evaluate formulation strategies and application methods for probiotic-based control of *Alternaria* in agriculture and post-harvest systems.
- Recognize current challenges, regulatory considerations, and emerging future directions for probiotics-based management approaches.

3.1 Introduction

Species belonging to *Alternaria* are among the most diverse fungal organisms in natural and agricultural settings. They are both saprophytes and opportunistic phytopathogens. Due to their melanized conidia and prolific spore formation, they can survive in wide environmental conditions, contributing to persistence and easy dissemination. *Alternaria* can infect a wide variety of plants, crops, vegetables, and fruits, resulting in deterioration of the agricultural products, loss of yield, and economic losses (Yang et al. 2024). Typically, infections in plants manifest as spots on leaves, blight rots and contamination of seeds and fruits (Cao et al. 2013). *Alternaria* spp. such as *A. alternata, A. tenuis* are also known to produce a wide range of toxic compounds known as mycotoxins. Examples include altenuenes, alternariols, tenuazonic acid, etc. (Lee et al. 2015). Often few species of this genus are known to infect immunocompromised humans. However, they are primarily recognized as plant pathogens (Xu 2022). Therefore, this chapter focuses on the role of *Alternaria* in plant health and crop disease.

Probiotics are beneficial microorganisms that are known to have beneficial properties when consumed by humans. In recent years, probiotics have also been emphasized as biocontrol agents and eco-friendly alternatives to chemical fungicides. Bacteria such as *Bacillus, Lactoplantibacillus, Pseudomonas*, and yeast such as *Saccharomyces* are documented to exert antagonistic mechanisms (Deepthi et al. 2022; Divyashree et al. 2024; Adithi et al. 2024, 2025; Shruthi et al. 2025). It could be competition for nutrients, production of metabolites, or induction of plant systemic resistance (Deepa et al. 2022; Vasundaradevi et al. 2024, 2025). These bacteria produce bioactive metabolites such as organic acids, lipoproteins, and bacteriocins that inhibit the growth of mycelium, suppress spore germination while promoting plant vigour (Garcia et al. 2016). In contrast to chemical fungicides, probiotics-based interventions promote soil microbial networks and enhance stability for sustainable agricultural practices.

This chapter presents the current understanding of the use of probiotic microorganisms for the management of *Alternaria*-associated plant infections. The chapter brings together the infection biology, mode of action of these probiotic microorganisms in inhibiting *Alternaria* infections, stimulating systemic resistance in plants with results from lab and field studies. Additionally, the key challenges in strategies, formulations, and delivery for field applications are discussed.

3.2 Pathogen Biology of *Alternaria* Infections

3.2.1 Taxonomic Diversity and Importance

The genus *Alternaria* is taxonomically placed within the phylum Ascomycota, class Dothideomycetes, and family Pleosporaceae. Previously, *Alternaria* was grouped under Deuteromycetes, the category lacking sexual stages. However, with the recent discovery of sexual morphs in several taxa, it is now reclassified. At present, more than 300 species of *Alternaria* are recognized, many of which are saprophytic and mild pathogens, but the subset includes highly aggressive pathogens (Woudenberg et al. 2013).

With advancements in molecular phylogenetics, multi-locus sequence analyses have refined species boundaries and delineated major species such as *A. alternata, A. solani, A. tenuissima*, and *A. brassicicola* (Dettman et al. 2023). Often, several species exhibit similar morphological features, making it challenging to identify. In such cases, the molecular makers such as internal transcribed spacer (ITS), translation elongation factor 1-alpha (TEF1-α), glyceraldehyde-3-phosphate dehydrogenase (GAPDH) are helpful when resolving at the species-level (Meena et al. 2017; Aloi et al. 2021). The ecological plasticity and adaptive capacity of *Alternaria* spp. enable them to colonize a wide range of hosts under variable environmental conditions. Thus, making the *Alternaria* genus a useful model in understanding fungal pathogenicity.

Alternaria-associated diseases impose substantial agricultural losses, particularly on important crops such as tomato, potato, brassicas, sunflower, and several fruits. Due to which the overall yield can reduce up to 80% depending on the environmental conditions and the susceptibility of the host. In addition to a reduction in crop productivity, *Alternaria* infections adversely affect the seed quality, post- harvest storage ability, and overall value of the product. As stated earlier, many species are capable of producing toxins, such as alternariol monoethyl ether, alternariol, etc., that enter through food or feed, posing health risks in humans and animals. The presence of these toxins has led to increased regulatory and trade restrictions further constraining the economy (Sreenivasa et al. 2008; Cao et al. 2013; Lee et al. 2015; Deepa et al. 2016; Fernandes et al. 2023).

3.2.2 Mechanism of *Alternaria* Infection

Alternaria spp. are predominantly necrotrophic, which derive nutrients for host plants by inducing cell death during infections. Infection begins when air-borne conidia settle on the plant surface, where adhesion is facilitated by the production of mucilaginous extracellular material. Under favourable conditions, typically 20–28 °C and high humidity, the conidia germinate to form germ tubes. Germ tubes penetrate the host tissues either directly through epidermis or indirectly via openings such as stomata or wounds (Singh et al. 2021). Following this, the pathogen colonizes host tissue by secreting cell-wall degrading enzymes such as cellulases, pectinases, and xylanases. These enzymes act directly by degrading the host cell wall and releasing nutrients. Among the cell-wall degrading enzymes, pectin-targeting enzymes play an important role in tissue maceration and lesion expansion (Ma et al. 2019) (Fig 3.1).

A key determinant of *Alternaria* pathogenicity is the production of host-specific toxins which largely define host range and disease severity. For example, the *A. alternata* sub sp. *lycopersici* produces AAL-toxin in tomato, sp. *mali* produces

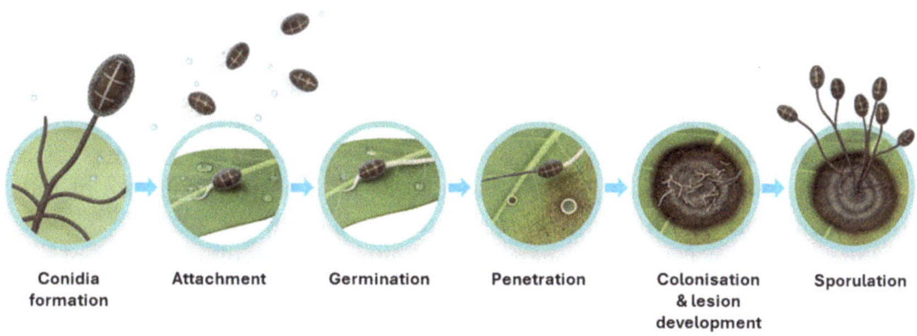

Fig. 3.1 Overview of the infection cycle of *Alternaria* spp.

AM toxin in apples, and the AK toxin sp. *kawasaki* in Brassica, etc. (Meena and Samal 2019; Ma et al. 2019). These toxins disrupt cell membranes, induce necrotic lesions and trigger apoptosis-like cell death, thereby creating a favourable environment for fungal colonization. During infection, *Alternaria* spp. also manipulate host reactive oxygen species (ROS) dynamics, promoting programmed cell death (Ma et al. 2019). In addition, melanin deposition in the fungal cell wall contributes to enhanced tolerance against environmental stress, protection from ultraviolet radiation, and the generation of sufficient turgor pressure required for effective host penetration.

Reproduction in *Alternaria* occurs predominantly through asexual conidia, enabling rapid dissemination and development of disease epidemics under conducive conditions. Although sexual reproduction is rare, it occurs through the formation of ascospores within pseudothecia (Salotti et al. 2024). The regulation of virulence, stress tolerance, and toxin biosynthesis is mediated by conserved molecular signalling pathways such as mitogen-activated protein kinase (MAPK) cascades, such as Hog1, cyclic AMP-protein kinase A (cAMP-PKA) pathway. Whereas genes such as Hog1 and CTT have been implicated in osmotic stress response and host-specific toxin production (Lin and Chung 2010; Lin et al. 2010). Collectively, these mechanisms demonstrate the versatility of *Alternaria* spp. in infection.

3.3 Characteristics of Probiotics Used in Fungal Pathogen Management

Probiotics are traditionally defined as live microorganisms that confer health benefits to the host when administered in adequate amounts (FAO/WHO 2002; Hill et al. 2014; Divyashree et al. 2024; Vasundaradevi et al. 2025). Probiotics have extended applications beyond human and animal health into plant and soil ecosystems (Vasundaradevi et al. 2024). In the phytopathological context, probiotic microorganisms refer to beneficial bacterial or yeast strains that promote plant growth, enhance immunity, and suppress pathogens through diverse ecological interactions (Hill et al. 2014). Mainly reported probiotic groups used in the management of fungal phytopathogens include lactic acid bacteria (LAB), such as *Lactobacillus plantarum*, *L. rhamnosus*, *Pediococcus pentosaceus*, *Bacillus* spp., including *B. subtilis*, *B. amyloliquefaciens*, and *B. velezensis*; and yeasts like *Saccharomyces cerevisiae* and *Pichia kudriavzevii* (Elshaghabee et al. 2017; Fira et al. 2018; Vasundaradevi et al. 2025; Shruthi et al. 2025a). These microorganisms demonstrate multifunctional plant-probiotic traits, such as the production of antimicrobial metabolites, biofilm formation, siderophore production, etc.

In the context of fungal management, probiotics act as biological control agents by reducing spore germination, mycelial growth, and toxin biosynthesis. For instance, *Lactobacillus plantarum* strains isolated from phyllosphere or fruit surfaces have shown strong antagonism against *A. alternata* via organic acid and bacteriocin production. Similarly, *Bacillus subtilis* and *B. amyloliquefaciens* are known to induce systemic resistance in host plants, effectively limiting *Alternaria*-induced necrosis (Liang et al. 2025; Radhakrishnan et al. 2017). Compared to chemical fungicides, probiotic formulations are environmentally safe, biodegradable, and compatible with sustainable agriculture. Besides, probiotics mitigate fungicide resistance and improve the soil microbial ecosystem.

3.4 Action of Probiotics Against *Alternaria* Infection

The antagonistic activity of the probiotic microorganisms involves a complex interplay of biological, ecological, and molecular processes. Instead of relying on a single mode of action, probiotic microorganisms often inhibit through multiple but overlapping interactions. These interactions include production of metabolites, competition for nutrients and space, enzymatic action against the cell wall, biofilm-associated exclusion, and induction of plant defence responses (Fig. 3.2).

(i) Antifungal metabolites and volatile compounds: Several microorganisms, such as *Lactobacillus,* exert antifungal activity through the synthesis of diverse secondary metabolites that interfere with spore germination and hyphal development. These can be organic acids such as lactic acids, acetic acid, butyric acid, that locally reduce the pH, or lipopeptides such as iturin, fengycin, and surfactin that can disrupt the membrane integrity. Bacteriocins cause cell leakage or inhibit protein synthesis (Fira et al. 2018; Vasundaradevi et al. 2024). Beyond these organic acids, several microorganisms also release volatile organic acids such as 2,3-butanediol, acetoin, and dimethyl sulphide (Raza and Shen 2020).

(ii) Competitive exclusion and nutrient limitation: Probiotic microorganisms rapidly colonize parts of the leaves' roots surface due to high hydrophobicity index and restrict other microorganisms from colonizing. By occupying the space, these microorganisms utilize carbon and other available nutrients (Dai et al. 2025). Surface colonization near the sites such as stomata also limits the occupancy and germination of conidia. This pre-emptive occupation exclusion

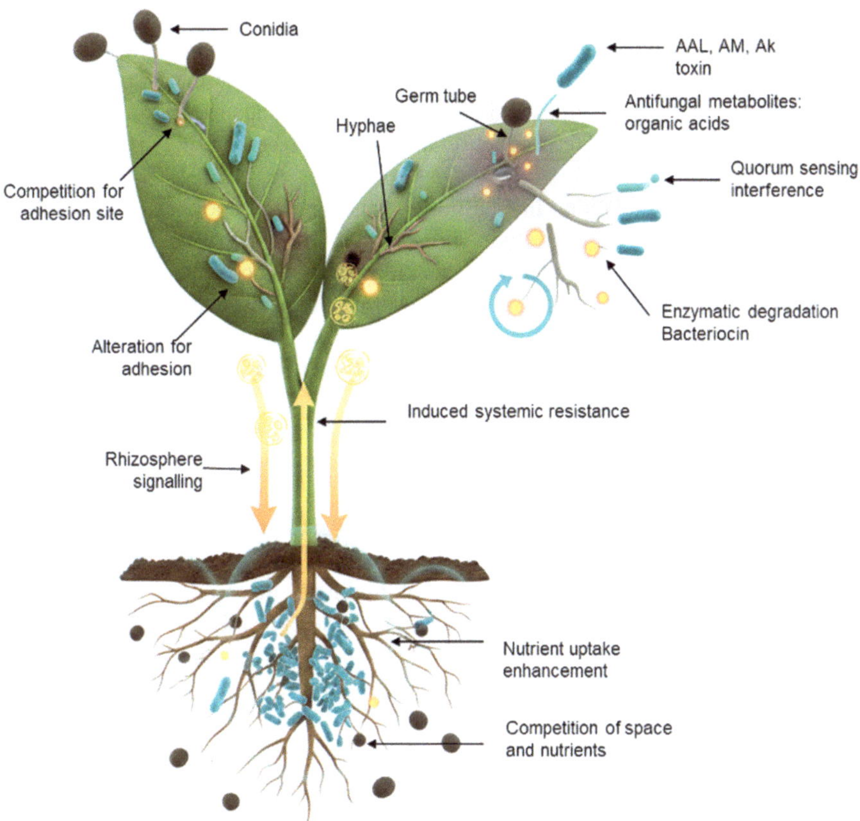

Fig. 3.2 Overview of the infection cycle of *Alternaria* and its inhibition in the presence of probiotic microorganisms

is beneficial under field conditions, where environmental fluctuations can enhance pathogen competitiveness.

(iii) Resistance in plants: Several microorganisms can stimulate induced systemic resistance (ISR), a plant-wide defence response activated by non-pathogenic microorganisms such as probiotics (Pieterse et al. 2014). ISR is commonly modulated by jasmonic acid and ethylene signalling pathways, which result in increased synthesis of phenolic compounds, lignin deposition (Flors et al. 2008; Fontenelle et al. 2011). As a result, probiotics associated with plants exhibit enhanced defensive readiness that limits pathogen infection but also strengthens overall plant immunity.

(iv) Biofilm formation and physical penetration: Several species, such as *Bacillus and Lactobacillus plantarum,* are known to form biofilm structures in roots and leaf surfaces, generating a stable microbial layer that acts as both a physical and biochemical barrier against pathogens. The biofilm enhances the persistence of beneficial populations and buffers the plant-microbes interface against oxidative stress or any other associated stress. The extracellular matrix of biofilms is rich in exopolysaccharides and bioactive peptides that support long-term colonization and metabolite production and release (Islam et al. 2016; Sarti et al. 2023).

(v) Enzymatic degradation: In addition to chemical compounds, the probiotic microorganisms also produce extracellular hydrolytic enzymes such as chitinases and β-1,3-glucanases capable of degrading structural components of the fungal hyphae resulting in cell wall destabilization and loss of viability. The degradation products can also function as elicitors of plant signalling further reinforcing ISR signalling responses (Dai et al. 2025).

3.5 Experimental Observations

The efficacy of probiotic microorganisms is explored in several in vitro assays, greenhouse experiments, and also at field evaluations. These studies assess multiple antifungal mechanisms such as production of inhibitory metabolites, competition for nutrients, colonization sites, and activation of plant defence mechanisms. In addition to these, several studies have reported development of formulations such as cell-free supernatants, encapsulated spores of probiotic microorganisms, and biofilm-based preparations, to improve shelf life, stability, and performance under field conditions.

Table 3.1 summarizes representative studies illustrating probiotic microorganism interactions with *Alternaria* spp. across different crop systems. Figure 3.3 summarizes in vitro assay used for strain evaluation. All these experimental studies consistently demonstrate inhibition of *Alternaria* in different environmental conditions. Initial screening is commonly carried out in methods such as dual culture confrontation assays (Fig. 3.3), exposure to cell-free filtrate and volatile compounds. Several studies have reported acidification, enzymatic degradation and production of secondary metabolites, etc. (Vasundaradevi et al. 2024). Microscopic evaluation revealed hyphal distortion and cytoplasmic leakage in *Alternaria* exposed to cell-free filtrates of *Bacillus subtilis*, attributable to cyclic lipopeptides such as iturin, surfactin, and fengycin (Zhang et al. 2022). Further, greenhouse and field trials have validated these findings under realistic conditions. In greenhouse trials, *B. velezensis* HN-Q-8 reduced *Alternaria solani* lesion formation in potato by more than 70%, alongside elevated activities of superoxide dismutase (SOD), peroxidase (POD), and catalase (CAT) (Zhang et al. 2020; Bai et al. 2023). Similarly, *B. velezensis* YXDHD1–7 decreased tomato early blight incidence by 68%, mediated by VOC production (2,3-butanediol, acetoin) and enhanced plant vigour through spermidine signalling (Raza and Shen 2020).

Table 3.1 Summary of selected studies on biocontrol of *Alternaria* spp. by probiotic microorganisms

Probiotic strain	Target *Alternaria* species	Key mechanism	Outcome	References
Bacillus velezensis HN-Q-8	*A. solani*	Lipopeptide production, induced systemic resistance (ISR)	70% lesion reduction	Bai et al. (2023)
Bacillus velezensis YXDHD1-7	*A. solani*	Volatile organic compounds (VOCs), Spermidine-mediated plant vigour	68% disease reduction	Li et al. (2024)
Trichoderma harzianum + *Trichoderma viride,* + *Bacillus subtilis* + *Pseudomonas fluorescens*	*A. porri*	unknown	23% disease management, 5.6% increase in seed yield	Saini et al. (2024)
Bacillus subtilis + *Pseudomonas fluorescens* + *Trichoderma harzianum*	*Alternaria* spp.	Microbial consortia, ISR, nutrient competition	98% disease reduction; 15% yield improvement	Giorni et al. (2024)
Bacillus subtilis BS01	*A. solani*	Lipopeptide production, volatile organic acids	–	Awan et al. (2023)
Leuconostoc mesenteroides GY-2	*A. alternata*	Volatile organic acids	91% lesion reduction	Dai et al. (2025)

3.6 Strategies and Applications

Successful field or post-harvest application of probiotics depends on formulation stability and efficient delivery. Formulations enhance microbial survival during storage, improve adherence and colonization on plant surfaces, and facilitate metabolite release. Common types include wettable powders (WP) with inert carriers such as talc or kaolin for seed coating and foliar sprays, liquid formulations in protective buffers or oils for short-term applications, slow-release granules or pellets (peat, vermiculite, biochar) for rhizosphere establishment, and microencapsulation or polymer coatings (alginate, chitosan) that protect cells from UV radiation, desiccation, and temperature fluctuations while allowing gradual metabolite diffusion (Gao et al. 2025). *Bacillus* spp., due to their spore-forming ability, achieve long shelf life (6–12 months) and resistance to desiccation, whereas LAB can be stabilized via freeze-drying or encapsulation in polysaccharide matrices (Montel Mendoza et al.

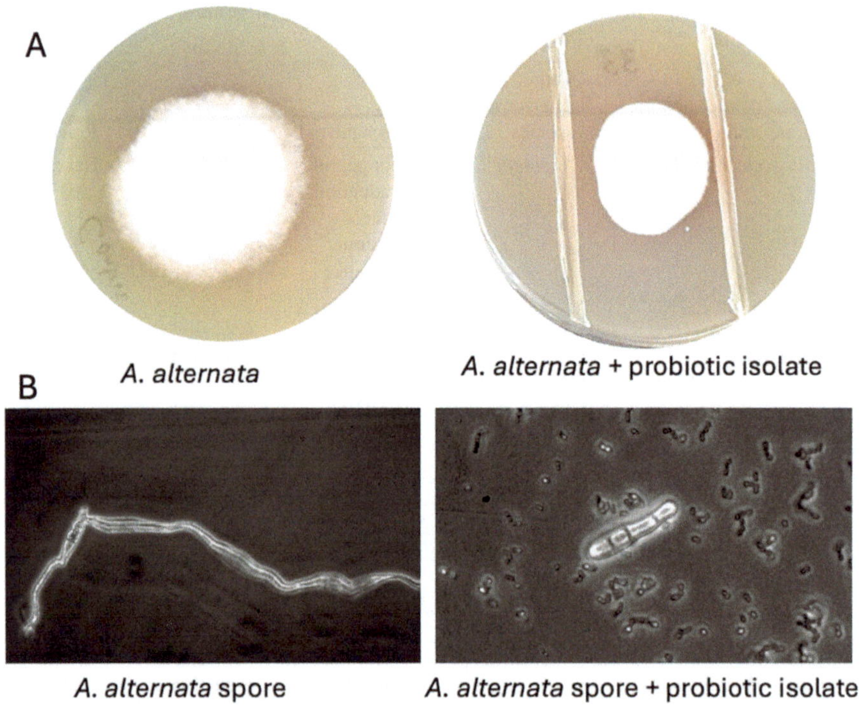

Fig. 3.3 In vitro inhibition of *A. alternata* by *Lactiplantibacillus* sp. Panel **A** depicts the growth of *A. alternata* on the modified de Man, Rogosa, and Sharpe (MRS) agar plates and its suppression in the presence of *Lactiplantibacillus* sp. Panel **B** shows the germination of *A. alternata* conidia and the inhibition of conidial germination when exposed to *Lactiplantibacillus* sp.

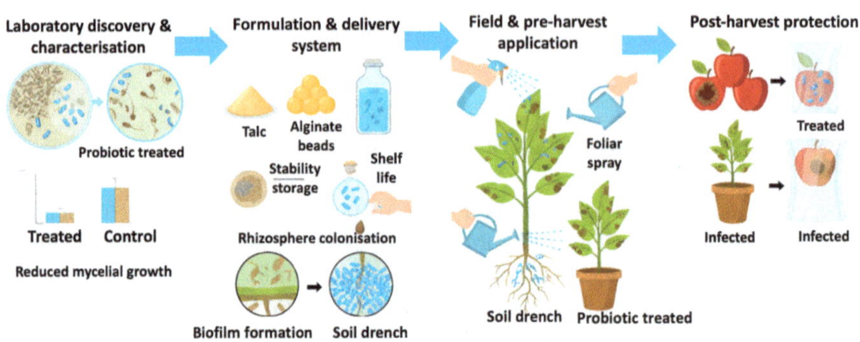

Fig. 3.4 Application pipeline of probiotic strain as biocontrol agents

2014). Advanced formulations may include bio-stimulants such as humic acids, amino acids, or prebiotic sugars to enhance probiotic establishment. Encapsulated VOC-emitting strains, can be a potential biofumigation-based post-harvest storage system for bridging laboratory efficacy with field reliability (Fig. 3.4).

3.6.1 Applications in Food Preservation and Storage

A. alternata is a common post-harvest pathogen, causing black rot and mycotoxin contamination in fruits and vegetables. With growing interest for cell-free management of fungal disease, probiotic based solution offers an ecofriendly alternative. In recent study, *Lactiplantibacillus plantarum* MYSVB1 isolated from banyan tree fruit exhibited antifungal activity by production to organic acids, bacteriocins, resulting in a 65% reduction in visible rot in tomato and apple fruits during 10-day ambient storage (Vasundaradevi et al. 2024). Similarly, yeasts such as *Saccharomycopsis fibuligera* KN-1 produce volatile organic compounds that inhibit both spores and mycelium and they are used as fumigants (Chen et al. 2025). The efficacy of these compounds can be enhanced with edible coatings and modified packaging, such that the incidence of infection can be reduced.

3.6.2 Application Strategies

Since *Alternaria* is known to infect seeds and plants during post-harvest storage, the effective control of *Alternaria* infections in plants involves intervention at multiple stages. These include seed treatment, soil application, foliar spray, and post-harvest management. Although specific studies for *Alternaria* are limited, broadly, there are several strategies developed that can be extended to manage *Alternaria* control.

Seed treatment, seed coating, and biopriming with probiotic microorganisms offer an effective strategy to establish beneficial populations at the early stage of plant development. Inoculation with spores of these microorganisms, for example, spores of *Bacillus,* rhizosphere colonization, reduces seed-borne infections and improves seedling vigour, nutrient uptake, etc. (Sarkar et al. 2021). Carrier materials such as talc or peat-based formulations are also commonly used to enhance the survival of these microorganisms and uniform application.

Soil-based applications represent another important strategy. The incorporation of probiotic granules, compost-derived formulations into soils can promote the development of a suppressive rhizosphere, indirectly reducing secondary foliar infections. Beneficial strains influence root exudation, increasing phenolic compounds and siderophores, as detailed earlier, which restrict the availability of nutrients for propagation of infection. Studies have also shown that integration with biochar improves microbial survival.

Foliar applications of probiotic formulations allow direct interaction with pathogen conidia or mycelium. The application strategy involves the addition of live spore suspension of probiotic microorganisms or fermentation-derived cell-free extracts applied as sprays. This is mostly followed by repetitive applications during the early and pre-infection stages. In studies under greenhouse conditions, foliar spray containing *B. velezensis* reduced the early blight severity by up to 70% (Bai et al. 2023).

Post-harvest management strategies extend the protective role of probiotics beyond field applications. Surface dipping, edible coatings, or vapour phase applications incorporating probiotic formulations can significantly reduce fungal proliferation. Lactic acid bacteria-based formulations containing *Lactiplantibacillus plantarum, Leuconostoc mesenteroides* maintain acidic microenvironments and produce antifungal peptides that inhibit mycelial growth (Vasundaradevi et al. 2024). This can greatly reduce post-harvest losses, especially when integrated with cold storage.

3.7 Challenges and Limitations

Several studies have demonstrated the benefits of using probiotic microorganisms against phytopathogens. Yet most of the applications in the field are constrained by technological and regulatory limitations. Variability in environmental conditions, inherent complexity of microbial interactions, and lack of defined standards for the development and approval of microbial agents limit the overall progress of large-scale adoption.

3.7.1 Environmental Variations

One of the main constraints in using probiotic microorganisms is the inconsistent performance under field conditions. Often, strong inhibition and antifungal activity are observed in laboratory conditions; greenhouse studies may not be achieved in real field conditions. This variability could be due to one or more of the following reasons: (i) temperature, humidity, and ultraviolet radiation, soil properties directly affect the survival, colonization and metabolic activity of the probiotic microorganisms. Sometimes, strain-specific variations may also influence the survival. For instance, *Bacillus* can sustain the harsh conditions by spore formation, while Lactobacillus is mainly vegetative cells, making it more sensitive to changes in the ambient environment and field conditions (Compant et al. 2019). (ii) Microbial competition from the indigenous microbial community of soil and the plant itself (endophytic microorganisms). They can limit the colonization and establishment of the probiotic microorganisms. (iii) Host plant pathogen plays an important role as differences in cultivars influence the characteristics surface properties that are important for probiotic microbial attachment and colonization. (iv) Pathogen diversity within the genus, such as *A. alternata, A. solani,* and *A. tenuissima* leads to variable susceptibility to probiotic antagonists. This creates hurdles in formulation strategies for uniform disease control.

3.7.2 Formulations and Delivery

Several technical limitations continue to resist the laboratory's transition to commercially viable agricultural practices. (i) Shelf life and microbial viability remain a greater concern, particularly for non-spore-forming strains such as *Lactobacillus* spp., which show rapid decline in viability during storage. In order to extend the shelf life, encapsulation-based approaches are often required. (ii) Carrier selection and compatibility also play a crucial role. Commonly used inert materials such as talc, bentonite, or starch must also support microbial survival (Khan et al. 2023). (iii) Achieving uniform microbial density on seeds, plant surfaces, or soil matrix is challenging, and it directly influences the colonization and disease suppression. (iv) Co-formulation with multiple strains adds further complexity, although the microbial consortia provide additive benefits. Differences in growth characteristics and metabolic interactions require careful optimization (Saini et al. 2024; Giorni et al. 2024). (v) In addition, stability of antifungal compounds, including volatile organic compounds, peptides, or lipopeptides, is often limited due to their degradation under field conditions (Aslan et al. 2023). In spite of these limitations, microencapsulation and nanoformulations and biochar-based carriers are used to enhance the survival and controlled release of microorganisms or compounds.

3.7.3 Safety and Regulatory Considerations

The regulatory framework remains one of the most significant non-technical barriers to the adoption of probiotic and bacterial formulations. Unlike synthetic pesticides, microbial products are evaluated under regulatory systems and differ based on regions and clear categorization. In many jurisdictions, probiotic strains may be assessed under food safety, environmental release, plant protection, or human health regulations in prolonged approval timelines (Merenstein et al. 2023). In addition, regulatory agencies require extensive safety documentation, including toxicological profiling, environmental risk assessment, and whole genome sequencing to verify the absence of virulence factors and toxin biosynthesis pathways, or even transferable antibiotic resistance genes (Marrone 2019). These evaluations are time-consuming and expensive for academic and small-scale developers. This is further complicated by the lack of international harmonization; regulatory standards and approval pathways differ substantially across regions, and in several Asian countries, formal guidelines governing agricultural probiotics remain limited or poorly defined (Arora and Baldi 2015). From a commercial perspective, protection of strain identity presents additional challenges as maintaining proprietary ownership requires rigorous genomic characterization, secure long-term storage, and traceability throughout production. Most of the probiotic microorganisms are classified as generally recognized as safe (GRAS); their deliberate release on a large scale continues to raise ecological dilemma mainly due to horizontal gene transfer and impact on the indigenous beneficial microbes.

3.7.4 Knowledge and Technology Gaps

Despite encouraging experimental outcomes, several knowledge gaps limit the deployability of probiotic microorganisms in agricultural settings. The major concern is the limited field validations, as most of the studies are small-scale or short-term or isolated to certain regions and limited seasonal validation. Besides, the actual mechanisms of inhibition, the interactions or probiotic microorganisms with the host is limited that strongly influence the outcome.

Translational challenges also arise during scale-up. The production of high-density, metabolically active cell concentration is often poorly optimized. This can result in variability in efficacy between batches. From economic perspective, production cost, formulation requirements, and cold-chain logistics can outweigh perceived benefits. This is true in most of the low-input resource-limited farming practices. Finally, adoption at the farmer level remains constrained by limited awareness. Many cultivators are unfamiliar with microbial biocontrol strategies and tend to favour conventional fungicides. This is due to the unpredictable and delayed response of probiotic-based formulations.

3.8 Future Perspectives

Future progress in sustainable management of *Alternaria* spp. is likely to depend on the close integration of molecular, microbial biotechnology, and delivery technologies. Instead of relying solely on single microbial inoculants, recent studies have increasingly suggested the microbial community approach. Developments in gene technology and synthetic biology have opened up new opportunities to improve the antifungal efficacy and robustness. Targeted modifications of biosynthetic pathways have the potential to enhance the production of antifungal lipopeptides such as surfactin, fengycin, or organic acids. Quorum sensing could allow beneficial microorganisms to regulate antifungal activity in response to pathogen-derived signals or ques. For non-spore-forming lactic acid bacteria, improving tolerance to ultraviolet radiation, desiccation, and temperature stress remains a key objective in improving field persistence and storage. Any such genetic enhancements must be accompanied by careful evaluation of safety and regulatory oversight. A useful extension of this strategy for field deployment is the use of mixed microbial consortia. By combining several approaches, such as the synthesis of volatile compounds, siderophore-mediated nutrient competition, and promotion of plant growth and defence responses, assemblages of *Bacillus, Pseudomonas*, and lactic acid bacteria can provide more comprehensive protection. Persistence under changing environmental conditions is enhanced by occupying specific ecological niches, such as the rhizosphere, phyllosphere, and endosphere. When compared to single-strain applications, metabolic cooperation and cross-feeding among consortium members have frequently led to greater suppression of the pathogen. Further advancements in microbiome characterization and computational modelling could significantly improve consortium design.

Recent microbiome-based approaches focus on strengthening native plant-associated microbial communities to achieve more stable and long-lasting suppression. Instead of introducing external agents alone, these strategies seek to selectively favour indigenous beneficial microorganisms that naturally limit pathogen establishment. For example, prebiotic inputs such as root exudate analogues or targeted nutrient amendments can shift microbial community structure toward antagonistic populations. Phage-assisted approaches offer an additional layer of specificity by selectively regulating phytopathogens or competing microbes without broadly disrupting the microbiome. Metagenomic and metatranscriptomic analyses have further enabled the identification of microbial groups consistently associated with disease suppression, providing a rational basis for microbiome manipulation. Continued progress in formulation and delivery method is equally important for maintaining probiotic performance under practical field and post-harvest conditions. Environmental stresses often limit microbial survival and activity after application, making formulation design a critical factor. Emerging delivery systems that respond to factors such as moisture availability or infection pressure may help synchronize probiotic activity with periods of highest disease risk. When combined with precision agriculture tools, such as disease forecasting models and remote sensing platforms, the success rate can be amplified.

Beyond technical considerations, regulatory and economic frameworks strongly influence the adoption of biocontrol agents. Clear and harmonized regulations, along with well-defined quality standards for microbial products, are necessary to build confidence among growers and manufacturers. Support for decentralized or regional production systems may further reduce costs and improve accessibility. Collaboration between academic researchers, industry stakeholders, and policymakers will be critical for translating laboratory advances into reliable commercial products. In the longer term, probiotic biocontrol strategies have the potential to support the circular bio-economy.

3.9 Conclusion

Probiotic microorganisms offer a promising, environmentally compatible approach for managing *Alternaria*-associated diseases in crop systems. Through a combination of niche occupation, production of antifungal enzymes and metabolites, and stimulation of plant defence responses, these beneficial microbes can limit fungal establishment while supporting overall plant health. Several experimental evidences demonstrate promising biocontrol strategies. However, challenges related to formulation stability, regulatory clearance, and performance under variable environmental conditions continue to restrict wider adoption. Addressing these limitations will require genomics-guided strain selection, carefully designed pilot and field trials across diverse agro-climatic regions, and regulatory frameworks that enable safe and transparent commercialization of microbial products.

3.10 Chapter Summary

This chapter provides a comprehensive overview of the use of probiotic microorganisms in mitigating the infection of *Alternaria* spp. It describes the biology and ecological relevance of *Alternaria*, followed by detailed mechanisms through which probiotics suppress fungal growth via metabolite production, competitive exclusion, and activation of plant defences. Few studies demonstrate practical applications in seed treatment, soil amendment, and post-harvest protection. Despite promising results, formulation stability, environmental variability, and regulatory inconsistencies remain major obstacles. Future advancements, including genetically optimized strains, biofilm-based delivery systems, and microbial consortia, hold promise for precision biocontrol. Overall, probiotics are emerging as a sustainable, eco-friendly alternative to chemical fungicides but need greater validation to meet modern goals of resilience and microbiome-conscious agriculture.

Competing Interest None.

Author Contribution MS: Conceptualization; MS and RV: Writing original and review editing; MYS: Review editing.

References

Adithi G, Divyashree S, Shruthi B, Deepa N, Sreenivasa MY (2024) Evaluation of *Limosilactobacillus fermentum* MYSAGAM1 isolated from herbal amla juice as a probiotic candidate with antifungal characteristics against *Fusarium equiseti*. Food Biosci 58:103843. https://doi.org/10.1016/j.fbio.2024.103843

Adithi G, Rakesh S, Shruthi B, Divyashree S, Deepa N, Vanitha PR, Vasundaradevi R, Deepthi BV, Sreenivasa MY (2025) Biological decontamination of myco-toxigenic *Fusarium verticillioides* using *Lactiplantibacillus plantarum* MYSN128 isolated from traditional vellappam: implications for food safety and global mycotoxin management. Curr Microbiol 82:404. https://doi.org/10.1007/s00284-025-04401-z

Aloi F, Riolo M, Sanzani SM, Mincuzzi A, Ippolito A, Siciliano I, Pane A, Gullino ML, Cacciola SO (2021) Characterization of *Alternaria* species associated with heart rot of pomegranate fruit. J Fungi 7(3):172. https://doi.org/10.3390/jof7030172

Arora M, Baldi A (2015) Regulatory categories of probiotics across the globe: a review representing existing and recommended categorization. Indian J Med Microbiol 33(Suppl):S2–S10. https://doi.org/10.4103/0255-0857.150868

Aslan M, Ertaş N, Demir MK (2023) Storage stability, heat stability, controlled release and antifungal activity of liposomes as alternative fungal preservation agents. Food Biosci 51:102281. https://doi.org/10.1016/j.fbio.2022.102281

Awan ZA, Shoaib A, Schenk PM, Ahmad A, Alansi S, Paray BA (2023) Antifungal potential of volatiles produced by *Bacillus subtilis* BS-01 against *Alternaria solani* in *Solanum lycopersicum*. Front Plant Sci 13:1089562. https://doi.org/10.3389/fpls.2022.1089562

Bai X, Li Q, Zhang D, Zhao Y, Zhao D, Pan Y, Wang J, Yang Z, Zhu J (2023) *Bacillus velezensis* strain HN-Q-8 induced resistance to *Alternaria solani* and stimulated growth of potato plants. Biology 12(6):856. https://doi.org/10.3390/biology12060856

Cao J, Yan J, Zhao Y, Jiang W (2013) Effects of postharvest salicylic acid dipping on *Alternaria* rot and disease resistance of jujube fruit during storage. J Sci Food Agric 93(13):3252–3258. https://doi.org/10.1002/jsfa.6167

Chen R, Sun R, Liang J, Sun L, Yang M, Qin Y (2025) Investigation into the antagonistic effects and mechanisms of VOCs produced by *Saccharomycopsis fibuligera* KN-1 against *Alternaria alternata*. Food Biosci 69:106867. https://doi.org/10.1016/j.fbio.2025.106867

Compant S, Samad A, Faist H, Sessitsch A (2019) A review on the plant microbiome: ecology, functions, and emerging trends in microbial application. J Adv Res 19:29–37. https://doi.org/10.1016/j.jare.2019.03.004

Dai P, Li B, Li Y, Wang L, Hu T, Wang Y, Meng X, Cao K, Wang S, Sun M (2025) Biocontrol efficiency of *Leuconostoc mesenteroides* GY-2 against postharvest black rot caused by *Alternaria alternata* and mechanisms of action. J Fungi 11(10):705. https://doi.org/10.3390/jof11100705

Deepa N, Charith Raj AP, Sreenivasa MY (2016) Nested PCR method for the early detection of fumonisin-producing *Fusarium verticillioides* in pure cultures, cereal samples and plant parts. Food Biotechnol 30(1):18–29

Deepa N, Chennappa G, Deepthi BV et al (2022) Antifungal potential of *Azotobacter* species and its metabolites against *Fusarium verticillioides* and biodegradation of fumonisin. J Appl Microbiol 133(4):2430–2444

Deepthi BV, Deepa N, Vanitha PR, Sreenivasa MY (2022) Stress responses on the growth and mycotoxin biosynthesis of *Fusarium proliferatum* associated with stored poultry feeds. Appl Food Res 2(1):1000U1. https://doi.org/10.1016/j.afres.2022.100091

Dettman JR, Eggertson QA, Kim NE (2023) Species diversity and molecular characterization of *Alternaria* section *Alternaria* isolates from cereal crops in Canada. Front Microbiol 14:1194911. https://doi.org/10.3389/fmicb.2023.1194911

Divyashree S, Ramu R, Sreenivasa MY (2024) Evaluation of new candidate probiotic *Lactobacillus* strains isolated from a traditional fermented food multigrain-millet dosa batter. Food Biosci 57: 103450. https://doi.org/10.1016/j.fbio.2023.103450

Elshaghabee FMF, Rokana N, Gulhane RD, Sharma C, Panwar H (2017) *Bacillus* as potential probiotics: status, concerns, and future perspectives. Front Microbiol 8:1490. https://doi.org/10.3389/fmicb.2017.01490

Fernandes C, Casadevall A, Gonçalves T (2023) Mechanisms of *Alternaria* pathogenesis in animals and plants. FEMS Microbiol Rev 47(6):fuad061. https://doi.org/10.1093/femsre/fuad061

Fira D, Dimkić I, Berić T, Lozo J, Stanković S (2018) Biological control of plant pathogens by *Bacillus* species. J Biotechnol 285:44–55. https://doi.org/10.1016/j.jbiotec.2018.07.044

Flors V, Ton J, Van Doorn R, Jakab G, García-Agustín P, Mauch-Mani B (2008) Interplay between JA, SA and ABA signalling during resistance against *Pseudomonas syringae* and *Alternaria brassicicola*. Plant J 54(1):81–92. https://doi.org/10.1111/j.1365-313X.2007.03397.x

Fontenelle ADB, Guzzo SD, Lucon CMM, Harakava R (2011) Growth promotion and induced resistance in tomato against *Xanthomonas euvesicatoria* and *Alternaria solani* by Trichoderma spp. Crop Prot 30(11):1492–1500. https://doi.org/10.1016/j.cropro.2011.07.019

Food and Agriculture Organization of the United Nations, World Health Organization (2002) Joint FAO/WHO working group report on drafting guidelines for the evaluation of probiotics in food. FAO, Rome

Gao R, Zhao D, Zhou X, Wan Z, Wang X, Rao H, Liu X, Gao X, Hao J (2025) Advances in polysaccharide-based biopolymers for probiotic encapsulation. Curr Res Food Sci 11:101186. https://doi.org/10.1016/j.crfs.2025.101186

Garcia EF, Luciano WA, Xavier DE et al (2016) Identification of lactic acid bacteria in fruit pulp by-products and probiotic potential. Front Microbiol 7:1371. https://doi.org/10.3389/fmicb.2016.01371

Giorni P, Bulla G, Bellotti G et al (2024) In planta evaluation of bacterial consortia against *Alternaria* spp. Front Hortic 3:1447425. https://doi.org/10.3389/fhort.2024.1447425

Hill C, Guarner F, Reid G et al (2014) ISAPP consensus statement on the term probiotic. Nat Rev Gastroenterol Hepatol 11(8):506–514. https://doi.org/10.1038/nrgastro.2014.66

Islam S, Akanda AM, Prova A, Islam MT, Hossain MM (2016) PGPR isolation and disease suppression. Front Microbiol 6:1360. https://doi.org/10.3389/fmicb.2015.01360

Khan A, Singh AV, Gautam SS et al (2023) Microbial bioformulation for sustainable agriculture. Front Plant Sci 14:1270039. https://doi.org/10.3389/fpls.2023.1270039

Lee HB, Patriarca A, Magan N (2015) *Alternaria* in food: ecophysiology and toxicology. Mycobiology 43(2):93–106. https://doi.org/10.5941/MYCO.2015.43.2.93

Li W, Sun L, Wu H et al (2024) *Bacillus velezensis* YXDHD1-7 suppresses early blight in tomato. Microorganisms 12(5):921. https://doi.org/10.3390/microorganisms12050921

Liang Q, Zhou W, Peng S et al (2025) Bacteriocin-producing LAB in food industry. Curr Res Food Sci 10:100997. https://doi.org/10.1016/j.crfs.2025.100997

Lin CH, Chung KR (2010) Histidine kinase and HOG1 MAPK signaling in *Alternaria alternata*. Fungal Genet Biol 47(10):818–827. https://doi.org/10.1016/j.fgb.2010.06.009

Lin CH, Yang SL, Wang NY, Chung KR (2010) FUS3 MAPK signaling in *Alternaria alternata*. Fungal Genet Biol 47(4):381–391. https://doi.org/10.1016/j.fgb.2009.12.009

Ma H, Zhang B, Gai Y et al (2019) Cell-wall-degrading enzymes in *Alternaria alternata*. Front Microbiol 10:2514. https://doi.org/10.3389/fmicb.2019.02514

Marrone PG (2019) Pesticidal natural products. Pest Manag Sci 75(9):2325–2340. https://doi.org/10.1002/ps.5433

Meena M, Samal S (2019) *Alternaria* host-specific toxins. Toxicol Rep 6:745–758. https://doi.org/10.1016/j.toxrep.2019.06.021

Meena M, Gupta SK, Swapnil P et al (2017) *Alternaria* toxins and virulence genes. Front Microbiol 8:1451. https://doi.org/10.3389/fmicb.2017.01451

Merenstein D, Pot B, Leyer G et al (2023) Emerging issues in probiotic safety. Gut Microbes 15(1): 2185034. https://doi.org/10.1080/19490976.2023.2185034

Montel Mendoza G, Pasteris SE, Otero MC, Nader-Macías MEF (2014) LAB survival after freeze-drying. J Appl Microbiol 116(1):157–166. https://doi.org/10.1111/jam.12359

Pieterse CMJ, Zamioudis C, Berendsen RL et al (2014) Induced systemic resistance by beneficial microbes. Annu Rev Phytopathol 52:347–375. https://doi.org/10.1146/annurev-phyto-082712-102340

Radhakrishnan R, Hashem A, Abd-Allah EF (2017) *Bacillus* as a crop improvement tool. Front Physiol 8:667. https://doi.org/10.3389/fphys.2017.00667

Raza W, Shen Q (2020) VOC-mediated plant–microbe interactions. In: Molecular aspects of plant beneficial microbes. Elsevier, pp 209–219

Saini S, Raj K, Kumar R, Saini AK (2024) Biocontrol agents against onion purple blotch. J Plant Pathol. https://doi.org/10.1007/s42161-024-01744-7

Salotti I, Giorni P, Battilani P (2024) Biology and epidemiology of *Alternaria* in tomato. Front Plant Sci 15:1430965. https://doi.org/10.3389/fpls.2024.1430965

Sarkar D, Singh S, Parihar M, Rakshit A (2021) Seed bio-priming with microbial inoculants. Curr Res Environ Sustain 3:100093. https://doi.org/10.1016/j.crsust.2021.100093

Sarti GC, Galelli ME, Arreghini S et al (2023) *Bacillus subtilis* biofilm promotes lettuce growth. Sustainability 15(21):15406. https://doi.org/10.3390/su152115406

Shruthi B, Deepthi BV, Adithi G, Vanitha PR, Sreenivasa MY (2025) *Saccharomyces cerevisiae* MYSY1 as a promising probiotic isolate with antifungal properties against foodborne *Aspergillus* species. Biocatal Agric Biotechnol G7:10374U. https://doi.org/10.1016/j.bcab.2025.103649

Shruthi B, Rakesh S, Adithi G, Divyashree S, Mahadevakumar S, Ajithkumar M, Sreenivasa MY (2025a) Probiotic characterization and antifungal potentials of *Pichia kudriavzevii* MYSSBYPS10 isolated from traditionally fermented green gram dosa batter. Food Biotechnol 39(1):1–27. https://doi.org/10.1080/08U0543G.2025.2457G78

Singh KP, Kumari P, Rai PK (2021) Disease-resistant genes in *Brassica juncea*. Front Plant Sci 12: 617405. https://doi.org/10.3389/fpls.2021.617405

Sreenivasa MY, Dass RS, Charith Raj AP, Janardhana GR (2008) PCR-based detection of genus *Fusarium* and fumonisin-producing isolates from freshly harvested sorghum grains grown in Karnataka, India. J Food Saf 28:236–237

Vasundaradevi R, Sarvajith M, Somashekaraiah R, Gunduraj A, Sreenivasa MY (2024) *Lactiplantibacillus plantarum* MYSVB1 against *Alternaria alternata*. Front Microbiol 15: 1322758. https://doi.org/10.3389/fmicb.2024.1322758

Vasundaradevi R, Sarvajith M, Divyashree S, Deepa N, Achar PN, Sreenivasa MY (2025) Tropical fruit-derived *Lactiplantibacillus* against *Fusarium oxysporum*. Sci Rep 15:2144. https://doi.org/10.1038/s41598-025-85190-0

Woudenberg JHC, Groenewald JZ, Binder M, Crous PW (2013) *Alternaria* redefined. Stud Mycol 75:171–212. https://doi.org/10.3114/sim0015

Xu J (2022) Assessing global fungal threats to humans. mLife 1(3):223–240. https://doi.org/10.1002/mlf2.12036

Yang G, Cui S, Huang W et al (2024) Greenhouses as evolutionary niches for *Alternaria alternata*. Microbiol Spectr 12(6):e0039024. https://doi.org/10.1128/spectrum.00390-24

Zhang D, Yu S, Yang Y et al (2020) Volatile-mediated inhibition of *Alternaria solani* by *Bacillus subtilis*. Front Microbiol 11:1196. https://doi.org/10.3389/fmicb.2020.01196

Zhang D, Qiang R, Zhou Z et al (2022) Fengycin-mediated biocontrol of *Alternaria solani*. Front Microbiol 13:861113. https://doi.org/10.3389/fmicb.2022.861113

Probiotics: An Innovative Approach to Combat Postharvest Anthracnose

4

Ata Aditya Wardana, Vincensius Marcellino, Fumina Tanaka, and Fumihiko Tanaka

Abstract

Postharvest diseases pose significant challenges to the handling and storage of fresh produce, as their quality and appearance directly influence marketability and consumer preference. Anthracnose, caused by *Colletotrichum* fungi, leads to lesions that degrade the visual appeal and edibility of affected commodities. Conventional antifungal treatments often pose risks to human health and the environment, highlighting the urgent need for safe, cost-effective, and broad-spectrum alternatives. Probiotics, beneficial microorganisms known for enhancing human immunity and resisting pathogen colonization, have emerged as promising biofungicides. Studies demonstrate their efficacy in controlling anthracnose on fruits and vegetables through antimicrobial compound production, nutrient and space competition, and activation of plant defense mechanisms. This review explores the advancements in utilizing probiotics as innovative solutions for controlling postharvest anthracnose.

A. A. Wardana (✉)
Department of Food Technology, Faculty of Engineering, Bina Nusantara University, Jakarta, Indonesia
e-mail: ata.wardana@binus.edu

V. Marcellino · F. Tanaka · F. Tanaka
Laboratory of Postharvest Science, Faculty of Agriculture, Kyushu University, Fukuoka, Japan

M. Y. Sreenivasa et al. (eds.), *Probiotics as Antifungals*,
https://doi.org/10.1007/978-981-92-0002-3_4

Graphical Abstract

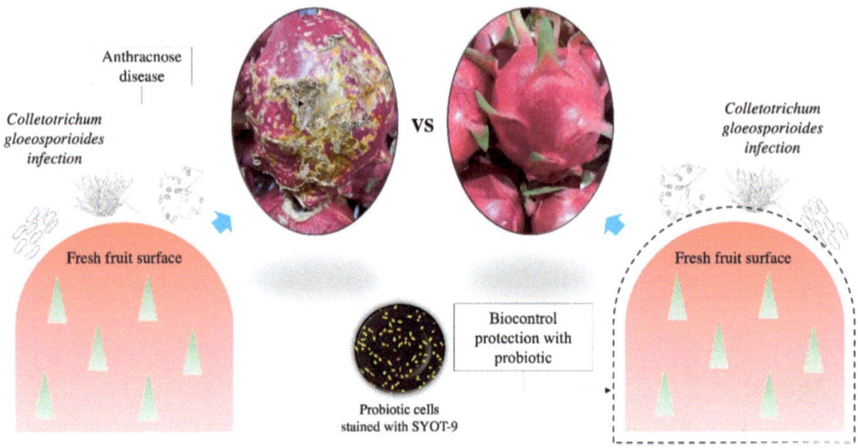

Conceptual framework of probiotics as a biocontrol agent for controlling postharvest anthracnose

Keywords

Food · Safety · Quality · Pathogen · Shelf-life

Learning Objectives

- To understand the recent updates on probiotics to combat postharvest anthracnose.
- To explain the conventional and emerging strategies for anthracnose prevention using probiotic-based biocontrol.
- To explore future prospects of probiotic-based anti-anthracnose agents.

4.1 Introduction

Anthracnose is a devastating fungal disease affecting a wide range of crops, leading to significant economic losses in agriculture. This disease, caused by various species of *Colletotrichum*, anthracnose, manifests in symptoms such as fruit rot, leaf blight, and stem lesions, ultimately reducing crop yield and quality. Conventional management strategies, including chemical fungicides, have been widely used to control this disease. However, concerns over environmental safety, fungicide resistance, and consumer health have driven the search for sustainable and eco-friendly alternatives.

Biological control is an effective strategy for managing fungal diseases caused by phytopathogens. It involves the use of live microorganisms or their derivatives to suppress plant pathogens, thereby minimizing economic losses (Carmona-Hernandez et al. 2019; Wardana et al. 2023). The mode of action, as described in numerous studies on antagonists, primarily involves competition for space and essential nutrients such as carbohydrates or amino acids (Saravanakumar et al. 2008; Carmona-Hernandez et al. 2019). Additionally, the production of antifungal compounds represents the second most significant mechanism (Romero et al. 2022).

Probiotics, primarily known for their health benefits, have gained attention in plant disease management. The World Health Organization (WHO) defined probiotics as live microorganisms that, when administered in adequate amounts, confer health benefits to the host (FAO and WHO 2002). Probiotics can be classified into various categories, including pharmaceutical probiotics, food-based probiotics, probiotics for animal feed, and genetically engineered probiotics. Among these, pharmaceutical probiotics are considered the most significant (Venugopalan et al. 2010). Beneficial microorganisms, such as lactic acid bacteria (LAB) including *Lactobacillus, Pediococcus*, and certain *Bacillus* species, exhibit antagonistic properties against plant pathogens through mechanisms such as competitive exclusion, enzyme production, and induced systemic resistance.

Recent studies suggest that probiotics could serve as effective biocontrol agents against anthracnose, offering an environmentally friendly alternative to synthetic fungicides (Mpeluza et al. 2023; Wang et al. 2024; Huang et al. 2024). This article explores the potential of probiotics as anti-anthracnose agents, highlighting their mechanisms of action, effectiveness, and potential applications in sustainable agriculture (El-Mabrok et al. 2013; Tefa et al. 2016; Greeshma et al. 2021; Thi et al. 2023; Mpeluza et al. 2023; Wang et al. 2024; Huang et al. 2024). From this perspective, probiotics can be considered as an innovative approach to combat postharvest anthracnose.

4.2 Anthracnose in Postharvest Commodities

Fruit losses occur throughout production, postharvest, storage, and transport phases, significantly reducing available food for human consumption. The FAO reported that 14% of global food is lost between harvesting and retail (FAO 2019). The substantial fruit loss in Asian countries is due to mishandling, inefficient transport, and fungal and bacterial contamination (Udayanga et al. 2013). Compliance with strict phytosanitary standards for export to developed countries impacts agricultural trade in tropical fruit-producing developing nations. Diseases like anthracnose, caused by *Colletotrichum* species (de Capdeville et al. 2007; Bahia 2021), highlighted the need for technological investments to mitigate economic losses.

The contamination of healthy fruits and vegetables is significantly influenced by the environmental conditions of tropical regions. Typically, elevated temperatures (around 27 °C) and high humidity levels (approximately 80%) create optimal conditions for anthracnose development in tropical fruits (Botelho et al. 2016; Bahia 2021). Persistent moisture on leaf surfaces facilitates fungal infection and growth. In fruit cultivation, a significant portion of produce is lost due to contamination during production and ripening (Bahia 2021). Anthracnose is particularly concerning for producers, as the responsible fungus has severely impacted crops such as mango, papaya, avocado, banana, and other fruits (Botelho et al. 2016; Bahia 2021).

Anthracnose primarily manifests during fruit ripening, particularly postharvest, in climacteric species. Dead leaves, infected branches, and fruits serve as contamination sources, with infection possible from fruiting to harvest (de Oliveira et al. 2023). Wind or water disperses conidiospores, which quickly adhere to fruit surfaces, germinate, and penetrate the cuticle. The hyphae then colonize the fruit wall, forming small, circular, dark brown to black lesions that become water-soaked and expand over time (de Oliveira et al. 2023). As the fruit ripens, lesions enlarge, darken, and develop gelatinous pink or orange spore masses, eventually leading to necrosis and fruit loss (Udayanga et al. 2013; Perkins et al. 2019). The visual appearance of this disease is exhibited in Fig. 4.1.

Fig. 4.1 Visual appearance of anthracnose disease on postharvest commodities: (**a**, **b**) chili, (**c**) eggplant, (**d**) tomato, (**e**) papaya, (**f**) duku fruit, (**g**) dragon fruit, (**h**) avocado

4.3 Conventional Treatments for Anthracnose Prevention

Fruit and vegetable commodities, historically produced for local consumption, have become global commodities due to significant advancements in technology, transportation, and storage. However, challenges persist in their commercialization, including inadequate infrastructure and technology, pest infestations, microbial growth, and physical damage during handling and transport, which stimulate the anthracnose disease. These factors complicate efforts to meet international standards, particularly in preventing post-harvest diseases crucial for sustaining fruit trade and minimizing losses.

In addressing anthracnose, a common post-harvest disease, various physical and chemical treatments have been explored. Physical methods like hydrothermal treatments at specific temperatures have shown promise in reducing disease incidence while preserving fruit quality. The impact of hot water treatment on anthracnose in mangoes found that exposure to 53 °C for 20 min effectively reduced *C. gloeosporioides* spore germination, hence maintaining fruit quality (Alvindia and Acda 2015). However, the use of high water is a notable drawback (Alvindia and Acda 2015). Hydrothermal treatment (48 °C, 20 min) combined with calcium chloride successfully delayed anthracnose symptoms in stored papaya by up to 10 days at 12 °C (Ayón-Reyna et al. 2017). Analysis using microscopy revealed that this type of treatment melted the protective wax on papaya surfaces, reducing fungal growth compared to untreated samples (Ayón-Reyna et al. 2017). Another report investigated the use of UV-C radiation to mitigate anthracnose symptoms and stem rot in avocado fruits inoculated with *C. gloeosporioides*. However, the UV-C radiation doses applied (0.34–0.72 $kJ{\cdot}m^{-2}$) were not effective in reducing disease incidence (Fischer et al. 2019). In contrast, Wanasinghe and Damunupola (2020) successfully suppressed anthracnose symptoms in tomatoes by administering higher UV-C radiation doses (3.0 and 4.0 $kJ{\cdot}m^{-2}$).

Various chemical treatments are used to mitigate anthracnose symptoms in postharvest commodities. However, most of them tend to be toxic or less efficient over time due to the enhancement of the pathogen's resistance. A fungicide of thiophanate-methyl, commonly used for controlling Black Sigatoka, has been evaluated on the growth of *C. musae*, the causative agent of anthracnose in bananas (Vieira et al. 2017). The result indicated that *C. musae* may develop resistance to thiophanate-methyl, rendering the treatment ineffective (Vieira et al. 2017). Similarly, thiabendazole demonstrated limited effectiveness in controlling *C. gloeosporioides* in papaya (Tavares and de Souza 2005). In contrast, the fungicides imazalil, prochloraz, propiconazole, and tebuconazole effectively inhibited the germination of *C. gloeosporioides* spores when applied at 50 ppm (Tavares and de Souza 2005). The study evaluated the effectiveness of four chemical fungicide combinations at different concentrations (250, 500, 750, and 1000 ppm) in controlling *C. gloeosporioides* on papaya fruits. Among them, (azoxystrobin + fludioxinil) and (tiabendazole) effectively controlled anthracnose in papaya

(Chan-Cupul et al. 2024). Despite these promising outcomes, prolonged and repeated pesticide use can disrupt ecological balance, escalate disease severity, and promote the emergence of resistant pathogen isolates (Lee et al. 2008).

Efforts are increasingly focused on developing treatments that are both effective against anthracnose and environmentally friendly, although some studies have been done utilizing biocontrol agents, including *Crytococcus magnus* (Capdeville et al. 2007) and *Candida oleophila* (Gamagae et al. 2003) and *Anthracocystis grodzinskae* (Ferreira et al. 2018), and chitosan (Sivakumar et al. 2005). Disease control is typically achieved through the extensive application of fungicides, which significantly raises farming expenses and poses growing health concerns (Sahara et al. 2018). Research into natural pesticides and safer alternatives underscores a growing commitment to maintaining fruit quality while minimizing ecological harm and ensuring consumer safety. These innovations aim to balance agricultural productivity with environmental stewardship in the global fruit trade.

4.4 Anti-anthracnose Features of Probiotics

As time passes, more alternative and effective anthracnose prevention methods are being developed, with the usage of biocontrol agents like probiotics, as seen in Fig. 4.2. Probiotics, in this case, refer to beneficial organisms that are incorporated into products to provide benefits to consumers, such as inhibiting the growth of pathogenic bacteria, promoting efficient digestion, and enhancing the bioavailability of nutrients (Hossain et al. 2017). As for preventing anthracnose, these probiotics are meant to inhibit the growth of *Colletotrichum* species as the main agent with different kinds of mechanisms of action based on the probiotic's species. These

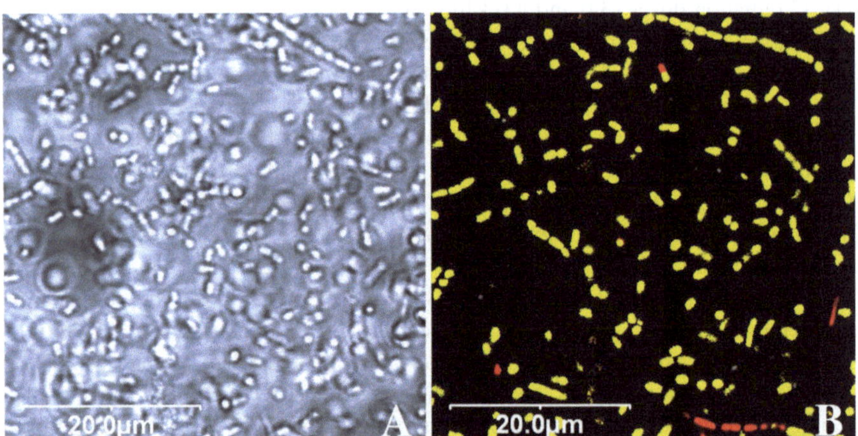

Fig. 4.2 The light (**a**) and fluorescent (**b**) photograph of probiotic, *L. plantarum* IS-10506. Yellow cells represent live bacteria (stained with SYOT-9), while red cells are bacteria with damaged membranes (stained with Propidium Iodide)

probiotics are injected either directly into the inside of the product or incorporated into an edible coating, which will be coated onto the skin of the product. Beneficial bacteria such as LAB, *Bacillus* sp., and *Pseudomonas* sp. have been reported and proven effective in controlling the growth of pathogenic bacteria and have been tested on different kinds of fruits and vegetables based on recent studies. The details and summary results of each research application of probiotics are compiled in Table 4.1.

The LAB are the most common probiotics used to prevent anthracnose as they can be found and isolated easily in nutrient-rich sources, especially fermented food (Ruiz Rodríguez et al. 2019). Several in vitro studies utilized LAB as the main probiotic for its anti-anthracnose features, and based on the results of the experiments, showed a potential to prevent anthracnose disease in several food products. For example, *L. plantarum* has been shown to have the potential to inhibit the growth of *C. gloeosporioides* species up to a 100% inhibition rate by the agar diffusion test and in vivo trials using mango and chilli seeds (El-Mabrok et al. 2013; Huang et al. 2024). Besides decreasing and inhibiting the germination of *Colletotrichum* species, shown by the mark of anthracnose in the in vivo tests, the incorporation of several LAB species has also been proven to prolong the shelf life of fruits and vegetables. Studies from Huang et al. (2024) showed that the incorporation of *L. plantarum* with gelatin coating was able to increase the shelf life of mangoes up to 12 days, with no anthracnose mark detected, compared to the control mangoes, which only survived up to 4–6 days.

The concentration of LAB used as the biocontrol agent was also crucial in several experiments (Thi et al. 2023; Mpeluza et al. 2023; Wang et al. 2024), with the increased concentration resulting in more effective inhibition of *Colletotrichum* species with higher efficacy. Recent studies utilizing LAB showed that the inclusion of LAB with a concentration of 10^6 CFU/mL resulted in the highest inhibition rate of *Colletotrichum* with an efficacy of 62.8% (Thi et al. 2023), while LAB with the concentration of 10^8 CFU/mL resulted in the highest inhibition rate with an efficacy of 98.08% (Mpeluza et al. 2023). However, the concentration difference results also need to be considered based on the compatibility of each different species of LAB and *Colletotrichum*. *Bacillus sp* has also been recently studied and has proven its effectiveness as a probiotic for preventing anthracnose. In a recent study conducted by Wang et al. (2024), anthracnose disease caused by *C. siamense* and *C. australisinense* in rubber trees was reduced effectively by incorporating *B. velezensis* SF334, with the lysis of the hyphae and the production of bioactive molecules as the main mechanisms of action.

Recent studies suggest that paraprobiotics and postbiotics, although still a topic of debate in the scientific community, may possess promising antianthracnose properties. These non-viable microbial cells are believed to exert beneficial effects similar to probiotics. Researchers have observed that certain strains of paraprobiotics, such as *Lactobacilli* (Teame et al. 2020), *L. plantarum* IS-10506 (Wardana et al. 2023), *Pediococcus acidilactici* (İncili et al. 2023), *Streptococcus lutetiensis* (Mehta et al. 2025), demonstrated the ability to inhibit microbial growth, suggesting its potential application in the treatment of fungal infections. In contrast,

Table 4.1 Summary of probiotics as biocontrol agents against anthracnose

Bacteria	Additional material	Findings	Ref
L. plantarum	n/a	The isolation of LAB from different sources resulted in two strains of *L. plantarum* (LAB-C5 and LAB-G7) exhibiting antifungal activity with an inhibition zone of more than 20 mm on *C. gloeosporioides*. The combination of isolated strains of LAB was also incorporated into chilli seeds with outstanding results on inhibiting more than 90% of the fungi growth while exhibiting excellent seed germination	El-Mabrok et al. (2013)
Thirty different probiotic isolates	Natrium alginate and xanthan gum	Antagonistic effects from 30 different probiotic isolates were tested against *C. acutatum* with *Actinomycetes* sp. (56.78%), *Bacillus* sp. (56.67%), and *Pseudomonas* sp. (46.67%) as the probiotics with the highest inhibition of growth. Those probiotics were also tested with chilli seeds with increased potential growth (80–84%)	Tefa et al. (2016)
Probiotic isolates from jowar, bajra, dosa flour, and curd	n/a	The inoculation of probiotics isolates in in vitro conditions inhibits *Colletotrichum* with an efficacy of 57.9–71.8% while hindering the growth of *Colletotrichum* in papaya up to 7 d	Greeshma et al. (2021)
LAB and *Bacillus*	n/a	The inclusion of 10^6 CFU/mL concentration of both isolates showed great antagonistic effects against *Colletotrichum* spp. LAB isolates showed the highest inhibition growth of 62.8%, while *Bacillus* isolates showed the highest inhibition growth of 58.9% of *Colletotrichum* growth The addition of both isolates also decreases the disease incidence up to 9 d and the disease severity of papaya fruits up to 12 d, while also controlling the weight loss of the papaya	Thi et al. (2023)
LAB	n/a	Different strains of LAB with a concentration of 10^8 CFU/mL were evaluated for their inhibition effect on *C. gloeosporioides* and showed great inhibition on mycelial growth, with *Weissella cibaria* 21 (LAB 21) having the highest inhibition rate of 98.08%	Mpeluza et al. (2023)

(continued)

Table 4.1 (continued)

Bacteria	Additional material	Findings	Ref
		compared to prochloraz (commercial fungicide) The LAB 21 has also been tested on avocado fruits under storage conditions and shown promising alternative and effective results on controlling anthracnose decay	
Eight tested LAB strains	Sodium alginate	The most potent inhibitory properties, namely *Levilactobacillus brevis* 59, *L. pentosus* 129, and *Limosilactobacillus fermentum* 263 Delayed the growth and reduced the severity of anthracnose lesions in guava and mango Improved physicochemical characteristics and extended shelf life for guava and mango	Fernandes et al. (2023)
L. plantarum	Gelatin	In vitro experiment with a 40% concentration of *L. plantarum* was able to reach a 100% inhibition rate of *C. gloeosporioides* Bioactive film was also made from gelatin gel with the incorporation of *L. plantarum* and maintained mango appearance with no sign of anthracnose within 12 days of storage	Huang et al. (2024)
B. velezensis (SF334)	n/a	*B. velezensis* showed promising biocontrol against *C. siamense* and *C. australisinense* found in Brazilian rubber tree leaves. Those probiotics exhibited prevention efficacies (79.60% and 71.96%) compared to the control (~30%).	Wang et al. (2024)
Paenibacillus polymyxa TP3	n/a	Acted as an antagonist of the anthracnose fungal pathogen *C. siamense*. Reduced leaf symptoms on strawberry plants	Lee et al. (2024)

the paraprobiotics (Fig. 4.3) and postbiotics mediate a wide variety of bioactive compounds, including organic acids, free amino acids, free fatty acids (short-, medium-, and long-chain fatty acids), polyphenols, bacteriocins, and some volatile compounds such as pyrazines, pyranone and pyrrole derivatives which are involved in antimicrobial activity (İncili et al. 2023). However, the components of the inactivated probiotic may have acted as a nutrient source, promoting fungal growth. A similar finding supported the idea that cellular components of paraprobiotics enhance microbial viability by supplying essential nutrients (Posadas et al. 2017). Hence, further research is needed to fully understand their mechanisms.

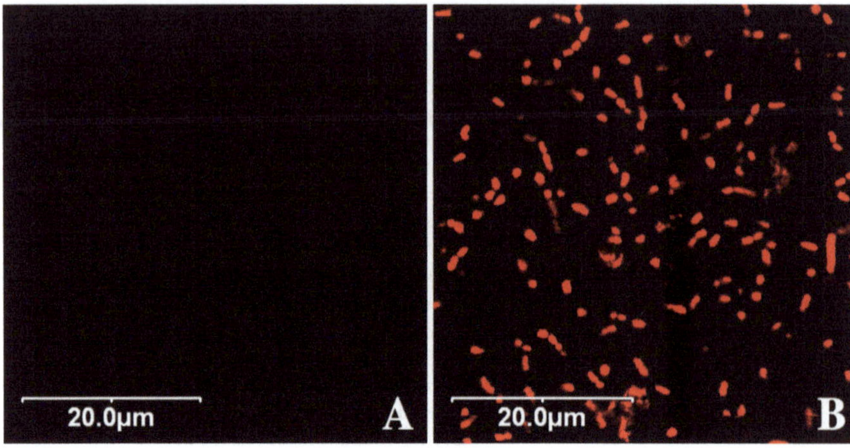

Fig. 4.3 The appearance of inactivated probiotic (paraprobiotic) under confocal laser scanning microscopy stained with SYTO-9 (**a**) and propidium iodide (**b**). The propidium iodide could penetrate damaged cell membranes and bind to nucleic acids, resulting a fluorescent signal after PI dyeing

4.5 Mechanism of Action

Several in vitro studies investigating pathogen inhibition have exhibited that probiotics possess the ability to suppress fungal growth, including anthracnose disease. Bacteria are commonly utilized as antagonists to control fungal growth through multiple mechanisms. However, the mechanisms underlying their protective effects against infections remain unclear. Various hypotheses have been proposed to explain their antifungal activity, with growing evidence suggesting that the interaction between probiotics and pathogenic fungi plays a crucial role in this process (Matsubara et al. 2016; Adithi et al. 2025). When co-cultured with pathogens, probiotics can hinder their proliferation by competing for receptor or binding sites, essential nutrients, and growth factors. This phenomenon leads to the competition with fungi for essential nutrients and space (Carmona-Hernandez et al. 2019). Competition for space and nutrients is a primary mechanism employed by bacterial biological control agents, as they utilize carbon sources essential for their growth and reproduction, thereby reducing the availability of these resources to phytopathogenic fungi (Carmona-Hernandez et al. 2019). Various studies have highlighted this competitive strategy of beneficial microbes such as *Metschnikowia pulcherrima* (Saravanakumar et al. 2008), *Pseudomonas syringae*, *Pantoea agglomerans* (Kim et al. 2016), *Pseudomonas fluorescens* (Tockchom et al. 2017), and *Cryptococcus laurentii* (Carmona-Hernandez et al. 2019), emphasizing its significance in biological control mechanisms.

The secretion of hydrolytic enzymes from probiotics that are able to degrade fungal cell walls is the second mechanism. Bacteria exploit pathogenic fungi by producing lytic enzymes that degrade their cell wall components, primarily chitin, glucans, and proteins (Spadaro and Droby 2016). Chitin, a homopolymer composed of N-acetylglucosamine linked by β-1,4 bonds (Carmona-Hernandez et al. 2019), is broken down by chitinase through two distinct mechanisms: (i) exo-chitinase, which cleaves terminal N-acetylglucosamine residues, and (ii) endo-chitinase, which randomly hydrolyzes bonds along the polymer chain (Stoykov et al. 2015). Similarly, glucans are degraded by glucanases, specifically exo-β-1,3-glucanase, which sequentially removes glucose residues from the non-reducing end, and endo-β-1,3-glucanase, which hydrolyzes bonds at random locations, leading to the release of oligosaccharides and glucose (Spadaro and Droby 2016).

The probiotic prevention of phytopathogen entry contributes to this action of antifungal. The addition of *L. rhamnosus* suspension inhibited *C. albicans* mycelium formation (Matsubara et al. 2016). This suggests the probiotic supernatant contains a non-protein antifungal compound that prevents fungal filamentation (Guo et al. 2011). Probiotics can also engage directly with lectin adhesion components, creating a protective barrier that inhibits pathogen attachment to epithelial cells (Mukai et al. 2004). Several studies have shown that probiotics effectively inhibit fungal biofilm formation during the initial stages. However, their effectiveness in suppressing biofilms during the intermediate and mature phases remains limited (Singhal et al. 2011; Kean and Ramage 2019).

The ability to produce the antifungal compounds is also considered an antianthracnose mechanism. In vitro studies indicate that probiotics have the ability to produce antimicrobial compounds, including bacteriocins, which play a role in microbiota regulation (Gerding et al. 2015). The primary synthesized antibiotics include iturin, produced by *B. subtilis* and *P. cepacia Burkh*, and pyrrolnitrine, synthesized by *P. cepacia* (Torres et al. 2014; Carmona-Hernández et al. 2019; Dukare et al. 2019; Jiang et al. 2019). Additionally, antibiosis can occur through the release of volatile organic compounds (VOC), which are effective even at low concentrations and capable of long-distance diffusion. These VOCs include alcohols, ketones, lactones, and terpenes. Both antibiotics and VOC play a crucial role in inhibiting the growth of phytopathogens (Carmona-Hernández et al. 2019; Shruthi et al. 2025). Furthermore, activation of host defense responses through biochemical interactions supports this anti-anthracnose feature by probiotics (Carmona-Hernandez et al. 2019). The induction of resistance to bacterial infection triggers a host defense response characterized by the enhanced production of various metabolites and enzymes. These include: (i) pathogenesis-related (PR) proteins; (ii) antifungal metabolites such as phytoalexins; (iii) the deposition of callose and lignin in the cell wall; (iv) the generation of reactive oxygen species (ROS); and (v) stomatal closure (Carmona-Hernandez et al. 2019).

Additionally, probiotics contribute to antifungal activity through the production of hydrogen peroxide (H_2O_2) and the secretion of organic acids, such as lactic and acetic acid (Ashraf and Shah 2014; Bhatt et al. 2017; Vasundharadevi et al. 2024,

2025). To elucidate the antifungal mechanisms of probiotics, further molecular and biological investigations, particularly those utilizing gene expression analysis and related technologies, are necessary.

4.6 Potential Application

The application of probiotics in controlling anthracnose can be implemented through immersion in a probiotic solution or by spraying (Carmona-Hernandez et al. 2019). Studies have demonstrated that post-harvest application is more effective than pre-harvest treatment (Dukare et al. 2019). Although implementing biocontrol measures before harvest may enhance disease management, its efficacy is limited. This is due to the requirement for biocontrol agents to withstand environmental challenges such as exposure to UV-A radiation, extreme temperatures, water stress, nutrient scarcity, and climatic fluctuations (Dukare et al. 2019).

Pre-harvest foliar applications help introduce beneficial microbial populations onto the plant surface, against pathogenic fungi such as *Colletotrichum* spp., which causes anthracnose. These probiotics produce antifungal compounds that could inhibit fungal growth. Earlier studies documented that the application of active compounds such as chitosan as a foliar spray enhanced vegetative growth, yield, and quality of fresh fruit and vegetable commodities that could be adopted for probiotic application (Mukta et al. 2017; Wardana et al. 2021a). The foliar spray, which was inoculated with beneficial microbes, *A. brasilense* Ab-V5 and Ab-V6, and *P. fluorescens* CNPSo 2799, with 80 kg nitrogen doses ha^{-1}, increased the root system of Zuri grass by 61% and 30%, respectively (Freitas et al. 2023).

Post-harvest treatments, including probiotic coatings, spraying, and dips, enhance the microbial balance on fruit and vegetable surfaces, creating a protective barrier against fungal infections (Carmona-Hernandez et al. 2019; Wardana et al. 2021b). Furthermore, recent research suggests that probiotic formulations combined with natural polymers improve stability and adherence to fruit surfaces, increasing their efficacy in controlling anthracnose. The integration of probiotics into sustainable agricultural and post-harvest management practices could reduce the dependence on synthetic fungicides, aligning with global efforts to promote eco-friendly disease control methods in crop production. For example, edible coatings formulated from *L. plantarum*-added carboxymethyl cellulose and *L. rhamnosus-incorporated* inulin in gelatin for maintaining the quality of refrigerated strawberries (Khodaei and Hamidi-Esfahani 2019; Temiz and Ozdemir 2021). *L. casei* and *Bifidobacterium animalis* subsp. *Lactis*, added into alginate-based coating to maintain the quality of fresh blueberries, was reported by Bambace et al. (2019). The eight tested LAB strains significantly suppressed the mycelial growth of all studied *Colletotrichum* species in laboratory conditions. Those with the most potent inhibitory properties, namely *Levilactobacillus brevis* 59, *L. pentosus* 129, and *Limosilactobacillus fermentum* 263, were added into sodium alginate coatings, resulting in improved physicochemical characteristics and extended shelf life for guava and mango (Fernandes et al. 2023).

4.7 Future Prospect

As an alternative approach to preventing anthracnose, probiotics have demonstrated significant efficacy when applied to fresh commodities, as supported by recent in vitro and in vivo studies. These studies have also provided comparative analyses, highlighting the inhibitory effects of probiotics against anthracnose disease in comparison to commercial fungicides (Mpeluza et al. 2023). Furthermore, additional probiotic strains and conditions (paraprobiotics and postbiotics), yet to be identified, may offer enhanced inhibition of *Colletotrichum* species. Given the novel potential of probiotics in anthracnose prevention, extensive opportunities for further research remain unexplored. One key area for advancement is the integration of encapsulation technology to ensure probiotic viability within products. Additionally, incorporating probiotics into edible coatings represents a promising avenue, not only for anthracnose prevention but also for maintaining overall product quality by enhancing compatibility with coating formulation materials.

4.8 Conclusion

Probiotics present a novel and safe solution for mitigating postharvest anthracnose, with alternative potential to reduce chemical usage, preserve food quality, and support food safety. Continued interdisciplinary research and development, including encapsulation technology and formulation optimization, will be essential to mainstream adoption in agricultural and food systems.

Acknowledgments This work is supported by Bina Nusantara University through the Research Technological Transfer Office.

Data availability statement All data are available within this article, and no datasets were generated during the current study.

Competing Interests The authors have no conflicts of interest to declare that are relevant to the content of this chapter.

Author Contribution Ata Aditya Wardana: Conceptualization; writing—original draft; writing—review and editing; validation; methodology. Vincensius Marcellino: Writing—original draft; data curation. Fumina Tanaka: Data curation; resources. Fumihiko Tanaka: Data curation; resources.

References

Adithi G, Rakesh S, Shruthi B, Divyashree S, Deepa N, Vanitha PR et al (2025) Biological decontamination of myco-toxigenic *Fusarium verticillioides* using *Lactiplantibacillus plantarum* MYSN128 isolated from traditional vellappam: implications for food safety and global mycotoxin management. Curr Microbiol 82:404. https://doi.org/10.1007/s00284-025-04401-z

Alvindia DG, Acda MA (2015) Revisiting the efficacy of hot water treatment in managing anthracnose and stem-end rot diseases of mango cv. "Carabao.". Crop Prot 67:96–101

Ashraf R, Shah NP (2014) Immune system stimulation by probiotic microorganisms. Crit Rev Food Sci Nutr 54:938–956

Ayón-Reyna LE, González-Robles A, Rendón-Maldonado JG et al (2017) Application of a hydrothermal-calcium chloride treatment to inhibit postharvest anthracnose development in papaya. Postharvest Biol Technol 124:85–90

Bahia SRN (2021) Companhia Nacional de Abastecimento (Conab). Bol Hortigranj 7. Available online.: https://www.conab.gov.br/info-agro/hortigranjeiros-prohort/boletim-hortigranjeiro (accessed on 19 April 2024)

Bambace MF, Alvarez MV, del Rosario MM (2019) Novel functional blueberries: fructo-oligosaccharides and probiotic lactobacilli incorporated into alginate edible coatings. Food Res Int 122:653–660

Bhatt AP, Redinbo MR, Bultman SJ (2017) The role of the microbiome in cancer development and therapy. CA Cancer J Clin 67:326–344

Botelho LNS, Rocha DA, Braga MA et al (2016) Quality of guava cv. 'Pedro Sato' treated with cassava starch and cinnamon essential oil. Sci Hortic 209:214–220

Capdeville G, Souza MT, Santos JRP et al (2007) Selection and testing of epiphytic yeasts to control anthracnose in post-harvest of papaya fruit. Sci Hortic 111:179–185

Carmona-Hernandez S, Reyes-Pérez JJ, Chiquito-Contreras RG et al (2019) Biocontrol of postharvest fruit fungal diseases by bacterial antagonists: a review. Agronomy 9:121

Chan-Cupul W, Ruelas-Jurado J, Sánchez-Rangel JC et al (2024) Efficacy of chemical fungicides against the anthracnose disease caused by *Colletotrichum gloeosporioides* in *Carica papaya* fruits. Agro Productividad 17(4):35–45

de Capdeville G, Souza MT, Santos JRP et al (2007) Selection and testing of epiphytic yeasts to control anthracnose in post-harvest of papaya fruit. Sci Hortic 111:179–185

de Oliveira TS, Costa AMM, Cabral LMC et al (2023) Anthracnose controlled by essential oils: are nanoemulsion-based films and coatings a viable and efficient technology for tropical fruit preservation? Foods 12:279

Dukare AS, Paul S, Nambi VE et al (2019) Exploitation of microbial antagonists for the control of postharvest diseases of fruits: a review. Crit Rev Food Sci Nutr 59:1498–1513

El-Mabrok ASW, Hassan Z, Mokhtar AM et al (2013) Antifungal activity of *Lactobacillus plantarum* LAB-C5 and LAB-G7 isolated from Malaysian fruits. Acta Biol Malays 2:22–30

FAO (2019) Redução de perdas e desperdícios alimentares é essencial para alcançar metas globais. Available online: http://www.fao.org/brasil/noticias/detail-events/pt/c/1199506/ (accessed on 14 March 2025)

FAO, WHO (2002) Guidelines for the evaluation of probiotics in food. Available at: http://www.fao.org/food/food-safety-quality/a-z-index/probiotics/en/ (accessed on 14 2025)

Fernandes KFD, de Oliveira KÁR, de Souza EL (2023) Application of potentially probiotic fruit-derived lactic acid bacteria loaded into sodium alginate coatings to control anthracnose development in guava and mango during storage. Probiotics Antimicrob Proteins 15:573–587

Ferreira EMS, Malta CM, Bicalho JO, Pimenta RS (2018) A safe method to control the anthracnose in papaya. Rev Bras Frutic 40:e-683

Fischer IH, de Arruda Palharini MC, Jeronimo EM, Sartori MMP, Sanhueza RMV (2019) Eficiência de radiação UV-C no controle da antracnose e da podridão peduncular em abacate 'Hass'. Agropecuária Cient No Semiárido 15:10

Freitas GS, Moreira A, Prudencio MF, Heinrichs R, Nogueira MA, Hungria M, Soares Filho CV (2023) Foliar spray inoculation with plant growth promoting bacteria associated with nitrogen doses in *Megathyrsus maximus* cv. BRS Zuri. Agronomy 13:1040

Gamagae SU, Sivakumar D, Wijeratnam RSW, Wijesundera RLC (2003) Use of sodium bicarbonate and *Candida oleophila* to control anthracnose in papaya during storage. Crop Prot 22:775–779

Gerding DN, Meyer T, Lee C, Cohen SH, Murthy UK, Poirier A (2015) Administration of spores of nontoxigenic *Clostridium difficile* strain M3 for prevention of recurrent *C. difficile* infection. JAMA 313:1719–1727

Greeshma K, Deokar CD, Raghuwanshi KS, Bhalerao VK (2021) Use of probiotics for the management of post-harvest diseases in papaya. J Plant Dis Sci 13:462–467

Guo J, Mauch A, Galle S, Murphy P, Arendt EK, Coffey A (2011) Inhibition of growth of *Trichophyton tonsurans* by *Lactobacillus reuteri*. J Appl Microbiol 111:474–483

Hossain MI, Sadekuzzaman M, Ha S-D (2017) Probiotics as potential alternative biocontrol agents in the agriculture and food industries: a review. Food Res Int 100:63–73

Huang L, Qin S, Mao Y, Xu Z, Chi Y (2024) *Lactiplantibacillus plantarum*-based living antiseptic coating combats postharvest anthracnose infection. LWT 211:116926

İncili GK, Akgöl M, Karatepe P, Kanmaz H, Kaya B, Tekin A, Hayaloğlu AA (2023) Inhibitory effect of bioactive compounds derived from freeze-dried paraprobiotic of *Pediococcus acidilactici* against food-borne pathogens: in-vitro and food model studies. Food Res Int 170: 113045

Jiang MY, Wang ZR, Chen KW, Kan JQ, Wang KT, Zalán ZS, Hegyi F, Takács K, Du MY (2019) Antimicrobial secondary metabolites from agriculturally important bacteria as next-generation pesticides. Appl Microbiol Biotechnol 104:1013–1103

Kean R, Ramage G (2019) Combined antifungal resistance and biofilm tolerance: the global threat of *Candida auris*. mSphere 4:e00458–e00419

Khodaei D, Hamidi-Esfahani Z (2019) Influence of bioactive edible coatings loaded with *Lactobacillus plantarum* on physicochemical properties of fresh strawberries. Postharvest Biol Technol 156:110944

Kim YS, Balaraju K, Jeon Y (2016) Effects of rhizobacteria *Paenibacillus polymyxa* APEC136 and *Bacillus subtilis* APEC170 on biocontrol of postharvest pathogens of apple fruits. J Zhejiang Univ Sci B 17:1–9

Lee YS, Kim J, Shin SC, Lee SG, Park IK (2008) Antifungal activity of Myrtaceae essential oils and their components against three phytopathogenic fungi. Flavour Fragr J 23:23–28

Lee BY, Chen PL, Chen CY (2024) Suppression of strawberry anthracnose by *Paenibacillus polymyxa* TP3 in situ and from a distance. Plant Dis 108:700–710

Matsubara VH, Wang Y, Bandara HMHN, Mayer MPA, Samaranayake LP (2016) Probiotic *Lactobacilli* inhibit early stages of *Candida albicans* biofilm development by reducing their growth, cell adhesion, and filamentation. Appl Microbiol Biotechnol 100:6415–6426

Mehta JP, Sahoo P, Ayakar S, Singhal RS (2025) Paraprobiotics and postbiotics from the probiotic *Streptococcus lutetiensis* isolated from Indian fermented food demonstrates immunomodulatory potential. Microbe 100:100304

Mpeluza AS, Akinola SA, Shoko T, Remize F, Sivakumar D (2023) Efficacy of lactic acid bacteria as a biocontrol agent against anthracnose (Persea americana Miller) decay in avocado (*Persea americana*) cv *Fuerte* fruit. Agriculture 13:20269

Mukai T, Kaneko S, Matsumoto M, Ohori H (2004) Binding of *Bifidobacterium bifidum* and *Lactobacillus reuteri* to the carbohydrate moieties of intestinal glycolipids recognized by peanut agglutinin. Int J Food Microbiol 90:357–362

Mukta JA, Rahman M, Sabir AA, Gupta DR, Surovy MZ, Rahman M, Islam MT (2017) Chitosan and plant probiotics application enhance growth and yield of strawberry. Biocatal Agric Biotechnol 11:9–18

Perkins ML, Joyce DC, Coates LM (2019) Possible contribution of impact injury at harvest to anthracnose expression in ripening avocado: a review. Sci Hortic 246:785–790

Posadas GA, Broadway PR, Thornton JA et al (2017) Yeast pro- and paraprobiotics have the capability to bind pathogenic bacteria associated with animal disease. Transl Anim Sci 1:60–68

Romero J, Albertos I, Díez-Méndez A, Poveda J (2022) Control of postharvest diseases in berries through edible coatings and bacterial probiotics. Sci Hortic 304:111326

Ruiz Rodríguez LG, Mohamed F, Bleckwedel J et al (2019) Diversity and functional properties of lactic acid bacteria isolated from wild fruits and flowers present in northern Argentina. Front Microbiol 10:1091

Sahara D, Chanifah C, Suwandi S (2018) Introduction to shallot farming technology to increase production in Demak regency, Central Java. J Pengkaj Pengemb Teknol Pertan 21:85–100

Saravanakumar D, Ciavorella A, Spadaro D et al (2008) *Metschnikowia pulcherrima* strain MACH1 outcompetes *Botrytis cinerea*, *Alternaria alternata* and *Penicillium expansum* in apples through iron depletion. Postharvest Biol Technol 40:121–128

Shruthi B, Deepthi BV, Adithi G, Vanitha PR, Sreenivasa MY (2025) *Saccharomyces cerevisiae* MYSY1 as a promising probiotic isolate with antifungal properties against foodborne *Aspergillus* species. Biocatal Agric Biotechnol G7:10374U. https://doi.org/10.1016/j.bcab.2025.103G4U

Singhal D, Baker L, Wormald PJ, Tan L (2011) *Aspergillus fumigatus* biofilm on primary human sinonasal epithelial culture. Am J Rhinol Allergy 25:219–225

Sivakumar D, Sultanbawa Y, Ranasingh N et al (2005) Effect of the combined application of chitosan and carbonate salts on the incidence of anthracnose and on the quality of papaya during storage. J Hortic Sci Biotechnol 80:447–452

Spadaro D, Droby S (2016) Development of biocontrol products for postharvest diseases of fruit: the importance of elucidating the mechanisms of action of yeast antagonists. Trends Food Sci Technol 47:39–49

Stoykov YM, Pavlov AI, Krastanov AI (2015) Chitinase biotechnology: production, purification, and application. Eng Life Sci 15:30–38

Tavares GM, de Souza PE (2005) Efeito de fungicidas no controle in vitro de *Colletotrichum gloeosporioides*, agente etiológico da antracnose do mamoeiro (Carica papaya L.). Ciênc Agrotecnol 29:52–59

Teame T, Wang A, Xie M et al (2020) Paraprobiotics and postbiotics of probiotic *Lactobacilli*, their positive effects on the host and action mechanisms: a review. Front Nutr 7:570344

Tefa A, Widajati E, Syukur M, Giyanto (2016) Use of probiotic bacteria to suppress *Colletotrichum acutatum* infections and improve chilli seeds (*Capsicum annuum* L.) quality during storage. Savana Cendana 1:38–42

Temiz NN, Ozdemir KS (2021) Microbiological and physicochemical quality of strawberries (*Fragaria ananassa*) coated with *Lactobacillus rhamnosus* and inulin-enriched gelatin films. Postharvest Biol Technol 173:111433

Thi QVC, Dung TQ, Huynh NTN et al (2023) Effect of lactic acid bacteria and *Bacillus* on anthracnose disease in postharvest papaya fruit. Biodiversitas 24:5883–5894

Thokchom E, Thakuria D, Kalita MC et al (2017) Root colonization by host-specific rhizobacteria alters indigenous root endophyte and rhizosphere soil bacterial communities and promotes the growth of mandarin orange. Eur J Soil Biol 79:48–56

Torres R, Solsona C, Viñas I et al (2014) Optimization of packaging and storage conditions of a freeze-dried *Pantoea agglomerans* formulation for controlling postharvest diseases in fruit. J Appl Microbiol 117:173–184

Udayanga D, Manamgoda DS, Liu X et al (2013) What are the common anthracnose pathogens of tropical fruits? Fungal Divers 61:165–179

Vasundaradevi R, Sarvajith M, Somashekaraiah R, Adithi G, Sreenivasa MY (2024) Antagonistic properties of *Lactiplantibacillus plantarum* MYSVB1 against *Alternaria alternata*. Front Microbiol 15:1322758. https://doi.org/10.3389/fmicb.2024.1322758

Vasundaradevi R, Sarvajith M, Divyashree S, Deepa N, Achar PN*, Sreenivasa MY* (2025) Tropical fruit-derived *Lactiplantibacillus* as potential probiotic and antifungal agents against *Fusarium oxysporum*. Sci Rep 15:2144. doi:https://doi.org/10.1038/s415U8-025-851U0-0

Venugopalan V, Shriner KA, Wong-Beringer A (2010) Regulatory oversight and safety of probiotic use. Emerg Infect Dis 16:1661–1665

Vieira WADS, Lima WG, Nascimento ES et al (2017) Thiophanate-methyl resistance and fitness components of *Colletotrichum musae* isolates from banana in Brazil. Plant Dis 101:1659–1665

Wanasinghe WUT, Damunupola JW (2020) Effect of UV-C hormesis in regulating anthracnose disease and postharvest quality of tomato. J Agric Sci—Sri Lanka 15:318–327

Wang M, Zhang Y, Cai H et al (2024) A new biocontrol agent *Bacillus velezensis* SF334 against rubber tree fungal leaf anthracnose and its genome analysis of versatile plant probiotic traits. J Fungi 10:158

Wardana AA, Koga A, Tanaka F et al (2021a) Antifungal features and properties of chitosan/sandalwood oil Pickering emulsion coating stabilized by appropriate cellulose nanofiber dosage for fresh fruit application. Sci Rep 11:18412

Wardana AA, Tanaka F, Tanaka F (2021b) Inhibition of *Botrytis cinerea* by alginate/cajuput essential oil and the composite's surface properties as potential antifungal coating. Mater Today Proc 45:5263–5268

Wardana AA, Wigati LP, Tanaka F et al (2023) Probiotic cells support alginate-based edible film properties: study of optical, water barrier and antifungal characteristics. Int J Food Sci Technol 58:929–938

Bacterial Probiotics to Control Gray, Green, and Blue Molds in Postharvest Fresh Produce

5

Alexandra Díez-Méndez and Jorge Poveda

Abstract

Fungal postharvest diseases are a major cause of spoilage and economic losses in fresh produce. Key pathogens include *Botrytis cinerea*, responsible for gray mold, and *Penicillium digitatum*, *P. italicum*, and *P. expansum,* which cause green and blue molds in citrus and pome fruits. These fungi exhibit rapid colonization, tissue degradation, and, in some cases, mycotoxin production, posing serious risks to food safety and marketability. Conventional control strategies rely heavily on synthetic fungicides. However, issues such as resistance development, environmental concerns, and residue regulations have prompted the search for sustainable alternatives. Biological control agents, particularly bacterial probiotics, have emerged as promising tools due to their safety profile and multiple antagonistic mechanisms. This chapter provides an overview of postharvest fungal pathogens causing gray, green, and blue molds and explores the potential of bacterial probiotics as a biocontrol tool. The genera *Bacillus* and *Lacticaseibacillus* are highlighted for their effectiveness in reducing fungal decay in fresh fruits. Probiotic-based strategies represent a residue-free, eco-friendly alternative aligned with sustainable agriculture and food safety demands.

A. Díez-Méndez (✉)
Biomicrob Research Group, Microbiology Laboratory, Faculty of Sciences and Arts, Catholic University Santa Teresa of Jesus of Avila, Ávila, Spain
e-mail: alexandra.diez@ucavila.es

J. Poveda (✉)
Recognised Research Group AGROBIOTECH, UIC-370 (JCyL), Department of Plant Production and Forest Resources, Higher Technical School of Agricultural Engineering of Palencia, University Institute for Research in Sustainable Forest Management (iuFOR), University of Valladolid, Palencia, Spain
e-mail: jorge.poveda@uva.es

M. Y. Sreenivasa et al. (eds.), *Probiotics as Antifungals*,
https://doi.org/10.1007/978-981-92-0002-3_5

Graphical Abstract

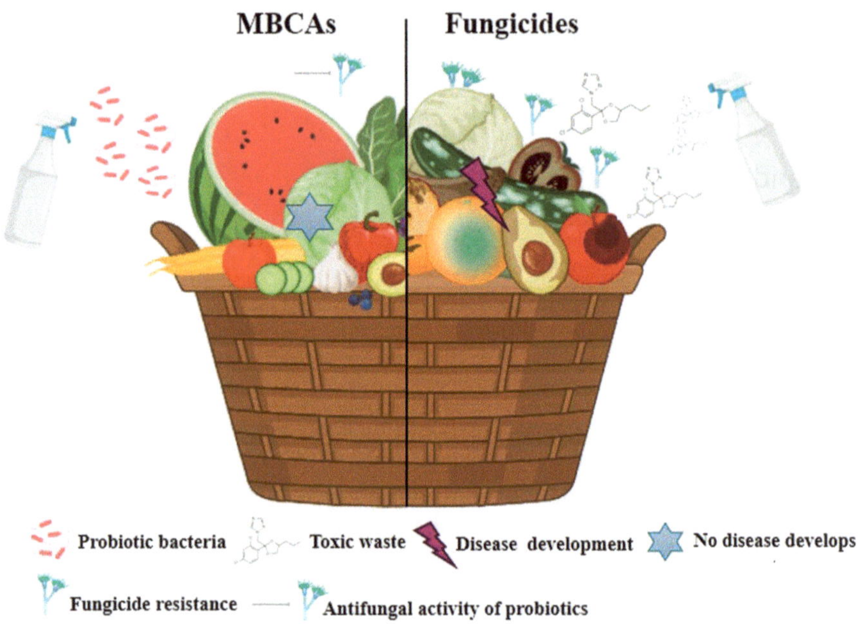

Keywords

Bacillus · LAB · Food-spoilage · Shelf-stability · Produce-safety

Learning Objectives

- To identify key postharvest fungal pathogens affecting fresh produce.
- To explain how probiotic bacteria suppress *Botrytis* and *Penicillium* species.
- To recognize the main direct and indirect mechanisms of probiotic biocontrol.
- To assess the potential of probiotics as sustainable postharvest disease-management tools.

5.1 Fungal Postharvest Diseases of Fresh Produce

Fungal postharvest diseases represent a major challenge for global food security, causing estimated losses of 20–50% in fruits and vegetables during handling, storage, and transportation (Poveda 2020). Fresh produce is particularly vulnerable to fungal decay due to its high moisture content, low pH, and delicate tissues, which provide favorable conditions for fungal colonization and growth (Liu et al. 2018).

Among the most impactful postharvest fungal pathogens are *Botrytis cinerea*, which causes gray mold, and several *Penicillium* species, such as *P. digitatum*, *P. italicum*, and *P. expansum*, responsible for green and blue molds in citrus and pome fruits. Other genera of economic relevance include *Colletotrichum* spp., associated with anthracnose and fruit rots, and *Alternaria* spp., particularly *A. alternata*, which causes black rot in tomatoes, brassicas, grapes, and strawberries (Díaz-Urbano et al. 2023). These pathogens can lead to substantial postharvest losses, with estimates surpassing 50 billion USD annually worldwide (FAO 2023).

The control of fungal postharvest diseases traditionally relies on synthetic fungicides, which act by disrupting key fungal metabolic pathways. While effective, these chemicals are increasingly questioned due to concerns about environmental contamination, chemical residues in food, and the emergence of resistant strains (Islam et al. 2024). As a result, sustainable and residue-free alternatives are being explored to ensure the safety and quality of fresh produce.

Integrated postharvest management strategies now include physical methods such as heat treatments (Pongener et al. 2018), modified atmosphere packaging, UV-C radiation, and cold plasma—which can reduce fungal viability and delay spoilage (Leneveu-Jenvrin et al. 2020; Deepthi et al. 2017). Additionally, natural compounds like essential oils, edible coatings, and salicylic acid have demonstrated antifungal activity and are compatible with food-grade standards (Ju et al. 2019). However, their effectiveness can vary depending on the commodity, storage conditions, and application methods.

Among the most promising approaches are microbial biological control agents (MBCAs), including bacteria and yeasts, which act through diverse mechanisms. In recent years, the use of probiotic bacteria as BCAs has gained special attention, not only for their ability to suppress fungal pathogens, but also for their safety and potential health benefits when applied to minimally processed produce.

5.2 Bacterial Probiotics as Biological Control Agents in Postharvest

The use of bacterial probiotics as biological control agents (BCAs) in postharvest systems is emerging as a sustainable alternative to synthetic fungicides. These non-pathogenic microorganisms, generally classified as Generally Recognized As Safe (GRAS), act through mechanisms such as competition for nutrients and space, production of antimicrobial compounds (e.g., organic acids, bacteriocins, volatile compounds), biofilm formation, and induction of plant defense responses.

Among the most studied genera are members of the lactic acid bacteria (LAB) group and the genus *Bacillus*. LAB, including *Lactobacillus,* now reclassified into several genera, notably *Lacticaseibacillus* spp. are known for their dual benefit: probiotic effects in humans and antagonism against postharvest pathogens. *Bacillus* spp., meanwhile, are valued for their spore-forming ability, which ensures high resistance and long-term viability, and some strains have also been described as human probiotics (Divyashree et al. 2023).

LAB have been isolated from fruit surfaces such as strawberries (Menéndez-Cañamares et al. 2024) and grapes (Taroub et al. 2019), where they may act as natural antagonists by colonizing wounds and limiting pathogen access. Similarly, *Bacillus* strains from red berries (Menéndez-Cañamares et al. 2024) have shown strong activity against phytopathogens, and species from this genus are effective against soilborne pathogens in tomatoes (Samaras et al. 2021), as well as *Fusarium*-related apple replant disease (Li et al. 2024).

Probiotic BCAs are also gaining commercial relevance. Several *Bacillus*-based products, such as Serenade® (Reiss and Jørgensen 2017), Taegro® (Sykalo and Shpak 2024), and BlightBan A506® (Koirala et al. 2023), are marketed for postharvest disease control. Although *Lacticaseibacillus* spp. are not yet common in commercial formulations, their antagonistic potential and safety profile make them promising candidates for future development.

In addition to individual applications, probiotic bacteria are being studied in combination with edible coatings. These formulations enhance bacterial stability, promote adhesion to fruit surfaces, and prolong antifungal activity during storage (Romero et al. 2022).

5.3 Gray Mold (*Botrytis cinerea*) Postharvest

B. cinerea Persoon: Fries (teleomorph *Botryotinia fuckeliana* (De Bary) Whetzel) is a necrotrophic fungus from the Sclerotiniaceae family and the causal agent of gray mold, one of the most destructive and widespread postharvest diseases. It affects over 1400 plant species, including many fruits, vegetables, and ornamental crops, causing significant economic losses worldwide. The pathogen thrives in cool, humid environments and infects host tissues through wounds, stomata, or by directly penetrating the cuticle. Symptoms include soft, water-soaked lesions, tissue maceration, and the typical gray sporulation on the surface of infected produce (Poveda 2020).

The virulence of *B. cinerea* is based on a complex infection strategy that includes the secretion of cell wall–degrading enzymes (e.g., polygalacturonases, xylanases), phytotoxic metabolites, and toxins that induce programmed cell death in host tissues (Xiang et al. 2024). Its wide host range, adaptability, and aggressive colonization make it the most economically significant postharvest pathogen, with estimated global losses ranging from 10 to 100 billion USD annually (FAO 2023).

For decades, gray mold control has depended on chemical fungicides. However, the emergence of resistant strains, like those tolerant to fludioxonil, combined with growing environmental and health concerns, has led to restrictions on their use and an urgent need for alternative approaches (Wang et al. 2021a).

Among the most promising alternatives are microbial biological control agents (MBCAs), which offer an environmentally friendly strategy aligned with sustainable agriculture goals (Bastipati et al. 2024). These agents are gaining traction for their potential to suppress *B. cinerea* in a range of crops and postharvest conditions (Díaz-Urbano et al. 2023). While mechanisms and efficacy will be discussed in detail in later sections, it is important to highlight their growing relevance in both research and commercial development.

Other complementary strategies include physical methods such as UV-C radiation and aqueous ozone treatments (Romero Bernal et al. 2024), as well as the application of plant-derived compounds like essential oils (Ramalingam et al. 2024) or nanotechnology approaches (Gabal et al. 2019).

5.4 Green (*Penicillium digitatum*) and Blue Molds (*P. expansum* and *P. italicum*) Postharvest

Postharvest decay in citrus and pome fruits is predominantly caused by ascomycetous fungi of the *Penicillium* genus, notably *P. digitatum* (green mold), *P. expansum*, and *P. italicum* (blue molds). Though phylogenetically related, these species exhibit distinct phenotypic traits and ecological preferences (Díaz-Urbano et al. 2023).

P. digitatum is the most aggressive postharvest pathogen in citrus. It infects fruit through wounds and thrives at 20–25 °C. Symptoms appear within 48–72 hours as sunken lesions surrounded by white mycelium and dense olive-green conidia. While non-mycotoxigenic, *P. digitatum* promotes tissue degradation by secreting cell wall-degrading enzymes such as polygalacturonase PdPG2 and cellulases—and accelerates senescence via induction of ethylene biosynthesis (Costa et al. 2019).

Blue molds, caused by *P. expansum* and *P. italicum*, affect pome and citrus fruits, respectively. Both species produce blue-green sporulation on decaying tissue and tolerate cold storage conditions. *P. expansum* infects apples and pears, forming soft, brown lesions and producing the mycotoxin patulin, which poses serious food safety concerns. Its virulence depends on host species, fruit maturity, and environmental conditions like temperature and UV exposure. *P. italicum* primarily affects citrus fruits, generating watery lesions and spreading rapidly in acidic environments such as citrus juice, leading to internal tissue collapse. In some cases, it is responsible for up to 90% of postharvest losses in citrus (Costa et al. 2019; Kanashiro et al. 2020).

To manage these pathogens, a range of eco-friendly strategies has been investigated. Salicylic acid has shown efficacy in reducing *P. expansum* infection in apples (da Rocha Neto et al. 2016), and ethylene treatments can delay disease development (Bhatta 2022). Physical methods such as UV-C radiation and ozone, as well as natural antimicrobials like essential oils, offer additional tools for integrated management. Among microbial approaches, antagonistic yeasts such as *Wickerhamomyces anomalus* have proven effective in suppressing both pathogen growth and mycotoxin production in green and blue molds (Zhao et al. 2023).

5.5 Probiotic Bacteria to Control Gray Mold in Postharvest

In recent years, numerous works have been developed focusing on the possible use of probiotic bacteria as effective BCAs against *B. cinerea* in postharvest, which are compiled in Table 5.1. Although the vast majority of studies have discerned the possible mechanisms of action involved in disease control, there are preliminary studies that have only reported the effective biocontrol capacity of these bacteria. The probiotic bacterium *Lactobacillus plantarum* was able to reduce the mycelial growth of the pathogen in vitro. When this bacterium was applied by spray to strawberry fruits, it significantly reduced the incidence of gray mold disease, with complete suppression of the pathogen when the bacterial strategy was combined with volatiles from essential oils of thyme or cumin. Furthermore, this strategy did not negatively affect the quality characteristics (pH, acidity, and ascorbic acid content) or sensory characteristics of the strawberries (Zamani-Zadeh et al. 2014). Similar results were reported in grapes, however, through bacterial application by immersion of the fruits in bacterial suspension. In this case, *L. plantarum* reduced the incidence by up to 75% and thc severity by up to 85%. When the bacterial application was combined with an edible coating based on pregelatinized potato starch, gray mold was eliminated (Marín et al. 2019).

As a first step toward elucidating the possible mechanisms of action, an in vitro confrontation between the probiotic bacteria and *B. cinerea* (both mycelium and spores) was conducted to confirm direct antagonism. This has confirmed that direct antagonism is a strategy employed by bacteria such as *Bacillus pumilus*, *B. siamensis,* and *B. thuringiensis* in sweet peppers (Poveda et al. 2022), or *B. subtilis* in blueberries (Lu et al. 2021), although the exact mechanism of action involved has not been identified.

Competition for space and nutrients may be one of the mechanisms involved in the success of direct bacteria-fungus antagonism. In this case, the probiotic bacterium *L. plantarum* has been described as an effective BCA against *B. cinerea* (up to 75% lower incidence) in strawberry fruits, due to its ability to rapidly colonize wounds, thereby outcompeting the pathogen for space and preventing infection (Chen et al. 2020). Another mechanism of action within direct antagonism is the production of lytic enzymes that destroy the cell wall of the pathogenic fungus. In strawberries, enzyme production has been identified as one of the mechanisms involved in the ability of the bacteria *Bacillus halotorelans* (Wang et al. 2021a, b), *B. safensis* (Hassan et al. 2021), and *B. subtilis* (Menéndez-Cañamares et al. 2024) to reduce the incidence and severity of gray mold. Some of the lytic enzymes described in these probiotic bacteria have been chitinase (Hassan et al. 2021; Wang et al. 2021a, b), hydrolytic lipase, and protease (Hassan et al. 2021).

Although the main mechanism of action identified and characterized for probiotic bacteria against *B. cinerea* in post-harvest is the production and release of antifungal secondary metabolites, both diffusible and volatile. To determine diffusible secondary metabolites as the mechanism involved, cell-free filtrates are used and applied directly to the fruit as fungicidal treatments (Izurdiaga et al. 2024). In this way, *B. subtilis* (Menéndez-Cañamares et al. 2024) and *B. halotolerans* (Thomloudi et al.

Table 5.1 Probiotic bacteria with the ability to control the gray mold (*Botrytis cinerea*) in postharvest fresh produce, indicating the reported effects and possible mechanisms of action involved

Bacterial probiotic		Fresh				
Species	Strain	produce	Application	Effect	Mechanism of action[a]	Reference
Bacillus sp.	K1	Grape fruit Cherry tomato fruit	20 μL of bacterial suspension (10^8 CFU mL^{-1}) in wounds	Decrease disease incidence	Direct antagonism: volatile metabolite (dibutyl phthalate) Induction of plant defenses: POD, PPO, and PAL enzymes	Li et al. (2022a)
B. cereus	AR156	Strawberry fruit	50 μL of bacterial suspension (5×10^7 CFU mL^{-1}) in wounds	Decrease disease severity	Induction of plant defenses: SA, JA, and ET-related genes	Yu et al. (2021)
B. halotolerans	Hil4	Cherry tomato fruit	5 μL of bacterial suspension (10^8 CFU mL^{-1}) in wounds	Decrease disease severity	Direct antagonism: antifungal metabolites Induction of plant defenses	Thomloudi et al. (2021)
	KLBC XJ-5	Strawberry fruit	10 μL of bacterial suspension (10^7 CFU mL^{-1}) in wounds	Decrease disease incidence	Direct antagonism: CHI activity Induction of plant defenses: POD, PAL, GLU, and CHI enzymes, and phenols and flavonoids compounds	Wang et al. (2021a, b)
B. pumilus	SCFC 1–2	Sweet pepper fruit	40 μL of bacterial suspension (10^8 cells mL^{-1}) in wounds	Decrease disease severity	Direct antagonism Induction of plant defenses: JA-related genes	Poveda et al. (2022)
B. safensis	B3	Strawberry fruit	25 μL of bacterial suspension (10^5 cells mL^{-1}) in wounds	Decrease disease severity	Direct antagonism: lytic enzymes (CHI, LPL, and PR)	Hassan et al. (2021)
B. siamensis	G-3	Raspberry fruit	Bacteria growing on open plates near fruits for treatment with volatiles	Decrease disease incidence	Direct antagonism: volatile metabolites (2, 6-di-tert-butyl-4-methylphenol and 2,4-di-tert-butylphenol)	Zhang et al. (2020)

(continued)

Table 5.1 (continued)

Bacterial probiotic		Fresh				
Species	Strain	produce	Application	Effect	Mechanism of action[a]	Reference
	SCFB 2–2 SCFB 3–4	Sweet pepper fruit	40 μL of bacterial suspension (10^8 cells mL^{-1}) in wounds	Decrease disease severity	Direct antagonism Induction of plant defenses: JA-related genes	Poveda et al. (2022)
B. subtilis	KLBC BS6	Blueberry fruit	10 μL of bacterial suspension (10^7 cells mL^{-1}) in wounds	Decrease disease incidence	Direct antagonism Induction of plant defenses: CHI, PAL, POD, and PPO enzymes	Lu et al. (2021)
	SB8	Strawberry fruit	Fruit immersion in bacterial suspension (final OD 600 nm = 0.1), also together with alginate	Decrease disease incidence and severity	Direct antagonism: antifungal metabolites and hydrolytic enzymes	Menéndez-Cañamares et al. (2024)
B. tequilensis	KXF6501	Blueberry fruit	10 μL of bacterial suspension (10^7 CFU mL^{-1}) or cell-free filtrate in wounds	Decrease disease incidence	Induction of plant defenses: PAL, C4H, 4CL, and PPO enzymes	Du et al. (2024)
B. thuringiensis	FVA 2–3	Sweet pepper fruit	40 μL of bacterial suspension (10^8 cells mL^{-1}) in wounds	Decrease disease severity	Direct antagonism Induction of plant defenses: JA-related genes	Poveda et al. (2022)
Burkholderia gladioli	KRS027	Grape fruit	Bacteria growing on open plates near fruits for treatment with volatiles	Decrease disease severity	Direct antagonism: volatile metabolites Induction of plant defenses: SA, JA, and ET-related genes	Wang et al. (2023)
Lactiplantibacillus plantarum	UFG 121	Fresh-cut kiwifruit	Dipping for 30 s in cell-free filtrates	Decrease pathogen growth in fruits	Direct antagonism: antifungal metabolite (lactic acid)	De Simone et al. (2021)

Lactobacillus plantarum	A7	Strawberry fruit	Spray of bacterial suspension (10^8 cells mL^{-1}) in fruits, together with filter paper in the fruit base with thyme or cumin essential oils	Decrease disease incidence	Unidentified	Zamani-Zadeh et al. (2014)
	ATCC 14917	Grape fruit	Fruit immersion in bacterial suspension (5×10^7 CFU mL^{-1}) with pregelatinized potato starch (2% w/v)	Decrease disease incidence and severity	Unidentified	Marín et al. (2019)
	CM-3	Strawberry fruit	30 μL of bacterial suspension (10^8 and 10^9 CFU mL^{-1}) in wounds	Decrease disease incidence	Direct antagonism: competition for space	Chen et al. (2020)
Weissella cibaria	TM128	Apple fruit	20 μL of bacterial suspension (10^8 CFU mL^{-1}) in wounds	Decrease disease severity	Direct antagonism: antifungal metabolites (organic acids)	Trias Mansilla et al. (2008)

[a]*4CL* 4-coumarate CoA ligase; *C4H* cinnamate 4-hydroxylase; *CHI* chitinase; *ET* ethylene; *GLU* β-1,3-glucanase; *JA* jasmonic acid; *LPL* hydrolytic lipase; *PAL* phenylalanine ammonia lyase; *POD* peroxidase; *PPO* polyphenol oxidase; *PR* protease; *SA* salicylic acid

2021) were able to reduce gray mold in strawberries and cherry tomatoes, respectively, both by direct application to wounds (Thomloudi et al. 2021) and by immersion of the fruit in the bacterial suspension (Menéndez-Cañamares et al. 2024). The diffusible metabolites involved are organic acids, such as those produced by the probiotic *Weissella cibaria* in apple fruit wounds (Trias Mansilla et al. 2008). For example, *Lactiplantibacillus plantarum* was capable of producing lactic acid that prevents the development of *B. cienerea* on fresh-cut kiwifruits (De Simone et al. 2021).

On the other hand, to determine whether volatile metabolites are involved in the bacterium's antagonistic capacity against the post-harvest pathogen, open Petri dishes containing the actively growing bacteria are normally placed in contact with the fruit in a closed container, so that these volatile compounds can act in a controlled manner (Poveda 2021). This determined that the volatile metabolites of the bacterium *Burkholderia gladioli* were involved in the reduced impact of *B. cinerea* on grapes. This effect was due to these antifungal compounds affecting colony extension and hyphal development by downregulating melanin biosynthesis and upregulating vesicle transport, G protein subunit 1, mitochondrial oxidative phosphorylation, disturbing the autophagy process, and degrading the cell wall of the pathogen (Wang et al. 2023). Identifying exactly which volatile metabolites have antifungal activity requires complex mass spectrometry techniques, but several have been identified. For example, *Bacillus* sp. was capable of reducing the incidence of gray mold in grapes and cherry tomatoes by releasing the volatile metabolite dibutyl phthalate (Li et al. 2022a). Meanwhile, *Bacillus siamensis* acts on raspberries by releasing 2,6-di-tert-butyl-4-methylphenol and 2,4-di-tert-butylphenol (Zhang et al. 2020).

But in addition to direct antagonism, probiotic bacteria act as effective BCAs against *B. cinerea* in post-harvest by inducing plant defenses in fruits. Cell receptors in fruits can recognize bacteria or compounds produced by them and activate plant defenses through hormonal signaling. In this regard, in sweet pepper fruits, recognition of the bacteria *B. pumilus*, *B. siamensis,* or *B. thuringiensis* in wounds induced jasmonic acid (JA)-mediated defence signaling that reduced the severity of the disease caused by *B. cinerea* (Poveda et al. 2022). Similarly, the recognition of volatiles produced by the probiotic *B. gladioli* led to the induction of grape fruit defenses against the pathogen, mediated by the hormones salicylic acid (SA), jasmonic acid (JA), and ethylene (ET) (Wang et al. 2023). These hormonal signals trigger the local and/or systemic activation of plant antifungal defenses. This occurred in strawberry fruits upon recognition of *B. cereus* in wounds, activating SA, JA, and ET hormonal pathways and locally accumulating benzoxazinoids and flavonoids (Yu et al. 2021).

The main induced plant defense response that has been studied as a mechanism of action of probiotic bacteria against *B. cinerea* has been the activity of defense enzymes by probiotics of the genus *Bacillus*. The defensive enzymes induced in blueberry, grape, and cherry tomato fruits by these bacteria include chitinase (CHI), phenylalanine ammonia-lyase (PAL), peroxidase (POD), polyphenol oxidase (PPO), cinnamate 4-hydroxylase (C4H), and 4-coumarate CoA ligase (4CL) (Lu et al. 2021;

Li et al. 2022a; Du et al. 2024). Furthermore, these probiotics can induce the activity of plant defense enzymes and the accumulation of antifungal compounds in the fruit in combination. This is the case of *B. halotorelans* in strawberry fruit, which is capable of reducing the incidence of gray mold by inducing β-1, 3-glucanase (GLU), CHI, PAL, and POD enzyme activity and the accumulation of phenols and flavonoid compounds. However, these induced defensive responses did not cause differences in nutritional quality in these fruits with respect to ascorbic acid, titratable acidity, and total soluble solids (Wang et al. 2021a, b).

5.6 Probiotic Bacteria to Control Green and Blue Molds in Postharvest

Other postharvest diseases for which probiotic bacteria have been used are green (*P. digitatum*) and blue molds (*P. expansum and P. italicum*), which are compiled in Tables 5.2 and 5.3, respectively.

In the case of *P. digitatum*, direct antagonism has been described as one of the strategies developed by this group of bacteria when acting as BCAs. In this regard, *Bacillus sonorensis* is capable of reducing the incidence of green mold in grape fruits through its ability, characterized in vitro, to inhibit the growth of mycelium and conidial germination of the pathogen (Deng et al. 2020). To date, the mechanism of action developed by these probiotic bacteria within the most studied direct antagonism is the production of diffusible metabolites. These compounds, produced by the probiotics *Lactobacillus paraplantarum* and *L. plantarum*, were able to reduce the incidence of green mold by up to 80% and severity by nearly 100% in different citrus fruits (Jahantigh et al. 2023; Volentini et al. 2023). These antifungal metabolites have been described in greater detail in probiotics of the genus *Bacillus*. For example, in mandarin orange fruits, the harmful effect of *P. digitatum* is reduced due to the production of a 30.2 kDa lipoprotein by *B. velezensis* (Wang et al. 2024), or surfactin, fengycin, bacillaene, and bacilysin by *B. subtilis* (Li et al. 2022b). In addition, in the case of the latter probiotic bacterium, it was also possible to identify, by labeling with green fluorescent protein (GFP), competition for space with the pathogen as the mechanism of action involved, since it easily colonized the citrus fruit peel and pulp (Li et al. 2022b).

In the case of *P. expansum*, some probiotic bacteria capable of reducing blue mold have been identified, although the mechanisms of action involved have not been characterized. Through direct application to apple and grape wounds, the bacteria *Bacillus* sp., *Lacticaseibacillus rhamnosus,* and *Lactiplantibacillus plantarum* were able to reduce the incidence of the disease by up to 40% and its severity by 60% (Zhimo et al. 2020; Hou et al. 2025). However, with the spray application of *Lactobacillus plantarum* and *L. paracasei* bacteria, the disease was completely inhibited in apples (Matei et al. 2016). In addition, bacteria belonging to the genera *Lacticaseibacillus* and *Lactiplantibacillus* are very good candidates as probiotics, as they exhibit high survival rates under extreme gastrointestinal conditions, enhanced auto-aggregation, co-aggregation, and hydrophobicity (Hou et al. 2025).

Table 5.2 Probiotic bacteria with the ability to control the green mold (*Penicillium digitatum*) in postharvest fresh produce, indicating the reported effects and possible mechanisms of action involved

Bacterial probiotic		Fresh				
Species	Strain	produce	Application	Effect	Mechanism of action[a]	Reference
Bacillus sonorensis	KLBC GS-3	Grape fruit	30 μL of bacterial suspension (10^7 CFU mL^{-1}) in wounds	Decrease disease incidence	Direct antagonism Induction of plant defenses: PPO, POD, and PAL enzymes	Deng et al. (2020)
B. subtilis	L1–21	Mandarin fruit	10 μL of bacterial suspension (10^8 CFU mL^{-1}) in wounds	Decrease disease incidence	Direct antagonism: antifungal metabolites (surfactin, fengycin, bacillaene, and bacilysin) and competition for space	Li et al. (2022b)
B. velezensis	wr8	Mandarin orange fruit	20 μL of bacterial secondary metabolites in wounds	Decrease disease severity	Direct antagonism: antifungal metabolite (lipoprotein)	Wang et al. (2024)
Lactobacillus paraplantarum	CRL 1905	Lemon fruits	10 μL of bacterial secondary metabolites in wounds	Decrease disease incidence	Direct antagonism: antifungal metabolites	Volentini et al. (2023)
L. plantarum	–	Sweet orange fruit	Fruit immersion in bacterial suspension (10^8 CFU mL^{-1}) or 30% cell-free supernatant	Decrease disease severity	Direct antagonism: antifungal metabolites	Jahantigh et al. (2023)

[a]*PAL* phenylalanine ammonia lyase; *POD* peroxidase; *PPO* polyphenol oxidase

Table 5.3 Probiotic bacteria with the ability to control the blue mold (*Penicillium expansum* and *P. italicum*) in postharvest fresh produce, indicating the reported effects and possible mechanisms of action involved

Bacterial probiotic						
Species	Strain	Fresh produce	Application	Effect	Mechanism of action[a]	Reference
Bacillus sp.	3	Apple (*P. expansum*)	20 μL of bacterial suspension (10^8 CFU mL^{-1}) in wounds	Decrease disease incidence	Unidentified	Zhimo et al. (2020)
Bacillus amyloliquefaciens	B4	Loquat fruit (*P. expansum*)	20 μL of bacterial suspension (10^8 CFU mL^{-1}) in wounds	Decrease disease incidence	Direct antagonism: competition for space	Ye et al. (2021)
B. tequilensis	Yh	Bingtang orange fruit (*P. italicum*)	5 μL of bacterial suspension (10^7 CFU mL^{-1}) in wounds	Decrease disease incidence	Direct antagonism: CHI and GLU activity Modification of fruit microbiota Induction of plant defenses: POD, PPO, PAL, CHI, and GLU enzymes	Zhou et al. (2025)
Lacticaseibacillus rhamnosus	SQ63	Grapefruit (*P. expansum*)	10 μL of bacterial suspension (10^{10} CFU mL^{-1}) in wounds	Decrease disease incidence	Unidentified	Hou et al. (2025)
Lactiplantibacillus plantarum	LR5–2	Grapefruit (*P. expansum*)	10 μL of bacterial suspension (10^{10} CFU mL^{-1}) in wounds	Decrease disease incidence	Unidentified	Hou et al. (2025)
Lactobacillus paraplantarum	CRL 1905	Lemon fruit (*P. italicum*)	10 μL of bacterial secondary metabolites in wounds	Decrease disease incidence	Direct antagonism: antifungal metabolites	Volentini et al. (2023)

(continued)

Table 5.3 (continued)

Bacterial probiotic						
Species	Strain	Fresh produce	Application	Effect	Mechanism of action[a]	Reference
L. plantarum	ATCC8014	Apple fruit (*P. expansum*)	Spray of bacterial suspension on fruits	Decrease disease severity	Unidentified	Matei et al. (2016)
	TR7 TR71	Tomato fruit (*P. expansum*)	Surface treatment with 10 mL of cell-free filtrate per fruit	Decrease disease incidence	Direct antagonism: antifungal diffusible (lactic acid, acetic acid, and phenyllactic acid) and volatile (pyrazine) metabolites	Luz et al. (2020)
L. paracasei	IC13239	Apple fruit (*P. expansum*)	Spray of bacterial suspension on fruits	Decrease disease severity	Unidentified	Matei et al. (2016)

[a]*CHI* chitinase; *GLU* β-1,3-glucanase; *PAL* phenylalanine ammonia lyase; *POD* peroxidase; *PPO* polyphenol oxidase

The mechanisms of action described to date for probiotic bacteria against *P. expansum* include competition for space and the production of antifungal metabolites, with no indirect mechanisms of action having been described. In the loquat fruit, the bacterium *Bacillus amyloliquefaciens* reduced the incidence of blue mold by 40% due to rapid and effective colonization of wounds. In addition, oral toxicity test results suggested that this bacterium was safe for human consumption (Ye et al. 2021). On the other hand, the probiotic bacterium *L. plantarum* was able to reduce the harmful effect of *P. expansum* on tomato fruits through the production and release of different diffusible metabolites, such as lactic acid, acetic acid, and phenyllactic acid, and volatiles, such as pyrazine (Luz et al. 2020).

Finally, against *P. italicum* on different citrus fruits, different direct and indirect mechanisms of action effective in post-harvest have been identified in probiotic bacteria. The direct mechanism of action described is the ability of *Bacillus tequilensis* to produce enzymes that degrade the fungal cell wall, such as CHI and GLU, which cause increased cell membrane damage and leakage of cellular compounds in *P. italicum* (Zhou et al. 2025). The production of antifungal secondary metabolites in the form of organic acids by the bacterium *L. paraplantarum* also inhibits conidial germination of the pathogen and is implicated in an up to 80% reduction in the incidence of green mold in lemon fruit (Volentini et al. 2023). On the other hand, different indirect mechanisms of action have been described in Bingtang orange fruits by the probiotic bacterium *B. tequilensis*. Specifically, this bacterium reduces the incidence of the disease by modifying the bacterial and fungal community structures on the fruit surface and inducing plant defenses, with increased activity of the enzymes POD, PPO, PAL, CHI, and GLU and accumulation of total phenolics (Zhou et al. 2025).

5.7 Conclusions and Future Perspectives

Gray, green, and blue molds represent important postharvest diseases of fresh vegetable products worldwide. In the current context of continued global population growth and the search for less use of chemical pesticides, which are harmful to the environment and health, biological control strategies represent an effective and sustainable alternative. Some probiotic bacteria, in addition to their benefits for human health, can also be used as effective BCAs against the pathogens *B. cinerea*, *P. digitatum*, *P. expansum,* and *P. italicum*, as discussed in this chapter and summarized in the infographic (Fig. 5.1).

The use of probiotic bacteria to effectively control these postharvest diseases is a very recent area of study, with the vast majority of studies having been conducted in the last five years. It is, therefore, a field of knowledge that is still to be explored and has enormous basic and applied scientific potential. These bacteria exert antifungal activity via multiple direct mechanisms of action (competition, antibiosis, and/or lytic enzymes) and indirect mechanisms (activation of plant defenses and/or modification of the fruit microbiota), which allows for greater effectiveness in reducing the incidence and severity of the disease in post-harvest.

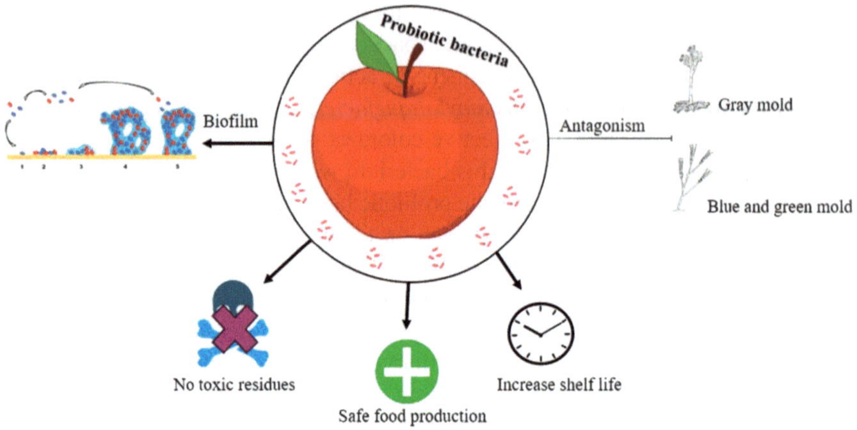

Fig. 5.1 Infographic illustrating how probiotic bacteria act as biological control agents against common postharvest molds, showing effects and mechanisms of action

Of the four pathogens included in this chapter, *B. cinerea* is the most widely studied for postharvest control with probiotic bacteria, characterizing different mechanisms of action. Less work has been done to control *Penicillium* diseases in these vegetable products. In the case of *P. digitatum* and *P. expansum*, indirect mechanisms of action developed by these probiotic bacteria have not even been characterized. Therefore, more work is needed to focus on the control of *Penicillium*, especially on the induction of plant defenses or changes in the plant microbiota after the application of probiotic bacteria.

Among the indirect mechanisms of action, only one study has been conducted on the modification of fruit microbiota after the use of a probiotic bacterium such as BCA, specifically for the control of *P. italicum*. This is a field of research that has developed exponentially in recent years and whose study techniques are becoming increasingly accessible. Future lines of research should study in depth the changes in epiphytic and endophytic microbiota in fruit after treatment with probiotic bacteria and their relationship with the control of decay caused by molds.

Another major area for further work is the industrial application of these probiotic bacteria as BCAs in post-harvest. The vast majority of studies focus on bacterial treatment of artificially made wounds in fruit. Although scientifically very useful, this methodology does not address the actual application of bioinoculants. This is an area of study that requires considerable effort on the part of the scientific community, in close collaboration with industry. In addition, in this industrial application, the use of other health control strategies combined with probiotic bacteria, such as edible coatings or the use of essential oils, should be considered, as preliminary studies are very promising.

Finally, as always when working with food products, there is an important aspect that must be considered in post-harvest fruits after the application of these probiotic bacteria, namely the quality of the final food product. This aspect has been largely overlooked in studies conducted to date, but future research should also analyze the effects of probiotic bacteria on the nutritional composition and organoleptic characteristics of fresh plant products, including during storage and under different temperatures.

5.8 Summary

Fungal postharvest diseases cause major spoilage and economic losses in fresh produce, particularly gray mold (*B. cinerea*) and green and blue molds caused by *Penicillium* spp. Reliance on synthetic fungicides is increasingly limited by resistance, environmental concerns, and residue regulations, driving demand for sustainable alternatives. Bacterial probiotics, especially *Bacillus* and *Lacticaseibacillus* spp., offer promising biocontrol through multiple antagonistic mechanisms and strong safety profiles. This chapter reviews these key pathogens and highlights probiotic-based strategies as eco-friendly, residue-free tools aligned with sustainable agriculture and food-safety requirements.

References

Bastipati SB, Vaishnavi DS, Prasad S, Deepu PG, Reddy KV (2024) Soil microbiomes and biocontrol of pathogens. In: Aransiola SA, Atta HI, Maddela NR (eds) Soil microbiome in green technology sustainability. Springer, Cham. https://doi.org/10.1007/978-3-031-71844-1_13

Bhatta UK (2022) Alternative management approaches of citrus diseases caused by *Penicillium digitatum* (green mold) and *Penicillium italicum* (blue mold). Front Plant Sci. https://doi.org/10.3389/FPLS.2021.833328

Chen C, Cao Z, Li J, Tao C, Feng Y, Han Y (2020) A novel endophytic strain *of Lactobacillus plantarum* CM-3 with antagonistic activity against *Botrytis cinerea* on strawberry fruit. Biol Control. https://doi.org/10.1016/J.BIOCONTROL.2020.104306

Costa JH, Wassano CI, Angolini CFF, Scherlach K, Hertweck C, Pacheco Fill T (2019) Antifungal potential of secondary metabolites involved in the interaction between citrus pathogens. Sci Rep 9:1–11. https://doi.org/10.1038/S41598-019-55204-9

da Rocha Neto AC, Luiz C, Maraschin M, Di Piero RM (2016) Efficacy of salicylic acid to reduce *Penicillium expansum* inoculum and preserve apple fruits. Int J Food Microbiol 221:54–60. https://doi.org/10.1016/J.IJFOODMICRO.2016.01.007

De Simone N, Capozzi V, de Chiara MLV, Amodio ML, Brahimi S, Colelli G, Drider D, Spano G, Russo P (2021) Screening of lactic acid bacteria for the biocontrol of *Botrytis cinerea* and the potential of *Lactiplantibacillus plantarum* for eco-friendly preservation of fresh-cut kiwifruit. Microorganisms. https://doi.org/10.3390/MICROORGANISMS9040773

Deepthi BV, Gnanaprakash AP, Sreenivasa MY (2017) Effect of γ-irradiation on fumonisin-producing *Fusarium* associated with animal and poultry feed mixtures. 3 Biotech 7:57. https://doi.org/10.1007/s13205-017-0693-x

Deng J, Kong S, Wang F, Liu Y, Jiao J, Lu Y, Zhang F, Wu J, Wang L, Li X (2020) Identification of a new *Bacillus sonorensis* strain KLBC GS-3 as a biocontrol agent for postharvest green mould in grapefruit. Biol Control. https://doi.org/10.1016/J.BIOCONTROL.2020.104393

Díaz-Urbano M, Velasco P, Rodríguez VM, Poveda J (2023) Endophytic fungi in postharvest disease management in fresh produce. Postharvest management of fresh produce:81–112. https://doi.org/10.1016/B978-0-323-91132-0.00004-6

Divyashree S, Shruthi B, Vanitha PR, Sreenivasa MY (2023) Probiotics and their postbiotics for the control of opportunistic fungal pathogens: a review. Biotechnol Rep. https://doi.org/10.1016/j.btre.2023.e00800

Du Q, Li R, Liu L, Chen L, Tang J, Deng J, Wang F (2024) Application of *Bacillus tequilensis* for the control of gray mold caused by *Botrytis cinerea* in blueberry and mechanisms of action: inducing phenylpropanoid pathway metabolism. Front Microbiol. https://doi.org/10.3389/FMICB.2024.1455008

FAO (2023) The state of food and agriculture 2023: food loss and waste reduction for improved food security and environmental sustainability. Food and Agriculture Organization of the United Nations

Gabal E, Amal-Asran Mohamed MA, Abd-Elsalam KA (2019) *Botrytis* gray mold nano- or biocontrol: present status and future prospects. Nanotechnol Life Sci 85–118. https://doi.org/10.1007/978-3-030-13296-5_5

Hassan EA, Mostafa YS, Alamri S, Hashem M, Nafady NA (2021) Biosafe management of *Botrytis* grey mold of strawberry fruit by novel bioagents. Plants 10(12):2737. https://doi.org/10.3390/PLANTS10122737

Hou Y, Duan Y, Wu G, Zhang J, Luo X, Zhang M, Pang H, Hao Y, Wang Y, Cai Y, Wang L, Tan Z (2025) Antibacterial activity, probiotic potential, and biocontrol efficacy of two lactic acid bacteria against *Penicillium expansum* on fresh grapes. Foods 14(3):493. https://doi.org/10.3390/FOODS14030493

Islam T, Danishuddin-Tamanna NT, Matin MN, Barai HR, Haque MA (2024) Resistance mechanisms of plant pathogenic fungi to fungicide, environmental impacts of fungicides, and sustainable solutions. Plants 13(19):2737. https://doi.org/10.3390/PLANTS13192737

Izurdiaga D, Sánchez-López ÁM, Fernández-San Millán A, Poveda J (2024) Cell-free filtrates from plant pathogens: potential new sources of bioactive molecules to improve plant health. Crop Prot. https://doi.org/10.1016/J.CROPRO.2023.106477

Jahantigh S, Oghaz NA, Rahnama K, Hatamzadeh S (2023) Application of *Lactobacillus* spp. for the biological management of green mold (*Penicillium digitatum*) on sweet orange fruit under *in vitro* and storehouse conditions. Biocontrol Sci Technol 33(6):567–581. https://doi.org/10.1080/09583157.2023.2208781

Ju J, Xie Y, Guo Y, Cheng Y, Qian H, Yao W (2019) Application of edible coating with essential oil in food preservation. Crit Rev Food Sci Nutr 59(15):2467–2480. https://doi.org/10.1080/10408398.2018.1456402

Kanashiro AM, Akiyama DY, Kupper KC, Fill TP (2020) *Penicillium italicum:* an underexplored postharvest pathogen. Front Microbiol. https://doi.org/10.3389/FMICB.2020.606852

Koirala S, Myers B, Shin GY, Gitaitis R, Kvitko BH, Dutta B (2023) Evaluating options to increase the efficacy of biocontrol agents for the management of *Pantoea* spp. under field conditions. Plant Dis 107(9):2701–2708. https://doi.org/10.1094/PDIS-11-22-2710-RE

Leneveu-Jenvrin C, Charles F, Barba FJ, Remize F (2020) Role of biological control agents and physical treatments in maintaining the quality of fresh and minimally processed fruit and vegetables. Crit Rev Food Sci Nutr 60(17):2837–2855. https://doi.org/10.1080/10408398.2019.1664979

Li P, Feng B, Yao Z, Wei B, Zhao Y, Shi S (2022a) Antifungal activity of endophytic *Bacillus* K1 against *Botrytis cinerea*. Front Microbiol 13:935675. https://doi.org/10.3389/FMICB.2022.935675

Li Y, Xia M, He P, Yang Q, Wu Y, He P, Ahmed A, Li X, Wang Y, Munir S, He Y (2022b) Developing *Penicillium digitatum* management strategies on post-harvest citrus fruits with metabolic components and colonization of *Bacillus subtilis* L1-21. J Fungi 8(1):80. https://doi.org/10.3390/jof8010080

Li B, He X, Guo S, Li D, Wang Y, Meng X, Dai P, Hu T, Cao K, Wang S (2024) Characterization of *Bacillus amyloliquefaciens* BA-4 and its biocontrol potential against *Fusarium*-related apple replant disease. Front Plant Sci 15:1370440. https://doi.org/10.3389/FPLS.2024.1370440

Liu J, Sui Y, Wisniewski M, Xie Z, Liu Y, Wang Q (2018) The impact of the postharvest environment on the viability and virulence of decay fungi. Crit Rev Food Sci Nutr 58(10): 1681–1687. https://doi.org/10.1080/10408398.2017.1279122

Lu Y, Ma D, He X, Wang F, Wu J, Liu Y, Jiao J, Deng J (2021) *Bacillus subtilis* KLBC BS6 induces resistance and defence-related responses against *Botrytis cinerea* in blueberry fruit. Physiol Mol Plant Pathol 114:101599. https://doi.org/10.1016/J.PMPP.2020.101599

Luz C, D'Opazo V, Quiles JM, Romano R, Mañes J, Meca G (2020) Biopreservation of tomatoes using fermented media by lactic acid bacteria. LWT. https://doi.org/10.1016/J.LWT.2020.109618

Marín A, Plotto A, Atarés L, Chiralt A (2019) Lactic acid bacteria incorporated into edible coatings to control fungal growth and maintain postharvest quality of grapes. HortScience 54(2): 337–343. https://doi.org/10.21273/HORTSCI13661-18

Matei GM, Matei S, Matei A, Cornea CP, Draghici EM, Jerca IO (2016) Bioprotection of fresh food products against blue mold using lactic acid bacteria with antifungal properties. Rom Biotechnol Lett 21(1):11201–11208

Menéndez-Cañamares S, Blázquez A, Albertos I, Poveda J, Díez-Méndez A (2024) Probiotic *Bacillus subtilis* SB8 and edible coatings for sustainable fungal disease management in strawberry. Biol Control. https://doi.org/10.1016/J.BIOCONTROL.2024.105572

Pongener A, Sharma S, Purbey SK (2018) Heat treatment of fruits and vegetables. In: Postharvest disinfection of fruits and vegetables, pp 179–196. https://doi.org/10.1016/B978-0-12-812698-1.00009-1

Poveda J (2020) Use of plant-defense hormones against pathogen diseases of postharvest fresh produce. Physiol Mol Plant Pathol. https://doi.org/10.1016/J.PMPP.2020.101521

Poveda J (2021) Beneficial effects of microbial volatile organic compounds in plants. Appl Soil Ecol. https://doi.org/10.1016/J.APSOIL.2021.104118

Poveda J, Calvo J, Barquero M, González-Andrés F (2022) Activation of sweet pepper defence responses by biocontrol agents of the genus *Bacillus*. Eur J Plant Pathol 164(4):507–524. https://doi.org/10.1007/S10658-022-02575-x

Ramalingam S, Le Myint Z, Ahn SY, Ryu JA, Lee SM, Yun HK (2024) UV-C treatment elicits resistant responses against *Botrytis cinerea* and improves fruit characteristics in grapevines. Hortic Environ Biotechnol 65(4):707–724. https://doi.org/10.1007/S13580-024-00602-W

Reiss A, Jørgensen LN (2017) Biological control of yellow rust of wheat with *Bacillus subtilis* strain QST713. Crop Prot 93:1–8. https://doi.org/10.1016/j.cropro.2016.11.009

Romero Bernal A, Raffellini S, Jaramillo-Sánchez G, Alzamora SM, Gómez PL (2024) Examining the response of *Botrytis cinerea* to aqueous ozone. J Food. https://doi.org/10.1080/19476337.2024.2339446

Romero J, Albertos I, Díez-Méndez A, Poveda J (2022) Control of postharvest diseases in berries through edible coatings and bacterial probiotics. Sci Hortic. https://doi.org/10.1016/J.SCIENTA.2022.111326

Samaras A, Roumeliotis E, Ntasiou P, Karaoglanidis G (2021) *Bacillus subtilis* MBI600 promotes tomato growth and induces systemic resistance. Plants 10:1113. https://doi.org/10.3390/plants10061113

Sykalo O, Shpak B (2024) Study and assessment of Taegro WP application on strawberries. Biol Syst Theory Innov 2(15):76–83. https://doi.org/10.31548/biologiya15(2).2024.007

Taroub B, Salma L, Manel Z, Ouzari HI, Hamdi Z, Moktar H (2019) Isolation of lactic acid bacteria from grapefruit. Ann Microbiol 69(1):17–27. https://doi.org/10.1007/S13213-018-1359-6

Thomloudi EE, Tsalgatidou PC, Baira E, Papadimitriou K, Venieraki A, Katinakis P (2021) Genomic and metabolomic insights into secondary metabolites of the novel *Bacillus halotolerans* hil4, an endophyte with promising antagonistic activity against gray mold and plant growth promoting potential. Microorganisms. https://doi.org/10.3390/MICROORGANISMS9122508

Trias Mansilla R, Bañeras Vives L, Montesinos Seguí E, Badosa Romañó E (2008) Lactic acid bacteria from fresh fruit and vegetables as biocontrol agents of phytopathogenic bacteria and fungi. https://doi.org/10.2436/20.1501.01.66

Volentini SI, Olmedo GM, Grillo-Puertas M, Rapisarda VA, Hebert EM, Cerioni L, Villegas JM (2023) Biological control of green and blue molds on postharvest lemon by lactic acid bacteria. Biol Control. https://doi.org/10.1016/J.BIOCONTROL.2023.105303

Wang W, Fang Y, Imran M, Hu Z, Zhang S, Huang Z, Liu X (2021a) Characterization of fludioxonil resistance in *Botrytis cinerea*. Microorganisms. https://doi.org/10.3390/microorganisms9020266

Wang F, Xiao J, Zhang Y, Li R, Liu L, Deng J (2021b) Biocontrol ability and action mechanism of *Bacillus halotolerans* against grey mould in strawberry. Postharvest Biol Technol. https://doi.org/10.1016/J.POSTHARVBIO.2020.111456

Wang D, Luo WZ, Zhang DD, Li R, Kong ZQ, Song J, Dai XF, Alkan N, Chen JY (2023) Insights into the biocontrol function of *Burkholderia gladioli* against *Botrytis cinerea*. Microbiol Spectrum. https://doi.org/10.1128/spectrum.04805-22

Wang X, Wang R, He S, Weng Y, Lan B, Zhou L, Yang H (2024) Biocontrol potential of *Bacillus velezensis* WR8 secondary metabolites against *Penicillium* sp. Gene. https://doi.org/10.1016/J.GENE.2023.147872

Xiang Y, Yuan H, Mao M, Hu Q, Dong Y, Wang L, Wu B, Luo Z, Li L (2024) Reciprocal inhibition of autophagy and *Botrytis cinerea*-induced programmed cell death. Food Chem:https://doi.org/10.1016/J.FOODCHEM.2024.140512

Ye WQ, Sun YF, Tang YJ, Zhou WW (2021) Biocontrol potential of a broad spectrum antifungal strain *Bacillus amyloliquefaciens* B4 for postharvest loquat fruit storage. Postharvest Biol Technol. https://doi.org/10.1016/J.POSTHARVBIO.2020.111439

Yu YY, Dou GX, Sun XX, Chen L, Zheng Y, Xiao HM, Wang YP, Li HY, Guo JH, Jiang CH (2021) Transcriptome and biochemical analysis reveal the effects of AR156 on postharvest strawberry gray mold and fruit quality. Front Plant Sci. https://doi.org/10.3389/FPLS.2021.700446

Zamani-Zadeh M, Soleimanian-Zad S, Sheikh-Zeinoddin M, Goli SAH (2014) Integration of *Lactobacillus plantarum* A7 with thyme and cumin essential oils for gray mold control in strawberry. Postharvest Biol Technol 92:149–156. https://doi.org/10.1016/J.POSTHARVBIO.2014.01.019

Zhang X, Gao Z, Zhang X, Bai W, Zhang L, Pei H, Zhang Y (2020) Control effects of *Bacillus siamensis* G-3 volatile compounds on raspberry postharvest diseases caused by *Botrytis cinerea* and *Rhizopus stolonifer*. Biol Control. https://doi.org/10.1016/J.BIOCONTROL.2019.104135

Zhao Q, Shi Y, Xu C, Jiang Z, Liu J, Sui Y, Zhang H (2023) Control of postharvest blue and gray mold in kiwifruit by *Wickerhamomyces anomalus*. Postharvest Biol Technol. https://doi.org/10.1016/J.POSTHARVBIO.2023.112345

Zhimo VY, Biasi A, Kumar A, Feygenberg O, Salim S, Vero S, Droby S (2020) Yeasts and bacterial consortia from kefir grains as effective biocontrol agents of postharvest fruit diseases. Microorganisms 8(3):428. https://doi.org/10.3390/microorganisms8030428

Zhou A, Zhou H, Yang K, Sun Y, Deng J, Wang F (2025) *Cryptococcus laurentii* combined with *Bacillus tequilensis* reduces blue mould by enhancing disease resistance and modulating the microbial community structure in Bingtang orange peel. Int J Food Microbiol. https://doi.org/10.1016/J.IJFOODMICRO.2025.111192

Probiotic Measures: Mildews, Smuts, and Rots

6

K. Ajithkumar, A. S. Savitha, M. Renuka, and S. Mahadevkumar

Abstract

Plant probiotics are beneficial microbes capable of inducing systemic resistance in hosts, exhibiting great efficiency in niche colonization and demonstrating antagonistic action against a broad spectrum of pathogens. The concept of plant probiotics has garnered significant attention recently in agriculture due to its substantial application in the biological management of plant diseases and bio-fertilization. Compared to conventional chemical pesticides, these probiotics are safer, less harmful to the environment and to human health, have much more targeted activity, have more potential in small amounts, decompose more quickly, prevent the development of resistance through a number of mechanisms and can be used in both conventional and integrated pest management systems. As for plant diseases are concerned, these probiotics are pioneers in the race to effectively manage the powdery mildew, smut, and rot caused by an array of fungal genera. Henceforth, this book chapter is mainly focused on the various probiotics entangled with the management of fungal plant pathogens, as well as their mode of action involved in tackling them.

K. Ajithkumar (✉) · M. Renuka
Main Agricultural Research Station, College of Agriculture, University of Agricultural Sciences, Raichur, Karnataka, India
e-mail: ajithk.path@uasraichur.edu.in

A. S. Savitha
Department of Plant Pathology, College of Agriculture, University of Agricultural Sciences, Raichur, Karnataka, India

S. Mahadevkumar
Botanical Survey of India, ANRC, Sri Vijayapuram, Andaman & Nicobar Island, India

M. Y. Sreenivasa et al. (eds.), *Probiotics as Antifungals*,
https://doi.org/10.1007/978-981-92-0002-3_6

Graphical Abstract

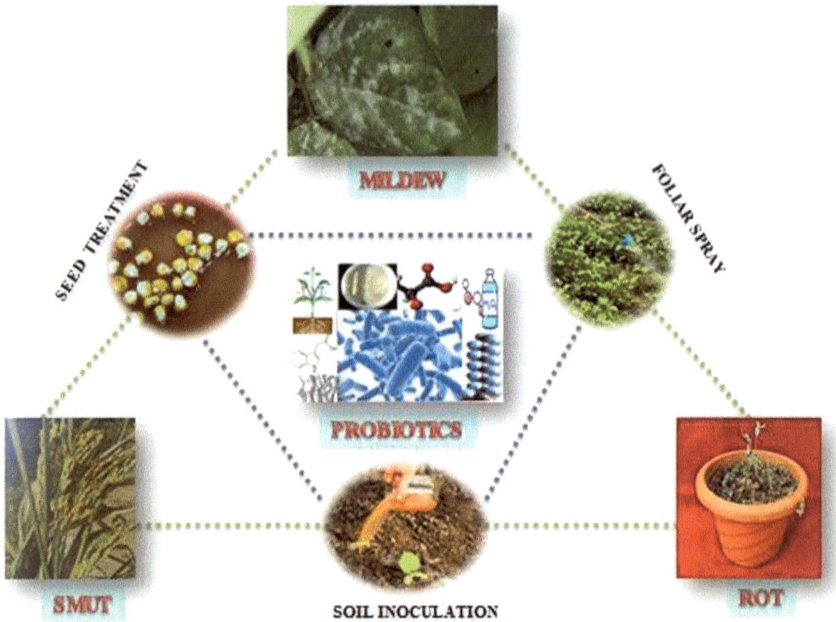

Keywords

Antifungal activity · Biocontrol agent · ISR · Lactic acid bacteria · Plant probiotics

Learning Objectives

- By the end of this chapter, readers will have a keen knowledge of the various kinds of symptoms produced by the mildews, smut, and rot diseases.
- This chapter enlightens the readers to understand the probiotics utilized as weapons for tackling the plant pathogens.
- This chapter will help readers to analyze the effectiveness of different modes of action of probiotics employed for managing the mildews, smut, and rot diseases.

6.1 Introduction

Probiotics are living organisms that play a vital role in promoting a wide range of health outcomes. With the growing prominence of functional foods, probiotics, whether in the form of food products or supplements, have become the most popular component. Probiotics have long been a crucial element and a lucrative target for possible human health benefits (Hamad et al. 2022). Furthermore, the anticipated 9.5 billion people on the planet by 2050 put significant pressure on the agricultural revolution and improvements to produce high food yields while maintaining environmental safety. Plant probiotics have therefore gained popularity since they offer an unmatched means of achieving sustainable agriculture without endangering biodiversity and the natural ecology.

Dieter Haas and Christoph Keel (2003) first coined the term "plant probiotics" to denote beneficial microorganisms that can induce systemic resistance in their hosts, high niche colonization efficiency and antagonistic action against a broad range of pathogens. Also, by releasing extracellular metabolites, these microorganisms improve the health of the soil and help in the breakdown of complex organic materials and insoluble nutrients into simpler forms that plants can easily absorb (Herrmann and Lesueur 2013). Probiotics' prospective uses in the agricultural industry, however, have not yet been formally acknowledged. The selection of novel probiotic strains and the creation of novel applications have become increasingly significant in light of the growing interest in probiotics' use in agriculture.

The application of probiotics across various agricultural sectors has steadily expanded due to their promising benefits as they offer improved nutrient absorption, disease resistance, and overall health in plants, fish, and animals (Song et al. 2012). The plant probiotic products can be utilized as phytostimulators, biopesticides, plant strengtheners, and biofertilizers, depending on their effects and mode of action. According to Berg (2009), plant probiotics offer several benefits over chemical pesticides, including improved safety, reduced environmental and human health risks, greater target specificity, effectiveness at low concentrations, the ability to proliferate under regulation by the host plant and its native microbiota, faster degradation compared to conventional pesticides, a lower resistance development through multiple mechanisms, and compatibility with conventional as well as integrated pest management systems.

6.2 Mildews

The pathogen causing this disease manifests as white, grey, brownish, or purplish growths on the surface of the host. In downy mildew, white cottony mycelial growth is seen on the lower surface of the leaf; the corresponding upper surface shows chlorotic patches, which, with advancement, become necrotic patches and finally result in death of the infected tissue. Most devastating diseases include sorghum downy mildew caused by *Perenosclerospora sorghi*, which results in the characteristic symptoms, including the lower surface of the leaf having abundant downy white growth consisting of sporangiophore and sporangia, while the corresponding upper surface shows chlorosis. As the infected, bleached leaves mature, they gradually become necrotic, and the interveinal tissues disintegrate, releasing resting spores (oospores). This results in loosely connected vascular bundles, producing the characteristic leaf-shredding symptom. In the case of green ear disease of bajra caused by *Sclerospora graminicola*, the infected inflorescence becomes partially or completely malformed, with florets transformed into leafy structures, giving rise to the typical "green ear" appearance. At later stages, these tissues turn necrotic and contain abundant masses of oospores. In maize, crazy top downy mildew caused by *Sclerophthora macrospora* is characterized primarily by the proliferation of auxiliary buds on the tassel stalk and cob, resulting in the distinctive' crazy top' symptom.

Grape downy mildew (*Plasmopara viticola*) caused heavy losses to the wine industry in France and led to the discovery of Bordeaux mixture by Prof. P. M. A. Millardet in 1885. The pathogen infects all aerial parts of the plant, including leaves, young shoots, tendrils, and immature berries. Initial symptoms appear as irregular, yellowish, translucent spots on the upper surface of the leaves. Corresponding to these lesions, a dirty white, cottony fungal growth develops on the lower surface. As the disease progresses, affected leaves turn yellow and brown, undergo necrosis, and eventually exhibit premature defoliation. Infected leaves, shoots, and tendrils become covered with a whitish fungal growth. Berries also develop a white fungal coating, after which they become leathery and shrivel. Diseased berries turn hard, bluish-green, and later brown, rendering them unsuitable for wine production. In sunflowers, downy mildew caused by *Plasmopara halstedii* results in chlorosis along the leaf midribs. As the disease advances, leaves become abnormally thick, curl downward, and display prominent yellow and green mottling on the upper (epiphyllous) surface. A characteristic downy fungal growth develops on the lower (hypophyllous) surface. Infected plants often produce sterile flower heads, and basal gall symptoms may also occur independently of foliar infection. In crucifers, downy mildew caused by *Peronospora parasitica* leads to discoloration of young seedlings, and in severe cases, complete plant death. Symptoms include purplish or yellow-brown spots on the upper leaf surface, with a fluffy downy fungal growth appearing on the lower surface. Downy mildew of cucurbits, caused by *Pseudoperonospora cubensis,* is characterized by yellow, angular spots on the upper surface. Under moist conditions, the corresponding lower surface develops a purplish downy growth. As the disease progresses, leaves turn yellow and fall prematurely. Infected plants become stunted and may die, while fruits that develop often fail to mature properly and exhibit poor taste.

Whereas in the case of powdery mildew, white powdery-like patches covering both upper and lower surfaces of the leaf, leading to powdery or dusty appearance, yellowing, and premature defoliation (chilli and pigeonpea powdery mildew incited by *Leveillula taurica*). This powdery patch consists of an enormous number of spores born on an oidiophore tangled with mycelia (Singh 2002). In roses, powdery mildew caused by *Sphaerotheca pannosa* var. *rosae* is characterized by a greyish powdery fungal growth on the leaves, accompanied by rolling of the leaf margins. Infected flower buds often fail to open, and the floral parts become discolored, dwarfed, and desiccated. Grey or areolate mildew of cotton, incited by *Ramularia areola*, appears as irregular to angular, pale translucent lesions on the lower surface, typically delimited by veinlets. On the upper surface, these lesions manifest as light green or yellow-green specks. A whitish-grey or frosty powdery growth, consisting of fungal conidiophores, develops on the lower surface. As lesions coalesce, the entire leaf may become covered with a white to grey powdery growth. The infection can spread to upper leaves, eventually affecting the whole plant. Severely infected leaves dry from the margins, turn yellowish-brown, and fall prematurely. In tobacco, powdery mildew caused by *Erysiphe cichoracearum* var. *nicotianae* initially appears as small, isolated white patches on the upper leaf surface. As the disease progresses, affected leaves turn brown, wither, and develop a scorched appearance. Severe infections result in defoliation and a significant reduction in both the yield and quality of curable leaves.

Mango powdery mildew, caused by *Oidium mangiferae*, is characterized by a whitish to greyish powdery growth on the inflorescence and tender leaves. The infection typically begins on the inflorescence and gradually spreads downward, affecting the floral axis, young leaves, and stems. Infected floral parts are severely damaged and often abscise prematurely. Even when fruit set occurs, the fruits fail to enlarge and may drop before reaching pea size. In some cases, fruits become malformed and discolored due to severe infection. Owing to poor fruit set and substantial flower and fruit drop, yield losses may reach as high as 70–80 percent. While in the case of green gram, black gram, and bean powdery mildew (*Erysiphe polygoni*), white powdery mass patches are seen on pods, which later result in distorted, malformed pods with few small, ill-filled seeds. *Erysiphe cichoracearum* affecting okra and cucurbits results in underdeveloped, deformed fruits. Even in the case of mustard powdery mildew (*E. cruciferarum*) appears as dirty white circular patches on siliquae and results in small and shriveled seeds. In pea, powdery mildew caused by *Erysiphe pisi* produces white, floury patches on the pods, consisting of a dense powdery fungal mass. These patches gradually change from white to light brown and finally dark brown, eventually becoming necrotic. Infected pods are generally unmarketable or fetch only a minimal price in the open market. Powdery mildew of grapes, incited by *Uncinula necator*, leads to cracking of the berry skin. Infected berries often develop a characteristic net-like pattern of scar tissue and may suffer yield losses of 40–60 percent. Such berries typically have more acidity than healthy fruits and are unsuitable for wine production. In apple, powdery mildew caused by *Podosphaera leucotricha* appears as small white or grey powdery patches on the undersurface of the leaves. Infected leaves become elongated, narrower than normal, and exhibit curled margins. Affected fruits remain small, deformed and frequently develop a roughened surface.

6.3 Smut

Smuts accurately refers to sooty or charcoal-like powder in plants. The disease is known as smut; the affected plant parts, such as florets, grains, stem, leaves, and roots, show a black or purplish-black dusty mass of spores. In some smuts, this mass of spores may be compact and held under the epidermis, appearing as black streaks or blotches on the host surface. The symptoms of smut usually appear on the floral organs, particularly the ovary, and the pustules are usually considerably larger than those of rusts (Singh 2002). In case of "false smut" of rice caused by *Ustilaginoidea virens* (Syn: *Claviceps oryzae - sativa*), the pathogen transforms individual ovaries/ grains into greenish spore balls of velvety appearance, which later become black in color. In sorghum, four distinct types of smut diseases have been reported. Head smut, caused by *Sphacelotheca reiliana*, is characterized by the complete replacement of the ear head with large sori. Grain smut, incited by *Sphacelotheca sorghi*, involves the substitution of individual grains with smut sori. Loose smut, caused by *Sphacelotheca cruenta*, results in plants that are shorter than healthy ones, with thinner stalks and pronounced tillering. The glumes become hypertrophied, and the ear head appears looser compared to healthy plants. Long smut, caused by *Tolyposporium ehrenbergii*, produces elongated, cylindrical to slightly curved sori enclosed by a relatively thick, creamy-brown covering membrane (peridium).

Loose smut of wheat, caused by *Ustilago nuda tritici*, is a major disease responsible for approximately 8–10 percent yield loss across the country. Symptoms become apparent only at the time of panicle emergence from the boot leaf. All spikelets within the panicle are transformed into a mass of black powdery spores. The infected panicle typically emerges about two days earlier than healthy ones and is initially covered by a thin, silvery membrane enclosing the spores. Corn smut, incited by *Ustilago maydis*, produces large smut sori that replace the tassel and ear. In some cases, the tassel may be partially or completely converted into a smut sorus. Whip smut of sugarcane, caused by *Ustilago scitaminea*, is characterized by the development of elongated, whip-like, dusty black structures in place of the normal central shoot. The whip may range from a few inches to several feet in length and is initially enclosed by a thin, white papery membrane. All shoots emerging from an infected clump typically produce whip-like structures. Diseased clumps may also form mummified arrows, where the lower portion bears a normal inflorescence with typical flowers, while the upper rachis is transformed into a characteristic smutted whip. Occasionally, smut sori may also develop on leaves and stems.

6.4 Rot

The infected tissue undergoes necrosis, progressively decomposes, and eventually turns brown. This condition is caused by fungi and bacteria that secrete enzymes capable of degrading the middle lamella between cell walls, leading to tissue maceration and disintegration. (Singh 2002). Sheath rot of rice caused by *Sarocladium oryzae* (Syn: *Acrocylindrium oryzae*) shows oblong or irregular greyish brown spots on the flag leaf sheath. They enlarge and develop a grey center and brown margins covering major portions of the leaf sheath. The panicles rot and abundant whitish powdery fungal growth is seen inside the leaf sheath. Stem rot of rice (*Sclerotium oryzae*) exhibits small black lesions formed on the outer leaf sheath. Later, the infested tissues are subjected to rot and abundant small black sclerotia are seen in the rotting tissues.

Charcoal rot of maize, caused by *Macrophomina phaseolina* and *Rhizoctonia bataticola,* is characterized by pronounced wilting of affected plants. Infected stalks display greyish streaks, and the pith becomes shredded. Numerous minute, greyish-black sclerotia develop along the vascular bundles. The internal shredding of the stalk often leads to breakage near the crown region, which typically turns dark in color. Additionally, shredding of the root bark and disintegration of the root system are common symptoms of the disease. The same results can also be seen in the case of dry root of chickpea, pigeon pea, black gram, green gram, soybean, groundnut, and cotton. Stem rot/collar rot incited by *S. rolfsii* affecting groundnut, sunflower, and brinjal shows a peculiar symptom of white mycelia growth on the collar region with mustard seed-like sclerotia. *Sclerotinia* stem rot (*Sclerotinia sclerotiorum*), also called white mold, affects sunflowers and mustard. The disease initially appears as water-soaked lesions on the stem, which later become covered with a cottony white fungal growth. As it advances, the affected stem portions develop a bleached appearance, and the tissues gradually shred. Girdling of the stem leads to premature ripening and lodging of plants. Hard, black sclerotia are formed within the stem and occasionally on its surface. Red rot of sugarcane, caused by *Colletotrichum falcatum* (Sexual stage: *Physalospora tucumanensis* or *Glomerella tucumanensis*), is characterized by reddening of the internal stalk tissues, particularly the vascular bundles, which appear elongated at right angles to the long axis of the stalk. A key diagnostic feature is the presence of crosswise white patches interrupting the reddened tissue. When split open, infected stems emit a characteristic acidic-sour odor. Another sett-borne disease, sett rot or pineapple disease, caused by *Ceratocystis paradoxa*, affects the internal tissues of the stalk, which initially develop a reddish discoloration and later turn brownish-black. Severely infected setts exhibit internodal cavities lined with fungal mycelium and abundant spores. Accompanied by a distinct pineapple-like odor.

Anabe roga/root rot of coconut and arecanut caused by *Ganoderma lucidum* results in the brittle, discolored, and dried roots. While in the case of bud rot of arecanut (*Phytophthora palmivora*), as well as *P. meadii,* causing koleroga in bunches, they also pass on to the buds and cause rotting. Secondary organisms colonize the rotting bud and convert it into a slimy mass, which would emit a foetid smell. *P. palmivora* causes black pod disease or pod rot of cocoa, where the infected pod and beans become brown colored and show rotting. *P. meadii* causes capsule rot of cardamom, leading to a crop loss of up to 30 per cent. Here, immature infected capsules rot and fall off by emitting a foul smell, but mature capsules become shriveled after infection. Rhizome rot of ginger and turmeric, caused by *Pythium* spp., primarily affects the rhizomes, which begin to rot and soften. The characteristic bright orange color of the rhizome gradually changes to various shades of brown, followed by progressive decomposition. Eventually, the tissues break down into a watery mass of putrefied material, often enclosed within the tough outer ring of the rhizome.

6.5 Lactic Acid Bacteria

The idea of lactic acid bacteria (LAB) first surfaced in the twentieth century and these bacteria now have appropriate biological potential in a variety of businesses. LAB can be found in fermented foods such as meat, dairy, veggies, and baked goods. Since ancient times, lactic acid bacteria (LAB) have been utilized for food fermentation. They can perform two tasks: they can operate as a food fermenting agent and possibly provide health advantages. Not all the LAB are automatically considered as GRAS (generally recognized as safe) or granted as QPS (Qualified presumption of safety) status. Although the group has a long history of use in food fermentation and is generally considered safe, regulatory agencies like the European Food Safety Authority (EFSA) and the US Food and Drug Administration (FDA) usually assign the official status at the species or even strain level. LAB are facultative anaerobic, Gram-positive, rod- or spherical, non-spore forming, and catalase-negative bacteria that include 6 families and 38 genera in the *Lactobacillales* order (Abubakr and Al-Adiwish 2017), which usually grow on substrates high in carbohydrates that they ferment to produce organic acids (Jaffar et al. 2023). LAB neutralizes harmful gases, improves soil fertility, aeration, and solubility, enhances seed germination, and reduces a variety of abiotic stressors. LAB plant growth-promoting qualities, however, have not been thoroughly investigated and have little supporting data in the literature (Lamont et al. 2017).

When considering lactic acid bacteria (LAB) for probiotic applications, several desirable attributes must be taken into account, including cost-effectiveness, maintenance of viability during processing and storage, ease of incorporation into products, and resilience to physicochemical stresses encountered during processing. Since these bacteria may prevent the growth of pathogens in crops, they appear to be promising candidates for bio-pesticides.

There are three recognized ways that LAB functions as a biocontrol agent:

(a) Production of bacteriocins, reactive oxygen species, and antimicrobial compounds.
(b) Prevention of pathogens from infecting plant tissues by colonizing them beforehand.
(c) Modifying the plant immune response (Konappa et al. 2016).

The ability of these microbes to produce a variety of metabolites, including lactic, propionic, acetic and formic acids; hydrogen peroxide; diacetyl; bacteriocins and bacteriocin like inhibitory substances (BLIS) (Jahatigh et al. 2023) and other metabolites, including diketopiperazines, hydroxy derivatives of fatty acids, phenyl lactic acid, phenyl lactate, hydroxyphenyl lactate, reuterin, cyclic dipeptides, pyrrolidone-5-carboxylic acid, 3-hydroxy fatty acids and amines are attributing towards the antifungal activity of *Lactobacillus* spp. (Gerez et al. 2009; Divyashree et al. 2024).

6.6 Mode of Action

6.6.1 Regulating Nutrient Absorption and Nitrogen Fixation

LAB strains can increase the amount of nutrients that plants can access from compost and other organic or inorganic materials. Phosphorus and potassium are solubilized by the acidity produced by LAB through the formation of organic acids, thereby making these nutrients available for plant uptake. According to Giassi et al. (2016), several LAB strains not only demonstrated phosphate-solubilizing ability but were also capable of fixing atmospheric nitrogen for plant utilization. Furthermore, Higdon et al. (2020) reported that *Lactococcus lactis*, isolated from the Sierra Mixe maize mucilage microbiota, was identified as a diazotroph capable of performing biological nitrogen fixation (BNF).

6.6.2 Production of Phytohormones

Plant root nutrition and water uptake are improved by these phytohormones, which also increase the length and surface area of root hairs (Kumar et al. 2022). Abiotic stress management, defense, and regular cell function are all aided by the enhanced metabolic activity brought on by the synthesis of phytohormones. Numerous LAB species can secrete auxins like indole-3-acetic acid (IAA) and phytohormones like gibberellins (GA), which have a variety of roles in promoting plant growth (Lamont et al. 2017).

6.6.3 Organic Acids

Numerous studies have suggested that one of the main ways LAB exhibits its antimicrobial action against a broad range of target phytopathogens is through organic acids. Lactic acid is the predominant metabolite produced by LAB; however, they also secrete other organic acids, including acetic. Propionic, formic, benzoic, and phenyllactic (PLA) acids. Lactic acid is thought to have an antimicrobial effect by disrupting the pathogen's membrane functions, preventing active transport, lowering intracellular pH, and inhibiting a number of metabolic processes, all of which result in the death of the target microbe (Rattanachaikunsopon and Phumkhachorn 2010).

6.6.4 Hydrogen Peroxide

In the presence of oxygen, LAB such as *Lactobacillus plantarum* P1–2 and *L. plantarum* 2142 produce the reactive oxygen species, i.e., hydrogen peroxide (H_2O_2) as a result of the action of flavoprotein oxidases or NADH peroxidase. The antimicrobial activity of hydrogen peroxide may stem from the oxidation of sulfhydryl groups, which denaturate many enzymes, and from the peroxidation of membrane lipids, which increases membrane permeability. Hydrogen peroxide serves as a precursor for the generation of bactericidal free radicals, including superoxide (O^{2-}) and hydroxyl (OH^-) radicals, which can inflict damage on DNA and prevent the growth of pathogenic and psychoactive microbes (Amenu 2013). However, more recent studies suggest that the antimicrobial efficacy of H_2O_2 is likely limited, as bacteria produce it only in small quantities, and its activity is largely exerted synergistically with other antifungal compounds (Le Lay et al. 2016).

6.6.5 Fatty Acids

As demonstrated in *Lactobacillus plantarum* MiLAB 14, hydroxy fatty acids (FAs) exhibit detergent-like properties that disrupt (partition) the lipid bilayers of fungal membranes. This disruption increases membrane permeability, leading to leakage of intracellular proteins and electrolytes and ultimately resulting in cytoplasmic disintegration of fungal cells (Sjogren et al. 2003).

6.6.6 Reuterin

The presence of glycerol in anaerobic conditions can either directly or indirectly increase the formation of reuterin, an antibacterial molecule derived from glycerol, which is produced by a number of lactobacilli. Numerous species of *Fusarium, Penicillium,* and *Aspergillus* are among the fungi for which reuterin has been shown to be effective. It has also been associated with the inhibition of mycotoxin production in fermented foods. Vimont et al. (2019) explained that reuterin produced by *L. reuteri* ATCC 53608 in yogurt showed a fungistatic effect at a concentration of 1.38 mM, while a fungicidal effect was obtained at 6.9 mM against *Penicillium chrysogenum* LMA-212 and *Mucor racemosus* LMA-722. Therefore, reuterin has a high potential as a food preservative, particularly owing to its biochemical properties and antibacterial and antifungal activities. Sun et al. (2022) reported that reuterin is a complex, multi-component antimicrobial system produced by *Limosilactobacillus reuteri* during glycerol metabolism. Owing to its functional versatility, reuterin holds significant potential for applications in both the food and medical industries. According to the reports, the reuterin system has anti-inflammatory, anti-cancer, and anti-infection properties in addition to controlling intestinal flora. Strong broad-spectrum antibacterial effects are typically exhibited by the reuterin system.

6.6.7 Cyclic Dipeptides

The smallest cyclic peptides are called cyclic dipeptides or cyclodipeptides (CDPs), also referred to as 2, 5-diketopiperazines. Based on the available data, bacteria produce almost 90 per cent of CDPs. Among the cyclic dipeptides that were extracted from LAB and have been shown to have antimicrobial properties are cyclo (GlyLeu), cyclo (Phe-Pro), cyclo (L-Phe-L-Pro), and cyclo (L-Phe-trans-4-OH-L-Pro), which were isolated from *L. plantarum* MiLAB 393 and showed broad-spectrum antifungal activity against *Fusarium sporotrichioides* and *Aspergillus fumigates* (Strom et al. 2002). Cyclo (L-Val-L-Pro) and cis-cyclo (L-Phe-L-Pro) obtained from *L. plantarum* LBP-K10 are significantly effective against *G. boninense*. Cyclo (L-Leu-L-Pro) isolated from *L. brevis* exhibits antifungal activity against various fungi and can also inhibit the production of harmful mycotoxins like aflatoxin. The antifungal compound Cyclo (Gly-Leu) from *L. plantarum* VTT E-78076 was found to have antifungal activity against the plant pathogen *F. avenaceum.* Cyclic peptides have drawn a lot of attention because of their stability in a variety of conditions, including heat, pH, and enzymes (Silpa and Rupachandra 2022).

6.7 Isolation of Lactic Acid Bacteria

Probiotics were isolated from fermented food, milk products, fruits, and vegetables, where 10 g of each sample was blended with 90 ml of sterilized water and serially diluted 10^{-1} to 10^{-5} in sterilized water to form a microbial suspension. Around 0.5 ml of the diluted sample was spread on De Man, Rogosa, and Sharpe (MRS) agar media and incubated at 30 °C for 24 hrs for bacterial growth. The plates were observed for the appearance of colonies, and further culture was purified by the streak plate method on MRS agar and incubated at 30 °C for 24 hrs. For long-term storage, the isolates will be preserved in broth solution with glycerol (30%) and maintained at −20 to −80 °C. Only fresh cultures will be used for the analysis throughout the experiment (Akter et al. 2021; Adithi et al. 2025; Vanitha et al. 2024; Shruthi et al. 2025). The bacteria were further identified based on morphological characters, biochemical tests, and molecular methods to confirm their antimicrobial activity and safety measures.

6.8 Probiotic Application Methodology

- *Seed dressing*: The seeds are submerged in bacterial suspension and subsequently dried before seed germination (known as bio-priming), or a liquid cell suspension or adhesive may be utilized to coat the seed with bacterial cells (referred to as film coating).
- *Seedling/root dip method*: The root of seedlings is dipped in the bacterial suspension for 5–10 min, allowing the probiotic to colonize roots thoroughly.

- *Foliar spray*: It is possible to manage fungal infections by directly spraying diluted probiotic cultures onto plant foliage. This technique can cause plants to develop systemic resistance while also responding quickly to disease symptoms.
- *Soil inoculation*: As a soil amendment, probiotics can be added to the soil to enhance its health and lower the prevalence of diseases. This technique encourages a robust rhizosphere microbial community, which can inhibit infections and boost plant development.

6.9 Probiotics in Management of Mildews, Smuts, and Rots

The effects of milk and whey as possible substitutes for synthetic fungicides in the management of grapevine powdery mildew incited by *Uncinula necator* were investigated by Crisp and his colleagues (2006). Within 24 hrs of treatment, different milk components created oxygen radicals under natural light, which led to the collapse of *U. necator* hyphae and ruptured oidia by lactoferrin, although hyphae damage did not become apparent until 48 hrs following treatment. The findings provide credence to the theory that lactoferrin's action and the generation of free radicals are linked to milk's ability to prevent powdery mildew. Ferrandino and Smith (2007) showed that foliar spray of cow's milk and water was about 50–70 per cent as effective in reducing foliar symptoms of powdery mildew (*Podosphaera xanthii*) of pumpkin and post-harvest fruit rot, and 40–50 per cent as effective in increasing marketable yield. DeBacco (2011) observed that milk and other dairy products fermented with *Lactobacillus were* used as a natural management for powdery mildew of cucurbits. Meena et al. (2024) found that butter milk spray at 10 per cent concentration has reduced 41.99 per cent disease intensity of pea powdery mildew over the untreated check.

Gajbhiye (2016) isolated lactic acid bacteria, i.e., *Lactobacillus lactis* subsp. *cremoris* PFL9 and *Lactobacillus brevis* PFR77 from aerial parts of the pomegranate plant. These exhibited strong antifungal activity, particularly against pathogens causing fruit rot like *Alternaria alternata, Fusarium graminearum, Corynespora cassiicola,* and *Phomopsis varsoniana.* They demonstrated significant biocontrol potential, surviving on fruit surfaces and delaying rot development comparably to mancozeb fungicide. Barrios-Roblero et al. (2019) demonstrated that *L. plantarum, L. paracasei,* and *L. pentosus* reduce the mycelial growth and spore germination of *Colletotrichum gloesporioides,* causing anthracnose in papaya. Steglinska et al. (2022) found that *Lactiplantibacillus plantarum* KB2 LAB 03 was able to reduce the disease caused by *Pectobacterium carotovorum* and *Rhizoctonia solani* in vivo up to 40–90 per cent. Riolo et al. (2024) isolated 13 *Lactiplantibacillus plantarum* from the rind of citrus fruits. The two isolates N3B2 and M2B2 exhibited strong antifungal activity against *Penicillium italicum, Alternaria alternata, Fusarium oxysporum,* and *Colletotrichum* spp., which caused rots in citrus fruits. The antifungal effect was mainly due to organic acids (lactic, acetic) and phenolic compounds (DL-3-phenyllactic acid, salicylic acid). Iparraguirre et al. (2024) reported that lactic acid bacteria, including *Lactobacillus brevis* and *Lactococcus lactis*, significantly reduced the damage caused by the phytopathogens *Uncinula necator* and *Phytophthora viticola* in grapevines cultivated under field conditions in the Ica Valley. Their findings showed that a 10-litre application of the microbial consortium was comparable to treatment with *Lecanicillium lecanii* for the curative control of *E. necator.* Furthermore, in developing plants of the Autumn Crips variety, a 1 litre dose of the microbial consortium proved more effective than the chemical treatment in managing *Plasmopara viticola.* Gwiazdowski et al. (2024) had conducted a mini-plot experiment and found that the application of LAB (*Lactiplantibacillus plantarum, Lacticaseibacillus paracasei,* and *Lentilactobacillus buchneri*) led to the reduced stem smut of rye and wheat common bunt incidence in the range of 34.5–94.7 per cent and 24.8–99.6 per cent, respectively.

6.10 Merits and Demerits of Probiotics

Merits

- LAB has achieved the status of generally regarded as safe (GRAS), with few exceptions, indicating its safe use in the cultivation of edible crops without raising health concerns for humans.
- Plant probiotics can induce systemic resistance in their host plants, exhibit strong niche colonization capacity, and exert antagonistic activity against a wide range of pathogens.
- Plant probiotic formulations can be employed as phytostimulators, biopesticides, plant strengtheners, and bio-fertilizers.

Demerits

- There is a clear lack of evidence correlating the antagonistic effects of probiotics observed under in vitro conditions with their actual efficacy in pathogen management under field conditions.
- It is important to exercise caution when forming mixed consortia of probiotics with other plant growth-promoting microbes (PGPM) to prevent any potential incompatibility issues.
- It is important to confirm the safety of the selected probiotic strains before they are used in industry in order to reduce possible impacts on plant, animal, and human health.

6.11 Summary

Plant probiotics are advantageous microbes that enhance plant defense mechanisms, effectively colonize ecological niches, and suppress a diverse array of plant pathogens through their antagonistic properties. Plant probiotics have therefore gained popularity since they offer an unmatched means of achieving sustainable agriculture without endangering biodiversity and the natural ecology. These probiotics are equipped with organic acids, enzymes, hydrogen peroxide, diacetyl, bacteriocins, siderophore, hydroxy derivatives of fatty acids, phenyl lactic acid, reuterin, cyclic dipeptides, and amines, which contribute to the antifungal activity against mildews, smuts, and rot pathogens. By competitively colonizing ecological niches and utilizing available nutrients, these strains limit pathogen establishment and proliferation, while simultaneously stimulating induced systemic resistance in plants.

6.12 Future Directions and Challenges

- To fully harness the potential of plant probiotics, research should prioritize the identification and understanding of existing plant-microbe signaling mechanisms that facilitate effective colonization by selected microbial strains.
- Investigating the effects of candidate plant probiotics on native microbial communities and overall ecosystem functions is essential to ensure their large-scale application with minimal non-target impacts.
- Developing a consistent bio-pesticide employing plant probiotics requires a comprehensive approach that considers delivery modalities, formulation modifications, agricultural methods, and plant-microbe symbioses.
- Although many promising probiotics were investigated, there are incredibly few commercially accessible probiotics. Thus, the development of probiotic products is crucial to their extensive application in the management of plant diseases.
- Examine the potential synergistic effects of mixing probiotics with fungicides or integrating multiple probiotics. We can expect to achieve the highest potential management efficacy of plant diseases by optimizing the ratio of each component.
- Investigate the molecular mechanism by looking at the transcriptome of probiotics and examining the genes that are expressed differently in them during the management of plant pathogens.

Conflict of Interest The authors confirm that they have no conflict of interest.

Authors Contributions Conceptualization and editing: A. K. and S. A. S.; Writing and review: R. M. and A. K; Supervision: A. K. and M. S.

References

Abubakr MAS, Al-Adiwish WM (2017) Isolation and identification of lactic acid bacteria from different fruits with proteolytic activity. Int J Microbiol Biotechnol 2:58–64

Adithi G, Rakesh S, Shruthi B, Divyashree S, Deepa N, Vanitha PR, Vasundaradevi R, Deepthi BV, Sreenivasa MY (2025) Biological decontamination of myco-toxigenic *Fusarium verticillioides* using *Lactiplantibacillus plantarum* MYSN128 isolated from traditional vellappam: implications for food safety and global mycotoxin management. Curr Microbiol 82:404. https://doi.org/10.1007/s00284-025-04401-z

Akter B, Salleh RM, Chun NWC, Bakar AMH, Furusawa G (2021) Identification and growth study of potential probiotic isolated from pineapple, watermelon and banana peels. J Fd Process Preserv 45(12):1–10

Amenu D (2013) Antimicrobial activity of lactic acid bacteria isolated from "Ergo", Ethiopian traditional fermented milk. Curr Res Microbiol Biotechnol 1:278–284

Barrios-Roblero C, Rosas-Quijano R, Salvador-Figueroa M, Galvez-lopez D, Vazquez-Ovando A (2019) Antifungal lactic acid bacteria isolated from fermented beverages with activity against *Colletotrichum gloeosporioides*. Fd Biosci 29:47–54

Berg G (2009) Plant-microbe interactions promoting plant growth and health: perspectives for controlled use of microorganisms in agriculture. Appl Microbiol Biotechnol 84(1):11–18

Crisp P, Wicks TJ, Troup G, Scott ES (2006) Mode of action of milk and whey in the control of grapevine powdery mildew. Australasian Plant Pathol 35(5):487–493

DeBacco MJ (2011) Compost tea and milk to suppress powdery mildew (*Podosphaera xanthii*) on pumpkins and evaluation of horticultural pots made from recyclable fibers under field conditions. Master's Thesis, University of Connecticut, Storrs. pp 1–64

Divyashree S, Ramu R, Sreenivasa MY (2024) Evaluation of new candidate probiotic *Lactobacillus* strains isolated from a traditional fermented food multigrain-millet dosa batter. Food Biosci:103450. https://doi.org/10.1016/j.fbio.2023.103450

Ferrandino F, Smith VL (2007) The effect of milk-based foliar sprays on yield components of field pumpkins with powdery mildew. Crop Prot 26(4):657–663

Gajbhiye MH (2016) Probiotic traits of lactic acid bacteria isolated from aerial surfaces of pomegranate. Int J Bioassays 5:4733–4738

Gerez CL, Torino MI, Rollan G, de Valdez GF (2009) Prevention of bread mould spoilage by using lactic acid bacteria with antifungal properties. Fd Control 20(2):144–148

Giassi V, Kiritani C, Kupper KC (2016) Bacteria as growth-promoting agents for citrus rootstocks. Microbiol Res 190:46–54

Gwiazdowski R, Kubiak K, Jus K, Marchwinska K, Gwiazdowska D (2024) The biocontrol of plant pathogenic fungi by selected lactic acid bacteria: from laboratory to field study. Agriculture 14(1):61–79

Hamad GM, Amer A, el-Nogoumy B, Ibrahim M, Hassan S, Siddiqui SA, El-Gazzar A, Khalifa E, Omar SA, Abou-Alella SA, Ibrahim SA, Esatbeyoglu T, Mehany T (2022) Evaluation of the effectiveness of charcoal, *Lactobacillus rhamnosus* and Saccharomyces cerevisiae as aflatoxin adsorbents in chocolate. Toxins 15(1): 21–38

Herrmann L, Lesueur D (2013) Challenges of formulation and quality of biofertilizers for successful inoculation. Appl Microbiol Biotechnol 97(20):8859–8873

Higdon SM, Huang BC, Bennett AB, Weimer BC (2020) Identification of nitrogen fixation genes in *Lactococcus* isolated from maize using population genomics and machine learning. Microorganisms 8:1–25

Iparraguirre HC, Huaman LEB, Baldino JLS, Huaman-Flores EJ (2024) Microbial consortium of yeasts and lactic acid bacteria as biological control agents for powdery mildew and downy mildew under field conditions. Trop Subtrop Agroecosyst 27:1–18

Jaffar NS, Jawan R, Chong KP (2023) The potential of lactic acid bacteria in mediating the control of plant diseases and plant growth stimulation in crop production - a mini review. Front Plant Sci 13:1047945

Jahatigh S, Oghaz NA, Rahnama K, Hatamzadeh S (2023) Application of *Lactobacillus* spp. for the biological management of green mold (*Penicillium digitatum*) on sweet orange fruit under in vitro and storehouse conditions. Biocontrol Sci Technol 33(6):567–581

Konappa NM, Maria M, Uzma F, Krishnamurthy S, Nayaka SC, Niranjana SR, Chowdappa S (2016) Lactic acid bacteria mediated induction of defense enzymes to enhance the resistance in tomato against *Ralstonia solanacearum* causing bacterial wilt. Sci Hort 207:183–192

Kumar S, Diksha, Sindhu SS, Kumar R (2022) Biofertilizers: an ecofriendly technology for nutrient recycling and environmental sustainability. Curr Res Microb Sci 3:100094

Lamont JR, Wilkins O, Bywater-Ekegard M, Smith DL (2017) From yogurt to yield: potential applications of lactic acid bacteria in plant production. Soil Biol Biochem 111:1–9

le Lay C, Coton E, le Blay G, Chobert JM, Haertlee T, Choiset Y, Van Long NN, MesletCladiere L, Mounier J (2016) Identification and quantification of antifungal compounds produced by lactic acid bacteria and propionibacteria. Int J Fd Microbiol 239:79–85

Meena DK, Singh J, Bhujangrao WM, Sharma PK (2024) Eco-friendly management of powdery mildew of pea (*Pisum sativum* L.). J Adv Biol Biotechnol 27(10):703–707

Rattanachaikunsopon P, Phumkhachorn P (2010) Lactic acid bacteria: their antimicrobial compounds and their uses in food production. Ann Biol Res 1:218–228

Riolo M, Villena AM, Calpe J, Luz C, Meca G, Tuccitto N, Cacciola SO (2024) A circular economy approach: a new formulation based on a lemon peel medium activated with lactobacilli for sustainable control of post-harvest fungal rots in fresh citrus fruit. Biol Control 189:105443

Shruthi B, Deepthi BV, Adithi G, Vanitha PR, Sreenivasa MY (2025) *Saccharomyces cerevisiae* MYSY1 as a promising probiotic isolate with antifungal properties against foodborne *Aspergillus* species. Biocatal Agric Biotechnol 67:10374U. https://doi.org/10.1016/j.bcab.2025.103649

Silpa S, Rupachandra S (2022) Cyclic peptide production from lactic acid bacteria (LAB) and their diverse applications. Crit Rev Fd Sci Nutr 62:2909–2927

Singh RS (2002) Introduction to principles of plant pathology 4th ed. Oxford and IBH Publishing Co Pvt Ltd pp. 41–45

Sjogren J, Magnusson J, Broberg A, Schnurer J, Kennc L (2003) Antifungal 3- hydroxy fatty acids from *Lactobacillus plantarum* MiLAB 14. Appl Environ Microbiol 69:7554

Song D, Ibrahim S, Haydek S (2012) Recent application of probiotics in food and agricultural science. In: Probiotics. In Tech, pp 1–36

Steglinska A, Kołtuniak A, Motyl I, Berłowska J, Czyżowska A, Cieciura-Włoch W, Okrasa M, Kregiel D, Gutarowska B (2022) Lactic acid bacteria as biocontrol agents against potato (*Solanum tuberosum* L.) pathogens. Appl Sci 12(15):7763

Strom K, Sjogren J, Broberg A, Schnurer J (2002) *Lactobacillus plantarum* MiLAB 393 produces the antifungal cyclic dipeptides cyclo (L-Phe-L-Pro) and cyclo (L-Phe-trans-4-OH-L-Pro) and 3-phenyllactic acid. Appl Environ Microbiol 68:4322–4327

Sun MC, Hu ZY, Li DD, Chen YX, Xi JH, Zhao CH (2022) Application of the reuterin system as food preservative or health-promoting agent: a critical review. Foods 11(24):4000

Vanitha PR, Divyashree S, Shruthi B, Adithi G, Deepthi BV, Sreenivasa MY (2024) *Limosilactobacillus fermentum* MYSY8, a potential probiotic isolate from fermented rice beverage for the control of *Microsporum canis*. Food Biotechnol 38(3):314–342. https://doi.org/10.1080/08U0543G.2024.2383743

Vimont A, Fernandez B, Ahmed G, Fortina H, Flissa I (2019) Quantitative antifungal activity of reuterin against food isolates of yeasts and moulds and its potential application in yogurt. Int J Fd Microbiol 289:182–188

Plant Defence Mechanisms Triggered by Probiotic Bacteria

7

Pruthviraj, M. K. Naik, T. N. Vijay Kumar, Manjunath Madhukar Mopagar, and M. L. Sachin

Abstract

Lactic acid bacteria (LAB) are gaining recognition as effective plant growth promoters and bio-control agents due to their multifaceted roles in enhancing plant health and productivity. These bacteria can produce important plant growth regulators such as indole-3-acetic acid (IAA) and gibberellic acid (GA3), which stimulate plant growth by modulating cellular processes. Additionally, LAB are involved in the solubilization of essential nutrients like phosphorus, potassium, zinc and silicon from soil rhizosphere, thereby enhancing their availability to plants and promoting better root development and overall plant vigor. LAB also plays a significant role in protecting plants from pathogenic fungi and bacteria by producing bioactive compounds such as proteases and siderophores, which inhibit the growth of harmful microorganisms. The application of LAB as seed treatment agents is a promising strategy for improving plant growth and protecting against soil-borne diseases. LAB's ability to produce IAA, GA3 and

Pruthviraj (✉)
Department of Plant Pathology, Keladi Shivappa Nayaka University of Agricultural and Horticultural Sciences, Shivamogga, Karnataka, India

M. K. Naik
Department of Plant Pathology, University of Agricultural Sciences, Raichur, Karnataka, India

T. N. Vijay Kumar
Research and Development Executive, Entomology, Reckitt Pvt Ltd, Gurugram, Haryana, India

M. M. Mopagar
AICRP on Sunflower, Zonal Agricultural Research Station, University of Agricultural Sciences, Bengaluru, Karnataka, India

M. L. Sachin
Department of Agronomy, College of Agriculture, Mandya, University of Agricultural Sciences, Bengaluru, Karnataka, India

M. Y. Sreenivasa et al. (eds.), *Probiotics as Antifungals*,
https://doi.org/10.1007/978-981-92-0002-3_7

efficacy in solubilizing critical nutrients can significantly improve seed germination, root colonization and plant development. This chapter delves into the mechanisms by which LAB promote plant growth and combat diseases, highlighting their ability to produce IAA and GA3, solubilize nutrients and inhibit pathogens through protease and siderophore production. These traits position LAB as a sustainable alternative to chemical fertilizers and pesticides.

Graphical Abstract

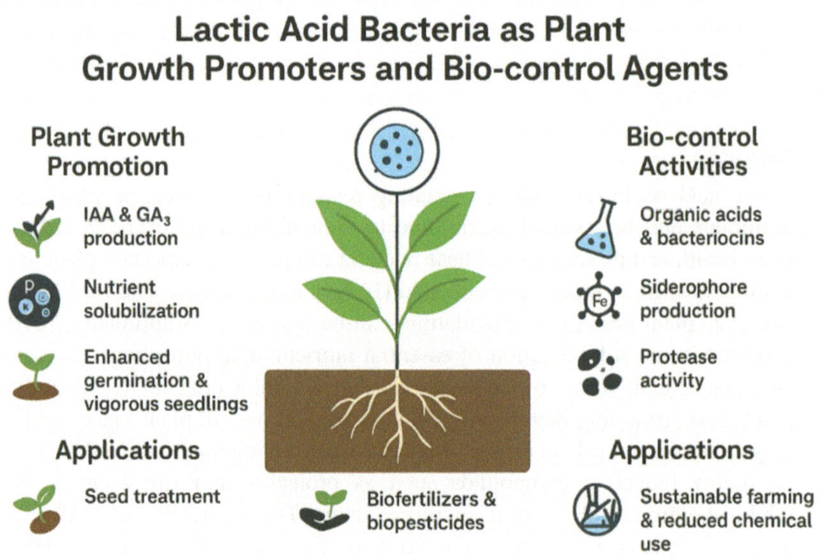

Keywords

Lactic Acid Bacteria (LAB) · Plant Growth Promotion · Nutrient Solubilization · Biocontrol Agents · IAA and GA3 Production

Learning Objectives

- To understand the diversity, classification and agricultural relevance of lactic acid bacteria (LAB).
- To explain the key mechanisms by which LAB enhance plant growth.
- To evaluate the biocontrol potential of LAB against plant pathogens.
- To assess the applications, limitations, and future prospects of LAB in sustainable agriculture.

7.1 Introduction to LAB and Their Role in Agriculture

LAB are a group of beneficial microorganisms characterized as Gram-positive, non-spore-forming and acid-tolerant. These bacteria are predominantly anaerobic and are known for their ability to ferment carbohydrates, producing lactic acid as the main metabolic end product. Common genera within this group include *Lactobacillus*, *Lactococcus*, *Leuconostoc*, *Pediococcus* and *Enterococcus* (Divyashree et al. 2024). LAB occur widely in nature and can be isolated from environments such as fermented food products, decaying plant residues and the digestive systems of animals (Zheng et al. 2023).

One of the most prominent roles of LAB in agriculture is in the production of silage. Silage is preserved forage used as animal feed, especially during times when fresh pasture is not available. LAB ferment the sugars present in chopped forage, producing lactic acid, which quickly lowers the pH. This acidic environment prevents the growth of spoilage organisms and preserves the nutritional content of the feed. The use of LAB as silage inoculants has become a common practice to improve the efficiency and quality of silage fermentation (Bai et al. 2024).

Beyond their role in feed preservation, LAB also enhances soil quality and supports plant health. Certain LAB strains function as biocontrol agents by producing antimicrobial compounds, including bacteriocins, hydrogen peroxide and organic acids that inhibit soil-borne pathogens (Deepa et al. 2022; Deepthi et al. 2022; Adithi et al. 2024, 2025; Vasundharadevi et al. 2024, 2025). They further contribute to nutrient cycling and promote plant growth by improving nutrient absorption and activating plant defense mechanisms. These combined benefits position LAB as an important tool for reducing reliance on chemical pesticides and advancing sustainable agricultural practices (Lamont et al. 2023).

In conclusion, LAB is beneficial in agricultural waste management. When used in composting or manure treatment, LAB help accelerate decomposition and suppress the activity of odor-causing and pathogenic microbes (Nishio et al. 2018). This results in improved quality compost and thereby reduces the environmental impact. Additionally, LAB have been studied for their potential in producing lactic acid for biodegradable plastics and other bio-based materials, highlighting their value in agricultural biotechnology and environmental sustainability (Wee et al. 2006; Abdel-Rahman et al. 2013).

7.1.1 Definition and Classification of LAB

LAB are a group of Gram-positive, non-spore-forming, typically non-motile bacteria that are known for their ability to ferment carbohydrates primarily into lactic acid. This metabolic trait not only defines the group but also gives it its name. LAB are generally facultative anaerobes or aerotolerant, meaning they can survive in the presence or absence of oxygen, although they typically grow best in anaerobic conditions. They have low GC (guanine–cytosine) content in their DNA and are acid-tolerant, enabling them to thrive in environments where other bacteria may be inhibited by low pH (Wang and Zhang 2023).

Taxonomically, LAB are divided into several genera within the phylum *Firmicutes*, class *Bacilli* and order *Lactobacillales*. The major genera include:

- *Lactobacillus*—rod-shaped bacteria, widely used in food fermentation and probiotics.
- *Lactococcus*—primarily cocci, important in dairy fermentations such as cheese and buttermilk.
- *Leuconostoc*—heterofermentative cocci involved in vegetable fermentations and flavor development.
- *Pediococcus*—cocci, often used in the fermentation of meats and vegetables.
- *Enterococcus*—cocci that inhabit the gastrointestinal tract and are sometimes used in probiotics, though some species are opportunistic pathogens.

Advances in molecular biology, particularly 16S rRNA gene sequencing, have significantly refined the taxonomy of LAB, leading to the reclassification of many species and the identification of new genera. For example, the genus *Lactobacillus* has recently been restructured into over 20 distinct genera to better reflect phylogenetic relationships.

7.1.2 LAB in the Context of Plant Growth Promotion and Bio-control

LAB have emerged as important microbial agents in promoting plant growth through several direct and indirect mechanisms. Some LAB strains can enhance nutrient availability in the soil by solubilizing phosphate, producing siderophores that bind and transport iron and synthesizing plant growth-promoting substances like indole-3-acetic acid (IAA), a common plant hormone (Li et al. 2024). These functions help improve root development, nutrient uptake and overall plant vigor. LAB also contributes to improved soil structure and microbial balance, fostering a healthier rhizosphere environment that supports sustainable crop production without heavy reliance on chemical fertilizers (Fig. 7.1).

7.2 Plant Growth Promoting Mechanisms of LAB

7.2.1 Production of Plant Growth Regulators (IAA and GA3)

LABs, traditionally recognized for their applications in food fermentation and animal health, have recently attracted attention for their plant growth-promoting potential. One of the key mechanisms involves the biosynthesis of plant growth regulators (PGRs), particularly indole-3-acetic acid (IAA) and gibberellic acid (GA3). These phytohormones regulate vital plant physiological processes such as

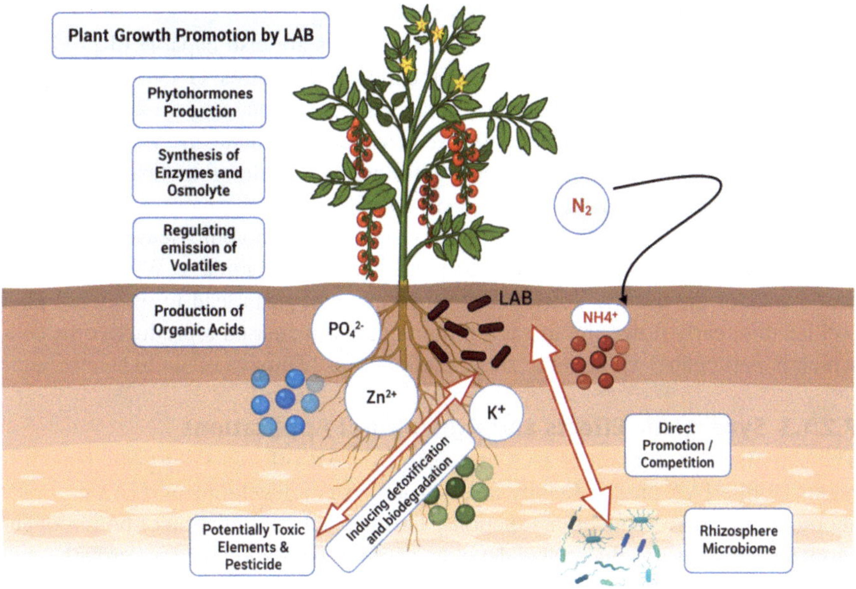

Fig. 7.1 Schematic representation of the plant growth promotion, bioprotection and biodegradation potential of lactic acid bacteria

root and shoot development, cell elongation, seed germination and flowering. In addition, LAB can enhance nutrient solubilization, improve plant stress tolerance, and stimulate systemic resistance against pathogens, thereby functioning as effective biofertilizers and bioprotectants (Raman et al. 2023; Sharma et al. 2020). The multifunctional nature of LAB highlights their potential role in sustainable agriculture as eco-friendly alternatives to synthetic agrochemicals.

7.2.1.1 Indole-3-Acetic Acid (IAA) Production

IAA is the most common and well-studied auxin, a group of plant hormones essential for plant growth and development. LAB strains, particularly those in the genera *Lactobacillus*, *Pediococcus*, and *Leuconostoc*, are known to produce IAA when cultivated in media containing L-tryptophan, a precursor amino acid commonly found in the rhizosphere due to plant root exudates. These bacteria convert tryptophan into IAA through several pathways, such as the indole-3-pyruvate pathway, via intermediates like indole-3-acetamide or indole-3-acetaldehyde. The indole-3-acetic acid produced by LAB in the rhizosphere contributes to improved plant development by encouraging longer primary roots, greater lateral root branching, and enhanced root hair proliferation. An enlarged root system enhances the plant's ability to absorb water and essential nutrients, especially nitrogen and phosphorus. This improvement supports greater plant vigor, stress tolerance and higher yield potential, even in less-than-ideal soil conditions (Nishida et al. 2025).

7.2.1.2 Gibberellic Acid (GA3) Production

In addition to auxins like IAA, certain LAB strains are also capable of producing gibberellic acids, with GA3 being one of the most active forms. GA3 is a diterpenoid compound that regulates seed germination, stem elongation, enzyme activation, and flowering. LAB-mediated production of GA3 can stimulate the mobilization of stored food reserves in seeds during germination, resulting in faster and more uniform seedling emergence (Lahmamsi et al. 2024). Furthermore, GA3 enhances cell division and elongation in plant tissues, particularly in young leaves to budding leaves, which contributes to increased plant height and biomass. When LAB that produce GA3 are used as soil or seed inoculants, they can lead to improved plant architecture, early flowering to early blooming, and increased reproductive success, which is particularly valuable in short-season crops (Turaeva et al. 2021).

7.2.1.3 Synergistic Effects and Agricultural Applications

The combined production of IAA and GA3 by LAB offers a synergistic benefit to plants. While IAA predominantly affects root system development, GA3 promotes shoot growth and reproductive functions. This balanced stimulation of both root and shoot systems ensures that plants develop a strong foundation for root growth while also maximizing productivity above ground (Ghatage et al. 2024).

7.2.2 Solubilization of Nutrients

Even though LABs are traditionally associated with food fermentation and probiotics, they are significantly recognized as plant growth-promoting microorganisms (PGPMs) due to their ability to enhance the bioavailability of essential nutrients in the soil. Many of the nutrients required by plants exist in an insoluble or bound state that is unavailable for uptake. LAB facilitates the conversion of nutrients into soluble forms accessible for plant uptake, using mechanisms such as organic acid production, pH reduction, chelation and enzymatic activity.

7.2.2.1 Phosphorous Solubilization

Phosphorus (P) is one of the most critical macronutrients required by plants, involved in a wide array of physiological and biochemical functions. It is an integral component of nucleic acids, DNA and RNA, phospholipids in cell membranes and most importantly, adenosine triphosphate, which is the universal energy currency in all living cells. Despite its importance, phosphorus is often a limiting nutrient in many agricultural soils due to its poor solubility and strong tendency to form insoluble complexes. In most soils, phosphorus exists in fixed forms such as tricalcium phosphate ($Ca_3(PO_4)_2$) in calcareous or alkaline soils and iron phosphate ($FePO_4$) or aluminum phosphate ($AlPO_4$) in acidic soils. These mineral forms are highly stable and poorly soluble in water, meaning that phosphorus remains largely inaccessible to plant roots even when the total phosphorus content of the soil is high. As a result, plants suffer from phosphorus deficiency, which manifests as stunted growth, poor root development and reduced yield (Sharma et al. 2023).

To address this problem sustainably, scientists and agronomists have been investigating the role of beneficial soil microorganisms in phosphorus solubilization. LABs contribute to phosphorus solubilization primarily through the secretion of low molecular weight organic acids, including lactic acid, acetic acid, and citric acid, which are byproducts of their carbohydrate metabolism. These organic acids play a dual role in the solubilization process. Primarily, they lower the pH of the soil microenvironment (especially in the rhizosphere, where root and microbial interactions are concentrated). Acidification increases the solubility of phosphate minerals by destabilizing the phosphate-metal bonds, particularly those involving calcium, iron and aluminum. Furthermore, these organic acids act as chelating agents, binding to the metal cations (Ca^{2+}, Fe^{3+}, Al^{3+}) that are tightly associated with phosphate in the soil (Fig 7.2a).

7.2.2.2 Potassium and Zinc Solubilization

Potassium (K) is one of the three primary macronutrients essential for plant growth and productivity, alongside nitrogen and phosphorus. It plays a critical role in various physiological and biochemical processes in plant growth. Despite potassium's abundance in soil, a large portion of it exists in insoluble forms, primarily bound within silicate minerals such as feldspar, mica and illite. These forms are not readily available for plant uptake, leading to potassium deficiency in many agricultural soils.

To overcome this challenge in an environmentally sustainable manner, recent studies have explored the role of beneficial microorganisms, including lactic acid bacteria (LAB), in potassium solubilization. LAB have shown potential as potassium-solubilizing microorganisms (KSMs). These bacteria contribute to potassium mobilization primarily through the production of organic acids (such as lactic acid, acetic acid and citric acid) and the release of protons (H^+). These substances acidify the surrounding microenvironment in the soil, particularly in the rhizosphere, thereby facilitating the weathering of potassium-bearing silicate minerals. The acidification process helps to break the silicate lattice structure of minerals like feldspar and mica, releasing K^+ ions into the soil rhizosphere where they become bioavailable for plant absorption (Fig. 7.2b) (Basak and Biswas 2023).

Zinc (Zn) is a vital micronutrient required for enzyme activity, protein synthesis, and hormone production in plants. Unfortunately, zinc is often present in soils in insoluble forms such as zinc oxide (ZnO), zinc carbonate ($ZnCO_3$) and zinc sulfide (ZnS), especially in alkaline or calcareous soils (Singh et al. 2018).

Bacteria employ various biochemical mechanisms to solubilize zinc (Zn), transforming insoluble mineral forms into bioavailable forms. These mechanisms include the production of organic acids, chelation and complexation of zinc ions and the synthesis of siderophores. Each of these processes is influenced by environmental factors such as pH, soil composition, and microbial activity. Understanding these metabolic pathways is essential for optimizing the role of zinc-solubilizing bacteria (ZSB) in sustainable agriculture. Zinc solubilization is often achieved through the secretion of organic acids by bacteria. These acids lower the pH of the surrounding environment, facilitating the conversion of insoluble zinc compounds like zinc oxide

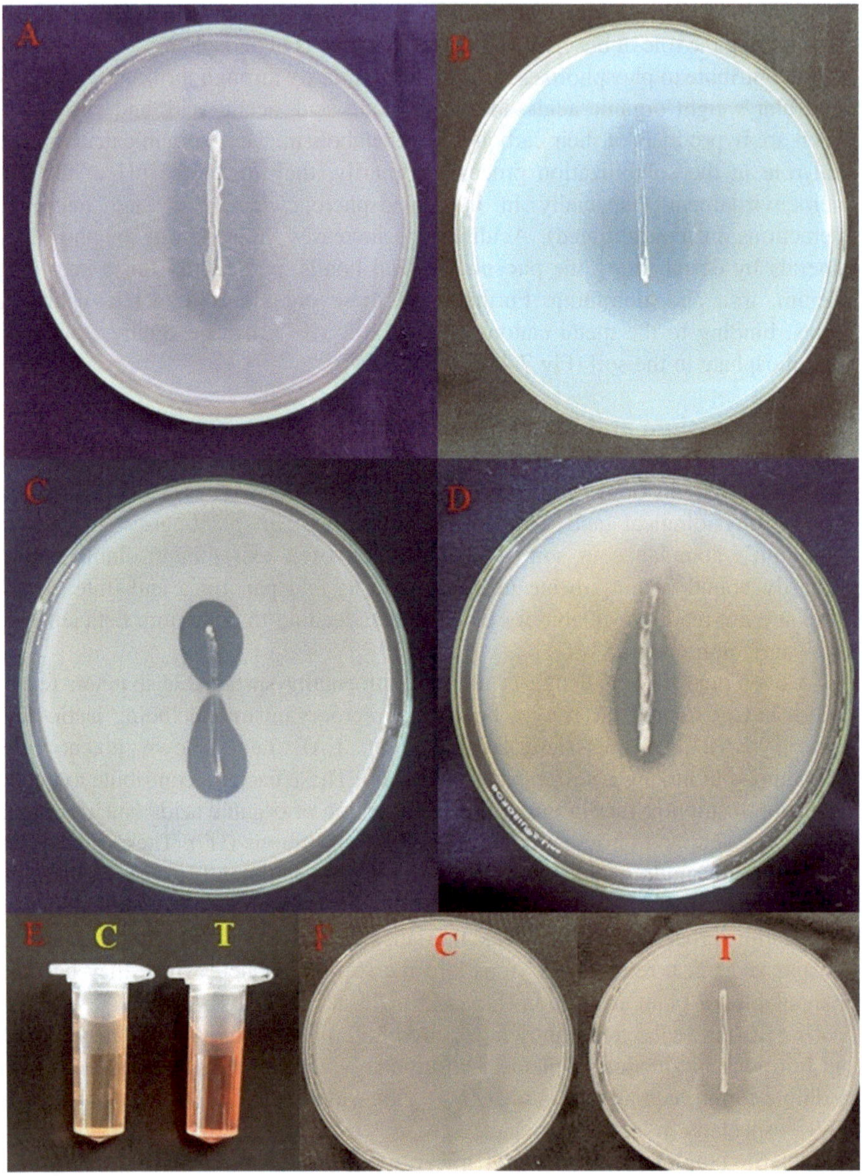

Fig. 7.2 Plant growth-promoting mechanisms of lactic acid bacteria involving nutrient solubilization and enzyme activities (**a**) Solubilization of inorganic phosphorus by lactic acid bacteria. (**b**) Lactic acid bacteria-mediated solubilization of potassium mineral. (**c**) Conversion of insoluble zinc compounds into soluble forms by LAB. (**d**) Solubilization of silicon substrates facilitated by LAB. (**e**) Production of siderophores as an iron-acquisition mechanism in LAB (**f**) Protease secretion and extracellular protein-degrading ability of LAB

and zinc phosphate into soluble forms such as Zn^{2+}, which plants can readily absorb from soil (Fig 7.2c) (Singh et al. 2018).

7.2.2.3 Silicon Solubilization

Silicon (Si), while not universally essential for plant growth, plays a key role in enhancing stress tolerance. It strengthens cell walls, improving plant structural integrity and reducing risks from mechanical stress, lodging and pathogen invasion. Nonetheless, its limited availability in many soils restricts its potential benefits in crop production (Ma and Yamaji 2015).

Recent studies have highlighted the role of soil microorganisms, particularly lactic acid bacteria (LAB), in enhancing the bioavailability of silicon through the process of silicon solubilization. LABs contribute to this process by secreting organic acids such as lactic, acetic, and citric acids that acidify the rhizosphere and facilitate the dissolution of silicate minerals (Etesami and Jeong 2018). This microbial-mediated dissolution releases soluble monosilicic acid into the soil solution, making it accessible for plant uptake (Fig. 7.2d).

7.3 Seed Treatment with LAB: Mechanisms and Benefits

7.3.1 Enhanced Germination and Root Development

LABs have garnered significant attention in recent years as effective biocontrol agents and plant growth promoters when applied as seed treatments. When seeds are treated with LAB before planting, these beneficial microorganisms can enhance seedling growth, improve soil health, and increase plant resilience to stress. Among the key benefits, enhanced seed germination and improved root development are among the most notable. Experimental evidence supports the positive effect of LAB on seed germination across various crops. For instance, certain strains of *Lactobacillus plantarum* and *Lactobacillus casei* have been reported to significantly increase germination rates and early seedling growth in cereals and vegetables by producing IAA and GA_3 (Fig. 7.3) (Naveed et al. 2014; Awasthi et al. 2020). Such findings underscore the potential of LAB-based bioinoculants in sustainable agriculture, particularly for improving crop establishment in challenging environments.

7.3.2 Interaction with Soil Microbiota and Plant Roots

Indole-3-acetic acid (IAA), the principal auxin produced by plants and certain beneficial microbes such as lactic acid bacteria (LAB), plays a critical role in regulating root development, including root elongation, lateral root initiation, and root hair formation. Several LAB strains, such as *Lactobacillus plantarum* and *Lactobacillus casei*, have been reported to synthesize IAA, particularly when supplemented with tryptophan in culture media (Awasthi et al. 2020). When seeds are inoculated with IAA-producing LAB, the exogenous supply of auxin in the

Fig. 7.3 Effect of seed treatment with LAB isolates on growth of tomato plants. C Control, MYSNRB20 *Limosilactobacillus fermentum*, MYSN7 *L. fermentum*, MYSN3 *Limosilactobacillus* sp., MYSJK8 *Lactiplantibacillus argentoratensis* and MYSPE4 *Lactiplantibacillus plantarum*

rhizosphere stimulates cell division and elongation in root meristems, leading to enhanced primary root growth. Simultaneously, increased auxin levels promote the formation of lateral roots and root hairs, expanding the total root surface area (Spaepen and Vanderleyden 2011). This improved root architecture enhances the seedling's capacity to absorb water and essential nutrients, including nitrogen, phosphorus, and potassium, thereby supporting overall plant vigor.

7.4 Bio-control Attributes of LAB

Lactic acid bacteria (LAB) possess notable biocontrol abilities, making them valuable in sustainable farming. They generate a range of antimicrobial compounds such as lactic and acetic acids, hydrogen peroxide, fatty acids, reuterin and bacteriocins that inhibit the growth of phytopathogens and spoilage organisms (Table 7.1).

Table 7.1 Lactic acid bacteria as a source of bio-stimulant and bio-control agent

Source	Isolate	Pathogen/crop	Mechanism	Reference
Lactic acid bacteria as a source of bio-stimulant				
Rhizosphere soil of wheat	*E. faecium* LB5, *Lactococcus lactis* LB6	*Fusarium graminearum*	Phosphate solubilization and biocontrol property	Strafella et al. (2021)
Blue Cherry fruit	*Enterococcus faecium* MYSBC14	*Ralstonia solanacearum and Xanthomonas oryzae* pv *oryzae*	Phosphate, potassium and zinc solubilization, IAA and GA3 production	Pruthviraj et al. (2023a)
Noni fruit	*Limosilactobacillus* sp. MYSN3	Tomato, Chilli, Okra	Phosphate, potassium and zinc solubilization, IAA and GA3 production	Pruthviraj et al. (2023b)
Jack fruit	*Lactiplantibacillus argentoratensis* MYSJK8	*Ralstonia solanacearum, Xanthomonas oryzae* pv *oryzae*	Peroxidase (POX), polyphenol oxidases (PPO), phenylalanine ammonia-lyase (PAL), β-1,3-glucanase and total phenols	Pruthviraj et al. (2024)
Lactic acid bacteria as a source of bio-control agent				
Herbal juice	*Lactiplantibacillus plantarum* MYSAGT3	*Aspergillus* sp.	Organic acids	Adithi et al. (2022)
Herbal Amla juice	*Limosilactobacillus fermentum* MYSAGAM1	*Fusarium equiseti*	Organic acids	Adithi et al. (2024)
Fermented green gram dosa batter	*Pichia kudriavzevii*	*Macrophomina phaseolina, Aspergillus niger and Fusarium oxysporum*	Organic acids	Shruthi et al. (2025a)
Ficus bengalensis	*Lactiplantibacillus plantarum* MYSVB1	*Alternaria alternata*	Organic acids	Vasundaradevi et al. (2024)
Ficus racemosa	*Lactiplantibacillus plantarum* MYSVCF3 and *Lactiplantibacillus* argentoratensis MYSVCF5	*Fusarium oxysporum*	Organic acids	Vasundaradevi et al. (2025)

7.4.1 Antifungal and Antibacterial Activity

LAB are a group of Gram-positive, non-sporulating microbes known for their ability to produce a wide range of bioactive compounds with antimicrobial properties. These include organic acids (such as lactic and acetic acid), hydrogen peroxide, bacteriocins and antifungal peptides. LABs, particularly *Lactiplantibacillus plantarum*, have demonstrated strong antifungal activity against a variety of plant pathogenic fungi, offering a natural and sustainable alternative to chemical fungicides. These strains have been shown to effectively suppress the growth of fungal genera such as *Fusarium*, *Aspergillus niger*, *Alternaria* and *Colletotrichum* spp. (Magnusson et al. 2003; Ström et al. 2002; Adithi et al. 2024, 2025; Shruthi et al. 2025a, b; Vasundharadevi et al. 2024, 2025).

The primary mechanism of antifungal activity involves the production of organic acids (e.g., lactic and acetic acids), which acidify the surrounding environment, leading to disruption of fungal cell membranes and inhibition of spore germination (Gänzle 2015). A lower pH creates hostile conditions for fungal growth and may enhance the permeability of fungal cell walls, facilitating the entry of antimicrobial agents. The antifungal efficacy of LAB has been validated in multiple studies. For instance, *L. plantarum* has shown inhibitory activity against *Colletotrichum gloeosporioides*, the causative agent of anthracnose in fruits, as well as against mycotoxin-producing fungi such as *Penicillium nordicum*, *P. commune* and *P. verrucosum* (Cheong et al. 2014; Brosnan et al. 2014). These findings underscore the biotechnological potential of LAB in managing plant diseases, particularly in postharvest treatments and organic farming systems. LABs possess potent antibacterial properties that make them effective biocontrol agents against a range of plant pathogenic bacteria. A major mechanism through which LAB exert their antagonistic effects is the production of organic acids (such as lactic and acetic acids), which acidify the surrounding environment and inhibit the growth of many bacterial pathogens by disrupting membrane integrity and enzyme activity (Gänzle 2015). Additionally, hydrogen peroxide, a metabolic byproduct of some LAB strains, contributes to bacterial inhibition by inducing oxidative stress in pathogens.

7.4.2 Protease Activity and Its Role in Disease Suppression

LAB are emerging as effective biocontrol agents in agriculture, particularly due to their protease activity, which plays a crucial role in suppressing plant diseases. These enzymes degrade complex proteins into simpler peptides and amino acids, disrupting the structural integrity of plant pathogens and enhancing plant health (Soomro et al. 2022).

LAB strains such as *Lactiplantibacillus plantarum* and *Pediococcus pentosaceus* have demonstrated extracellular proteolytic activity, particularly under acidic conditions. For instance, *P. pentosaceus* UP-2 exhibited high protease activity at pH 5, leading to substantial amino acid production, which is beneficial for both plant growth and pathogen suppression (Fig. 7.2f) (Yu et al. 2020).

7.4.3 Siderophore Production as a Mechanism of Bio-control

LABs contribute to plant nutrition by producing siderophores, which are low-molecular-weight chelating agents that bind to ferric iron (Fe^{3+}) in the soil, making it more soluble and bioavailable to plants (Ahmed and Holmström 2014). Iron is a vital micronutrient required for chlorophyll synthesis, electron transport, photosynthesis, and root development (Marschner 2012). Despite its abundance in soil, iron often exists in insoluble forms under aerobic or alkaline conditions, rendering it inaccessible to plants. LAB-produced siderophores compete with other soil microorganisms and chemical chelators for iron, forming stable Fe^{3+}-siderophore complexes that can be more readily absorbed by plant roots (Fig. 7.2e) (Saikia et al. 2011). This increased iron availability enhances seedling vigor, promotes healthy root elongation and supports optimal photosynthetic efficiency especially critical during the early stages of plant development.

7.5 Applications of LAB in Sustainable Agriculture

LABs are emerging as versatile contributors to sustainable agriculture, offering eco-friendly alternatives to conventional agrochemicals. These beneficial microbes enhance soil health by decomposing organic matter, producing organic acids, and facilitating nutrient cycling, thereby improving soil fertility and structure (Kumar et al. 2021). LAB also act as natural biocontrol agents, suppressing plant pathogens through the production of antimicrobial compounds such as bacteriocins, organic acids and hydrogen peroxide (Benitez-Cabello et al. 2020).

7.5.1 LAB as Biofertilizers

LAB are emerging as effective biofertilizers in sustainable agriculture due to their multifaceted roles in promoting plant growth, enhancing soil fertility and suppressing plant pathogens. These beneficial microbes offer an eco-friendly alternative to chemical fertilizers and pesticides, aligning with the principles of organic farming and integrated pest management (Kumar et al. 2021). LAB contribute to soil health by decomposing organic matter, producing organic acids and facilitating the nutrient cycle, thereby improving soil fertility and structure. They enhance plant growth by synthesizing phytohormones like indole-3-acetic acid (IAA), solubilizing essential minerals, and enhancing nutrient uptake (Benitez-Cabello et al. 2020). Additionally, their ability to detoxify heavy metals and degrade mycotoxins further supports plant health and productivity.

7.5.2 LAB as Biopesticides

LAB are increasingly recognized as effective biopesticides in sustainable agriculture due to their multifaceted roles in controlling plant pathogens and promoting plant health. These beneficial microorganisms produce a variety of antimicrobial compounds, including organic acids, hydrogen peroxide, bacteriocins, and reuterin, which inhibit the growth and spore germination of various phytopathogens such as *Fusarium*, *Aspergillus* and *Botrytis* species (Ogunbanwo et al. 2003). Additionally, LAB can outcompete pathogens for nutrients and space, further suppressing their proliferation (Mokoena 2017). Their ability to form biofilms enhances their persistence and efficacy in the rhizosphere, providing long-term protection against disease (Chowdhury et al. 2015).

7.5.3 Synergistic Effects with Other Microorganisms

The synergistic interactions between lactic acid bacteria (LAB) and other beneficial microorganisms, such as *Pseudomonas*, *Bacillus*, and *Trichoderma,* have garnered significant attention in sustainable agriculture for their enhanced biocontrol efficacy and plant growth-promoting capabilities. LAB, known for its production of antimicrobial compounds likc organic acids and hydrogen peroxide, can effectively suppress a wide range of plant pathogens. When combined with *Pseudomonas* species, which produce siderophores and antibiotics or with *Bacillus* species, known for their ability to produce lipopeptides and enzymes, the resulting microbial consortia exhibit a broader spectrum of antagonistic activities (Mendes et al. 2011; Radhakrishnan et al. 2017). Additionally, *Trichoderma* species contribute by producing chitinases and other enzymes that degrade fungal cell walls, further enhancing the biocontrol potential of these microbial combinations (Harman 2006).

7.6 Current Challenges and Emerging Directions for LAB in Agriculture

7.6.1 Limitations and Practical Considerations

Despite their recognized potential as biopesticides and plant growth-promoting agents, lactic acid bacteria (LAB) face several limitations and practical challenges that hinder their widespread application in agriculture. One of the primary issues is the strain-specific variability in efficacy. Not all LAB strains exhibit consistent antimicrobial or growth-promoting activity across different crops or environmental conditions. A strain that performs well under laboratory settings may fail to deliver similar results in the field due to variations in soil pH, temperature, moisture, and microbial competition (Giraud et al. 2010; Monteiro et al. 2020). Additionally, LAB are non-spore-forming and sensitive to environmental stresses such as ultraviolet light, desiccation and heat, which compromise their survivability and persistence in open-field applications when compared to spore-forming bacteria like *Bacillus* (Yuan et al. 2022).

Another major challenge lies in the formulation and delivery of LAB. Their sensitivity to processing and storage conditions often leads to a decline in viability over time, posing difficulties in developing stable and cost-effective commercial biopesticide products (Santos et al. 2019). Technologies such as microencapsulation or protective carriers are being explored to improve stability, but these add complexity and cost. Furthermore, even though many LAB strains are generally recognized as safe (GRAS) for food use, their approval for agricultural applications involves navigating complex regulatory frameworks, which differ by country and can be costly and time-consuming (FAO/WHO 2002; EFSA 2012).

7.6.2 Future Research and Development

The future success of lactic acid bacteria (LAB) as biopesticides and plant growth-promoting agents in sustainable agriculture depends heavily on targeted research and technological advancements. One critical area is the isolation and genomic characterization of novel LAB strains with broader antimicrobial activity and enhanced rhizosphere competence. Whole-genome sequencing and comparative genomics can help identify genes responsible for the production of antimicrobial compounds, stress tolerance, and plant-interactive traits, which are essential for strain selection and optimization.

In conclusion, the integration of LAB-based products into integrated pest management (IPM) programs, associated with farmer education and technology transfer initiatives, will be key to widespread adoption. Public-private partnerships, interdisciplinary collaboration and policy support will play pivotal roles in advancing LAB as a sustainable solution for crop protection and productivity enhancement.

7.7 Conclusion

LABs offer considerable potential for sustainable crop management due to their diverse roles in enhancing plant growth and health while minimizing the need for chemical inputs. These naturally occurring microbes produce antimicrobial substances, including organic acids, hydrogen peroxide and bacteriocins, which help control plant pathogens and lower disease incidence. When applied as biofertilizers and biopesticides, LAB improve soil microbial diversity and facilitate nutrient availability, particularly for nitrogen and phosphorus. Additionally, they strengthen plant tolerance to abiotic stresses such as drought and salinity by influencing hormone regulation and promoting root development. LAB also aids in breaking down organic matter and generating bioactive compounds, thereby improving compost quality and supporting eco-friendly waste recycling in agriculture. Integrating LAB into crop management systems supports eco-friendly farming practices, aligning with the goals of environmental conservation, soil health preservation and long-term agricultural sustainability.

7.8 Summary

Lactic acid bacteria (LAB) have emerged as promising plant growth-promoting and bio-control agents due to their diverse physiological and ecological functions in the rhizosphere. They enhance plant growth through the production of key phytohormones such as indole-3-acetic acid (IAA) and gibberellic acid (GA3), which stimulate root elongation, shoot development, and early seedling vigor. LAB also plays an important role in nutrient mobilization by solubilizing phosphorus, potassium, zinc and silicon, thereby increasing their availability to plants and improving nutrient uptake efficiency. Beyond growth promotion, LAB exhibit strong bio-control potential through the secretion of organic acids, hydrogen peroxide, proteases and siderophores, which collectively suppress fungal and bacterial pathogens. Application of LAB as seed treatments improves germination, strengthens root architecture and enhances plant resilience to biotic and abiotic stresses. Overall, LAB represents an eco-friendly, sustainable alternative to chemical fertilizers and pesticides, offering significant potential for improving crop productivity and promoting environmentally responsible agricultural practices.

Conflict of Interest The authors declare that they have no conflict of interest.

Author Contribution Statement PRU conceptualized the chapter, collected the literature, and prepared the initial draft. MKN provided critical guidance on structure, scientific content, and revisions. VK contributed to the sections related to microbial mechanisms and applications in sustainable agriculture. MMM assisted with data interpretation, figure preparation and refinement of plant growth–promotion sections. SML contributed to editing, formatting and cross-checking references. All authors reviewed, revised and approved the final version of the manuscript.

References

Abdel-Rahman MA, Tashiro Y, Sonomoto K (2013) Recent advances in lactic acid production by microbial fermentation processes. Biotechnol Adv 31:877–902. https://doi.org/10.1016/j.biotechadv.2013.04.002

Adithi G, Somashekaraiah R, Divyashree S, Shruthi B, Sreenivasa MY (2022) Assessment of probiotic and antifungal activity of *Lactiplantibacillus plantarum* MYSAGT3 isolated from locally available herbal juice against mycotoxigenic *Aspergillus* species. Food Bioscience Vol (50)

Adithi G, Divyashree S, Shruthi B, Deepa N, Sreenivasa MY (2024) Evaluation of *Limosilactobacillus fermentum* MYSAGAM1 isolated from herbal amla juice as a probiotic candidate with antifungal characteristics against *Fusarium equiseti*. Food Biosci 58:103843. https://doi.org/10.1016/j.fbio.2024.103843

Adithi G, Rakesh S, Shruthi B, Divyashree S, Deepa N, Vanitha PR, Vasundaradevi R, Deepthi BV, Sreenivasa MY (2025) Biological decontamination of myco-toxigenic *Fusarium verticillioides* using *Lactiplantibacillus plantarum* MYSN128 isolated from traditional vellappam: implications for food safety and global mycotoxin management. Curr Microbiol 82:404. https://doi.org/10.1007/s00284-025-04401-z

Ahmed E, Holmström SJM (2014) Siderophores in environmental research: roles and applications. Microb Biotechnol 7:196–208. https://doi.org/10.1111/1751-7915.12117

Awasthi A, Chauhan R, Pandey P, Tiwari R (2020) Enhancement in germination and seedling growth of wheat by lactic acid bacteria producing phytohormones. Biocatal Agric Biotechnol 26:101634

Bai J, Xu D, Xie D, Wang M, Li Z, Guo X (2024) Effects of lactic acid bacteria inoculants on fermentation quality, aerobic stability and bacterial community of whole-plant corn silage. Front Microbiol 15:1320735

Basak BB, Biswas DR (2023) Microbial solubilization of potassium from waste mica in presence of different carbon and nitrogen sources. Geoderma 424:116021

Benítez-Cabello A, Calero-Delgado B, Rodríguez-Gómez F, Garrido-Fernández A (2020) Lactic acid bacteria in table olives: Application and functional properties. Front Microbiol 11:584395

Brosnan B, Coffey A, Arendt EK, Furey A, Zannini E (2014) Antifungal activity of lactic acid bacteria and its application in food-based systems. Biotechnol Res Int 2014:146921. https://doi.org/10.1155/2014/146921

Cheong EYL, Sandhu A, Jayabalan J, Le TT, Ho TM (2014) Inhibition of *Colletotrichum gloeosporioides* and anthracnose disease of papaya by lactic acid bacteria. Biol Control 73:1–6. https://doi.org/10.1016/j.biocontrol.2014.02.006

Chowdhury SP, Hartmann A, Gao X, Borriss R (2015) Biocontrol mechanism by root-associated *Bacillus amyloliquefaciens* FZB42—a review. Front Microbiol 6:780. https://doi.org/10.3389/fmicb.2015.00780

Deepa N, Chennappa G, Deepthi BV et al (2022) Antifungal potential of *Azotobacter* species and its metabolites against *Fusarium verticillioides* and biodegradation of fumonisin. J App Microbiol 00:1–15

Deepthi BV, Deepa N, Vanitha PR, Sreenivasa MY (2022) Stress responses on the growth and mycotoxin biosynthesis of *Fusarium proliferatum* associated with stored poultry feeds. Appl Food Res 2(1):1000U1. https://doi.org/10.1016/j.afres.2022.1000U1

Divyashree S, Ramu R, Sreenivasa MY (2024) Evaluation of new candidate probiotic *Lactobacillus* strains isolated from a traditional fermented food multigrain-millet dosa batter. Food Biosci 57: 103450. https://doi.org/10.1016/j.fbio.2023.103450

EFSA Panel on Biological Hazards (BIOHAZ) (2012) Scientific opinion on the maintenance of the list of qualified presumption of safety (QPS) biological agents. EFSA J 10(12):3020

Etesami H, Jeong BR (2018) Silicon (Si): review and future prospects on the action mechanisms in alleviating abiotic and biotic stresses in plants. Ecotoxicol Environ Saf 147:881–896

FAO/WHO (2002) Guidelines for the evaluation of probiotics in food. Report of a joint FAO/WHO working group, London, Ontario, Canada

Gänzle MG (2015) Lactic metabolism revisited: metabolism of lactic acid bacteria in food fermentations and food spoilage. Curr Opin Food Sci 2:106–117

Ghatage A, Yamgar VM, Masurkar S (2024) Isolation and characterization of indole acetic acid, gibberellic acid, and phosphate solubilizing microorganisms from the rhizosphere of *Zingiber officinale*. J Appl Environ Microbiol 12:1–10

Giraud E, Lelong B, Raimbault M (2010) Influence of lactic acid bacteria on plant growth and disease suppression. J Appl Microbiol 109(3):859–868

Harman GE (2006) Overview of mechanisms and uses of *Trichoderma* spp. in plant disease control. Phytopathology 96(2):190–194

Kumar A, Verma JP, Singh B (2021) Lactic acid bacteria: effective eco-friendly tools for sustainable agriculture. Environ Sustain 4:123–134. https://doi.org/10.1007/s42398-021-00174-4

Lahmamsi H, Ananou S, Lahlali R, Tahiri A (2024) Lactic acid bacteria as an eco-friendly approach in plant production: current state and prospects. Folia Microbiol 69:465–489

Lamont JR, Wilkins O, Bywater-Ekegärd M, Smith DL (2023) From dairy to fields: lactic acid bacteria as plant growth–promoting rhizobacteria and biocontrol agents. Front Microbiol 14: 1133788

Li H, Li S, Zhang J, Zhang Q, Ren J (2024) Phosphate solubilization and plant growth properties are promoted by a lactic acid bacterium in calcareous soil. Appl Microbiol Biotechnol 108:24. https://doi.org/10.1007/s00253-023-12850-4

Ma JF, Yamaji N (2015) A cooperative system of silicon transport in plants. Trends Plant Sci 20: 435–442

Magnusson J, Jonsson H, Schnürer J, Roos S (2003) *Lactobacillus coryniformis* subsp. *coryniformis* produces antifungal compounds. Appl Environ Microbiol 69(12):7554–7560

Marschner H (2012) Mineral nutrition of higher plants, 3rd edn. Academic Press, London

Mendes R, Garbeva P, Raaijmakers JM (2011) The rhizosphere microbiome: significance of plant beneficial, plant pathogenic, and human pathogenic microorganisms. FEMS Microbiol Rev 37: 634–663

Mokoena MP (2017) Lactic acid bacteria and their bacteriocins: classification, biosynthesis and applications against uropathogens. Molecules 22:1255. https://doi.org/10.3390/molecules22081255

Monteiro CR, do Carmo MS, Melo BO, Alves MS, Dos Santos CI, Monteiro SG, Franco BDGM (2020) Lactic acid bacteria as potential probiotics for agricultural applications. Appl Microbiol Biotechnol 104:8297–8311

Naveed M, Mitter B, Reichenauer TG, Wieczorek K, Sessitsch A (2014) Increased drought stress resilience of maize through endophytic colonization by *Burkholderia phytofirmansstrain* PsJN and *Lactobacillus plantarum.* Appl Environ Microbiol 80:5884–5893

Nishida S, Kato T, Horiuchi H (2025) Emerging roles of lactic acid bacteria in sustainable agriculture and plant growth promotion. Microbiol Res 285:127405

Nishio M, Araki H, Arai S (2018) Effects of lactic acid bacteria on composting and manure management. J Environ Manag 210:23–30

Ogunbanwo ST, Sanni AI, Onilude AA (2003) Characterization of bacteriocin produced by *Lactobacillus plantarum* F1 and *Lactobacillus brevis* OG1. Afr J Biotechnol 2:219–227

Pruthviraj, Naik MK, Naik GR, Naik GB, Nandish M, Ekabote SD, Sreenivasa MY (2023a) Investigation on antibacterial, probiotic and plant growth promoting attributes of *Enterococcus faecium* MYSBC14 from Blue Cherry. J Saudi Soc Agric Sci. https://doi.org/10.1016/j.jssas.2023.04.003

Pruthviraj NMK, Naik GR, Nandish M, Ekabote SD, Sreenivasa MY (2023b) Antibacterial and plant growth promoting attributes of *Limosilactobacillus* sp. MYSN3 isolated from noni fruit. S Afr J Bot 162:559–567

Pruthviraj, Naik MK, Ekabote SD, Nandish MS, Ajithkumar K, Shruthi B, Adithi G, Divyashree S, Sreenivasa MY (2024) Evaluation of *Lactiplantibacillus argentoratensis* MYSJK8 from Jackfruit for its multifarious antibacterial, probiotic, plant growth promoting and bio-control attributes. J Crop Health. https://doi.org/10.1007/s10343-024-01050-3

Radhakrishnan R, Hashem A, Abd Allah EF (2017) *Bacillus*: a biological tool for crop improvement through bio-molecular changes in adverse environments. Front Physiol 8:667

Raman Y, Kim JS, Choi KR, Somasundaram S (2023) Plant growth–promoting lactic acid bacteria: prospects for sustainable agriculture. Front Microbiol 14:1145682

Saikia SP, Sarma RK, Deka M, Deka B (2011) Siderophore production by *Lactobacillus* spp. and its application in iron nutrition of crop plants. Indian J Microbiol 51:472–476

Santos TT, Varavallo MA, de Melo Pereira GV (2019) Application of lactic acid bacteria in agriculture: current status and future perspectives. World J Microbiol Biotechnol 35:138

Sharma R, Sanodiya BS, Thakur GS, Jaiswal P, Bisen PS (2020) Lactic acid bacteria as biofertilizers and biopesticides: an eco-friendly approach for sustainable agriculture. Biocatal Agric Biotechnol 27:101678

Sharma A, Khan N, Begum N, Singh N (2023) Role of phosphorus in plant growth and its uptake mechanisms under limited availability. J Soil Sci Plant Nutr 23:15–32

Shruthi B, Deepthi BV, Adithi G, Vanitha PR, Sreenivasa MY (2025a) *Saccharomyces cerevisiae* MYSY1 as a promising probiotic isolate with antifungal properties against foodborne *Aspergillus* species. Biocatal Agric Biotechnol G7:10374U. https://doi.org/10.1016/j.bcab.2025.103649

Shruthi B, Somashekaraiah R, Divyashree S, Adithi G, Ajithkumar K, Mahadevakumar S, Sreenivasa MY (2025b) Probiotic characterization and antifungal potentials of *Pichia kudriavzevii* MYSSBYPS10 isolated from traditionally fermented green gram dosa batter. Probiotics Antimicrob Proteins 15:857–870

Singh R, Rahi P, Dey P (2018) Zinc solubilization by bacteria enriched from rice rhizosphere and their growth-promoting potential. J Appl Biol Biotechnol 6:34–40

Soomro AH, Masud T, Anwaar K (2022) Role of lactic acid bacteria (LAB) in plant disease management. In: Khan MN et al (eds) Plant microbiome: stress response. Springer, pp 299–312

Spaepen S, Vanderleyden J (2011) Auxin and plant–microbe interactions. Cold Spring Harb Perspect Biol 3:a001438

Strafella S, Simpson DJ, Khanghahi MY, de Angelis M, Gänzle M, Minervini F et al (2021) Comparative genomics and in vitro plant growth promotion and biocontrol traits of lactic acid bacteria from the wheat rhizosphere. Microorganisms 9:78

Ström K, Sjögren J, Broberg A, Schnürer J (2002) *Lactobacillus plantarum* MiLAB 393 produces antifungal cyclic dipeptides cyclo(L-Phe-L-Pro), cyclo(L-Phe-trans-4-OH-L-Pro) and 3-phenyllactic acid. Appl Environ Microbiol 68:4322–4327

Turaeva BI, Kamolova KF, Soliev AB, Kutlieva GJ (2021) Gibberellic and indole acetic acids producing features of bacteria from various plants. Asian J Biol Life Sci 10:681–685

Vasundaradevi R, Sarvajith M, Somashekaraiah R, Gunduraj A, Sreenivasa MY (2024) Antagonistic properties of *Lactiplantibacillus plantarum* MYSVB1 against *Alternaria alternata*: a putative probiotic strain isolated from banyan tree fruit. Front Microbiol 15:1322758

Vasundaradevi R, Sarvajith M, Divyashree S, Deepa N, Premila NA, Achar MY, Sreenivasa MY (2025) Tropical fruit-derived *Lactiplantibacillus* as potential probiotic and antifungal agents against *Fusarium oxysporum*. Sci Rep 15:2144. https://doi.org/10.1038/s41598-025-85190-0

Wang Y, Zhang H (2023) Lactic acid bacteria: taxonomy, characteristic features, physiology, and probiotic applications. In: Lactic acid bacteria: from metabolism to applications. Springer, pp 1–18

Wee YJ, Kim JN, Ryu HW (2006) Biotechnological production of lactic acid and its recent applications. Food Technol Biotechnol 44(2):163–172

Yu Z, Zhang X, Li S, Li C, Yang Z (2020) Characterization of proteolytic activity and amino acid production by *Pediococcus pentosaceus* UP-2 isolated from fermented food. J Appl Microbiol 129:367–375

Yuan J, Raza W, Shen Q, Huang Q (2022) Antifungal activity of plant growth-promoting bacteria and their role in biological control of soil-borne pathogens. Appl Soil Ecol 170:104261

Zheng J, Wittouck S, Salvetti E, Franz CMAP, Harris HMB (2023) A taxonomic update for the bacterial genus *Lactobacillus*: reclassification and nomenclature updates. Int J Syst Evol Microbiol 73:005717

Part II

Probiotics for Combating Fungal Infections in Animals

Decoding the Gut Microbiome and Its Interactions in Animals: Mechanisms of Action

8

A. Anita Roy and M. Nandhini

Abstract

The host physiology, development, immunity, and behaviour in animals are influenced by their gut microbiome, which establishes a complex and dynamic ecosystem. The chapter elaborates on such mechanisms by which the microbial community networks with their animal hosts and emphasises their functional, molecular, and ecological dynamics. The chapter involves the overview of taxonomic and evolutionary patterns of host–microbe symbiosis, microbes across life stages, and the influence of the environment on microbial acquisition. The role of various key microbial taxa, functional guilds, and related species in maintaining gut homeostasis is discussed. Crosstalk between the molecular and neuroimmune system, between microbes and hosts such as nutrient assimilation, metabolic signalling, and immune modulation, was explored. The chapter also delves into dysbiosis-linked pathologies, host genomic and epigenomic contributions, and current analytical methodologies for microbiome research. Modern findings of gut microbiome as therapeutics, their nutritional interventions, comparative microbiome ecology, and emerging systems biology approaches are also highlighted with future directions and translational implications for animal health and welfare.

A. Anita Roy (✉)
Department of Biotechnology & Genetics, School of Sciences, Jain (Deemed-to-be) University, Bengaluru, Karnataka, India
e-mail: r.anitha@jainuniversity.ac.in

M. Nandhini
Department of Microbiology & Botany, School of Sciences, Jain (Deemed-to-be) University, Bengaluru, Karnataka, India

M. Y. Sreenivasa et al. (eds.), *Probiotics as Antifungals*,
https://doi.org/10.1007/978-981-92-0002-3_8

Graphical Abstract

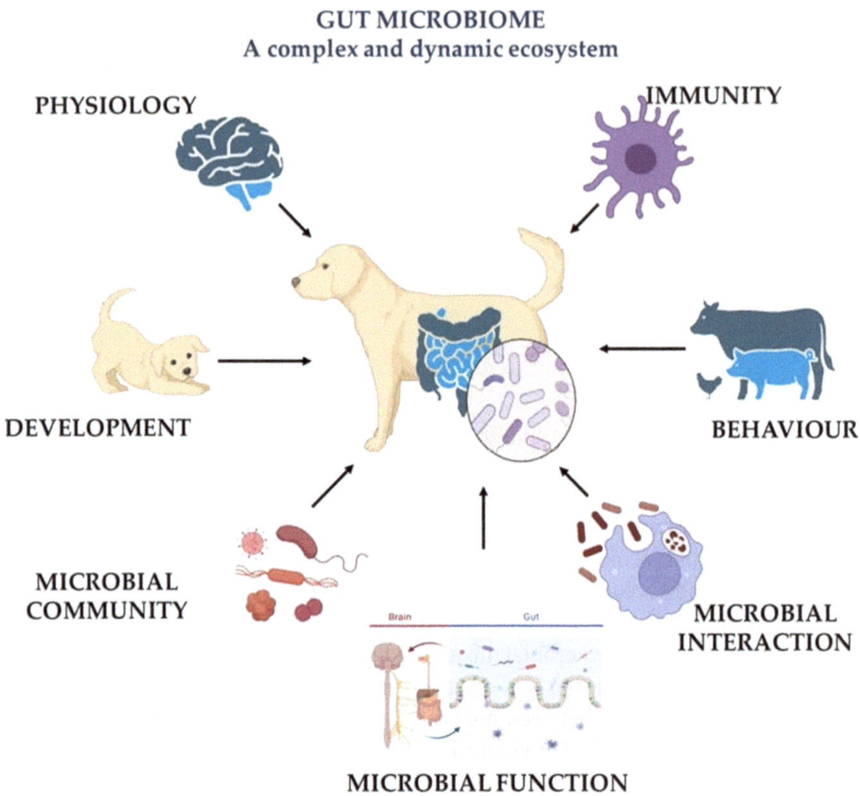

Keywords

Gut microbiome · Host–microbe interactions · Dysbiosis · Microbial ecology · Animal health

Learning Objectives

By the end of this chapter, readers will be able to:

- Describe the key microbial communities that reside in the animal gut and their variations across their communication with host biology, diet, and evolution.

(continued)

- Relate the mechanism of the formation, modification, and stability of gut microbial communities from birth to adulthood with the impact of maternal, dietary, and environmental conditions.
- Understand the communication of the gut microbiome to regulate the host biochemical, metabolic, and immunological pathways
- Compare various microbiome assemblies across diverse animal digestive systems and emerging research tools and interventions to support animal health and sustainable management.

8.1 Overview of Gut Microbiome in Animal Hosts

The gut microbiome refers to a diverse community of microorganisms that survives in the gastrointestinal tracts of animals. Such a microbial network comprises bacteria, archaea, viruses, and fungi, and plays a vital role in host health. They are involved in major biological processes such as digestion, synthesis of essential vitamins, immune responses, and protection against pathogenic organisms. Depending on factors such as diet, age, genetics, environment, and antibiotic exposure, there is a difference in the microbiome composition across species (Kamada et al. 2013; McFall-Ngai et al. 2013; Bahrndorff et al. 2018; Pickard et al. 2017). Fermentative bacteria are known to support the process of cellulose breakdown in herbivores, whereas in carnivores, the microbiome helps in protein and fat metabolism (Feiyun et al. 2026). Understanding the diverse microbiomes across animal models provides insights into health, disease, and potential therapeutic interventions for it.

8.1.1 Taxonomic Diversity and Ecological Distribution

There is a remarkable taxonomic diversity of the gut microbiome, which is majorly composed of bacteria from the phyla *Firmicutes*, *Bacteroidetes*, *Actinobacteria*, and *Proteobacteria* (Laterza et al. 2016). *Firmicutes* and *Bacteroidetes* are those that classically dominate in both humans and animal species, with vital contributions in carbohydrate fermentation and energy production (Flint et al. 2012). *Actinobacteria*, such as *Bifidobacterium*, are found to be more predominant in childhood, contributing to vitamin synthesis, while *Proteobacteria* are mainly found during dysbiosis (Kandpal et al. 2022). It is observed that factors such as pH, oxygen levels, nutrient availability, and host secretions influence the distribution of microbial taxa in the gastrointestinal tract (Ahn and Hayes 2021). Apart from these, external factors such as habitat, diet, use of antibiotics, and co-evolution with the host also influence both diversity and function of microbiome. Such ecological and taxonomic complexity highlights the adaptive nature of microbiomes across animal lineages and diverse environments.

8.1.2 Evolutionary Dynamics of Host–Microbial Symbiosis

The gut microbiome has an evolutionary relationship with its respective animal hosts, which is a dynamic and co-adaptive process formed by various mutual selection pressures. Phylogenomic studies reveal that microbial lineages have diversified simultaneously with their hosts, indicating long-term co-evolution. Host–microbe co-diversification in line with adaptive specialization is found in host-specific microbial taxa, such as *Bacteroides* in mammals and *Spirochaetes* in termites. The mechanism of such symbiosis is enabled by vertical transmission routes and host genetic regulation of microbial colonization. Such regulations can be regulated by pattern recognition receptors (e.g., TLRs, NOD-like receptors), and they mediate processes like microbial sensing and immune tolerance (Chu et al. 2013). The host traits like gut morphology, immune system maturation, and metabolic homeostasis are also influenced by the microbial communities. Host-mediated selective restrictions like immune responses and nutrient availability support the evolution of microbial populations. Therefore, the gut microbiome remains an evolving ecological and functional partner, essential for understanding the host physiology and adaptive evolution.

8.2 Ontogeny and Succession of Gut Microbial Communities

The process that involves the development and sequential changes of gut microbial communities in animals is a dynamic and regulated process, which initiates during birth and remains throughout the animal's lifetime. This progression is supported by several aspects like host genetics, diet, environmental exposure, and developmental stage. Early microbial colonizers play a crucial role in shaping the trajectory of gut ecosystem assembly (Maritan et al. 2024). In the beginning, the pioneer microbes, which are often facultative anaerobes, tend to dominate the gut environment, to generate a condition that allows the succession of more specialized and obligate anaerobes (Eggesbo et al. 2011). As the host matures, these microbial communities become highly complex and stable and contribute significantly to immune development, nutrient metabolism, and pathogen resistance (Pickard et al. 2017). Understanding this developmental timeline is essential to decode the gut microbiome's interaction with its animal hosts and its influence on health and disease across different stages of life.

8.2.1 Temporal Assembly from Neonatal to Adult Stage

The development of the gut microbiome from birth and throughout adulthood of the host is a sequential and time-dependent process. Such development and related changes in the gut microbiome of animals begin immediately after birth, even though the gastrointestinal tract is sterile during birth. Initially, the tract is colonized by the fibrolytic bacteria and methanogens, and the microbial density increases (Tanaka and Nakayama 2017). Both aerobic and facultative anaerobic microbes are found to be the early colonizers and they are later replaced majorly by anaerobic taxa. These microbes play key roles in the process of cellulose and hemicellulose degradation and for energy harvesting. The period between birth and weaning is critical for rumen development and during which the microbial colonies directly influence the maturation of rumen structures like the developing villi and the digestive capacity. This microbial succession is enhanced by factors such as diet, host–microbe interactions, environmental exposure, and genetic influences (Messer et al. 2017). As the host enters the post-weaning stage, the microbial ecosystem becomes stabilized and becomes resistant to change (Waters et al. 2025). Thus, the early-life interventions can be related to the important phase to shape a beneficial microbiome for improved health, productivity, and reduced methane emissions.

8.2.2 Environmental, Dietary, and Maternal Transmission Factors

The regulation of development and related changes in the gut microbiome in animals is influenced by various environmental, dietary, and maternal transmission factors. All these three factors are deeply interconnected and together influence the ontogeny and succession of gut microbial communities in animals (Amin and Seifert 2021). Maternal transmission introduces basic microbes such as *Lactobacillus* and *Bifidobacterium*, supporting early colonization throughout the process of birth and colostrum feeding (Ferretti et al. 2018). Environmental exposure, such as the microbiome interactions with conspecifics and surroundings, further supports the diversification of the microbiota, while geographic and ecological contexts contribute to the composition of the microbial community. Diet also has a significant impact, especially when there is a transition of feed from milk to solids, which is necessary for a shift to fibre-degrading and metabolically versatile taxa (Myer et al. 2017). Understanding these interactions across species provides valuable insight into microbiome-driven approaches for improving animal and human well-being.

8.3 Functional Microbial Taxa and Ecological Guilds

Functional microbial taxa are referred to as a set of microbes in the gut that are known for their metabolic roles rather than taxonomy (Wilmanski et al. 2021). These taxa form an ecological association, where microbial assemblies utilize similar resources or perform complementary functions within the gut ecosystem (Table 8.1). For example, *Ruminococcus* and *Fibrobacter* species are involved in cellulolytic guilds, specialized in degrading complex plant fibres, and producing fermentable substrates (Flint et al. 2008). *Bacteroides* and *Prevotella* species are associated with saccharolytic guilds, involved in metabolizing carbohydrates into short-chain fatty acids like acetate and propionate to maintain host energy and gut health. Proteins are degraded into amino acids and bioactive metabolites via proteolytic guilds (Diether and Willing 2019). Ecological guilds enhance nutrient

Table 8.1 Functional guilds of the gut microbiota and their metabolic roles

Functional guild	Microbial taxa involved	Primary substrates utilized	Key metabolic products	Major host benefits	References
Cellulolytic (fibre-degrading)	*Ruminococcus flavefaciens*, *Ruminococcus albus*, *Fibrobacter succinogenes*, *Bacteroidetes* spp.	Cellulose, hemicellulose, and resistant plant fibres	SCFAs (acetate, butyrate), oligosaccharides	Provides energy, enhances gut motility and herbivore digestion, and maintains epithelial health	Froidurot and Julliand (2022)
Saccharolytic (carbohydrate-fermenting)	*Bacteroides*, *Prevotella*, *Lactobacillus*, *Bifidobacterium*	Dietary carbohydrates, resistant starch, simple sugars	Acetate, propionate, butyrate	Regulates metabolism and immune homeostasis, strengthens gut barrier, and improves insulin sensitivity.	Baars et al. (2024)
Proteolytic (protein- and peptide-degrading)	*Clostridium*, *Bacteroides*, *Peptostreptococcus*	Dietary proteins, peptides, amino acids	Ammonia, branched-chain fatty acids, amines, phenols	Supports nitrogen recycling and manages inflammation	Adesogan (2009)
Mucin-degrading	*Akkermansia muciniphila*, *Bacteroides thetaiotaomicron*	Host-derived mucin glycoproteins	SCFAs (especially acetate & propionate)	Maintains mucus layer, enhances epithelial regeneration, and improves metabolic health	Aggarwal et al. (2022)
Methanogenic (Archaea)	*Methanobrevibacter smithii*, *Methanosphaera stadtmanae*	Hydrogen, CO_2, methyl compounds	Methane (CH_4)	Removes excess H_2 to optimize fermentation, improves the efficacy of fibre breakdown and influences energy harvest	De la Fuente et al. (2019)
Sulfate-reducing bacteria (SRB)	*Desulfovibrio*	Sulfate, lactate, H_2	Hydrogen sulfide (H_2S)	Supports H_2 disposal and maintains microbial balance	Rowan et al. (2009)
Lactic acid–producing bacteria (LAB)	*Lactobacillus*, *Streptococcus*, *Enterococcus*	Simple sugars	Lactic acid	Reduces gut pH, inhibits pathogens, promotes healthy fermentation, and supports immune tolerance	Pessione (2012), Divyashree et al. (2024)
Lactic acid–utilizing bacteria (LAU)	*Megasphaera*, *Selenomonas*, *Veillonella*	Lactic acid	Propionate, acetate	Prevents lactic acidosis, stabilizes fermentation, and volatile fatty acid (VFA) balance	Chen et al. (2020)

Methanotrophic/hydrogenotrophic fermenters	*Blautia*, *Ruminococcus*, *Christensenella*	Hydrogen, simple carbon substrates	SCFAs	Maintains low hydrogen pressure and thereby favours fermentation	Ungerfeld (2020)
Vitamin-producing microbes	*Bacteroides*, *Lactobacillus*, *Bifidobacterium*, *Clostridium* spp.	Dietary substrates, amino acids	B-vitamins (B1, B2, B6, B12), Vitamin K	Supports host metabolism, erythropoiesis, and immune function	Uebanso et al. (2020)
Bile acid-modifying microbes	*Clostridium scindens*, *Bacteroides*, *Lactobacillus*	Primary bile acids	Secondary bile acids	Regulates lipid absorption, modulates immune and metabolic signalling	Baars et al. (2015)
Short-chain fatty acid specialists	*Faecalibacterium prausnitzii*, *Butyrivibrio*, *Roseburia*	Complex polysaccharides, plant fibres	Butyrate (major), acetate	Strong anti-inflammatory effect, fuels colonocytes, and maintains tight junction integrity	Fernández et al. (2016)

utilization efficiency and ecosystem stability by creating metabolic networks via cross-feeding interactions. This guild-involved organization allows functional classification, where microbes inhabit distinct niches based on substrate preference, oxygen tolerance, and spatial localization in the gut (Tropini et al. 2017).

8.3.1 Keystone Species and Core Microbiota

Among the diversified set of gut microbiomes, certain microbial taxa play excessive roles to maintain their community structure and the host functionality and are termed as keystone species. Such species exert critical ecological influence by regulating microbial interactions, stabilizing community dynamics, and facilitating key metabolic pathways (Banerjee et al. 2018). For example, keystone taxa involved in anti-inflammatory activity and starch degradation are *Faecalibacterium prausnitzii* in humans and *Ruminococcus bromii* in ruminants, respectively. There are other sets of microbial communities referred to as the core microbiota and are consistently found across individuals of a host species, irrespective of external variables such as diet or geography. These core members contribute to other essential roles such as nutrient absorption, immune modulation, and protection against pathogens. Relating the roles of both keystone species and core microbiota is essential for translating the functional architecture of the gut ecosystem and thereby designing targeted microbiome-based interventions to improve animal health and productivity (Shade and Handelsman 2012).

8.3.2 Functional Redundancy and Niche Specialization

The gut ecosystem comprises numerous species present in the human microbiome that perform similar tasks as functional redundancies because they all produce short-chain fatty acids. Redundant microbiomes provide practical redundancy when faced with an ecological perturbation (e.g., antibiotics, dietary shifts, etc.). For instance, several species within the Firmicute and Bacteroidetes phyla ferment carbohydrate molecules that have been broken down into their constituent sugars by another group of microbes and are ultimately converted into SCFA for continued colonocyte energy supply, even as individual species have peaks and valleys in their population sizes (den Besten et al. 2013). On the other hand, some genera within the human microbiome perform specialized functions and can dominate unique niches. *Akkermansia muciniphila*, for example, thrives on mucin degradation at the interface between our intestinal mucosa and the gut lumen (Elzinga et al. 2024). Similarly, methanogenic archaea and sulphate-reducing bacteria also occupy their own niche, consuming fermentation metabolites, i.e., hydrogen and sulphate. In combination, specialised microbiota niches and redundancy permit environmental shift adaptability and retain core functional capabilities critical to the health of the host (Ramond et al. 2025).

8.4 Molecular Mechanisms Underpinning Host–Microbiome Crosstalk

Host-microbiome interactions are influenced by molecular signalling pathways that regulate immune responses, metabolism, and gut homeostasis. The microbial molecules, such as MAMPs-lipopolysaccharides, peptidoglycans, and flagellin, are identified by host PRRs, including TLRs and NLRs. Such recognition triggers a signalling cascade, which leads to varied inflammatory responses and displays immune tolerance against commensals. Microbial metabolites, especially SCFAs like butyrate, propionate, and acetate, promote differential host gene expression through their function as histone deacetylase inhibitors or ligands for GPCRs, thus influencing epithelial barrier function, immune cell differentiation, and energy metabolism. Microbial-derived tryptophan metabolites influence AhR signalling and shape mucosal immunity. The host epithelium and the immune system actively sculpt microbial communities through the secretion of antimicrobial peptides and mucins that create selective niches (Rooks and Garrett 2016).

8.4.1 Microbial-Derived Metabolites and Signalling Pathways

Microbial metabolites that are secreted by gut microbiota serve as signalling molecules that affect a wide range of host physiological functions and signalling pathways. They influence receptor molecules on immune cells, for example, in mucosal surfaces, microbial metabolites can change their immune cell behaviour, for instance, activation, differentiation, migration, and cytokine production. Apart from this, they also reach systemic circulation and initiate physiological responses, which may result in direct as well as indirect involvement of the central nervous system.

SCFAs are mainly acetate, propionate, and butyrate, which are the fermentative products of dietary fibres by gut bacteria such as Bacteroides and Firmicutes. SCFAs provide energy to colonocytes and also perform signalling functions when attached to G protein-coupled receptors (GPCRs) such as GPR41 and GPR43, thus controlling inflammatory reactions, gut barrier strength, and appetite. In addition, butyrate operates as a histone deacetylase (HDAC) inhibitor that changes the gene expression of the host on the epigenetic level, thereby activating anti-inflammatory pathways and epithelial cell differentiation. Moreover, various microbial metabolites like secondary bile acids, indole derivatives resulting from tryptophan metabolism, and polysaccharide A from Bacteroides fragilis bind to host receptors such as the aryl hydrocarbon receptor (AhR) and TLRs, thus affecting mucosal immunity and systemic inflammation (Gasaly et al. 2021). These metabolites are involved in complex signalling networks that regulate immune tolerance and defence, influence metabolic homeostasis, and even impact neurological functions via the gut–brain axis (Yang and Cong 2021). Knowledge of these mechanisms leads to the use of therapeutic agents that target microbial metabolites to treat inflammatory, metabolic, and neurodegenerative diseases. They can interact with the brain not only through direct but also through indirect pathways, which can have an effect on brain

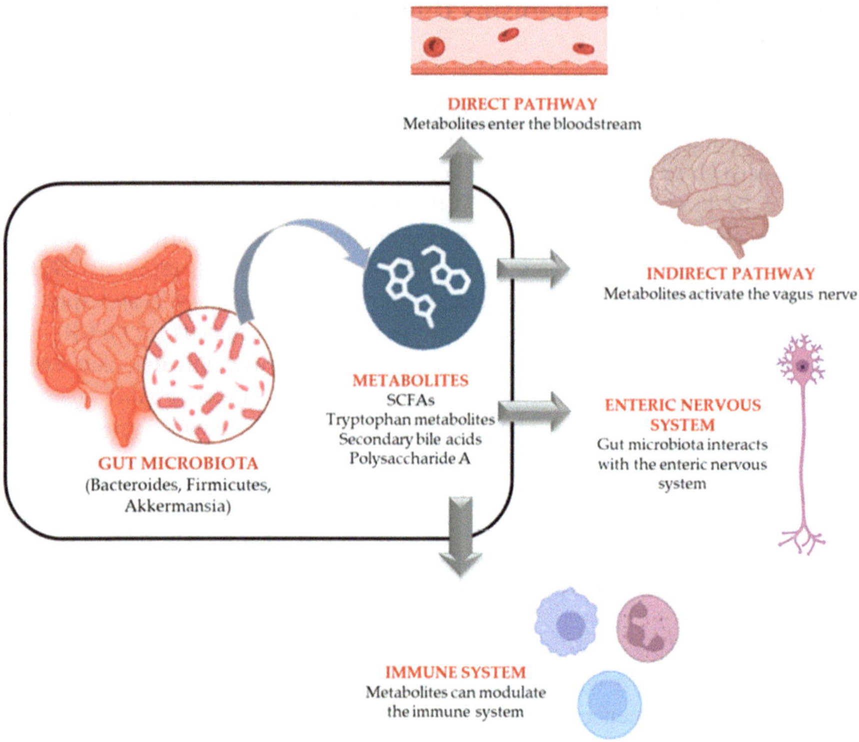

Fig. 8.1 Metabolite-induced pathway

function, neurodevelopment, and even the occurrence of diseases. Besides that, these metabolites have the power to influence immune responses, energy metabolism, and other physiological processes (Fig. 8.1).

(a) *Direct Pathway:* Microbial metabolites are described to reach the blood vessels, and also, they are capable of crossing the blood-brain barrier (BBB), thus they can directly alter brain functioning and, in this way, their involvement in the genesis of neurodegenerative or psychiatric disorders can be implied.
(b) *Indirect Pathway*: Metabolites have the potential to turn on the vagus nerve, which is a very important part of the autonomic nervous system, thus they can communicate with the brain.
(c) *Enteric Nervous System:* Gut microbiota is closely associated with the enteric nervous system, which, in turn, regulates digestive motility, permeability, and immune responses and thus has a great influence on the metabolites generated.
(d) *Immune System:* Microbial metabolites are capable of changing the immune system, which, in response, can have both local and systemic effects. Among these effects, there can be the production of neurotransmitters and gut hormones (Wang et al. 2024).

8.5 Host Pattern Recognition Receptors and Downstream Immune Signalling

The host pattern recognition receptors (PRRs) in host cells are a type of innate immune system that recognize certain conserved microbial structures known as MAMPs (microbe-associated molecular patterns) (Rooks et al. 2016). While there are various examples of PRRs, TLRs (Toll-like receptors), NLRs (NOD-like receptors), CLRs (C-type lectin receptors), and RLRs (RIG-I-like receptors) are the most prominent. All of these receptors are capable of sensing the presence of edible microbial structures, for example, lipopolysaccharides (LPS), peptidoglycans, flagellin, and microbial nucleic acids, leading to the stimulation of the immune system (Abe et al. 2019). The principal endogenous pathway for activating TLRs is mediated through ligand binding, followed by interactions with adaptor proteins that ultimately activate the NF-κB (nuclear factor kappa B) and MAPK (mitogen-activated protein kinase) pathways; this results in the secretion of cytokines, chemokines, and antimicrobial peptides, which coordinate the inflammatory and antimicrobial responses. NLRs, like NOD1 and NOD2, detect bacterial products in the cytosol (NLR) and activate the NF-κB pathway as well as the inflammasome for the regulation of interleukin-1β (IL-1β) and pyroptosis (Akira et al. 2006). The functions of these signalling pathways coordinate the balance between being tolerant to the microbiota and being protective against pathogens, ensuring homeostasis within the gut. Dysregulated signalling can lead to a state of chronic inflammation and/or the development of diseases such as inflammatory bowel disease (IBD). Therefore, the signalling pathways that involve PRRs are important for developing therapies aimed at moderating host immune responses through interactions with the gut's microbiota.

8.6 Nutrient Sensing and Metabolic Integration

Gut nutrient sensing is a complex interaction between the host and the microbiome. By integrating dietary sources with metabolic and immune responses, the gut's nutrient sensing enables the maintenance of homeostasis. The gut epithelial and immune cells identify nutrients and microbial-derived metabolites through specialized receptors, such as G-protein-coupled receptors (GPCRs) that include GPR41, GPR43, and GPR109A. These receptors sense SCFAs produced from microbially fermented foods (Kim et al. 2013, 2018; Digby et al. 2012) and therefore play a role in regulating energy metabolism, hormone secretion, and inflammatory signalling. In this way, nutrient status directly affects appetite and facilitates the development of immune tolerance to nutrients. Examples of nutrient-sensing pathways include the mTOR and AMPK pathways. These pathways integrate signals from microbial metabolites and the host nutrient status, regulating cellular growth, autophagy, and energy flux. Furthermore, microbial metabolic processes influence the composition of bile acids, which activate host nuclear receptors such as FXR and influence lipid and glucose metabolism. Therefore, through the interactions of several interconnected pathways, the gut's nutrient sensing supports a two-way communication system between diet, microbiota, and host physiology that promotes the maintenance of metabolic homeostasis and supports the development of an immune response. The failure of this integrated communication has been associated with metabolic diseases and may provide new targets for treatment with microbiome-based therapies.

8.7 Immunomodulatory Functions of the Gut Microbiome

The immune system plays an important role in establishing health and disease by interacting with self and foreign molecules encountered by the body. The gut microbiome with a more complex and diverse microbiota also takes a key role in providing defence mechanisms against microbial pathogens along with the immune system by educating it to distinguish between harmful and beneficial microbes, enhancing immune function. The gut microbiome triggers the host immune system both locally and systemically. These microorganisms interact with pattern recognition receptors, such as Toll-like receptors, on intestinal epithelial cells and immune cells and affect the development of innate and adaptive immunity in the host by building immune tolerance and preventing overactive immune responses (Kim et al. 2025). It also acts as a barrier against infections caused by outcompeting pathogens (Ramond et al. 2025).

Understanding the interactions between the host immune system and the gut microbiota is an evolving field of research. A study conducted by Geva-Zatorsky et al. (2017) mono-colonized each germ-free mouse with one of 53 human gut-derived bacterial strains belonging to all five major phyla (*Bacteroidetes; Firmicutes; Proteobacteria; Actinobacteria; and Fusobacteria*) to assess the modulation of innate and adaptive immune responses, and also focusing on the intestinal tissue transcriptome. The results showed a clear involvement of the colonized bacterial strain in influencing the immune system by expression of antimicrobial peptides, cytokine production, T-cell differentiation, and immunoglobulin A (IgA) responses.

Certain bacteria produce short-chain fatty acids comprising butyrate, propionate, and acetate, produced during fermentation, which influence the host immune cells and also provide a carbon source. These metabolites interfere with the cross-talk between the gut epithelium and immune cells (Yoo et al. 2020). Germ-free animals produce fewer Intraepithelial lymphocytes (IEL), IgA-secreting plasma cells, and T helper cells and regulatory T cells (Tregs).

8.7.1 Induction and Regulation of Mucosal Immune Responses

The gut microbiota along with their metabolites, host epithelial cells, mucus, IgA, antimicrobial peptides, and immune cells create a "mucosal firewall" that plays a critical role in combating pathogen translocation. Gut microbes provoke the development and maturation of gut-associated lymphoid tissues (GALT), interspersed with follicle-associated epithelia (FAE) and mucosa-associated lymphoid tissue (MALT), which influence the differentiation of T helper cells and regulatory T cells (Tregs) (Acheson and Luccioli 2004). During the initial stages of colonization of gut microbiota, the innate immune cells ascertain the microbial-associated molecular patterns (MAMPs) through pattern recognition receptors (PRRs) presented on the immune cells like MHC II-expressing antigen-presenting cells (APCs), dendritic cells, and mucosal epithelial cells. These signals initiate adaptive immune responses, namely the IL-12-associated differentiation of naïve T cells into T-helper cells. For example, Toll-like receptor (TLR) 2 recognizes MAMPs like lipoteichoic acid from Gram-positive bacteria, while TLR4 detects lipopolysaccharide (LPS) from Gram-negative bacteria in coordination with CD14 and MD-2 (Sjögren et al. 2009).

Certain gnotobiotic animal studies (Sonnenburg et al. 2006) have shown that metabolites secreted from microbes are essential for developmental pathways in the host and that different microbial strains elicit distinct host responses.

8.7.2 Role in Immune Homeostasis and Tolerance Mechanisms

The components of the innate and adaptive immune system, which influence the gut microbiome and contribute to immune homeostasis, are monocytes, dendritic cells, CD4+ T cells, macrophages, Treg cells, granulocytes, Th17 cells, cytotoxic natural killer cells, B cells and antibodies, lymphoid tissue inducer cells, non-cytotoxic innate lymphoid cells, and epithelial cells. The gut microbiota plays a fundamental role in maintaining immune homeostasis and promoting tolerance by secreting metabolites such as such as short-chain fatty acids (SCFAs), secondary bile acids, sphingolipids, tryptophan derivatives, and other bioactive compounds which interacts closely with the host's immune system, particularly through microbial-associated molecular patterns (MAMPs) that are recognized by pattern recognition receptors (PRRs) on innate immune cells.

The enormous diversity of gut microbiota continuously challenges the gut's immune system by presenting diverse antigenic determinants, which are recognised through pattern recognition receptors (PRRs), which associate with pathogen-associated molecular patterns (PAMPs) from microbes. Apart from these direct microbial signals, the immune system also responds through the secretion of cytokines, chemokines, and other inflammatory mediators against the pathogen. Factors such as host genetics, environmental exposures, diet, metabolic status, immune dysfunctions (either innate or acquired), and infections largely impact the interactions between gut microbes and immune cells. A study by Thompson et al. (2010) witnessed variation in community dynamics in mice, which suggests a separate individuality index for host and microbe-dependent phenotypic traits.

N-butyrate, a fermentation byproduct from microbes, deploys immunomodulatory effects on intestinal macrophages by reducing their reactivity towards the gut microflora present in the colon by histone deacetylase inhibition, which contributes to intestinal immune homeostasis. A study by suggested that the composition of the intestinal microbiota plays a crucial role in regulating the balance between Th17 (IL-17-producing T-helper) cells and regulatory T cells (Tregs) in the lamina propria (LP), thereby influencing intestinal immune responses, immune tolerance, and the risk of developing inflammatory bowel diseases.

Various papers have reported that the intestinal commensal microbiota play a critical role in regulating Treg-mediated immune responses, which are essential for mounting effective defence against pathogens and for preventing autoimmune disorders, food allergies, gut hypersensitivity to luminal antigens, and other harmful immunopathological conditions (Chistiakov et al. 2015).

8.8 Pathobiology of Dysbiosis and Microbiome-Associated Disorders

Dysbiosis is the disruption of the normal balance of gut microbiota and serves as an underlying factor in the development of numerous diseases. The factors that influence the imbalance of gut microbiota are diet, lifestyle, antibiotics, infections, inflammation, the genetic makeup of the host, and the diversity of intestinal flora. The pathogenic mechanisms that lead to dysbiosis are:

- Compromised intestinal mucosal barrier integrity
- Activation of inflammatory pathways
- Immune system imbalance
- Metabolic disturbances
- Disruptions in the gut–brain axis and gut–liver axis, which result in widespread systemic effects (Shen et al. 2025)

Several reports have reported the incidence of diseases as a result of an imbalance in the intestinal commensals. Loss of diversity is also a criterion for dysbiosis, which leads to lesser microbial service to the host. Microbial dysbiosis leads to metabolic dysbiosis, which is detected by metabolic concentrations in colon, blood, urine, or exhaled air, which are used as biomarkers that provide new diagnostic and therapeutic approaches to detect dysbiosis and disease (Sitkin et al. 2016).

8.8.1 Mechanistic Insights into Gastrointestinal and Systemic Pathologies

Host-associated factors influence the diversity and composition of gut microbiota in the host (Table 8.2). Conducted a study on the influence of antibiotics such as ampicillin, bacitracin, meropenem, neomycin, and vancomycin on the gut microbiome and confirmed that the antibiotics depleted the bacteria-derived metabolites in the colon and altered the lipid species, and also impaired the memory of antibiotic-treated mice. A lesser population of *Blautia* along with an increased

Table 8.2 Comparative overview of dysbiosis and host immune responses across different animal hosts

Sl. no	Host	Pathogen/causative agent	Disease	Effect on gut microbiome	The defence mechanism of the host	References
1.	Humans, farm animals, birds, rodents, reptiles, amphibians, fish	*Blastocystis*	Irritable bowel syndrome and inflammatory bowel disease	Decrease in population of beneficial *Bifidobacterium* and *Lactobacillus* spp.	Redox-mediated killing of the pathogen	Yason et al. (2019)
2.	Humans	Autoimmune	Coeliac disease	Targeted killing of intestinal epithelial cells by T cells	Production of short-chain fatty acids (SCFAs)	Di Cagno et al. (2011)
3.	Ruminants	Metabolic disorder causing diarrhoea, mastitis, and laminitis	Subacute ruminal acidosis (SARA)	Intestinal tissue damage, resulting in a "leaky gut"	Bacterial homeostasis, increased concentrations of VFA, acetate, propionate, and butyrate	Fu et al. (2022)
4.	Carnivores	Unknown	Intestinal Lymphoma	Decrease in bacterial diversity	Elevation in the level of Treg lymphocytes	Breczko et al. (2024)
5.	Cattle	*Mannheimia haemolytica*, *Histophilus somni*, *Mycoplasma bovis*, and *Pasteurella multocida*	Bovine respiratory disease	Imbalance in the bacterial community in the nasopharynx	Provide colonization resistance against BRD bacterial pathogens	Alexander et al. (2020)
6.	Cattle	*Mycobacterium avium* subsp. *paratuberculosis*	Johne's disease	Bacteroidetes and Firmicutes decreased	Intestinal inflammation	Fecteau et al. (2016)
7.	Chicken	*Eimeria maxima*	Coccidiosis	Reduction in *Lactobacillus, Ligilactobacillus,* and *Limosilactobacillus, Faecalibacterium* species	Production of SCFA	Liu et al. (2024)
8.	*Crocodylus intermedius*	*Helicobacter* spp.	Gastrointestinal disease	Significant decrease in microbial richness	–	Castelli et al. (2025)

population of *Ruminococcus gnavus* was identified as a susceptibility factor for Kawasaki disease (Teramoto et al. 2023).

8.9 Comparative Microbiome Ecology Across Taxonomic Lineages

Comparative studies of microbiomes across various animal taxa suggested prominent changes in diversity, composition of microbes, and functional capacity influenced by host-associated factors such as evolutionary history, diet, physiology, and habitat. For example, herbivores (ruminants) harboured cellulolytic microbes such as *Fibrobacter and Ruminococcus*, which enable efficient fibre degradation. Carnivores' gut microbiome was found to be dominated by proteolytic bacteria. Omnivores displayed considerably larger microbial diversity depending on their varied diets. In primates, host phylogeny strongly influenced microbial diversity, which is due to co-evolution and vertical transmission when compared to host diet. In contrast, in mammals, the host diet influenced the microbial community structure rather than host phylogeny (Degregori et al. 2025). However, the diversity and composition of the gut microbiome were altered depending on the environmental factors and lifestyle. For example, captive and experimental lab animals showed altered microbiomes when compared with their wild counterparts. This comparative framework aids in identifying phylosymbiosis (the extent of gut microbiome compositions relatable to specific hosts) and specific microbial functions, which improve the comprehension of host adaptation and help in microbiome-targeted diagnosis and treatment of disease in animals (McKenzie et al. 2017).

8.9.1 Ruminant, Monogastric, and Hindgut Fermenter Microbiota

Host diets shape distinct gut microbiota compositions, displaying their adaptations to diet and physiology. Martínez-Romero et al. (2021) reported the ingestion of endophytes by herbivores and humans, which modify the gut microbiome composition and diversity due to their fibres, flavonoids, carotenoids, alkaloids, bioactive metabolites, antimetabolites, and toxins. The rumen serves as a fermentation chamber that is home to dense populations of cellulolytic bacteria (*Fibrobacter, Ruminococcus*), archaea (methanogens), protozoa, and fungi in the stomach of the majority of ruminants (such as cows and sheep) (Chowdhury et al. 2015). In order to give the host energy, these microbes break down fibrous plant material into volatile fatty acids (VFAs) like butyrate, propionate, and acetate. For the breakdown of lignocellulose, the rumen microbiome has a high degree of diversity and functional specialization. The small and large intestines are the main organs used for digestion in monogastric animals, such as humans and pigs, which have a single-chambered stomach. Bacteroidetes and Firmicutes, which ferment dietary polysaccharides and generate SCFAs, predominate in their microbiota. Additionally, the gut microbiota supports pathogen resistance, immune modulation, and vitamin synthesis. Fibrous material is mostly fermented in the colon and cecum of hindgut fermenters, such as horses and rabbits. Their microbiota, which is frequently enriched in cellulolytic bacteria and anaerobic fungi adapted to the hindgut environment, is similar to that of ruminants in terms of fibre degradation but differs in terms of spatial location and microbial community structure (Wunderlich et al. 2025). Dietary management and health interventions specific to each type of digestive system are informed by an understanding of these microbial differences.

8.9.2 Laboratory Versus Wild-Type Animal Microbiome Dynamics

Laboratory animals and wild-type animals have markedly different microbiome composition and dynamics as a result of environmental exposure, diet, and lifestyle (Fig. 8.2). Laboratory animals are raised in a controlled, sanitized environment with standardized diets, which results in a less diverse microbiome and community structure compared to wild-type animals. The simplification often results in fewer commensals and overgrowth of competitors. Wild-type animals encounter more complex, changing environments, and they feed on a more diverse natural diet, which accounts for their more diverse microbiome (Couch et al. 2022). The diversity contributes to a higher resistance to pathogens as well as to a greater range of metabolic capacities that more closely reflect host–microbe co-evolution in the natural environment. This effect on microbiome composition has important consequences for immune development, metabolism, and behaviour and represents a challenge for translational research using laboratory models. Current efforts aiming to make the microbiome of laboratory animals more relevant by inoculating environmental microbes the microbiome of laboratory animals is expected to improve

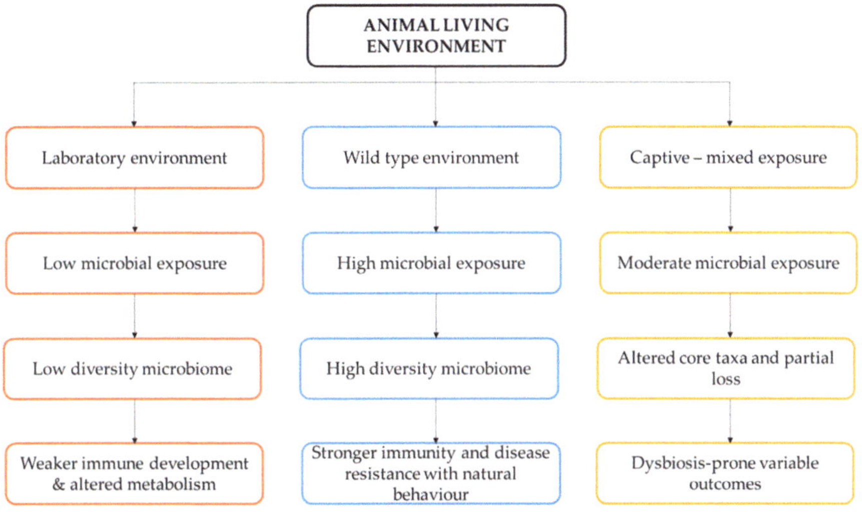

Fig. 8.2 Comparative microbiome dynamics in laboratory and wild-type animals

model relevance. The dynamics of these events are important for the interpretation of experimental results and the design of microbiome studies that have ecological validity. These results illustrate that laboratory animals have less diverse and compositionally distinct microbiomes than wild-type animals.

For example, studies by Rosshart et al. (2017) showed that laboratory mice have less microbial diversity and changed immune responses compared to wild mice. Wild mice have richer and more complex microbiota, which is linked to better disease resistance. Similarly, Clayton et al. (2016) found that the microbiomes of captive animals are different from those of wild populations. Captive animals often lack important microbial groups that help with nutrient metabolism and defending against pathogens. Controlled diets and sterile lab environments limit microbial exposure. This leads to simpler microbial communities dominated by Firmicutes and Bacteroidetes. In contrast, wild animals keep a wider range of types and functions because of their varied diets and exposure to environmental microbes. Attempts to "wildize" laboratory models of immunity, for instance, faecal microbiota transfer from wild organisms, have already been demonstrated to restore diversity and enhance physiological measures of fitness in laboratory models (Beura et al. 2016). Such data illustrate the essentiality of acknowledging host–microbe interaction contexts in experimental model design and interpretation.

8.10 Emerging Trends and Future Directions in Microbiome Mechanistic Research

Microbiome research is rapidly evolving, advancing, and shifting from simple microbial discovery and identification to uncovering their functional capability to influence human health. A multi-omics approach, which combines genomic, transcriptomic, proteomic, and metabolomic data to understand microbial physiology and functions, is becoming the major area of research (Schmidt et al. 2018). Tools like CRISPR, germ-free animal models, and synthetic microbiomes are used to understand the microbial roles in health and disease. Host-microbiome interactions play a major role in unravelling the immune modulation, metabolic control, and neurobiology in the host. Technologies like single-cell and spatial microbiomics enable the scientists to delve deep into understanding the microbial activity within specific tissues (Afrizal et al. 2022). The field is also moving toward personalized microbiome medicine, where individual microbial profiles guide interventions like tailored probiotics or dietary changes (Lee et al. 2021). Another exciting trend is microbiome engineering, where microbes are genetically modified to deliver therapeutic molecules or suppress harmful activity.

The future of microbial research will involve creating therapeutics from the microbiome. These include postbiotics, probiotics engineered for targeted therapies, and bacteriophage therapies. The development of standard protocols will be essential in ensuring scientific validity. The microbial populations in the skin, lungs, and reproductive systems are also being explored for involvement in systemic diseases. The application of artificial intelligence and computational simulations will likely be integral in predicting host responses and creating precise therapies. Ethical considerations will also be important in the future, especially in addressing issues in the handling of personalized microbiome data. Live microbial therapies will also require special considerations in terms of scientific approval.

8.10.1 Translational Implications for Animal Health, Productivity, and Welfare

The translational potential of host–microbiome interactions has been vast concerning enhanced animal health, productivity, and well-being. The manipulation of gut microbiota through probiotics, prebiotics, and diet has been useful in improving nutrient uptake, growth rates, and resistance to infections, thus making it an effective alternative to antibiotics. Thus, feeding animals a diet supplemented with *Lactobacillus-Bifidobacterium* has been demonstrated to increase growth and improve gastrointestinal health. Microbiome-based methods are also useful in disease prevention methods through enhanced barriers and reduced responses to pro-inflammatory agents. Also, methods using microbiota manipulation through diet manipulation, enhancing stress resistance and behaviours, are also useful. The development of microbiome analysis has provided for precision breeding based on species, breeds, and conditions to ensure maximum productivity alongside reduced impacts on the environment. Also, incorporating microbiota information into breeding has a vast translational approach, thus having a breeding criterion of animals characterized by a favourable composition of microbiota underlining superior health and performance (Ramond et al. 2025).

8.11 Future Perspectives

Despite the immense success achieved in deciphering the intricacies of the animal microbiota microbiome, there are a number of remaining gaps that could be explored in future research. Multi-omics, especially metagenomics, metatranscriptomics, metabolomics, and single-cell sequencing techniques at a high resolution, will enable a deeper understanding of microbial function and their interactions within a host. Moreover, machine learning will accelerate the development of microbial biomarkers, models of health and disease, and precision nutritional/therapeutic strategies. Moreover, longitudinal comparative research studies undertaken based on varied species will also be aided. The translational research will be based upon microbiome engineering, precision probiotics, postbiotics, and special diets that could improve both animal health, their immunity, and productivity. Moreover, translational research based upon ecologically relevant animal models of microbiota, along with microbiota-based diagnostic tools, could also revolutionize veterinary medicine. Moreover, early detection of diseases based on microbiota followed by personalized treatment strategies will also be a possibility. Importantly, future research will also need to address global challenges of animals like climate change, antibiotic resistance, and factory farming, which are known to be important influencers of gut microbial diversity.

8.12 Chapter Summary

This chapter is about the animal gut microbiome and how it affects the body. The gut microbiome is made up of different tiny living things that work together. We will look at what kinds of living things are in the gut and how they get there. The gut microbiome starts to form when animals are born and it changes as they get older. The food animals eat their genes and the world around them all play a role in what the gut microbiome looks like. The animal gut microbiome is very important because it helps animals digest their food, stay healthy, and keep their bodies in balance. The gut microbiome has different types of tiny living things that do different jobs, like helping animals break down the food they eat and keeping them from getting sick. The chapter also highlights major molecular pathways—including metabolite signalling, pattern-recognition receptor activation, and nutrient-sensing mechanisms—that govern host–microbe communication. Comparative perspectives across ruminants, monogastric species, and hindgut fermenters emphasize evolutionary and ecological diversity. Finally, emerging technologies and therapeutic strategies are discussed, underscoring the growing potential of microbiome-based approaches to enhance animal health, productivity, and welfare.

Conflict of Interest There is no conflict of interest regarding the publication of this chapter.

Author Contribution AR: Formulated the idea of the chapter, contributed to the literature review, and drafted the manuscript; NM: Contributed to content development, manuscript revision, and final editing.

References

Abe T, Marutani Y, Shoji I (2019) Cytosolic DNA-sensing immune response and viral infection. Microbiol Immunol 63:51–64

Acheson DW, Luccioli S (2004) Mucosal immune responses. Best Pract Res Clin Gastroenterol 18: 387–404

Adesogan AT (2009) Using dietary additives to manipulate rumen fermentation and improve nutrient utilization and animal performance. In: Proc 20th Florida Ruminant Nutrition Symp, Gainesville, pp 13–37

Afrizal A, Hitch TCA, Viehof A, Treichel N, Riedel T, Abt B et al (2022) Anaerobic single-cell dispensing facilitates the cultivation of human gut bacteria. Environ Microbiol 24:3861–3881

Aggarwal V, Sunder S, Verma SR (2022) Disease-associated dysbiosis and potential therapeutic role of *Akkermansia muciniphila*, a mucus-degrading bacterium of the gut microbiome. Folia Microbiol 67:811–824

Ahn J, Hayes RB (2021) Environmental influences on the human microbiome and implications for noncommunicable disease. Annu Rev Public Health 42:277–292

Akira S, Uematsu S, Takeuchi O (2006) Pathogen recognition and innate immunity. Cell 124:783–801

Alexander TW, Timsit E, Amat S (2020) Role of bovine respiratory microbiota in health and disease. Anim Health Res Rev 21:168–171
Amin N, Seifert J (2021) Dynamic progression of the calf microbiome and its influence on host health. Comput Struct Biotechnol J 19:989–1001
Ashraf R, Shah NP (2014) Immune system stimulation by probiotic microorganisms. Crit Rev Food Sci Nutr 54:938–956
Baars A, Oosting A, Knol J, Garssen J, Van Bergenhenegouwen J (2015) The gut microbiota as a therapeutic target in IBD and metabolic disease: a role for the bile acid receptors FXR and TGR5. Microorganisms 3:641–666
Baars DP, Fondevila MF, Meijnikman AS, Nieuwdorp M (2024) The central role of the gut microbiota in the pathophysiology and management of type 2 diabetes. Cell Host Microbe 32: 1280–1300
Bahrndorff S, De Jonge N, Hansen JK, Lauritzen JMS, Spanggaard LH, Sørensen MH et al (2018) Diversity and metabolic potential of the microbiota associated with a soil arthropod. Sci Rep 8: 2491
Bambace MF, Alvarez MV, del Rosario MM (2019) Novel functional blueberries: fructo-oligosaccharides and probiotic lactobacilli incorporated into alginate edible coatings. Food Res Int 122:653–660
Banerjee S, Schlaeppi K, van der Heijden MGA (2018) Keystone taxa as drivers of microbiome structure and functioning. Nat Rev Microbiol 16:567–576
Beura LK, Hamilton SE, Bi K, Schenkel JM, Odumade OA, Casey KA, Thompson EA, Fraser KA, Rosato PC, Filali-Mouhim A, Sekaly RP, Jenkins MK, Vezys V, Haining WN, Jameson SC, Masopust D (2016) Normalizing the environment recapitulates adult human immune traits in laboratory mice. Nature 532:512–516
Breczko WJ, Bubak J, Miszczak M (2024) Importance of intestinal microbiota and dysbiosis in intestinal lymphoma in dogs and cats. Cancers 16:2255
Castelli EM, Campos DL, English A, Nightingale KK, Brashears MM (2025) Dose-response effects of *Lactobacillus salivarius* on food-borne pathogens. J Food Prot 88:100502
Chen C, Cao Z, Li J, Tao C, Feng Y, Han Y (2020) A novel endophytic strain of *Lactobacillus plantarum* CM-3 with antagonistic activity against *Botrytis cinerea* on strawberry fruit. Biol Control. https://doi.org/10.1016/j.biocontrol.2020.104306
Chistiakov DA, Bobryshev YV, Kozarov E, Sobenin IA, Orekhov AN (2015) Intestinal mucosal tolerance and impact of gut microbiota to mucosal tolerance. Front Microbiol 5:781
Chowdhury SP, Hartmann A, Gao X, Borriss R (2015) Biocontrol mechanism by root-associated *Bacillus amyloliquefaciens* FZB42—a review. Front Microbiol 6:780
Chu H, Mazmanian SK (2013) Innate immune recognition of the microbiota promotes host-microbial symbiosis. Nat Immunol 14:668–675
Clayton JB, Vangay P, Huang H, Ward T, Hillmann BM, Al-Ghalith GA, Travis DA, Long HT, Tuan BV, Minh VV, Cabana F, Nadler T, Toddes B, Murphy T, Glander KE, Johnson TJ, Knights D (2016) Captivity humanizes the primate microbiome. Proc Natl Acad Sci USA 113: 10376–10381
Couch CE, McGee KM, Borgerhoff Mulder M, Tung J (2022) The gut microbiome in wild primates: ecological and evolutionary drivers of microbial diversity. Annu Rev Ecol Evol Syst 53:395–420
De la Fuente G, Yañez-Ruiz DR, Seradj AR, Balcells J, Belanche A (2019) Methanogenesis in animals with foregut and hindgut fermentation: a review. Anim Prod Sci 59:2109–2122
Degregori S, Wang X, Kommala A, Schulhof N, Moradi S, Mac Donald A, Eblen K, Jukovich S, Smith E, Kelleher E, Suzuki K (2025) Comparative gut microbiome research through the lens of ecology: theoretical considerations and best practices. Biol Rev 100:748–763
den Besten G, van Eunen K, Groen AK, Venema K, Reijngoud DJ, Bakker BM (2013) The role of short-chain fatty acids in the interplay between diet, gut microbiota, and host energy metabolism. J Lipid Res 54:2325–2340
Di Cagno R, De Angelis M, De Pasquale I, Ndagijimana M, Vernocchi P, Ricciuti P et al (2011) Duodenal and faecal microbiota of celiac children: molecular, phenotypic and metabolomic characterization. BMC Microbiol 11:219

Diether NE, Willing BP (2019) Microbial fermentation of dietary protein: an important factor in diet–microbe–host interaction. Microorganisms 7:19

Digby JE, Martinez F, Jefferson A, Ruparelia N, Chai J, Wamil M et al (2012) Anti-inflammatory effects of nicotinic acid in human monocytes are mediated by GPR109A dependent mechanisms. Arter Thromb Vasc Biol 32:669–676

Divyashree S, Ramu R, Sreenivasa MY (2024) Evaluation of new candidate probiotic Lactobacillus strains isolated from a traditional fermented food multigrainmillet dosa batter. Food Biosci 57:103450. https://doi.org/10.1016/j.fbio.2023.103450

Dong Z, Zhang S, Cheng Y, He X, Haase I, Liang Y, Jiang Y, Wu Y (2022) Comparative analysis of intestinal tract microbiota and feeding habits of five sympatric flycatchers. Avian Res 13: 100050

Eggesbø M, Moen B, Peddada S et al (2011) Development of gut microbiota in infants not exposed to medical interventions. APMIS 119:17–35. https://doi.org/10.1111/j.1600-0463.2010.02688.x

Elzinga J, Narimatsu Y, de Haan N et al (2024) Binding of *Akkermansia muciniphila* to mucin is O-glycan specific. Nat Commun 15:4582

Fecteau ME, Pitta DW, Vecchiarelli B, Indugu N, Kumar S, Gallagher SC et al (2016) Dysbiosis of fecal microbiota in cattle infected with Mycobacterium avium. PLoS One 11:e0160353

Feiyun F, Shi X, Chen P, Hu Q, Zhao Y, Chen Z, Zhang X (2026) Dietary drivers of gut microbiota diversity and function in wildlife of Wolong Nature Reserve: a metagenomic study. Curr Zool 72:14–29

Fernández CE, Giacaman RA, Tenuta LMA, Cury JA (2016) Effect of *Lactobacillus rhamnosus* on cariogenicity of Streptococcus mutans. Caries Res 49:583–590

Ferretti P, Pasolli E, Tett A et al (2018) Mother-to-infant microbial transmission from different body sites shapes the developing infant gut microbiome. Cell Host Microbe 24:133–145

Flint HJ, Bayer EA, Rincon MT, Lamed R, White BA (2008) Polysaccharide utilization by gut bacteria: potential for new insights from genomic analysis. Nat Rev Microbiol 6:121–131

Flint HJ, Scott KP, Duncan SH, Louis P, Forano E (2012) Microbial degradation of complex carbohydrates in the gut. Gut Microbes 3:289–306

Froidurot A, Julliand V (2022) Cellulolytic bacteria in the large intestine of mammals. Gut Microbes 14:2031694

Fu X, Lyu L, Wang Y, Zhang Y, Guo X, Chen Q, Liu C (2022) Safety assessment and probiotic characteristics of *Enterococcus lactis* JDM1. Microb Pathog 163:105380

Fülling C, Dinan TG, Cryan JF (2019) Gut microbe-to-brain signaling: what happens in vagus? Neuron 101:998–1002

Gasaly N, de Vos P, Hermoso MA (2021) Impact of bacterial metabolites on gut barrier function and host immunity: focus on bacterial metabolism and intestinal inflammation. Front Immunol 12:658354

Geva-Zatorsky N, Sefik E, Kua L, Pasman L, Tan TG, Ortiz-Lopez A et al (2017) Mining the human gut microbiota for immunomodulatory organisms. Cell 168:928–943

González A, Hall MN, Lin SC, Hardie DG (2020) AMPK and TOR: the yin and yang of cellular nutrient sensing and growth control. Cell Metab 31:472–492

Gutierrez Lopez DE, Lashinger LM, Weinstock GM, Bray MS (2021) Circadian rhythms and the gut microbiome synchronize the host metabolic response to diet. Cell Metab 33:873–887

Kamada N, Seo SU, Chen GY, Núñez G (2013) Role of the gut microbiota in immunity and inflammatory disease. Nat Rev Immunol 13:321–335

Kandpal M, Indari O, Baral B, Jakhmola S, Tiwari D, Bhandari V et al (2022) Dysbiosis of gut microbiota from the perspective of the gut-brain axis: role in neurological disorders. Metabolites 12:1064

Kim M, Friesen L, Park J, Kim HM, Kim CH (2018) Microbial metabolites, short-chain fatty acids, restrain tissue bacterial load, chronic inflammation, and associated cancer in the colon of mice. Eur J Immunol 48:1235–1247

Kim MH, Kang SG, Park JH, Yanagisawa M, Kim CH (2013) Short-chain fatty acids activate GPR41 and GPR43 on intestinal epithelial cells to promote inflammatory responses in mice. Gastroenterology 145:396–406

Kim S, Ndwandwe C, Devotta H, Kareem L, Yao L, O'Mahony L (2025) Role of the microbiome in regulation of the immune system. Allergol Int 74:187–196

Laterza L et al (2016) The gut microbiota and immune system relationship in human graft-versus-host disease. Mediterr J Hematol Infect Dis 8:2016025

Lee SH, Cho SY, Yoon Y, Park C, Sohn J, Jeong JJ et al (2021) *Bifidobacterium bifidum* strains synergize with immune checkpoint inhibitors to reduce tumour burden in mice. Nat Microbiol 6: 277–288

Liu C, Yang L, Wei W, Fu P (2024) Efficacy of probiotics/synbiotics supplementation in patients with chronic kidney disease: a systematic review and meta-analysis. Front Nutr 11:1434613

Maritan E, Quagliariello A, Frago E et al (2024) The role of animal hosts in shaping gut microbiome variation. Philos Trans R Soc B 379:20230071

Martínez-Romero E, Aguirre-Noyola JL, Bustamante-Brito R, González-Román P, Hernández-Oaxaca D, Higareda-Alvear V et al (2021) We and herbivores eat endophytes. Microb Biotechnol 14:1282–1299

McFall-Ngai M, Hadfield MG, Bosch TCG, Carey HV, Domazet-Lošo T, Douglas AE et al (2013) Animals in a bacterial world, a new imperative for the life sciences. Proc Natl Acad Sci USA 110:3229–3236

McKenzie VJ, Song SJ, Delsuc F et al (2017) Effects of captivity on the mammalian gut microbiome. Integr Comp Biol 57:690–704

Messer JS, Liechty ER, Vogel OA, Chang EB (2017) Evolutionary and ecological forces that shape the bacterial communities of the human gut. Mucosal Immunol 10:567–579

Myer PR, Freetly HC, Wells JE, Smith TPL, Kuehn LA (2017) Analysis of the gut bacterial communities in beef cattle and their association with feed intake, growth, and efficiency. J Anim Sci 95:3215–3224

Pessione E (2012) Lactic acid bacteria contribution to gut microbiota complexity: lights and shadows. Front Cell Infect Microbiol 2:86

Pickard JM, Zeng MY, Caruso R, Núñez G (2017) Gut microbiota: role in pathogen colonization, immune responses, and inflammatory disease. Immunol Rev 279:70–89

Ramond P, Galand PE, Logares R (2025) Microbial functional diversity and redundancy: moving forward. FEMS Microbiol Rev 49:fuae031

Rooks MG, Garrett WS (2016) Gut microbiota, metabolites and host immunity. Nat Rev Immunol 16:341–352

Rosshart SP, Vassallo BG, Angeletti D, Hutchinson DS, Morgan AP, Takeda K, Hickman HD, McCulloch JA, Badger JH, Ajami NJ, Trinchieri G, Pardo-Manuel de Villena F, Yewdell JW, Rehermann B (2017) Wild mouse gut microbiota promotes host fitness and improves disease resistance. Cell 171:1015–1028

Rowan FE, Docherty NG, Coffey JC, O'Connell PR (2009) Sulphate-reducing bacteria in ulcerative colitis. Br J Surg 96:151–158

Sarkar A, McInroy CJA, Harty S et al (2024) Microbial transmission in social microbiomes and host health. Cell 187:17–43

Schmidt TSB, Raes J, Bork P (2018) The human gut microbiome: from association to modulation. Cell 172:1198–1215

Shade A, Handelsman J (2012) Beyond the Venn diagram: the hunt for a core microbiome. Environ Microbiol 14:4–12

Shen Y, Fan N, Ma SX et al (2025) Gut microbiota dysbiosis: pathogenesis, diseases, prevention, and therapy. MedComm 6:e70168

Sitkin SI, Tkachenko EI, Vakhitov TY (2016) Metabolic dysbiosis of the gut microbiota and its biomarkers. Clin Exp Gastroenterol 12:6–29

Sjögren YM, Tomicic S, Lundberg A et al (2009) Influence of early gut microbiota on maturation of childhood immune responses. Clin Exp Allergy 39:1842–1851

Sonnenburg JL, Chen CTL, Gordon JI (2006) Genomic and metabolic studies of the impact of probiotics on a model gut symbiont and host. PLoS Biol 4:e413

Tanaka M, Nakayama J (2017) Development of the gut microbiota in infancy and its health impact. Allergol Int 66:515–522

Tasnim N, Abulizi N, Pither J, Hart MM, Gibson DL (2017) Linking gut microbial ecosystems with environment. Front Microbiol 8:1935

Teramoto Y, Akagawa S, Hori SI et al (2023) Dysbiosis of the gut microbiota as a susceptibility factor for Kawasaki disease. Front Immunol 14:1268453

Thomloudi EE, Tsalgatidou PC, Baira E et al (2021) Genomic and metabolomic insights into *Bacillus halotolerans* hil4. Microorganisms 9:2508

Tasnim N, Abulizi N, Pither J, Hart MM, Gibson DL (2017) Linking the gut microbial ecosystem with the environment: does gut health depend on where we live? Front Microbiol 8:1935

Thompson J, Gregory S, Plummer S, Shields RJ, Rowley AF (2010) Assessment of *Vibrio* spp. as probiotics for Pacific white shrimp. J Appl Microbiol 109:1177–1187

Tropini C, Earle KA, Huang KC, Sonnenburg JL (2017) The gut microbiome: connecting spatial organization to function. Cell Host Microbe 21:433–442

Uebanso T, Shimohata T, Mawatari K, Takahashi A (2020) Functional roles of B-vitamins in the gut microbiome. Mol Nutr Food Res 64:2000426

Ungerfeld EM (2020) Metabolic hydrogen flows in rumen fermentation. Front Microbiol 11:589

Ussar S, Fujisaka S, Kahn CR (2016) Interactions between host genetics and gut microbiome in diabetes and metabolic syndrome. Mol Metab 5:795–803

Wahlström A, Sayin SI, Marschall HU, Bäckhed F (2016) Intestinal crosstalk between bile acids and microbiota and its impact on host metabolism. Cell Metab 24:41–50

Wang X, Wang R, He S et al (2024) Biocontrol potential of *Bacillus velezensis* WR8 secondary metabolites. Gene 882:147872

Waters SM, Roskam E, Smith PE, Kenny DA, Popova M, Eugène M et al (2025) The role of rumen microbiome in the development of methane mitigation strategies for ruminant livestock. J Dairy Sci

Wilmanski T, Rappaport N, Diener C, Gibbons SM, Price ND (2021) From taxonomy to metabolic output: what factors define gut microbiome health? Gut Microbes 13:1–20

Wunderlich G, Bull M, McGilchrist N, Zhao C, Ross T, Rose M, Chapman B (2025) The horse gut bacteriome and anaerobic mycobiome are influenced by seasonal forages. J Appl Microbiol 136

Yang W, Cong Y (2021) Gut microbiota-derived metabolites in the regulation of host immune responses and inflammatory diseases. Cell Mol Immunol 18:866–877

Yason JA, Liang YR, Png CW, Zhang Y, Tan KSW (2019) Interactions between a pathogenic Blastocystis subtype and gut microbiota: in vitro and in vivo studies. Microbiome 7:30

Yoo JY, Groer M, Dutra SVO, Sarkar A, McSkimming DI (2020) Gut microbiota and immune system interactions. Microorganisms 8:1587

Probiotics Against Mycoses and Mycotoxin-Related Illness in Livestock and Other Domestic Animals

9

Mohamed E. Abd El-Hack, Mahmoud Moustafa, Islam M. Youssef, Ayman E. Taha, and Asmaa F. Khafaga

Abstract

Fungal infections and mycotoxin contamination are ongoing concerns in contemporary animal husbandry, impacting production, animal welfare, and food safety. Conventional antifungal medications frequently encounter resistance problems and residual difficulties, but probiotics are developing as a viable biological alternative. This chapter examines the function of probiotic bacteria in managing fungal infections (mycoses) and alleviating the harmful effects of mycotoxins in livestock and other agricultural animals. We examine the methods by which probiotics suppress fungal proliferation, bolster immunological response, and bind or break down mycotoxins in the gastrointestinal tract. Emphasis is directed towards their impacts on ruminants, swine, poultry, and aquaculture. This chapter

M. E. Abd El-Hack (✉)
Department of Industrial Pharmacy, College of Pharmaceutical Sciences and Drug Manufacturing, Misr University for Science and Technology (MUST), Giza, Egypt
e-mail: mohamed.abdelhaq@must.edu.eg

M. Moustafa
Department of Biology, College of Science, King Khalid University, Abha, Kingdom of Saudi Arabia

I. M. Youssef
Animal Production Systems Research Department, Animal Production Research Institute, Agricultural Research Center, Dokki, Giza, Egypt

A. E. Taha
Department of Public Health, College of Veterinary Medicine, King Faisal University, Al-Ahsa, Saudi Arabia

A. F. Khafaga (✉)
Department of Pathology, College of Veterinary Medicine, King Faisal University, Al-Ahsa, Saudi Arabia
e-mail: akhfajah@kfu.edu.sa

M. Y. Sreenivasa et al. (eds.), *Probiotics as Antifungals*,
https://doi.org/10.1007/978-981-92-0002-3_9

examines recent discoveries from laboratory and field studies, emphasizes probiotic strains exhibiting antifungal properties, and delineates practical aspects for their application in animal feed. In a time requiring safer and more sustainable agriculture, probiotics emerge as an effective approach to address fungal problems in livestock systems.

Graphical Abstract

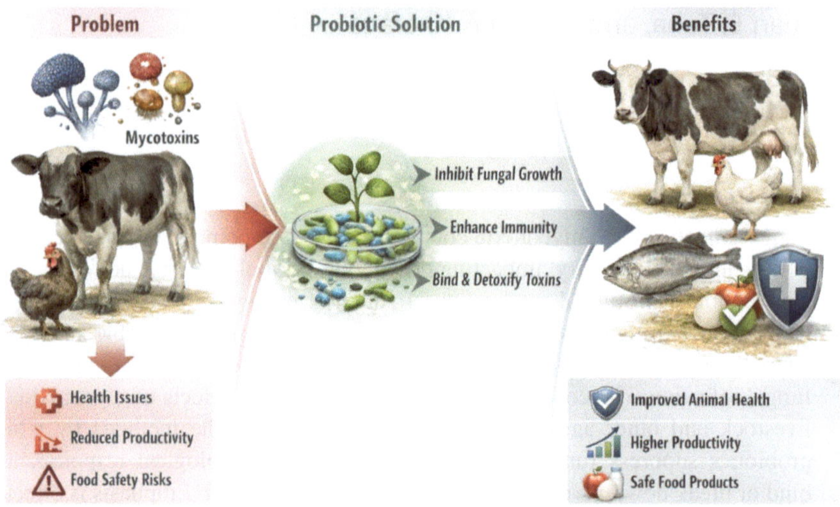

Keywords

Probiotics · Mycoses · Mycotoxins · Livestock health · Antifungal activity

Learning Objectives

By the end of the chapter, readers will be able to

- Understand the impact of mycoses and mycotoxin-related illnesses in livestock animals
- Describe the probiotic-mediated mycotoxin detoxification mechanisms
- Discuss the applications of probiotics in the livestock production system
- Promote sustainable and antibiotic-free disease management strategies

9.1 Introduction

Certain moulds generate toxic metabolic compounds known as mycotoxins. Numerous mycotoxins exhibit resilience to the severe circumstances encountered throughout the food supply chain due to their chemical stability. Aflatoxin, ochratoxins, citrinin, patulin, fumonisins, zearalenone, deoxynivalenol, nivalenol, and ergot alkaloids are the predominant mycotoxins detrimental to humans and animals. Certain mycotoxins are predominantly generated in the field, whereas others may be produced both in the field and during the postharvest stage. Mycotoxins can cause various health issues, including specific diseases, a compromised immune system, irritation, and allergic or lethal effects in both animals and people (Dass et al. 2007; Awuchi et al. 2022). Various fungi generate toxic secondary metabolites known as mycotoxins, which contaminate animal feed and pose significant threats to food safety, productivity, and animal health (Akinmoladun et al. 2025).

The assemblage of secondary metabolites is chemically diverse, with notable examples including aflatoxins (AFs) from *Aspergillus flavus*, *Penicillium verrucosum*, ochratoxin A (OTA) from *Aspergillus ochraceus*, and various toxins produced by *Fusarium* species, such as zearalenone (ZEN), deoxynivalenol (DON), trichothecenes, and fumonisins (FUM) (Maphaisa et al. 2025). Recently, newly identified mycotoxins such as enniatins (ENNs) and beauvericin (BEA) have garnered interest due to their sporadic presence in food and feed (Siri-Anusornsak et al. 2022).

Cereals, nuts, and their derivatives are prevalent sources of mycotoxins, which often accumulate in compounded animal feeds at escalating levels (Gruber-Dorninger et al. 2019). Mycotoxins primarily infiltrate animals and their byproducts (including eggs, organs, meat, and milk) via the ingestion of contaminated feed, presenting a significant threat to public health. Excessive ingestion of mycotoxins leads to mycotoxicosis, which results in diminished production, compromised immunity, increased disease susceptibility, and potentially mortality, contingent upon the dosage and duration of exposure (Njobeh et al. 2012).

In addition to toxicity issues, fungal infections present difficulties owing to biofilm formation, antifungal tolerance, and resistance, all of which elevate fungal morbidity and mortality (Garvey et al. 2022). Tolerance to antifungal agents denotes a partial growth that continues in the presence of inhibitory drug levels, whereas resistance indicates a total absence of response to antifungal therapy (Gow et al. 2022). Fungal species frequently exhibit resistance to at least one of the four principal classes of antifungal agents: polyenes, echinocandins, allylamines, and azoles (Fisher et al. 2022). Additional information regarding the chemical structures of these antifungal compounds has been encapsulated in reviewed papers (Casteli et al. 2016; Van Daele et al. 2019). Alongside fungal resistance, environmental pollution with antibiotic residues, resistant microbes, and resistance genes has emerged as a chronic global public health issue (Manyi-Loh et al. 2018). Researchers are currently concentrating on efforts to reduce antibiotic usage in cattle, specifically in the prevention of illnesses such as diarrhoea in heifers, by developing safer, sustainable alternatives (Agyare et al. 2018). Lactic acid bacteria (LAB) are a promising alternative due to their antimicrobial properties, ability to improve gut microbial equilibrium, and capacity to modulate immune responses by inhibiting Th2 cytokines (IL-4 and IL-5), augmenting CD4+ and CD8+ cell populations, increasing IL-12 and IFN-γ levels, and enhancing phagocytic activity (Markowiak and Śliżewska 2017; Grigas et al. 2021; Youssef et al. 2024).

Lactobacillus acidophilus, Bacillus spp., *Lactobacillus casei*, *Lactococcus lactis*, *Bifidobacterium bifidum*, and *Saccharomyces cerevisiae* have been employed as probiotics (Salamoura et al. 2014; Ashour et al. 2024; Cufadar et al. 2024). Biological approaches for mycotoxin breakdown and detoxification yield minimal or no end products and/or hazardous intermediates, providing an alternative approach to mycotoxin control (Awuchi et al. 2021). Fermentation is effective in decreasing and removing mycotoxins. Certain research indicates that the biological elimination of mycotoxin toxicity and its ability to modify their chemical structures merits increased attention (Afsah-Hejri et al. 2020). Microorganisms employed for this purpose include *Mycobacterium fluoranthenivorans*, *Flavobacterium aurantiacum*, *Nocardia corynebacterioides*, *Enterococcus faecium*, *Saccharomyces cerevisiae*, and *Lactobacillus rhamnosus* (Afsah-Hejri et al. 2020). To elaborate and distinguish from (Afsah-Hejri et al. 2020), briefly address dietary diversification and fermentation, including fungus, yeast, and bacteria (Awuchi et al. 2021). The primary objective of this chapter is to examine the detrimental effects of mycotoxins and the methods by which probiotics suppress fungal proliferation, bolster immunological response, and bind or break down mycotoxins in the gastrointestinal tract. Emphasis is placed on their advantageous effects in cattle species.

9.2 Fungal Pathogens and Mycotoxins in Livestock Systems

9.2.1 Common Fungal Pathogens in Animals

Aspergillus, *Candida* spp., and *Cryptococcus* are the primary fungal strains associated with gas generation that can affect the brain, bloodstream, and lungs (Gow et al. 2022). Infections that disseminate are typically caused by *Coccidioides*, *Blastomyces*, *Histoplasma*, *Cryptococcus* spp., and *Paracoccidioides* (Hossain et al. 2022). Patulin is derived from *Penicillium griseofluvum* and is regarded as poisonous to both plants and animals; however, this is not definitively established (de Rocha et al. 2014). Strains such as *Cryptococcus*, *Aspergillus*, *Candida*, and *Pneumocystis* are associated with immunocompromised individuals, whereas dimorphic protozoans like *Paracoccidioides*, *Coccidioides*, *Blastomyces*, and *Histoplasma* affect immunocompetent individuals (Alastruey-Izquierdo et al. 2015). Exposure to mycotoxins, typically through ingestion, leads to many disorders, including mycotoxicoses and mycoses, which can ultimately be lethal (Omotayo et al. 2019). Mycotoxins can enter the food chains of people and animals through direct or indirect contamination. Once an element has been contaminated by a toxigenic fungus, the mycotoxins persist in the final product even after the fungus has been eradicated (de Rocha et al. 2014). This is referred to as indirect contamination of food products and animal feed. In contrast, direct contamination transpires when a toxigenic fungus infects a product, food, or feed, resulting in the subsequent synthesis of mycotoxins (Awuchi et al. 2021). There are more than one hundred identified mycotoxins, primarily produced by species within the fungal genera *Penicillium*, *Aspergillus*, and *Fusarium* (Van Egmond 2013).

Classic mycoses are diseases in animals and humans caused by pathogenic fungi, specifically dermatophytes (Moretti et al. 2013; Moskaluk and VandeWoude 2022). Manoyan et al. (2024) assert that dermatophytes are characterized by their pronounced keratinophilicity and ability to degrade keratin proteolytically. This trait enables dermatophytes to degrade keratinized tissues such as the epidermis, stratum corneum, and its components (Moretti et al. 2013; Paryuni et al. 2020; Moskaluk and VandeWoude 2022; Sylvén et al. 2023). Multiple agents are responsible for dermatophytosis. Numerous species exist among animal taxa (rodents, canines, fur-bearing mammals, felines, etc.), mostly inhabiting soil (Pontes et al. 2013), which is notably abundant in substrates such as keratin (avian feathers and animal fur) (Jartarkar et al. 2021; Martinez-Rossi et al. 2021). Farm animals, including small cattle and larger cattle (Rambozzi et al. 2022), as well as fur-bearing animals such as minks, Arctic foxes, and nutria rabbits (Chang et al. 2022; Kottferová et al. 2022), serve as primary reservoirs of dermatophyte fungi in rural regions.

9.2.2 Mycotoxins of Concern in Animal Feed

The predominant mycotoxins, as illustrated in Fig. 9.1, found in cereal crops such as corn, barley, and wheat are aflatoxins (AFT), zearalenone (ZEA), deoxynivalenol (DON), fumonisins (FB), and ochratoxins (OTA) (Grace 2013). Aflatoxins (G1, G2, B1, and B2) are regarded as the most concerning class of mycotoxins due to their extreme toxicity (Bulgaru et al. 2025). Teratogenicity, diminished feed efficiency, heightened disease susceptibility, hepatic tumours and lesions, as well as compromised eggshell thickness and carcass quality have all been associated with prolonged exposure of animals to sub-acute levels of AFBs (Lizárraga-Paulín et al. 2011; Bakker et al. 2018; Sewunet et al. 2024). Chen et al. (2019) assert that FB, ZEA, and DON are the most significant mycotoxins produced by *Fusarium*, given their detrimental effects on animal health and the subsequent cost repercussions.

For instance, DON or vomitoxin (Pinton et al. 2008) is associated with nausea, diarrhoea, anorexia, emesis, and gastrointestinal tract haemorrhage (Pinilla 2015). Studies have demonstrated that prolonged exposure to low quantities of DON can have adverse effects on pigs (Wellington et al. 2020). Severe diseases such as pulmonary oedema in pigs (Machado et al. 2018) and leukoencephalomalacia in horses (Cendoya et al. 2018) are associated with feed contaminated by FB. Moreover, studies have demonstrated that FB may exert immunosuppressive,

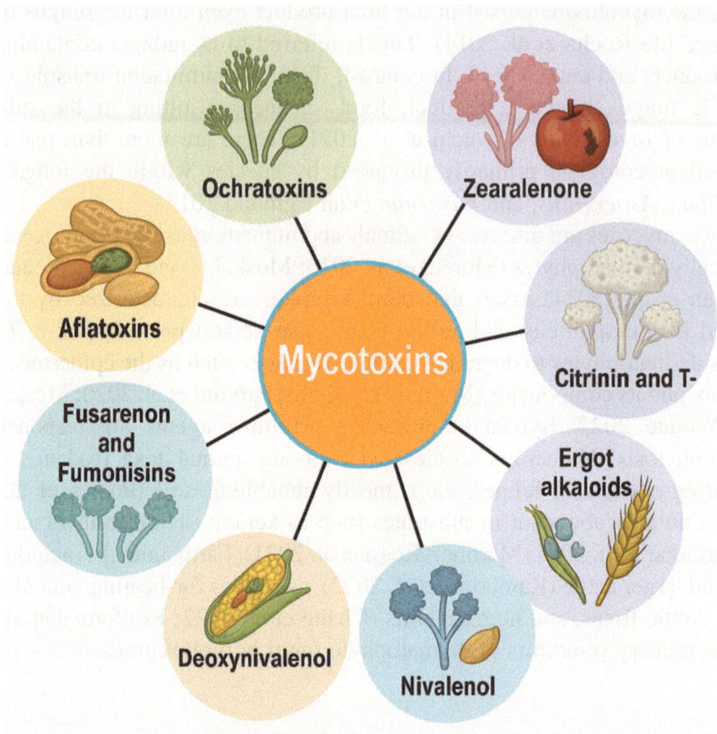

Fig. 9.1 Different types of mycotoxins produced by mycotoxigenic fungi

hepatotoxic, and nephrotoxic influences (Ruhnau et al. 2021). Data suggest that, unlike pigs, poultry doesn't demonstrate increased allergies to DON; instead, at high concentrations, it may compromise gut integrity (Awad et al. 2008), leading to reduced appetite, weight loss, and increased susceptibility to pathogenic bacteria (Aguzey et al. 2019; Bulgaru et al. 2021).

9.2.3 Clinical and Subclinical Effects in Farm Animals

Studies on animals indicate that OTA is immunotoxic, hepatotoxic, teratogenic, nephrotoxic, and neurotoxic. Pigs had the greatest incidence of allergies accompanied by nephropathy after being exposed to OTA-acute toxicity, primarily affecting the kidneys (Reddy and Bhoola 2010). Numerous animal investigations, including those involving chicks, quail, rabbits, hamsters, rats, and mice, have documented the teratogenic effects of mycotoxins, mostly characterized by reduced birth weight and craniofacial malformations (Malir et al. 2013; Heussner and Bingle 2015). In rats, the administration of a single dosage of T-2 toxin demonstrated dose-dependent metabolic changes. Even minimal concentrations (0.5 and 2 mg/kg body weight) elicited significant alterations in urinary metabolites, whereas elevated doses (4 mg/kg body weight) caused pronounced metabolic disruptions in the spleen, thymus, stomach, and liver, interfered with various metabolic pathways, and modified gut microbiota structure (Wan et al. 2015).

Exposure to T-2 toxin elicits oxidative stress as a primary mechanism of harm in vivo. In murine models, temporal elevations in oxidative damage to proteins and lipid peroxidation have been documented, suggesting that the formation of reactive oxygen species (ROS) is pivotal in cellular injury (Chaudhary et al. 2013). The suggested pathway can be illustrated as follows: Exposure to T-2 toxin leads to the formation of reactive oxygen species (ROS), resulting in oxidative stress, lipid peroxidation, protein oxidation, cellular and tissue damage, organ malfunction, and systemic toxicity. The percutaneous exposure route produces more severe consequences than the subcutaneous approach, indicating variations in toxin absorption and systemic distribution (Chaudhary et al. 2013). The extent of T-2 toxicity is affected by various parameters, such as species, age, dosage, and delivery method. T-2 toxin is regarded as one of the most perilous members of the trichothecene mycotoxin group due to its powerful cytotoxic, immunosuppressive, and oxidative properties (Li et al. 2011). *Fusarium* toxins, especially zearalenone (ZEA), demonstrate significant estrogenic action along with genetic, hepatic, and immunological toxicity (Zavistanaviciute et al. 2023). These mycotoxins primarily affect the reproductive system, interfering with endocrine functions and gonadal development in mammals, potentially causing abortion, stillbirth, and foetal malformations, thereby resulting in significant economic losses (Kowalska et al. 2016; Zheng et al. 2018; Zatecka et al. 2014; Tian et al. 2022). Likewise, ochratoxin A (OTA) has been linked to chronic hepatic injury and general health decline in mice (Awuchi et al. 2022).

While studies indicate that ZEA can elicit an immune response and affect various systems, including hepatic, digestive, and nephrological, the primary consequences of ZEA exposure in pigs are reproductive issues and hyperestrogenic symptoms, such as uterine enlargement and vulval swelling (Gerez et al. 2015). Further research on swine has demonstrated that a month of exposure to food contaminated with low concentrations of DON (2 ppm), ZEA (1.5%), and NIV (1.3 ppm) can lead to histological alterations in the colon, liver, and lymphoid organs, as well as impede the weight gain rate of piglets (Kalita et al. 2024). Furthermore, studies on poultry demonstrate that simultaneous exposure to OTA (1.8 ppb) and AFT (13 ppb) led to anorexia and developmental impairment, alongside necrotic lesions in the kidneys and liver (Grenier et al. 2016). DON and FB have been demonstrated in previous poultry studies to modify the immune system, exacerbating protozoan diseases in hens (Ying et al. 2019). OTA pollution has been linked to nephropathy infections in poultry and pigs, rendering kidney toxins the key target (Bulgaru et al. 2025).

Moreover, immunosuppression, diminished growth, and elevated mortality rates are correlated with OTA. In vitro studies have demonstrated that OTA can exacerbate intestinal permeability abnormalities induced by DON, even at non-toxic concentrations (van der Fels-Klerx 2010). Moreover, dietary contamination with

HT2 and T2 has been demonstrated to induce oral lesions in avians and impair the acquired immune response in swine (Wang et al. 2018).

9.3 Probiotic Mechanisms Action Against Mycotoxins and Fungi

9.3.1 Antifungal Effects of Probiotics

A critical functional attribute of lactic acid bacteria (LAB) is their capacity to reduce mycotoxin levels in contaminated feed and food items (Muhialdin et al. 2020). The detoxifying action, especially against aflatoxin B1 (AFB1), is reliant on the strain, indicating that efficacy differs among various LAB species and isolates (Afshar et al. 2020). The principal process entails the physical adsorption of AFB1 onto the bacterial cell surface, as opposed to enzymatic breakdown. The binding is primarily regulated by hydrophobic interactions, electrostatic attraction, hydrogen bonding, and van der Waals forces between mycotoxin molecules and functional groups on bacterial cell wall components, including peptidoglycans, polysaccharides, and teichoic acids (Afshar et al. 2020; Rämö et al. 2022). Upon binding, the toxin is sequestered on the bacterial surface, diminishing its bioavailability and absorption in the gastrointestinal system. In vitro studies have shown that *Lactobacillus plantarum* may eliminate roughly 21–60% of AFB1 within 30 minutes of incubation, contingent upon strain type and experimental conditions (Huang et al. 2017; Rämö et al. 2022). Fermentation by *Lactobacillus casei* and *Lactobacillus paracasei* can diminish AFB1 levels in wheat bran by an average of 52% and 46%, respectively, as reported by Zokaityte et al. (2021).

Strains of *Penicillium* and *Aspergillus* produce ochratoxin A (OTA), a secondary metabolite that ranks second in importance in feed after aflatoxin B1 (AFB1) (Khan Achakzai et al. 2017; González-López et al. 2021). Luz et al. (2018) reported that the capacity of LAB to reduce OTA depends on species characteristics and substrate pH. The *Lactobacillus plantarum* strain can achieve OTA reductions of 30–97% at pH 3.5 and 90–97% at pH 6.5. Luz et al. (2018) also demonstrated that the *Lactobacillus paracasei* strain reduces OTA by 21–16% at substrate pH levels of 3.5 and 6.5, respectively. Zokaityte et al. (2021) observed a 64% reduction in OTA levels in wheat bran fermented with *Lactobacillus paracasei* bacteria compared with untreated samples. The type of LAB cell wall protein and its structure, which differ among strains, determine the strain-specific reduction in ZEA (Adunphatcharaphon et al. 2021). Złoch et al. (2020) reported that the *Lactobacillus paracasei* strain diminishes ZEA through two mechanisms: (I) mycotoxin biosorption/binding within bacterial cells and (II) the metabolism of ZEA into less toxic ZOL and ά-ZOL forms. Adunphatcharaphon et al. (2021) reported that plant-derived *Lactobacillus plantarum* reduces ZEA by adsorption, resulting in a reduction of this mycotoxin by 0.5 to 21% (Fig. 9.2). Zokaityte et al. (2021) and Zavistanaviciute et al. (2023) reported that *Lactobacillus paracasei* reduces the T-2 toxin concentration in the outer layer of cereal. Reports indicate that LAB species are associated with the

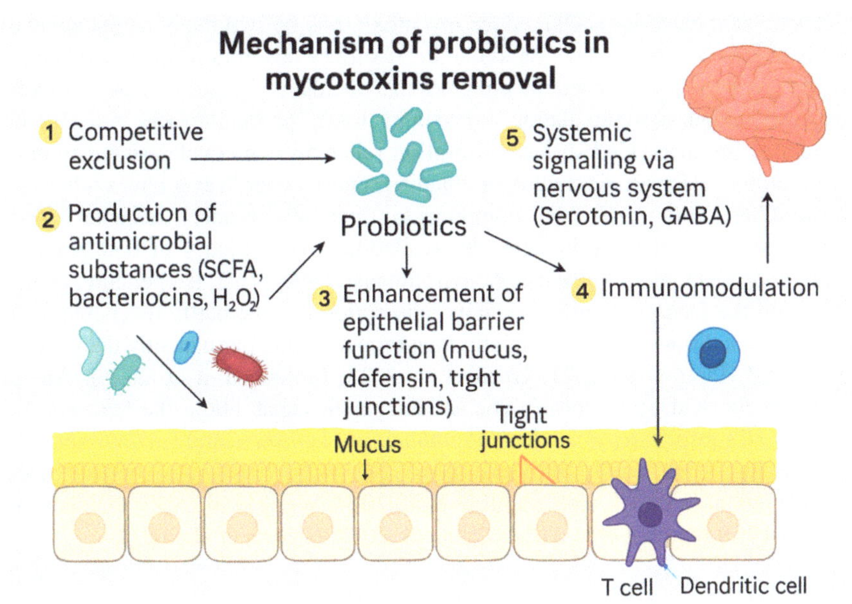

Fig. 9.2 Mechanism of probiotics in the removal of mycotoxins

12,13-epoxyring, which is crucial for reducing T-2 concentrations (Adhikari et al. 2017). Furthermore, T-2 can be converted to HT-2, and HT-2 may be partly responsible for the toxicity of T-2 (Arcella et al. 2017). Moreover, given the global prevalence of contamination by these deleterious compounds, one of the most vital aspects of LAB is its ability to suppress fungal growth and reduce mycotoxin levels in the substrate (Bartkiene et al. 2019; Luz et al. 2018; Guan et al. 2023). Aflatoxin B1 (AFB1) is recognised for causing hepatic issues and diminishing daily weight gain and production efficiency in livestock (Streit et al. 2012; Tian et al. 2022). AFM1, a toxic metabolite of AFB1, is excreted in the milk of dairy cows, raising significant concerns regarding AFB1 exposure in feed and presenting health risks to humans (Min et al. 2021). The removal of mycotoxins by lactic acid bacteria (LAB) is significantly influenced by culture parameters such as pH, temperature, incubation time, growth medium, and bacterial biomass, as indicated by several studies (Dalié et al. 2010). Desmosomes, adherens junctions, and tight junctions facilitate the mechanical interconnection of epithelial cells, establishing a barrier that excludes pathogens and large particles while permitting the passage of water, electrolytes, and dietary nutrients. A chemical barrier composed of cytokines, mucus, and antimicrobial peptides secreted by epithelial cells constitutes this barrier. The immunological barrier comprises capillary and gut-associated lymphoid tissue (GALT), located in the lamina propria. GALT comprises diverse immune system cells, including dendritic cells, macrophages, and lymphoid cells. To finalise the barrier established by epithelial cells, immune system cells secrete cytokines and immunoglobulin A.

Lactic acid bacteria (LAB) reduce mycotoxin risk by two main, complementary mechanisms: enzymatic degradation and cell-surface adsorption. Enzymatic degradation involves specific bacterial enzymes that chemically modify the toxin (hydrolysis, oxidation, de-epoxidation, or conjugation), producing less toxic, more excretable products. Cell-surface adsorption relies on non-covalent interactions—hydrophobic, electrostatic, hydrogen bonding, and van der Waals forces—between mycotoxins and cell-wall components (peptidoglycans, teichoic acids, polysaccharides), thereby lowering bioavailability. The efficacy of both routes is strain-dependent and influenced by environmental factors (pH, temperature, contact time) and the feed/gut matrix; crucially, adsorption may be partly reversible under digestive conditions, while enzymatic transformation can permanently detoxify the compound (Piotrowska 2021; Guan et al. 2023; Mohamed et al. 2025). Among them, species with strong proteolytic and enzymatic capabilities, like *Lactobacillus delbrueckii* subsp. *bulgaricus*, *Lactiplantibacillus plantarum*, and *Lacticaseibacillus paracasei*—are particularly recommended for use as feed additives to reduce mycotoxin levels in livestock diets (Aguirre et al. 2014; Wang et al. 2017; García-Cano et al. 2019; Raveschot et al. 2018; Luz et al. 2018; Nicosia et al. 2023).

9.3.2 Mycotoxin Detoxification and Removal

The hydrophobicity of the bacterial cell wall may be the principal factor responsible for toxin binding (Kumar et al. 2020). The binding of mycotoxins by bacterial cells was enhanced by cell wall components, chiefly polysaccharides, peptidoglycan, and teichoic and lipoteichoic acids (Niderkorn et al. 2009). Kumar et al. (2020) observed that after evaluating thirty LAB strains for their ability to bind OTA from solution, at least 95% of OTA removal occurred in liquid media containing *L. acidophilus*. Type A trichothecenes (such as T-2 toxin, diacetoxyscirpenol, and HT-2 toxin) pose a greater risk to humans and animals compared to other foodborne trichothecenes in the type B group (nivalenol and deoxynivalenol) (Janik et al. 2021). Juodeikiene et al. (2023) assert that the specific LAB strain utilized influences the degradation of trichothecenes, since different species may exhibit distinct cell densities and viabilities in contaminated feeds and meals with varying pH levels (Fig. 9.2). Trichothecene levels in contaminated medium are significant during the detoxification process (Zadeike et al. 2021; Juodeikiene et al. 2023). Given that HT-2 can induce neurological problems, emesis, and reduced feed consumption while also diminishing production, it is imperative to eliminate it from cattle feed (Ekwomadu et al. 2021). Moreover, it is established that the mycotoxins in vivo have a similar toxicity spectrum, and T-2 can be rapidly metabolized to HT-2 toxin in the gastrointestinal tract (Ekwomadu et al. 2021; Zadeike et al. 2021). Zavistanaviciute et al. (2023) reported that the pollution of L.pc244 and L.pl135 species could diminish levels of AFB1, ZEA, OTA, T-2, and HT-2. They proposed that a 3% concentration of both examined LAB species, as well as their combination, could serve as an environmentally sustainable and natural approach for mycotoxin reduction.

Detoxification of mycotoxins can be achieved through the metabolic or degradative activity of bacteria or enzymatic systems within the gastrointestinal tract. This biological process effectively neutralizes the toxins without leaving harmful residues or undesirable byproducts, making it an irreversible and environmentally friendly approach (Colović et al. 2019). Enhancing animal health through the inclusion of feed additives that lower exposure to mycotoxins has become an important strategy in livestock production (Garvey and Rowan 2023). In recent years, multidisciplinary research efforts have focused on developing efficient biological agents for mycotoxin detoxification (Hassan and Zhou 2018). Extensive evidence demonstrates that various microorganisms, particularly fungi and bacteria, can degrade or inactivate mycotoxins present in food and feed (Xia et al. 2017). LAB, including probiotics like *Bifidobacterium* and *Lactobacillus* strains, along with other bacterial species like *Micrococcus luteus* and *Bacillus subtilis*, can associate with fumonisins B1 and B2, presumably through peptidoglycan interactions (Scott 2012; Kabak and Var 2008). *Saccharomyces cerevisiae* is a recently identified microorganism that effectively binds to AFB1 and reduces its concentration in peanuts contaminated with *Aspergillus parasiticus* spores. This likely occurred due to the degradation of the mycotoxin or its binding to the yeast's cell wall (Giovati et al. 2014).

9.3.3 Immunomodulation

Cytokines, as principal markers in the intestinal immune system, are crucial for autoimmune diseases, inflammatory responses, and host defence (Huang et al. 2019). Various stimulators, such as mycotoxins, microbial infections, and additional factors, induce the release of IL-6, a crucial signalling molecule in the gut's immune and inflammatory response (Huang et al. 2018). AFB1 and ZEA, whether individually or in conjunction, have been demonstrated to elevate IL-6 mRNA expression, signifying that these mycotoxins provoke intestinal inflammation. Treatments including Z500 + A40 + CFSCP + MDE or the separate incorporation of CFSCP + MDE markedly decreased IL-6 mRNA levels, indicating that probiotics and mycotoxin-degrading enzymes are capable of reducing mycotoxin cytotoxicity (Fig. 9.3) (Huang et al. 2019). The intestine functions as the principal physical and immunological barrier, controlling nutrition and water absorption while obstructing the ingress of infections and toxins (Bouhet and Oswald 2007). The small intestine, as the initial organ to encounter and absorb AFB1 and ZEA, is especially vulnerable to oxidative and inflammatory damage (Pinton and Oswald 2014). ZEA has demonstrated the capacity to compromise epithelial cell integrity by affecting tight junction proteins and modifying intestinal permeability (Robert et al. 2017). Despite studies identifying these detrimental effects, additional studies are required to elucidate the mechanisms regulating the intestinal inflammatory response, barrier dysfunction, and nutritional absorption under concurrent AFB1 and ZEA exposure (Huang et al. 2019).

The immunomodulatory impacts of probiotics and mycotoxin-degrading enzymes (MDEs) are crucial in counteracting these detrimental consequences

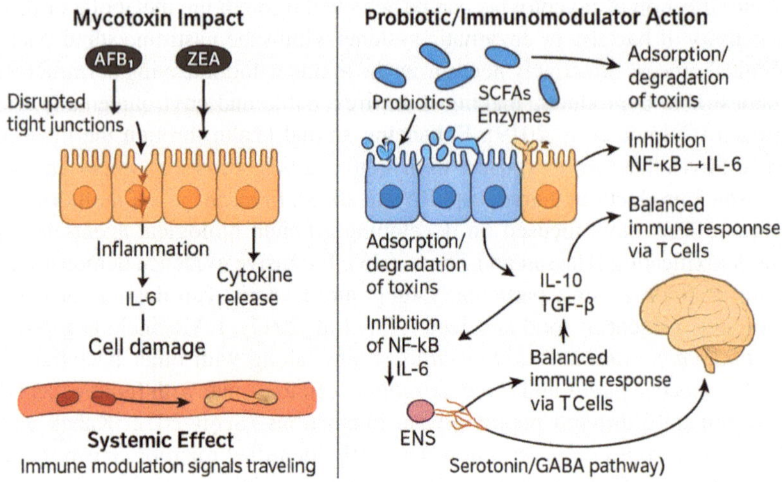

Fig. 9.3 Mechanistic pathway of probiotic and enzyme-based immunomodulation against mycotoxin-induced cytotoxicity

through many complementary pathways. Probiotics and MDEs enhance antioxidant enzyme activity (SOD, GPx, and CAT) and reduce pro-inflammatory cytokines (IL-6, TNF-α, and IL-1β) via inhibiting the activation of NF-κB and MAPK pathways (Fan et al. 2025). Furthermore, they enhance mucin production and reinstate tight junction proteins (occludin, claudins, and ZO-1), fortifying the intestinal barrier and preventing toxin transfer. Probiotics enhance immunological homeostasis and reduce tissue damage by increasing anti-inflammatory cytokines such as TGF-β and IL-10. Besides immunoregulation, probiotics can directly absorb or degrade mycotoxins into less toxic metabolites via enzymatic processes. Furthermore, microbial metabolites such as short-chain fatty acids (SCFAs) facilitate mucosal healing and modulate the gut–immune axis (Lázaro et al. 2024). These systems, when integrated, diminish inflammation, maintain intestinal and systemic homeostasis, and protect epithelial integrity in the presence of mycotoxins.

9.4 Applications in Different Animal Species

9.4.1 Poultry

Poultry production systems are equally vulnerable to the adverse effects of co-pollution and mycotoxins. Elevated levels of DON (0.1–0.9 mg/kg), ZEN (0.4–3.5 mg/kg), and FUM (1–33 mg/kg) markedly impaired the performance of chicken broilers in the research conducted by Shanmugasundaram et al. (2023). Concurrent exposure to mycotoxins results in substantial reductions in hatchability and egg production in laying hens. Verma et al. (2003) found that laying chickens subjected to simultaneous exposure to OTA (1–4 ppm) and AF (0.5–2 ppm) exhibited reduced egg production and feed efficiency. The interactions varied from cooperation to slightly less than additive, contingent upon the dosage. Ogido et al. (2004) discovered that FB1 (10 mg/kg) and AFB1 (50–200 μg/kg) exhibited an antagonistic interaction in quails, resulting in a less pronounced reduction in egg production when both toxins were combined compared to their individual administration. The research consistently indicates that exposure to various mycotoxins adversely affects laying performance. Moreover, Chowdhury et al. (2005) demonstrated that the consumption of grains often contaminated with *Fusarium* mycotoxins (DON) led to an increase in unmarketable eggs and a decline in egg production, mass, and weight of both eggs and eggshells. These findings present substantial concerns pertaining to food safety and productivity. Birds administered a diet contaminated with T-2 (4 ppm) and FB1 (20 ppm) exhibited a similar reduction in egg mass (Dazuk et al. 2020). The adverse effects of mycotoxins on animals are illustrated in Table 9.1.

The elimination of antibiotics as antibiotic growth promoters (AGPs) has adversely affected animal production systems economically (Laxminarayan et al. 2015). The chicken industry seeks alternatives to antibiotics that provide comparable benefits due to antibiotic resistance and the ban on antimicrobial growth promoters (Hernandez-Patlan et al. 2019). To mitigate antimicrobial resistance in the food chain and satisfy the increasing consumer demand for healthier products, novel antibiotic alternatives are necessary. Probiotics in chickens have shown several beneficial effects, including improved meat quality and flavour (Wang et al. 2017; Popova 2017), reduced pathogenic microorganisms (Kumar et al. 2018; Hernandez-Patlan et al. 2019), and enhanced growth characteristics (Alagawany et al. 2018; Yadav and Jha 2019). Salamoura et al. (2014) indicate that probiotics, including *Lactobacillus acidophilus*, *Bacillus* species, *Bifidobacterium bifidum*, *Lactobacillus casei*, *Lactococcus lactis*, and *Saccharomyces cerevisiae,* have been utilized. These strains positively influence intestinal microbiota equilibrium, suppress pathogens (Higgins et al. 2007; Kabir 2009), and enhance broiler performance (Kabir 2009; Vieco-Saiz et al. 2019).

9.4.2 Ruminants

Gruber-Dorninger et al. (2025) assert that multi-mycotoxin contamination adversely affects lactational performance in dairy cows. The intake of dry matter, milk production, energy-corrected milk, and 4% fat-corrected milk exhibited decreasing trends when nursing cows were administered mouldy maize meal (AF: 3 μg/kg; ZEN: 490 μg/kg) and cottonseed (AF: >150 μg/kg; ZEN: 84 μg/kg), as reported by Wu et al. (2022). The direct reduction of lactogenesis and indirect effects such as diminished dry matter intake (DMI), hepatic dysfunction, and modified rumen fermentation are among the mechanisms involved. A comprehensive mixed diet contaminated with various mycotoxins (AFB1 (38 μg/kg), OTA (501 μg/kg), ZEN (541 μg/kg), T-2 (270 μg/kg), FB1 (701 μg/kg), and DON (720 μg/kg)) markedly

Table 9.1 Adverse impact of mycotoxins on several livestock species

References	Livestock	Mycotoxin	Consequence
Kolawole and Tech (2020)	Pigs	DON+ZEN	Significantly reduced feed intake and weight gain compared to DON or ZEN administered individually.
Jia et al. (2020), Tessari et al. (2010)	Broilers	AFB1 + FB1 + OTA	↓ NDV antibody titres, significant hepatic damage, and complex immunological responses. Apoptosis and histological changes in the liver and kidney
Pierron et al. (2016)	Dairy goats	AFB1 + OTA+ ZEN (or FB1)	Enlarged liver enzyme concentrations (ALT, ALP), ↓ SOD, ↑ MDA, ↓ GSH-Px
Jovaisiene et al. (2016), Dänicke et al. (2017)	Dairy cows and calves	AFB1 + DON+ ZEN	↑ Somatic cell count, diminished immunological function ↓ WBC count, mild neutropenia
Kiyothong et al. (2012), Jia et al. (2020)	Ruminants	AFB1 + OTA + ZEN	Milk production and dry matter intake dramatically declined; the effects of certain mycotoxins were minimal
Prabu et al. (2013)	Rabbits	AFB1 + OTA	40% reduction in weight gain compared to 12% on an individual level; 25% mortality rate compared to 12.5%
Malekinejad et al. (2007), Chaytor et al. (2011)	Pig	AFB1 + DON + ZEN	Decreased macrophage activity, enhanced immunosuppression; reduced levels of IgG and IgA, diminished IL-2 and IFN-γ, decreased phagocytosis, reduced lymphocyte proliferation, and increased susceptibility to infection
Kiyothong et al. (2012)	Dairy cattle	AFB1 + DON + ZEN	Increased somatic cell count, decreased milk supply (14%), reduced milk fat and protein, diminished dry matter intake, and altered ruminal fermentation
Kolawole and Tech (2020)	Dairy goats, cows	AFB1 + DON + ZEN; AFB1 + OTA; AFB1 + OTA + ZEN	Reduction in milk yield (18%), synergistic effects of toxins, reduction in milk protein (12%)
Mohamed et al. (2019)	Sheep	AFB1 + OTA	↓ Fertility rate, lambing rate, fecundity rate, litter quantity and size, ↑ stillbirth

(continued)

Table 9.1 (continued)

References	Livestock	Mycotoxin	Consequence
Dazuk et al. (2020), Shanmugasundaram et al. (2023)	Broilers/ pullets	FUM + DON + ZEN; AF + OTA	Increased feed conversion ratio (20%), decreased average daily growth, reduced egg hatchability and production, diminished egg quality
Döll et al. (2003)	Pigs (sows)	ZEN + DON	Elevated levels of FSH and increased mean uterine weight
Malekinejad et al. (2007)	Pigs (grower–finishers)	DON + ZEN + FB1	Increased backfat thickness, decreased water retention capacity, reduced loin eye region, and darker coloration

diminished milk production, as reported by Kiyothong et al. (2012). Jiang et al. (2021) emphasized the compounded risks of multi-mycotoxin exposure, observing that cows administered contaminated aflatoxin mixtures produced less milk than those exposed to non-toxic substances.

Simultaneous mycotoxin exposure throughout pregnancy has been shown to impact the reproductive and production capabilities of animals, including their offspring. Mohamed et al. (2019) reported that pregnant Ossimi sheep consuming a diet contaminated with OTA (100 μg/kg DM) and AFB1 (50 μg/kg DM) exhibited reductions in pregnancy ratio, fertility rate, litter size, fecundity rate, lambing rate, and stillbirth by 37.5%, 44.6%, 44.12%, 47.93%, 39.29%, and 35.77%, respectively. Moreover, the composition of Ossimi ewes' milk, including fat, protein, total solids, and lactose, experienced a significant decline. Goats and dairy cattle appear to be significantly susceptible to multi-mycotoxin issues. Huang et al. (2017) demonstrated that AFB1 alone had minimal effects; however, dairy goats administered varying amounts of AFB1 (50 μg/kg DM), ZEN (500 μg/kg DM), and OTA (100 μg/kg DM) had substantial declines in milk production, suggesting a synergistic interaction. The integration of compound probiotics with mycotoxin-degrading enzymes for the concurrent detoxification of ZEA and AFB1 is an innovative and promising approach. The combination of *Candida utilis* and *Lactobacillus casei* (1:1:1) significantly enhanced the degradability of AFB1 and ZEA, achieving degradation rates of 45.49% and 44.90%, respectively, compared to Bacillus subtilis alone (Table 9.2, Liu et al. 2022).

9.4.3 Aquaculture

Bryden (2012) indicates that many ichthyological investigations have demonstrated diminished growth rates, particularly in Nile tilapia and channel catfish, when fed diets containing 1880 and 10,000 μg AFB1 per kg of feed, respectively. Published research indicates that dietary levels of DON below the recommended EU threshold (5 mg/kg) adversely affect the health of the liver, stomach, kidneys, gills, and spleen

Table 9.2 Summary of probiotic strains' effects on mycotoxin removal in livestock

Probiotic strain	Mycotoxins	Animal Species	Effect	References
Lactobacillus acidophilus	Aflatoxins	Cattle	Reduced toxicity, improved gut health	Jiang et al. (2021)
Lactobacillus acidophilus	Fumonisins	Pigs	↓ toxicity, improved gut health	Wang et al. (2023)
Bifidobacterium bifidum	Deoxynivalenol (DON)	Goats	↓ toxicity, improved gut health	Chen et al. (2021)
Bifidobacterium bifidum	Zearalenone	Poultry	↓ toxicity, improved reproductive health	Aalipanah et al. (2022)
Streptococcus thermophilus	Ochratoxin A	Cattle	↓ toxicity, improved kidney function	Ghadaksaz et al. (2022)
Streptococcus thermophilus	Fumonisins	Pigs	↓ toxicity, improved gut health	
Bacillus subtilis	Aflatoxins	Poultry	↓ toxicity, improved immune function	Gao et al. (2022)
Bacillus subtilis	Deoxynivalenol (DON)	Pigs	Reduced toxicity, improved gut health	Huang ct al. (2020)
Enterococcus faecium	Zearalenone	Cattle	↓ toxicity, improved reproductive health	Zavistanaviciute et al. (2023)
Enterococcus faecium	Ochratoxin A	Pigs	↓ toxicity, improved kidney function	Prasad et al. (2023)

in cyprinids and salmonids, in addition to impairing growth performance and feed efficiency (Gruber-Dorninger et al. 2025). Moreover, shrimp development performance was adversely affected by sub-regulatory levels of DON, which also led to histological changes in the intestines and hepatopancreas. Huang et al. 2018, 2019, and 2020 reported that the minimal detrimental effect level of DON in fish was 0.318 mg/kg feed, which resulted in decreased feed consumption and induced oxidative stress in grass carp. The reduced growth rate in Pacific white shrimp exhibited the least apparent adverse impact at 0.2 mg/kg DON in the diet (Trigo-Stockli et al. 2000; Gruber-Dorninger et al. 2025). The growth and health of fish and shrimp may be compromised, as a recent study indicates that 9.3% of tested fish feed samples and 8.2% of tested shrimp feed samples exceeded 0.318 mg/kg of deoxynivalenol (DON) (Gruber-Dorninger et al. 2025). *Lactobacillus amylovorus* demonstrated the capacity to bind food contaminants with an efficiency of up to 60% when associated with AFB1. *Lactobacillus fermentum* was utilized in the fermentation of whole-grain sorghum, leading to reductions of 84% in T-2 and 98% in FB1 (Adebo et al. 2019).

9.4.4 Rabbits and Pigs

The reproductive efficacy of swine is particularly susceptible to concurrent mycotoxin exposure. Co-exposure to ZEN and DON has been shown to modify reproductive physiology, diminishing piglet survival and herd fertility (Akinmoladun et al. 2025). Döll et al. (2003) discovered that female weaned piglets consuming diets contaminated with maize containing Fusarium toxins (ZEN-1.2 mg/kg and DON-8.6 mg/kg) exhibited reduced serum levels of follicle-stimulating hormone and an approximate 100% increase in the average uterine weight relative to body weight. Reproductive challenges for swine breeders lead to considerable financial losses. Mycotoxin combinations influence the carcass features and meat quality of grower-finisher pigs, as well as their reproductive performance (Akinmoladun et al. 2025). Additionally, Grenier and Oswald (2011) found that pigs exposed to DON, ZEN, and FB1 demonstrated increased back fat thickness, diminished loin eye regions (the muscular section of the loin along the pig's back), and inferior meat quality, marked by darker colouration and less water-holding ability.

Combination mycotoxin exposure significantly adversely affects rabbits (Akinmoladun et al. 2025). The synergistic effects of FB1 (10 mg/kg) and T-2 (2 mg/kg) on weaned rabbits had a detrimental influence on red blood cell NA(+)/K(+) ATPase activity, with a decrease observed in the T-2 group and an increase in the FB1 group, as reported in the haematotoxin study by Szabó et al. (2014). Furthermore, Prabu et al. (2013) documented increased mortality and synergistic liver damage resulting from AFB1 + OTA co-exposure, with a mortality rate of 25%, which is double that of individuals exposed to individual toxins.

9.5 Chosen Probiotic Strains Exhibiting Antifungal and Mycotoxin-Binding Capabilities

Probiotics are utilized in cattle breeding to mitigate adverse external effects due to their ability to modulate immune and nervous system functions, reduce blood levels of toxins, facilitate food digestion throughout the cow's digestive process, and restore microbiota to baseline levels (Wochner et al. 2018; Bren et al. 2023). Incorporating spore-forming probiotic cultures and probiotics from spore-forming bacteria into feed is a promising strategy. *Bacillus* strains in the spore stage exhibit resilience to elevated temperatures and endure the processes of extrusion, granulation, and expansion (Bren et al. 2023). A study investigating *Bacillus licheniformis* with ZEA-removal capability and *Bacillus amyloliquefaciens* with a similar function (Ju et al. 2019), an examination of *Bacillus natto* and *subtilis* (Lee et al. 2017), and a study on *Bacillus licheniformis* exhibiting activity against AFB1 constitute the in vitro research on *Bacillus* species. Huang et al. (2018) examined the elimination of AFB1 and ZEA from animal feed by the combination of *Bacillus subtilis*, *Candida utilis*, and *Lactobacillus casei*. Summary of the utilization of probiotics in the treatment of mycotoxins (Table 9.2).

9.6 Evidence from Experimental and Field Trials

According to previous studies (Li et al. 2018), ZEA may enhance the caecal microbial abundance in rabbits. Other research indicates that microbial richness (ACE and Chao1) was augmented upon the incorporation of probiotic Lactobacillus sp. into the diets of pigs and broilers (Nakphaichit et al. 2011; Shin et al. 2019). Considering the varied results, it can be asserted that probiotics will diminish the microbial populations that mycotoxins have increased (Chang et al. 2020). A potential method for mycotoxin decontamination is biodegradation. Khaniana et al. (2019) indicate that *Lactobacillus plantarum* can degrade AFB1, whereas *Bacillus pumilus* has shown effectiveness in degrading ZEA (Wang et al. 2017). Research revealed that *Streptomyces* sp. can degrade both ZEA and AFB1 (Harkai et al. 2016). Reports indicate that *Bacillus subtilis* alleviates the adverse effects of ZEA and AFB1 on feed consumption and egg production (Jia et al. 2016). Previous investigations conducted in our laboratory (Huang et al. 2018) suggest that AFB1 and ZEA may be concurrently degraded by the probiotic compounds *Bacillus subtilis*, *Candida utilis*, and *Lactobacillus casei*. Chang et al. (2020) assert that gut microbiota significantly influences the health and productivity of animals.

An optimal microbiome governs gastrointestinal development, modulates gut-associated systemic immunity, and prevents pathogen colonization of the intestinal epithelium (Mills et al. 2012). Enhancing the proliferation of beneficial bacteria such as Lactobacillus strains while diminishing the proliferation of harmful bacteria like *Terrisporobacter* and *Clostridium* (Fig. 9.4) indicates that the incorporation of

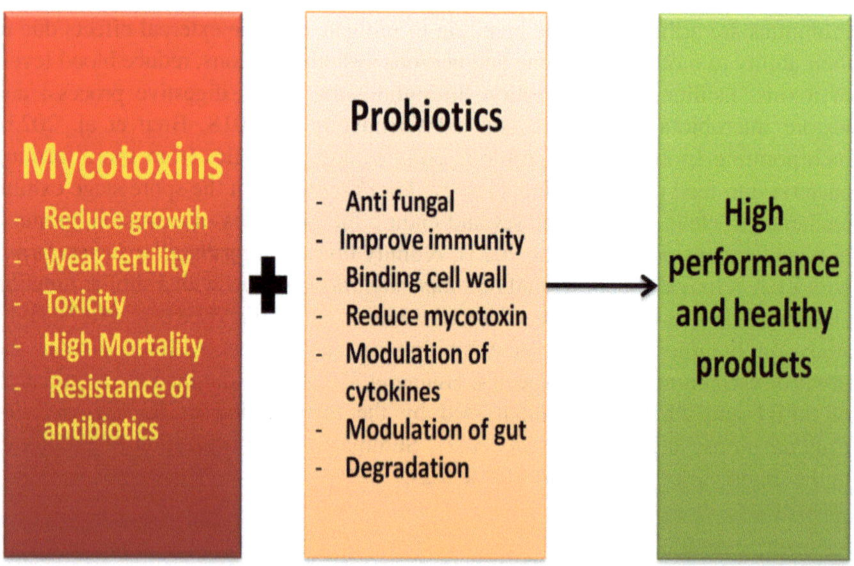

Fig. 9.4 Effect of probiotics to prevent the negative effects of mycotoxins

compound probiotics can mitigate ZEA and AFB1 residues and promote broiler growth (Chang et al. 2020).

Alterations in cell discharges and receptors were believed to modify the gut microbiome, as the microorganisms reside in proximity to epithelial cells (Cario 2005). Given the recent reviews on the impact of mycotoxins on the gastrointestinal system, we will enumerate the mechanisms via which mycotoxin exposure induces gut toxicity and alters the gut microbiota (Liew and Mohd-Redzwan 2018; Akbari et al. 2017; Bertero et al. 2020; Gao et al. 2020).

9.7 Practical Implementation and Limitations

Evidence suggests that synbiotic and probiotic formulations derived from *Bacillus* bacteria may alleviate this condition. Specific enzymes exhibiting lactonase activity can decompose mycotoxins due to the presence of a lactone ring. Liu et al. (2008) identify AHL lactonase and laccase as included entities. Bacteria often employ these enzymes to degrade the signalling molecules of competing species, which are utilized for intercellular communication and the development of quorum sensing, rather than to counteract mycotoxins (Rasch et al. 2004). The predominant autoinducer of quorum-sensing systems in gram-negative bacteria is N-acyl homoserine lactone (AHL), while probiotic Bacilli are recognized for harbouring various beneficial lactone-degrading enzymes (Tazehabadi et al. 2021). The mechanisms behind the beneficial effects of probiotic therapy in chickens remain unknown. Most research indicate that a probiotic species engages with the host's intestinal microbiome post-colonization by activating the gut immune system and producing biocidal chemicals such as bacteriocins and organic acids (Abd El-Hack et al. 2020).

9.8 Future Perspectives

Phytogenics, or plant-derived feed additives, are chemicals sourced from plants that confer health benefits to animals. They are utilized to augment digestion, enhance health, and promote growth performance in animals. Phytogenics demonstrate antimicrobial properties by suppressing the proliferation of pathogenic bacteria and fungi in the gastrointestinal tract, hence reducing the risk of infections and fostering a healthy gut microbiota (Austin and Ojebiyi 2025). The One Health idea recognizes the intricate interconnection of environmental integrity, human health, and animal health. Antimicrobial resistance (AMR) is a worldwide health issue affecting both human and veterinary medicine, exacerbated by the misuse of antibiotics in dairy production (Kiula and Makene 2025).

Lactobacillus casei and *Lactobacillus reuteri* (LAB) have shown effectiveness in binding aflatoxins in aqueous solutions. Other in vitro studies indicate that *Lactobacillus rhamnosus* and *amylovorus* can bind dietary contaminants with up to 60% effectiveness when associated with AFB1. *Lactobacillus fermentum* was employed in the fermentation of whole-grain sorghum, leading to reductions of 84% in T-2 and 98% in FB1 (Adebo et al. 2019). Ultrasound techniques, pulsed electric fields, and high-pressure processing (HPP) have been identified as novel and eco-friendly methods for controlling mycotoxins and fungi in food (Alizadeh et al. 2021). Given the vast spectrum of toxicities displayed, mycotoxins can, in fact, have harmful impacts on humans, animals, food, and plants. A primary challenge confronting several nations globally is the management of mycotoxin contamination in agricultural products, leading to the adoption of various preventive strategies (Awuchi et al. 2021). Pre-harvest and post-harvest preventive measures are influenced by mycotoxin decontamination, detoxification, and control protocols. It is apparent that techniques for mycotoxin detoxification and decontamination have diverse outcomes.

Future research should evaluate which specific harvesting and transportation methods more successfully mitigate mycotoxin toxicities along the food chain and the ecosystem, encompassing animal feed, human food, and plants. This contribution is deemed highly beneficial since it elucidates the properties of mycotoxins impacting animal, human, and plant consumption, hence affecting public health (Awuchi et al. 2021). Recent research has investigated the impact of gut microbiota in many diseases, demonstrating that probiotics, postbiotics (PPSP), prebiotics, and synbiotics can positively influence the microbiome (Lázaro et al. 2024).

9.9 Conclusion

Certain moulds generate toxic metabolic compounds known as mycotoxins. Numerous mycotoxins can endure the severe conditions encountered throughout the food supply chain due to their chemical stability. Aflatoxin, citrinin, fumonisins, patulin, ochratoxins, zearalenone, deoxynivalenol, nivalenol, and ergot alkaloids are the most common mycotoxins detrimental to humans and animals. Certain mycotoxins are predominantly generated in the field, whereas others may be produced both in the field and during the postharvest stage. Health effects of mycotoxins in animals and humans encompass various infections, compromised immunological responses to specific mycotoxins, death, and allergic or irritative reactions. Probiotics comprise *Lactobacillus acidophilus*, *Bacillus* species, *Lactococcus lactis*, *Bifidobacterium bifidum*, *Lactobacillus casei*, and *Saccharomyces cerevisiae*. Probiotics are employed in cattle breeding to mitigate adverse external effects due to their ability to modulate immune and nervous system functions, reduce blood levels of toxins, facilitate food digestion throughout the cow's digestive process, and restore microbiota to baseline levels. The integration of compound probiotics (CP) with mycotoxin-degrading enzymes for the concurrent detoxification of ZEA and AFB1 is an innovative and promising approach. Contemporary research has investigated the significant role of gut microbiota in several diseases, demonstrating that probiotics, postbiotics (PPSP), prebiotics, and synbiotics can enhance health by altering the microbiome. Researchers must concentrate on doing further studies to identify the appropriate probiotic strains for each type of livestock and to explore novel means of administration.

References

Aalipanah S, Fazeli MR, Akhavan Sepahi A et al (2022) Synergistic effects of probiotic *Bifidobacterium* isolated from chicken's intestine in combination with polyvinylpyrrolidone on reduction of aflatoxin B1. Lett Appl Microbiol 75(5):1160–1170

Abd El-Hack ME, El-Saadony MT, Shafi ME et al (2020) Probiotics in poultry feed: a comprehen sive review. J Anim Physiol Anim Nutr 104:1835–1850

Adebo OA, Kayitesi E, Njobeh PB (2019) Reduction of mycotoxins during fermentation of whole grain sorghum to whole grain ting (a Southern African food). Toxins 11:180

Adhikari M, Negi B, Kaushik N et al (2017) T-2 mycotoxin: toxicological effects and decontamination strategies. Oncotarget 8:33933–33952

Adunphatcharaphon S, Petchkongkaew A, Visessanguan W (2021) In vitro mechanism assessment of Zearalenone Remova by plant-derived lactobacillus plantarum BCC 47723. Toxins 13:286

Afsah-Hejri L, Hajeb P, Ehsani RJ (2020) Application of ozone for degradation of mycotoxins in food: a review. Compr Rev Food Sci Food Saf 19:1777–1808

Afshar P, Shokrzadeh M, Raeisi SN et al (2020) Aflatoxins biodetoxification strategies based on probiotic bacteria. Toxicon 178:50–58

Aguirre L, Hebert EM, Garro MS (2014) Proteolytic activity of *Lactobacillus* strains on soybean proteins. LWT Food Sci Technol 59:780–785

Aguzey HA, Gao Z, Haohao W et al (2019) The effects of Deoxynivalenol (DON) on the gut microbiota, morphology and immune system of chicken—a review. Ann Anim Sci 19:305–318

Agyare C, Boamah VE, Zumbi CN et al (2018) Antibiotic use in poultry production and its effects on bacterial resistance. IntechOpen, London UK. ISBN 978–784

Akbari P, Braber S, Varasteh S et al (2017) The intestinal barrier as an emerging target in the toxicological assessment of mycotoxins. Arch Toxicol 91:1007

Akinmoladun OF, Fon FN, Nji Q et al (2025) Multiple mycotoxin contamination in livestock feed: implications for animal health, productivity, and food safety. Toxins 17:365

Alagawany M, El-Hack ME, Farag MR et al (2018) The use of probiotics as eco-friendly alternatives for antibiotics in poultry nutrition. Environ Sci Pollut Res Int 25:10611–10618

Alastruey-Izquierdo A, Melhem MSC, Bonfietti LX et al (2015) Susceptibility test for fungi: clinical and laboratorial correlations in medical mycology. Rev Inst Med Trop Sao Paulo 57 (Suppl 19):57–64

Alizadeh AM, Hashempour-Baltork F, Khaneghah AM et al (2021) New perspective approaches in controlling fungi and mycotoxins in food using emerging and green technologies. Curr Opin Food Sci 39:7–15

Arcella D, Gergelova P, Innocenti ML et al (2017) Steinkellner H. Human and animal dietary exposure to T-2 and HT-2 toxin. EFSA J 15:e04972

Ashour EA, Youssef IM, Elsherbeni AI et al (2024) Influence of probiotic strains on growth performance, carcass characteristics and seasonal changes of growing New Zealand white rabbits. Trop Anim Health Prod 56(8):331

Austin K, Ojebiyi P (2025) Immune system development in poultry: role of early nutrition in enhancing immune responses in broilers

Awad WA, Ghareeb K, Bohm J et al (2008) The impact of the Fusarium toxin deoxynivalenol (DON) on poultry. Int J Poult Sci 7:827–842

Awuchi CG, Ondari EN, Ogbonna CU et al (2021) Mycotoxins affecting animals, foods, humans, and plants: types, occurrence, toxicities, action mechanisms, prevention, and detoxification strategies—a revisit. Foods 10(6):1279

Awuchi CG, Ondari EN, Nwozo S et al (2022) Mycotoxins' toxicological mechanisms involving humans, livestock and their associated health concerns: a review. Toxins 14(3):167

Bakker MG, Brown DW, Kelly AC et al (2018) Fusarium mycotoxins: a trans-disciplinary overview. Can J Plant Pathol 40:161–171

Bartkiene E, Bartkevics V, Mozuriene E et al (2019) The safety, technological, nutritional, and sensory challenges associated with lacto-fermentation of meat and meat products by using pure lactic acid bacteria strains and plant-lactic acid bacteria bioproducts. Front Microbiol 10:1036

Bertero A, Fossati P, Tedesco DEA et al (2020) Beauvericin and Enniatins: in vitro intestinal effects. Toxins 12(11):686

Bouhet S, Oswald IP (2007) The intestine as a possible target for fumonisin toxicity. Mol Nutr Food Res 51:925–931

Bren A, Denisenko Y, Prazdnova E et al (2023) Development of synbiotic preparations that restore the properties of cattle feed affected by toxin-forming micromycetes. Agriculture 13(3):523

Bryden WL (2012) Mycotoxin contamination of the feed supply chain: implications for animal productivity and feed security. Anim Feed Sci Technol 173(1–2):134–158

Bulgaru CV, Marin DE, Pistol GC, Taranu I (2021) Zearalenone and the immune response. Toxins 13:248

Bulgaru VC, Gras MA, Popa A et al (2025) Trends in mycotoxins co-occurrence in the complete feed for farm animals in southern Romania during 2021–2024 period. Toxins 17:201

Cario E (2005) Bacterial interactions with cells of the intestinal mucosa: toll-like receptors and nod2. Gut 54:1182–1193

Casteli MV, Derita MG, Lopez SN (2016) Novel antifungal agents: a patent review (2013–present). Expert Opin Ther Pat 27:415–426

Cendoya E, Chiotta ML, Zachetti V (2018) Fumonisins and Fumonisin-producing Fusarium occurrence in wheat and wheat by products: a review. J Cereal Sci 80:158–166

Chang J, Wang T, Wang P et al (2020) Compound probiotics alleviating aflatoxin B1 and zearalenone toxic effects on broiler production performance and gut microbiota. Ecotoxicol Environ Saf 194:110420

Chang CC, Wechtaisong W, Chen SY et al (2022) Prevalence and risk factors of zoonotic dermatophyte infection in pet rabbits in northern Taiwan. J Fungi 8(6):627

Chaudhary M, Bhaskar ASB, Rao PVL (2013) Differential effects of route of T-2 toxin exposure on hepatic oxidative damage in mice. Environ Toxicol 30:64–73

Chaytor AC, See MT, Hansen JA (2011) Effects of chronic exposure of diets with reduced concentrations of aflatoxin and deoxynivalenol on growth and immune status of pigs. J Anim Sci 89:124–135

Chen C, Turna NS, Wu F (2019) Risk assessment of dietary deoxynivalenol exposure in wheat products worldwide: are new codex DON guidelines adequately protective? Trends Food Sci Technol 89:11–25

Chen X, Su X, Li J et al (2021) Real-time monitoring of ruminal microbiota reveals their roles in dairy goats during subacute ruminal acidosis. NPJ Biofilms Microbiomes 7(1):45

Chowdhury SR, Smith TK, Boermans HJ et al (2005) Effects of feed-borne Fusarium mycotoxins on hematology and immunology in laying hens. Poult Sci 84:1698–1706

Colović R, Puvača N, Cheli F et al (2019) Decontamination of mycotoxin contaminated feedstuffs and compound feed. Toxins 11:617

Cufadar Y, Beyari EA, Hassoubah SA et al (2024) Impacts of Bacillus probiotics on productive performance and egg quality criteria in laying Japanese quails. J Appl Poult Res 33(3):100445

Dalié D, Deschamps A, Richard-Forget F (2010) Lactic acid bacteria potential for control of mould growth and mycotoxins: a review. Food Control 21:370–380

Dänicke S, Winkler J, Meyer U et al (2017) Haematological, clinical–chemical and immunological consequences of feeding Fusarium toxin contaminated diets to early lactating dairy cows. Mycotoxin Res 33:1–13

Dass RS, Sreenivasa MY, Janardhana GR (2007) High incidence of Fusarium verticillioides in animal and poultry feed mixtures produced in Karnataka, India. Plant Pathol J 6(2):174–178

Dazuk V, Boiago MM, Rolim G et al (2020) Laying hens fed mycotoxin-contaminated feed produced by Fusarium fungi (T-2 toxin and fumonisin B1) and *Saccharomyces cerevisiae* lysate: impacts on poultry health, productive efficiency, and egg quality. Microb Pathog 149: 104517

de Rocha MEB, da Chagas Oliveira Freire F, Maia FEF et al (2014) Mycotoxins and their effects on human 93. and animal health. Food Control 36:159–165
Döll S, Dänicke S, Ueberschär KH et al (2003) Effects of graded levels of Fusarium toxin contaminated maize in diets for female weaned piglets. Arch Anim Nutr 57:311–334
Ekwomadu TI, Akinola SA, Mwanza M (2021) Fusarium mycotoxins, their metabolites (free, emerging, and masked), food safety concerns, and health impacts. Int J Environ Res Public Health 18:11741
Fan Y, Liu Y, Shao C et al (2025) Gut microbiota-targeted therapeutics for metabolic disorders: mechanistic insights into the synergy of probiotic-fermented herbal bioactives. Int J Mol Sci 26: 5486
Fisher MC, Alastruey-Izquierdo A, Berman J et al (2022) Tackling the emerging threat of antifungal resistance to human health. Nat Rev Microbiol 20:557–571
Gao Y, Meng L, Liu H et al (2020) The compromised intestinal barrier induced by mycotoxins. Toxins 12:619
Gao S, Zhang L, Zhu D et al (2022) Effects of glucose oxidase and *Bacillus subtilis* on growth performance and serum biochemical indicexs of broilers exposed to aflatoxin B1 and endotoxin. Anim Feed Sci Technol 286:115186
García-Cano I, Rocha-Mendoza D, Ortega-Anaya J et al (2019) Lactic acid bacteria isolated from dairy products as potential producers of lipolytic, proteolytic and antibacterial proteins. Appl Microbiol Biotechnol 103:5243–5257
Garvey M, Rowan NJ (2023) Pathogenic drug resistant fungi: a review of mitigation strategies. Int J Mol Sci 24(2):1584
Garvey M, Meade E, Rowan NJ (2022) Effectiveness of front line and emerging fungal disease prevention and control interventions and opportunities to address appropriate eco-sustainable solutions. Sci Total Environ 851:158284
Gerez JR, Pinton P, Callu P et al (2015) Deoxynivalenol alone or in combination with nivalenol and zearalenone induce systemic histological changes in pigs. Exp Toxicol Pathol 67:89–98
Ghadaksaz A, Nodoushan SM, Sedighian H et al (2022) Evaluation of the role of probiotics as a new strategy to eliminate microbial toxins: a review. Probiotics Antimicrob Proteins 14(2): 224–237
Giovati L, Gallo A, Masoero F et al (2014) Vaccination of heifers with anaflatoxin improves the reduction of aflatoxin B1 carry over in milk of lactating dairy cows. PLoS One 9:e94440
González-López NM, Huertas-Ortiz KA, Leguizamon-Guerrero JE et al (2021) Omics in the detection and identification of biosynthetic pathways related to mycotoxin synthesis. Anal Methods 13:4038–4054
Gow NAR, Johnson C, Berman J et al (2022) The importance of antimicrobial resistance in medical mycology. Nat Commun 13:5352
Grace D (2013) Animals and aflatoxins. International Food Policy Research Institute, Washington DC. USA Brief 5: ISBN 9780896296763
Grenier BOIP, Oswald I (2011) Mycotoxin co-contamination of food and feed: meta-analysis of publications describing toxicological interactions. World Mycotoxin J 4:285–313
Grenier B, Dohnal I, Shanmugasundaram R et al (2016) Susceptibility of broiler chickens to coccidiosis when fed subclinical doses of deoxynivalenol and fumonisins—special emphasis on the immunological response and the mycotoxin interaction. Toxins 8:231
Grigas J, Ruzauskas M, Pautienius A et al (2021) Investigation of immunomodulatory and gut microbiota-altering properties of multicomponent nutraceutical prepared from lactic acid bacteria, bovine colostrum, apple production by-products and essential oils. Foods 10:1313
Gruber-Domninger C, Jenkins T, Schatzmayr G (2019) Global mycotoxin occurrence in feed: a ten-year survey. Toxins 11:375
Gruber-Domninger C, Müller A, Rosen R (2025) Multi-mycotoxin contamination of aquaculture feed: a global survey. Toxins 17(3):116
Guan Y, Lv H, Wu G et al (2023) Effects of lactic acid bacteria reducing the content of harmful fungi and mycotoxins on the quality of mixed fermented feed. Toxins 15:226

Harkai P, Szabo I, Cserhati M et al (2016) Biodegradation of aflatoxin-B1 and zearalenone by *Streptomyces* sp. collection. Int Biodeterior Biodegrad 108:48–56

Hassan Y, Zhou T (2018) Addressing the mycotoxin deoxynivalenol contamination with soil-derived bacterial and enzymatic transformations targeting the C3 carbon. World Mycotoxin J 11:101–112

Hernandez-Patlan D, Solis-Cruz B, Hargis BM et al (2019) The use of probiotics in poultry production for the control of bacterial infections and aflatoxins. In: Prebiotics and probiotics-potential benefits in human nutrition and health. IntechOpen, London

Heussner AH, Bingle LEH (2015) Comparative Ochratoxin toxicity: a review of the available data. Toxins 7:4253–4282

Higgins J, Higgins S, Vicente J et al (2007) Temporal effects of lactic acid bacteria probiotic culture on Salmonella in neonatal broilers. Poult Sci 86:1662–1666

Hossain CM, Ryan LK, Gera M et al (2022) Antifungal and drug resistance. Encyclopedia 2:1722–1737

Huang S, Zheng N, Fan C et al (2017) Effects of aflatoxin B1 combined with ochratoxin Aand/or zearalenone on metabolism, immune function, and antioxidant status in lactating dairy goats. Asian Australas J Anim Sci 31:505–513

Huang W, Chang J, Wang P et al (2018) Effect of the combined compound probiotics with mycotoxin–degradation enzyme on detoxifying aflatoxin B1 and zearalenone. J Toxicol Sci 43:377–385

Huang C, Feng L, Jiang WD et al (2019) Deoxynivalenol decreased intestinal immune function related to NF-kappaB and TOR signalling in juvenile grass carp (Ctenopharyngodon idella). Fish Shellfish Immunol 84:470–484

Huang C, Feng L, Liu XA et al (2020) The toxic effects and potential mechanisms of deoxynivalenol on the structural integrity of fish gill: oxidative damage, apoptosis and tight junctions disruption. Toxicon 174:32–42

Janik E, Niemcewicz M, Podogrocki M et al (2021) T-2 toxin the most toxic trichothecene mycotoxin: metabolism, toxicity, and decontamination strategies. Molecules 26:6868

Jartarkar SR, Patil A, Goldust Y et al (2021) Pathogenesis, immunology and management of dermatophytosis. J Fungi (Basel) 8(1):39

Jia R, Ma QG, Fan Y et al (2016) The toxic effects of combined aflatoxins and zearalenone in naturally contaminated diets on laying performance, egg quality and mycotoxins residues in eggs of layers and the protective effect of Bacillus subtilisbiodegradation product. Food Chem Toxicol 90:142–150

Jia R, Liu W, Zhao L et al (2020) Low doses of individual and combined deoxynivalenol and zearalenone in naturally moldy diets impair intestinal functions via inducing inflammation and disrupting epithelial barrier in the intestine of piglets. Toxicol Lett 333:159–169

Jiang Y, Ogunade IM, Vyas D et al (2021) Aflatoxin in dairy cows: toxicity, occurrence in feedstuffs and milk and dietary mitigation strategies. Toxins 13(4):283

Jovaisiene J, Bakutis B, Baliukoniene V, Gerulis G (2016) Fusarium and *Aspergillus* mycotoxin effects on dairy cow health, performance and the efficacy of anti-mycotoxin additive. Pol J Vet Sci 19:79–87

Ju J, Tinyiro SE, Yao W et al (2019) The ability of *Bacillus subtilis* and *Bacillus natto* to degrade zearalenone and its application in food. J Food Process Preserv 43:e14122

Juodeikiene G, Trakselyte-Rupsiene K, Reikertaite K et al (2023) Influence of biotreatment on Hordeum vulgare L. cereal wholemeal contamination and enzymatic activities. Foods 12:1050

Kabak B, Var I (2008) Factors affecting the removal of aflatoxin M1 from food model by *Lactobacillus* and *Bifidobacterium* strains. J Environ Sci Health B 43:617–624

Kabir S (2009) The role of probiotics in the poultry industry. Int J Mol Sci 10:3531–3546

Kalita JJ, Arya RS, Rajkhowa TK et al (2024) Pathology of co-contamination of mycotoxins in poultry farms of Aizawl district of Mizoram. Indian J Vet Sci Biotechnol 94:406–410

KhanAchakzai AK, Yaqoob M, Khan Barozai MY (2017) Chemical structure, occurrence and health hazard status of Ochratoxin A (OTA) in cereal food and feeds of Pakistan: a review. J Chem Soc Pak 39:867–878

Khaniana M, Karimi-Torshizia MA, Allamehb A (2019) Alleviation of aflatoxin-re lated oxidative damage to liver and improvement of growth performance in broiler chickens consumed *Lactobacillus plantarum* 299v for entire growth period. Toxicon 158:57–62

Kiula AH, Makene VA (2025) Phenotypic and genotypic resistance patterns of *Escherichia coli* in livestock from Dar es Salaam Tanzania. Biologia (Bratisl) 80:1897–1906

Kiyothong K, Rowlinson P, Wanapat M, Khampa S (2012) Effect of mycotoxin deactivator product supplementation on dairy cows. Anim Prod Sci 52:832–841

Kolawole O, Tech B (2020) Determination of Farm Animal Exposure to Multiple Mycotoxins and Assessment of Novel Mycotoxin Binders. Ph.D. Thesis, Queen's University Belfast, Belfast UK

Kottferová L, Molnár L, Čonková E et al (2022) Fungal flora in asymptomatic pet Guinea pigs and rabbits. Animals (Basel) 12(18):2387

Kowalska K, Habrowska-Górczy'nska DE, Piastowska-Ciesielska AW (2016) Zearalenone as an endocrine disruptor in humans. Environ Toxicol Pharmacol 48:141–149

Kumar S, Chen C, Indugu N et al (2018) Effect of antibiotic withdrawal in feed on chicken gut microbial dynamics, immunity, growth performance and prevalence of foodborne pathogens. PLoS One 13:e0192450

Kumar P, Mahato DK, Sharma B et al (2020) Ochratoxins in food and feed: occurrence and its impact on human health and management strategies. Toxicon 187:151–162

Laxminarayan R, Van Boeckel T, Teillant A (2015) The economic costs of withdrawing antimicrobial growth promoters from the livestock sector. OECD Food Agric Fish Pap 78

Lázaro Á, Vila-Donat P, Manyes L (2024) Emerging mycotoxins and preventive strategies related to gut microbiota changes: probiotics, prebiotics, and postbiotics–a systematic review. Food Funct 15:8998

Lee A, Cheng KC, Liu JR (2017) Isolation and characterization of a Bacillus amyloliquefaciens strain with zearalenone removal ability and its probiotic potential. PLoS One 12:e0182220

Li Y, Wang Z, Beier RC et al (2011) T-2 toxin, a trichothecene mycotoxin: review of toxicity, metabolism, and analytical methods. J Agric Food Chem 59:3441–3453

Li P, Yang S, Zhang X et al (2018) Zearalenone changes the diversity and composition of caecum microbiota in weaned rabbit. BioMed Res Int 2018:3623274

Liew WPP, Mohd-Redzwan S (2018) Mycotoxin: its impact on gut health and microbiota. Front Cell Infect Microbiol 8:60

Liu D, Momb J, Thomas PW et al (2008) Mechanism of the quorum-quenching lactonase (AiiA) from *Bacillus thuringiensis*. Product-bound structures. Biochemistry 47:7706–7714

Liu L, Xie M, Wei D (2022) Biological detoxification of mycotoxins: current status and future advances. Int J Mol Sci 23:1064

Lizárraga-Paulín EG, Moreno-Martínez E, Miranda-Castro SP (2011) Aflatoxins and their impact on human and animal health: an emerging problem. Aflatoxins Biochem Mol Biol 13:255–262

Luz C, Ferrer J, Mañes J, Meca G (2018) Toxicity reduction of Ochratoxin A by lactic acid bacteria. Food Chem Toxicol 112:60–66

Machado M, Andrade LR, Moraes DDA et al (2018) Concurrent ionophore poisoning and leukoencephalomalacia by consumption of contaminated commercial ration in horses. Acta Sci Vet 46:336

Malekinejad H, Schoevers EJ, Daemen IJ et al (2007) Exposure of oocytes to the Fusarium toxins zearalenone and deoxynivalenol causes aneuploidy and abnormal embryo development in pigs. Biol Reprod 77:840–847

Malir F, Ostry V, Pfohl-Leszkowicz A et al (2013) Ochratoxin A: developmental and reproductive toxicity—an overview. Birth Defects Res B Dev Reprod Toxicol 98:493–502

Manoyan MG, Gursheva AS, Gabuzyan NA, Panin AN (2024) Dermatophytoses of animals in the Russian regions: etiological structure and sensitivity of pathogens to antifungal drugs. Agric Biol 59(2):342–354

Manyi-Loh C, Mamphweli S, Meyer E, Okoh A (2018) Antibiotic use in agriculture and its consequential resistance in environmental sources: potential public health implications. Molecules 23:795

Maphaisa TC, Akinmoladun OF, Adelusi OA et al (2025) Advances in mycotoxin detection techniques and the crucial role of reference material in ensuring food safety. A review. Food Chem Toxicol 200:115387

Markowiak P, Śliżewska K (2017) Effects of probiotics, prebiotics, and synbiotics on human health. Nutrients 9(9):102

Martinez-Rossi NM, Peres NTA, Bitencourt TA et al (2021) State-of-the-art dermatophyte infections: epidemiology aspects, pathophysiology, and resistance mechanisms. J Fungi (Basel) 7(8):629

Mills S, Shanahan F, Stanton C et al (2012) Movers and shakers: influence of bacteriophages in shaping the mammalian gut microbiota. Gut Microbes 4:4–16

Min L, Fink-Gremmels J, Li D et al (2021) An overview of aflatoxin B1 biotransformation and aflatoxin M1 secretion in lactating dairy cows. Anim Nutr 7:42–48

Mohamed MY, Abd El-Hafeez AM, Ibrahim EMM et al (2019) Ameliorating effects of organic and inorganic mycotoxin binders on the performance of Ossimi sheep. Egypt J Sheep Goats Sci 14: 33–48

Mohamed MA, Anas H, Alduwish MA et al (2025) Influence of probiotic supplementation on growth, health and gut characteristics in growing rabbits. Ital J Anim Sci 24(1):1499–1514

Moretti A, Agnetti F, Mancianti F et al (2013) Dermatophytosis in animals: epidemiological, clinical and zoonotic aspects. G Ital Dermatol Venereol 148(6):563–572

Moskaluk AE, VandeWoude S (2022) Current topics in dermatophyte classification and clinical diagnosis. Pathogens 11(9):957

Muhialdin BJ, Saari N, Meor Hussin AS (2020) Review on the biological detoxification of mycotoxins using lactic acid bacteria to enhance the sustainability of foods supply. Molecules 25:2655

Nakphaichit M, Thanomwongwattana S, Phraephaisarn C et al (2011) The effect of including *Lactobacillus reuteri* KUB-AC5 during post-hatch feeding on the growth and ileum microbiota of broiler chickens. Poult Sci 90:2753–2765

Nicosia FD, Pino A, Maciel GLR et al (2023) Technological characterization of lactic acid bacteria strains for potential use in cheese manufacture. Foods 12:1154

Niderkorn V, Morgavi DP, Aboab B et al (2009) Cell wall component and mycotoxin moieties involved in the binding of fumonisin B1 and B2 by lactic acid bacteria. J Appl Microbiol 106(3): 977–985

Njobeh PB, Dutton MF, Åberg AT, Haggblom P (2012) Estimation of multi-mycotoxin contamination in South African compound feeds. Toxins 4:836–848

Ogido R, Oliveira CAF, Ledoux DR et al (2004) Effects of prolonged administration of aflatoxin B1 and fumonisin B1 in laying Japanese quail. Poult Sci 83:1953–1958

Omotayo OP, Omotayo AO, Mwanza M, Babalola OO (2019) Prevalence of mycotoxins and their consequences on human health. Toxicol Res 35:1–7

Paryuni AD, Indarjulianto S, Widyarini S (2020) Dermatophytosis in companion animals: a review. Vet World 13(6):1174–1181

Pierron A, Alassane-Kpembi I, Oswald IP (2016) Impact of two mycotoxins deoxynivalenol and fumonisin on pig intestinal health. Porc Health Manag 2:21

Pinilla V (2015) In vitro and in vivo effects of deoxynivalenol (DON) mycotoxin on porcine reproductive and respiratory syndrome virus (PRRSV) in piglets

Pinton P, Oswald IP (2014) Effect of deoxynivalenol and other type B trichothecenes on the intestine: a review. Toxins 6:1615–1643

Pinton P, Accensi F, Beauchamp E et al (2008) Ingestion of deoxynivalenol (DON) contaminated feed alters the pig vaccinal immune responses. Toxicol Lett 177:215–222

Piotrowska M (2021) Microbiological decontamination of mycotoxins: opportunities and limitations. Toxins 13:819

Pontes ZB, Oliveira AC, Guerra FQ et al (2013) Distribution of dermatophytes from soils of urban and rural areas of cities of Paraiba State, Brazil. Rev Inst Med Trop Sao Paulo 55(6):377–383

Popova T (2017) Effect of probiotics in poultry for improving meat quality. Curr Opin Food Sci 14:72–77

Prabu P, Dwivedi P, Sharma AK (2013) Toxicopathological studies on the effects of aflatoxin B1, ocharatoxin A and their interaction in New Zealand white rabbits. Exp Toxicol Pathol 65:277–286

Prasad S, Streit B, Gruber C, Gonaus C (2023) Enzymatic degradation of ochratoxin A in the gastrointestinal tract of piglets. J Anim Sci 101:skad171

Rambozzi L, Meneguz PG, Molinar Min AR et al (2022) Concurrent chorioptic mange and dermatophytosis in dairy goats: case report. Vet Sci 9(12):677

Rämö S, Kahala M, Joutsjoki V (2022) Aflatoxin B1 binding by lactic acid bacteria in protein-rich plant material fermentation. Appl Sci 12:12769

Rasch M, Buch C, Austin B et al (2004) An inhibitor of bacterial quorum sensing reduces mortalities caused by Vibriosis in Rainbow Trout (Oncorhynchus mykiss, Walbaum). Syst Appl Microbiol 27:350–359

Raveschot C, Cudennec B, Coutte F et al (2018) Production of bioactive peptides by *lactobacillus* species: from gene to application. Front Microbiol 9:2354

Reddy L, Bhoola K (2010) Ochratoxins—food contaminants: impact on human health. Toxins 2:771–779

Robert H, Payros D, Pinton P et al (2017) Impact of mycotoxins on the intestine: are mucus and microbiota new targets? J Toxicol Environ Health B 20:1–27

Ruhnau D, Hess C, Doupovec B et al (2021) Deepoxy-deoxynivalenol (DOM-1), a derivate of deoxynivalenol (DON), exhibits less toxicity on intestinal barrier function, campylobacter jejuni colonization and translocation in broiler chickens. Gut Pathog 13:44

Salamoura C, Kontogianni A, Katsipi D et al (2014) Probiotic fermented milks made of cows milk, goats milk and their mixture. J Biotechnol 185:125

Scott PM (2012) Recent research on fumonisins: a review. Food Addit Contam A 29:242–248

Sewunet SD, Kebede E, Melaku A et al (2024) Dairy farmers' knowledge, attitudes, and practices (KAP) towards aflatoxin contamination in milk and feeds in Bahir Dar, Ethiopia. Int J Microbiol 2024:5568286

Shanmugasundaram R, Lourenco J, Hakeem WA et al (2023) Subclinical doses of dietary fumonisins and deoxynivalenol cause cecal microbiota dysbiosis in broiler chickens challenged with *Clostridium perfringens*. Front Microbiol 14:1106604

Shin D, Chang SY, Bogere P et al (2019) Beneficial roles of probiotics on the modulation of gut microbiota and immune response in pigs. PLoS One 14:e0220843

Siri-Anusornsak W, Kolawole O, Mahakarnchanakul W et al (2022) The occurrence and co-occurrence of regulated, emerging, and masked mycotoxins in rice bran and maize from Southeast Asia. Toxins 14:567

Streit E, Schatzmayr G, Tassis P et al (2012) Current situation of mycotoxin contamination and co-occurrence in animal feed—focus on Europe. Toxins 4:788–809

Sylvén KR, Bergefur AL, Jacobson M et al (2023) Dermatophytosis caused by *Trichophyton mentagrophytes* complex in organic pigs. Acta Vet Scand 65(1):32

Szabó A, Szabó-Fodor J, Fébel H et al (2014) Individual and combined haematotoxic effects of fumonisin B1 and T-2 mycotoxins in rabbits. Food Chem Toxicol 72:257–264

Tazehabadi MH, Algburi A, Popov IV et al (2021) Probiotic Bacilli inhibit *Salmonella* biofilm formation without killing planktonic cells. Front Microbiol 12:242

Tessari EN, Kobashigawa E, Cardoso ALS et al (2010) Effects of aflatoxin B1 and fumonisin B1 on blood biochemical parameters in broilers. Toxins 2:453–460

Tian M, Feng Y, He X et al (2022) Mycotoxins in livestock feed in China—current status and future challenges. Toxicon 214:112–120

Trigo-Stockli DM, Obaldo LG, Dominy WG et al (2000) Utilization of deoxynivalenol-contaminated hard red winter wheat for shrimp feeds. J World Aquac Soc 31:247–254

Van Daele R, Spriet I, Wauters J et al (2019) Antifungal drugs: what brings the future? Med Mycol 57:S328–S343 van der Fels-Klerx HJ (2010) Occurrence data of trichothecene mycotoxins T-2 toxin and HT-2 toxin in food and feed. EFSA Support Publ 7:66E

Van Egmond HP (2013) Mycotoxins: risks, regulations and European co-operation. J Nat Sci Matica Srp Novi Sad 125:7–20

Verma J, Johri TS, Swain BK (2003) Effect of varying levels of aflatoxin, ochratoxin and their combinations on the performance and egg quality characteristics in laying hens. Asian Australas J Anim Sci 16:1015–1019

Vieco-Saiz N, Belguesmia Y, Raspoet R et al (2019) Benefits and inputs from lactic acid bacteria and their bacteriocins as alternatives to antibiotic growth promoters during food-animal production. Front Microbiol 10:57

Wan Q, Wu G, He Q et al (2015) The toxicity of acute exposure to T-2 toxin evaluated by the metabonomics technique. Mol BioSyst 11:882–891

Wang G, Yu MZ, Dong F et al (2017) Esterase activity inspired selection and characterization of zearalenone degrading bacteria Bacillus pumilus ES-21. Food Control 77:57–64

Wang J, Yang C, Yuan Z et al (2018) T-2 toxin exposure induces apoptosis in TM3 cells by inhibiting mammalian target of rapamycin/serine/threonine protein kinase (MTORC2/AKT) to promote Ca^{2+} production. Int J Mol Sci 19:3360

Wang J, Li S, Tang W et al (2023) Dietary complex probiotic supplementation changed the composition of intestinal short-chain fatty acids and improved the average daily gain of growing pigs. Vet Sci 10(2):79

Wellington MO, Bosompem MA, Petracek R et al (2020) Effect of long-term feeding of graded levels of deoxynivalenol (DON) on growth performance, nutrient utilization, and organ health in finishing pigs and DON content in biological samples. J Anim Sci 98(12):skaa378

Wochner KF, Becker-Algeri TA et al (2018) The action of probiotic microorganisms on chemical contaminants in milk. Crit Rev Microbiol 44:112–123

Wu X, Guo L, Huang G et al (2022) Effects of dietary natural mycotoxins exposure on performance, biochemical parameters and milk small molecule metabolic pathways of lactating cows. Agriculture 12:420

Xia X, Zhang Y, Li M et al (2017) Isolation and characterization of a *Bacillus subtilis* strain with aflatoxin B1 biodegradation capability. Food Control 75:92–98

Yadav S, Jha R (2019) Strategies to modulate the intestinal microbiota and their effects on nutrient utilization, performance, and health of poultry. J Anim Sci Biotechnol 10:1–11

Ying C, Hong W, Nianhui Z et al (2019) Nontoxic concentrations of OTA aggravate DON-induced intestinal barrier dysfunction in IPEC-J2 cells via activation of NF-κB signaling pathway. Toxicol Lett 311:114–124

Youssef IM, Elsherbeni AI, Almuraee AA et al (2024) Influence of using synbiotics by various routes on Mandarah male chicks: intestinal bacterial counts, gut morphology and histological status. Poult Sci 103(5):103601

Zadeike D, Vaitkeviciene R, Bartkevics V et al (2021) The expedient application of microbial fermentation after whole-wheat milling and fractionation to mitigate mycotoxins in wheat-based products. LWT 137:110440

Zatecka E, Ded L, Elzeinova F et al (2014) Effect of zearalenone on reproductive parameters and expression of selected testicular genes in mice. Reprod Toxicol 45:20–30

Zavistanaviciute P, Ruzauskas M, Antanaitis R et al (2023) Antimicrobial and mycotoxin reducing properties of lactic acid bacteria and their influence on blood and feces parameters of newborn calves. Animals 13(21):3345

Zheng W, Wang B, Li X et al (2018) Zearalenone promotes cell proliferation or causes cell death? Toxins 10:184

Złoch M, Rogowska A, Pomastowski P et al (2020) Use of *Lactobacillus paracasei* strain for Zearalenone binding and metabolization. Toxicon 181:9–18

Zokaityte E, Lele V, Starkute V et al (2021) The influence of combined extrusion and fermentation processes on the chemical and biosafety parameters of wheat bran. LWT 146:111498

Probiotics Against Fungal Diseases and Intoxications in Poultry and Other Avian Species

10

Mohamed E. Abd El-Hack, Ayman E. Taha, Asmaa F. Khafaga, and Mahmoud Kamal

Abstract

Fungal infections and mycotoxin exposure pose significant threats in chicken production, resulting in impaired health, diminished output, and considerable economic losses. The escalating limitations on synthetic antifungals and antibiotic growth promoters (AGPs) have necessitated the exploration of alternate strategies to protect avian health. Probiotic microorganisms have consequently gained attention as a safe and sustainable option to prevent and mitigate fungal infections and mycotoxin-associated toxicities in poultry. This chapter examines the predominant pathogenic fungi impacting birds, subsequently elucidated by the mechanisms through which probiotics impart antifungal and detoxifying properties. These mechanisms encompass the control of the immune system and the competitive exclusion of infections. Emphasis is placed on the efficacy and application of probiotics in diverse avian categories such as broilers, layers, turkeys, and ducks. By synthesizing current research, this chapter also outlines practical probiotic implementation strategies within commercial poultry

M. E. Abd El-Hack (✉)
Department of Industrial Pharmacy, College of Pharmaceutical Sciences and Drug Manufacturing, Misr University for Science and Technology (MUST), Giza, Egypt
e-mail: mohamed.abdelhaq@must.edu.eg

A. E. Taha (✉)
Department of Public Health, College of Veterinary Medicine, King Faisal University, Al-Ahsa, Saudi Arabia
e-mail: aetaha@kfu.edu.sa

A. F. Khafaga
Department of Pathology, College of Veterinary Medicine, King Faisal University, Al-Ahsa, Saudi Arabia

M. Kamal
Animal Production Research Institute, Agricultural Research Center, Dokki, Giza, Egypt

M. Y. Sreenivasa et al. (eds.), *Probiotics as Antifungals*,
https://doi.org/10.1007/978-981-92-0002-3_10

production systems. Collectively, these studies endorse the efficacy of probiotics as a significant instrument to improve avian health, alleviate fungal adversities, and foster sustainable antibiotic-free chicken production.

Graphical Abstract

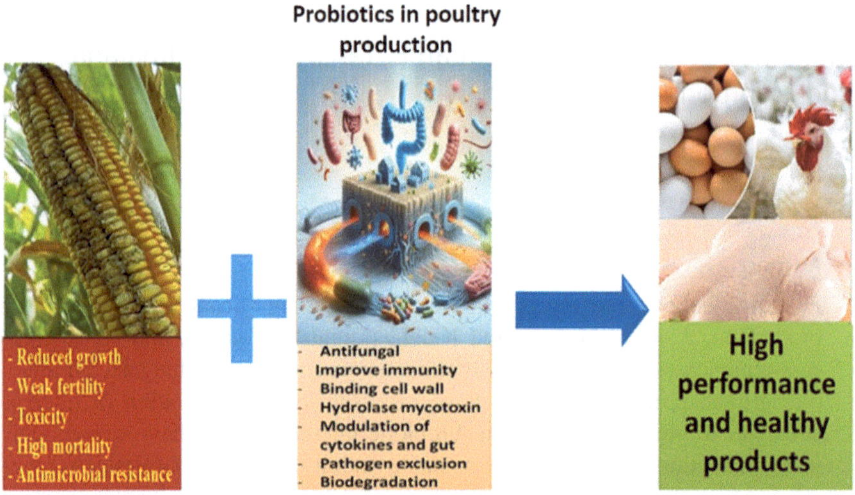

Keywords

Probiotics · Poultry · Mycotoxins · Avian mycoses · Antifungal strategies

Learning Objectives

By the end of the chapter, readers will be able to

- Describe clinical manifestations associated with fungal diseases and mycotoxicosis in aves
- Discuss probiotic-mediated mycotoxin mitigation and antifungal mechanisms
- Explore strategies for integrating probiotics into commercial poultry production

10.1 Introduction

Mycotoxins such as "zearalenone (ZEN), aflatoxins (AF), ochratoxin A (OTA), trichothecenes (TCN), fumonisins (FUM), T-2 toxin, and deoxynivalenol (DON)" adversely affect poultry production by diminishing performance traits, product quality, immunity, antioxidative activity, and welfare, leading to economic unsustainability. According to Attia et al. (2025), these mycotoxins particularly impact the kidney and liver, exhibiting carcinogenic, teratogenic, and mutagenic effects. There are over 500 mycotoxins identified, mainly produced by five fungal genera: *Fusarium* spp., *Penicillium* spp., *Aspergillus* spp., *Claviceps* spp., and *Alternaria* spp. (Haque et al. 2020). The most significant mycotoxins affecting poultry health include AF, ZEN, OTA, TCN, DON, FUM, and T-2 toxins (Murugesan et al. 2015).

Mycotoxin exposure in animal feed diminishes the quality, ecology, and safety of chicken products and adversely affects poultry productivity (Abd El-Hack et al. 2023a; Attia et al. 2025). The duration and dosage of mycotoxin exposure dictate its adverse consequences. Significant hepatic and renal impairment or death may occur within 72 hours due to elevated AF levels (Ochieng et al. 2021). Moreover, excessive AF induces immunosuppression, muscular hemorrhage, reproductive impairment, diminished development rate, reduced nutritional metabolism, gastrointestinal injury, hunger suppression, and immunosuppression (Tahir et al. 2022). *Aspergillus* species can penetrate eggshells and compromise the developing embryo, perhaps resulting in death or hatching with a fully formed lesion (Shaapan and Girh 2024).

The prohibition of AGPs has indirectly contributed to increased disease burden and economic loss in poultry systems (Laxminarayan et al. 2015; Abd El-Hack et al. 2023b). Probiotics in chickens have shown various beneficial effects, including improved meat quality and flavor (Wang et al. 2017; Popova 2017; Kamal et al. 2023), reduced pathogenic microorganisms (Kumar et al. 2018; Hernandez-Patlan et al. 2019), and improved growth efficiency (Alagawany et al. 2018; Yadav and Jha 2019). Additionally, Ouwehand et al. (2002) and Salamoura et al. (2014) identify probiotics as including "*Lactobacillus acidophilus*, *Lactobacillus casei*, *Bifidobacterium bifidum*, *Bacillus* spp., *Lactococcus lactis*, and *Saccharomyces cerevisiae*." These strains have shown beneficial effects on intestinal microbiota equilibrium, pathogen suppression (Higgins et al. 2007; Kabir 2009; Kamal et al. 2025), and broiler performance (Kabir 2009; Vieco-Saiz et al. 2019; Deepthi et al. 2017). The ability of probiotics to modify gut microbiota, enhance immunity, and mitigate pathogen proliferation, alongside their impact on poultry health, renders their use in disease management a promising strategy for improving animal productivity and health (Ayana and Kamutambuko 2024; Khan et al. 2025; Yushanaji et al. 2025). This chapter aims to examine the impact of probiotics as a potential means to improve avian health, mitigate fungal risks, minimize mycotoxins, and promote antibiotic-free chicken production.

10.2 Fungal Pathogens and Mycotoxins in Poultry and Avian Species

10.2.1 Major Fungal Pathogens in Birds

Pigeons, geese, and ducks, along with other domestic animals, are less susceptible to aspergillosis, a respiratory disease primarily affecting humans, turkeys, fowl, and mammals, caused by the *Aspergillus* fungus. The disease is often endemic in farm-raised turkeys and chickens, typically affecting individual birds, particularly juveniles aged 7 to 40 days. *Aspergillus* species are widely present in the environment, facilitating the disease's occurrence (Shaapan and Girh 2024).

Aspergillus fumigatus is the main cause of aspergillosis, but other molds like "*Rhizopus* sp., *Mucor* sp., *Aspergillus flavus*, and *A. niger*" can also play a role (Shaapan and Girh 2024). Effective management practices, including avoiding moldy feeds, maintaining cleanliness, and ensuring proper ventilation, are crucial for preventing the disease, as no strong treatment exists (Elfadaly et al. 2017). Aspergillosis significantly impacts the respiratory health of birds, leading to high morbidity and mortality, which points to improved management in poultry operations (Shaapan and Girh 2024). According to Shokat et al. (2018), lower respiratory tract infections (LRTI), which primarily affect the lower respiratory tract, are often caused by various viruses and bacteria.

Maciorowski et al. (2007) also say that the main types of fungi that get into storage are *Aspergillus, Rhizopus, Cladosporium*, and *Penicillium*. *Aspergillus* species (Fig. 10.1), particularly "*Aspergillus flavus*, *A. nomius*, *A. parasiticus*, *A. pseudotamarii*, and *A. tamarii*," are the primary strains responsible for producing aflatoxin and other toxins. *Fusarium* and *Penicillium* also produce toxins (Abo Nouh et al. 2020). Veterinary diets are made for sick animals and usually have wheat and

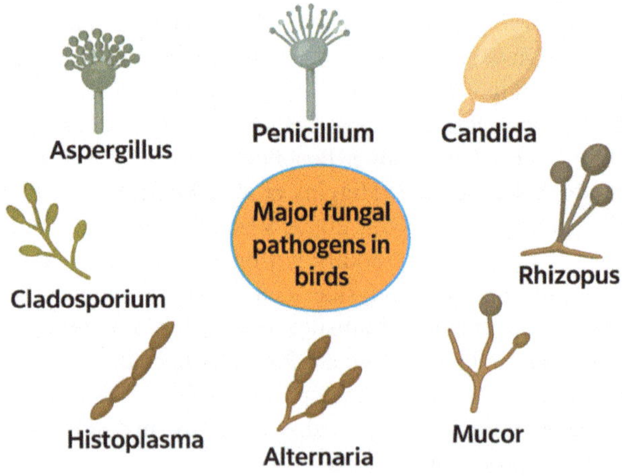

Fig. 10.1 Major fungal pathogens in birds and the mycotoxins they produce

corn as their main ingredients. *Aspergillus* and *Fusarium* pathogens, recognized for generating diverse deleterious mycotoxins (Witaszak et al. 2019), frequently infect these organisms.

Figure 10.1 provides a visual summary of the major fungal pathogens affecting birds, highlighting representative genera such as *Aspergillus, Penicillium, Candida, Rhizopus, Cladosporium, Histoplasma, Alternaria*, and *Mucor*. Each pathogen is illustrated to aid recognition and understanding of its diversity and relevance in avian health.

Filamentous fungi and yeasts have been isolated from the gastrointestinal tract of chickens, predominantly consisting of *Candida* species, followed by *Trichosporon* spp., *Saccharomyces cerevisiae*, *Rhodotorula* spp., and *Geotrichum* spp. (Ziółkowska and Tokarzewski 2007). Mikulski et al. (2008) indicate that specific fungi, such as *S. cerevisiae* and *S. boulardii*, are utilized as probiotics in animal husbandry. Poultry afflicted with candidiasis have also been identified to contain additional species, including "*Candida parapsilosis, Candida rugosa, Candida krusei, Candida famata,* and *Candida tropicalis*" (Singh et al. 2015). Non-specific clinical symptoms of candidiasis in birds include ruffled feathers, lethargy, anorexia, and stunted growth. Sokół et al. (2017) indicate that the symptoms may be mistaken for those of primary bacterial and viral infections. White diphtheritic patches and ulcers, or gray pseudomembranous lesions, are often observed, accompanied by diffuse or localized thickening of the affected mucosa (Sokół et al. 2015).

10.2.2 Mycotoxins in Poultry Feeds

Abdel-Azeem (2010) found 2281 fungal taxa across 755 taxonomic groups, including 57 myxomycete species. Toxigenic fungi that contaminate grain are conventionally categorized into two groups: "field" fungi, which affect seed crops (including *Fusarium*, *Cladosporium*, and *Alternaria* spp.) and presumably infiltrate seeds during plant development, and "storage" fungi, which proliferate during storage (such as *Penicillium* spp. and *Aspergillus*) (Legan 2000). Khalifa et al. (2022) reported that mycotoxins in animal feed samples, particularly soybean, varied from 1 to 14.8 parts per billion (ppb) for aflatoxins and from 1 to 14 parts/billion (ppb) for Ochratoxin A.

In sub-Saharan Africa, mycotoxins, including aflatoxins (AFs), trichothecenes, fumonisins (FBs), and zearalenone (ZEN), are commonly present in chicken diets and feed additives. These mycotoxins adversely affect hens' productivity and health, and they may also be present in chicken products, hence posing health risks to human consumers (Ochieng et al. 2021). Reports indicate the presence of over 400 mycotoxins in food and feed. Aflatoxins [B1 (AFB1), B2 (AFB2), G1 (AFG1), G2 (AFG2)], fumonisins (FBs), trichothecenes (including T-2 toxin (T-2) and deoxynivalenol (DON)), and ochratoxin A (OTA) are the most prevalent mycotoxins of global concern (Kemboi et al. 2020). The response of an animal to mycotoxin is determined by the specific type of mycotoxin, the amount and length of exposure, and the animal's age and species; substantial ingestion can result in acute poisoning, marked by mortality and specific clinical manifestations (Dass et al. 2007; Mngadi et al. 2008).

Due to their diverse origins, mycotoxins exhibit structural variations that result in minor differences in their physical, biological, and chemical properties (Karlovsky et al. 2016). Prevalent mycotoxins that often coexist in cereals and animal feed include aflatoxin B1 (AFB1), deoxynivalenol (DON), fumonisin B series (FBs), and zearalenone (ZEN). Mycotoxins, such as AFB1, DON, and ZEN, are commonly found in animal feed and agricultural products (Ma et al. 2018). Research indicates that poultry has significant susceptibility to AFB1. The effects of AFB1 on poultry are contingent upon various factors, including species, sex, age, and other relevant features of the population (Motbaynor et al. 2021; Wang et al. 2023).

10.2.3 Clinical Manifestations

Aflatoxins are associated with diminished immunity, impaired performance, organ damage, and reduced egg production (Hernández-Ramírez et al. 2020; Saleemi et al. 2020). Nonetheless, due to the structural similarities between FBs and sphingosine (So) as well as sphinganine (Sa), the toxicity of FBs has been associated with the disruption of sphingosine biosynthesis (Antonissen et al. 2020; Liu et al. 2020a). Furthermore, FBs have been demonstrated to adversely affect chicken organs, induce diarrhea and nephrotoxicity, and result in diminished body weight gain (Metayer et al. 2019). Additionally, Liu et al. (2020a) and Kolawole et al. (2020) assert that trichothecenes, such as DON, are predominantly recognized for their detrimental effects by obstructing protein synthesis and binding to sulfhydryl groups at cellular, subcellular, and systemic levels. Numerous studies indicate that animals, particularly chickens, are inherently subjected to elevated quantities of trichothecenes compared to humans. Chronic exposure has been demonstrated to induce immunological suppression and weight loss, whereas acute exposure has been associated with vomiting, nausea, irritation, and lesions (Chen et al. 2017; Yu et al. 2018). Reports indicate that zearalenone emulates estrogen's effects, and its toxicity has been associated with damage to reproductive organs (Chang et al. 2020; Chen et al. 2019).

According to Bhatti et al. (2019), OTA toxicity results from the production of DNA adducts, which impair protein synthesis, generate oxidative stress, and diminish mitochondrial activity. In animals, subacute OTA concentrations are teratogenic, immunosuppressive, and nephrotoxic (Bhatti et al. 2019; Vartiainen et al. 2020). According to Aikore et al. (2019), broiler chickens administered 500 g AFB1/kg feed exhibited the highest mortality rate in Nigeria. Given that AFB1 concentrations as high as 1067 g/kg have been documented in chicken feeds from the same country, the values employed in the subsequent study are pragmatically pertinent (Akinmusire et al. 2018). Furthermore, Naseem et al. (2018) reported that young layers exposed to chicken adenovirus 4 and administered 200 g AFB1/kg feed exhibited the highest mortality rate. In a comparable study, broiler hens that ingested AFB1 (750 or 1500 g/kg feed) in conjunction with the pathogen *Clostridium perfringens* exhibited the highest fatality rate (Cravens et al. 2013). Additionally, Manaf et al. (2011) reported that broiler chicks administered a diet containing 1000 g OTA/kg had the highest mortality rate due to OTA when infected with coccidia, whereas young birds receiving 800 g mycotoxins (OTA)/kg feed demonstrated a mortality rate of approximately 13% (Elaroussi et al. 2006).

Significant mycotoxins, including "AFs, OTA, ZEN, FB1, CPA, CIT," and patulin, have been demonstrated by Devegowda et al. (2008) to adversely affect egg quality, leading to the rejection of both table and hatching eggs, and resulting in substantial losses in layer production. Reports indicate that mycotoxicosis significantly jeopardizes commercial layer production, diminishing both the amount and quality of egg output (Salwa and Anwer 2009; Devegowda et al. 2008). Forty-eight hours post-exposure to mycotoxin (T-2) and IBV (infectious bronchitis virus), the avian subjects exhibited respiratory distress, characterized by recumbency, acute debilitation, and the dragging and extension of their necks on the ground (Yohannes et al. 2012). The administration of T-2 toxin to Japanese quails resulted in brown diarrheal feces, limb rigidity, and torticollis, alongside many trials conducted by Madheswaran et al. (2005). Cavanagh and Gelb (2008) reported that chicks infected with IBV had characteristic respiratory symptoms, including sneezing and gasping. The new component of this investigation is the detection of more severe clinical signs in birds exposed to both IBV and T-2 toxin. The T-2 toxin may have compromised the chicks' systemic and local defenses against the IBV infection, hence enhancing its pathogenicity (Yohannes et al. 2012). The disease primarily affects the respiratory systems of birds, manifesting as bronchopneumonia. In poultry, it occurs in both acute and chronic forms (Asfaw and Dawit 2017). Chronic aspergillosis predominantly affects adult birds, but the clinical indicators of acute aspergillosis are usually observed in young chickens, sometimes during outbreak events characterized by significant morbidity and mortality (Charlton et al. 2008).

10.3 Probiotic Mechanisms of Action

10.3.1 Antifungal Effects

The precise methods of action of numerous diverse and strain-specific probiotics remain unidentified. Mechanisms understood include pathogen exclusion, bacteriocin production, modulation of fecal enzymatic activity, production of branched-chain and short-chain fatty acids, cell adhesion and mucin production, immune system modulation, and interaction with the gut-brain axis through regulation of neurological and endocrine functions (Panitsidis et al. 2023). Figure 10.2 illustrates the mechanisms by which probiotics promote mycotoxin removal.

10.3.1.1 Competitive Exclusion

Competitive exclusion occurs when one strain of bacteria surpasses another in securing receptor loci within the digestive tract (Bermudez-Brito et al. 2012). Additionally, Collado et al. (2010) suggest that this phenomenon may result from

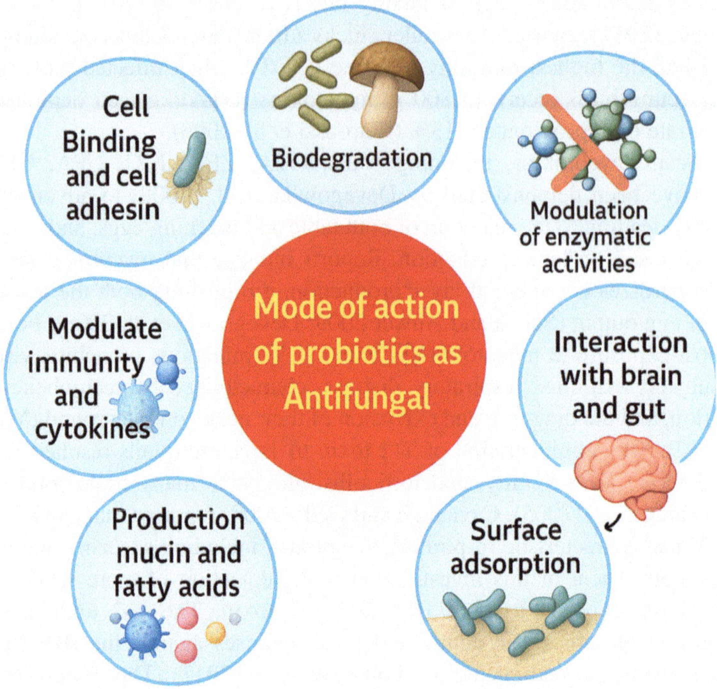

Fig. 10.2 Mechanisms of probiotics involved in mycotoxin removal. Diagram illustrating the multifaceted mechanisms by which probiotics exert antifungal impacts, including cell binding and adhesion, biodegradation, modulation of enzymatic activities, interaction with the brain–gut axis, surface adsorption, production of mucin and fatty acids, and modulation of host immunity and cytokine responses

a decrease in the surrounding pH, competition for nutrients, or the production of bacteriocins or similar compounds. Furthermore, probiotics may modify the metabolism of bile acids in the gut lumen, subsequently affecting cholesterol absorption (Panitsidis et al. 2023). Probiotics colonize and stick to the intestinal mucosa via surface molecules such as mucus-binding proteins, establishing a competitive environment that restricts pathogen adhesion and toxin absorption. This physical adsorption mechanism sequesters deleterious chemicals, diminishing their bioavailability and toxicity in the gastrointestinal tract. This method is facilitated by probiotics that enhance the intestinal barrier through the stimulation of mucus production and tight junction proteins, thereby impeding toxin transfer (Idowu et al. 2025).

10.3.1.2 Enzyme Production

Hydrolase bile salt, an enzyme potentially involved in the initial deconjugation of salts, is synthesized by most recognized bacterial strains and probiotics from various genera associated with the digestive tract (Pavlovic et al. 2012). Specific probiotic strains generate enzymes capable of biotransforming or degrading mycotoxins and other deleterious chemicals in the gastrointestinal tract. Enzymes produced by *Bacillus subtilis* and other probiotics can digest toxins, thereby cleansing the gut environment (Hamid et al. 2025). This enzyme activity promotes gut health by averting toxin-induced lesions and reinstating villus integrity, hence improving nutrition absorption (Opazo et al. 2025).

10.3.1.3 Immune Modulation

The gut microbiota modulates systemic immunity by producing compounds with anti-inflammatory properties and immunomodulatory capabilities, as well as by activating immune cells. To elicit these immunomodulatory effects, probiotic bacteria interact with dendritic cells, lymphocytes, macrophages, and epithelial cells (D'Amelio and Sassi 2017). Probiotics enhance the host immune system by amplifying both innate and adaptive immune responses. They activate immune cells, including monocytes, macrophages, dendritic cells, T lymphocytes, and B lymphocytes, modulating cytokine production to equilibrate inflammation and pathogen protection (Argañaraz-Martínez et al. 2025). Furthermore, probiotics enhance the release of secretory IgA, thereby strengthening mucosal defenses against pathogen colonization. They influence Toll-like receptor signaling, enhancing the production of anti-inflammatory cytokines such as IL-10, while diminishing pro-inflammatory markers like TNF-α and IFN-γ (Idowu et al. 2025). Furthermore, metabolic byproducts such as short-chain fatty acids derived from probiotics affect regulatory T cell differentiation via epigenetic pathways, thereby enhancing immunological homeostasis (Pewan et al. 2025).

Mycotoxins in feedstuff are detoxified using chemical and physical processes, including solvent extraction, washing, heat inactivation, chemical agents, and irradiation. Moreover, a prevalent method in chicken production involves the dietary substitution of adsorbents, including zeolites, hydrated sodium calcium aluminosilicate, bentonites, and glucomannan. Figure 10.2 illustrates techniques for the elimination of mycotoxins.

Moreover, many feed additives, including natural antioxidants, phytogenic additives, and microbes, may mitigate the adverse effects of mycotoxins and yield beneficial results in the poultry industry (Attia et al. 2025). Research indicates that high dosages of antioxidants, including vitamin E, vitamin B1, vitamin C, selenium (Se), flavonoids, and carotenoids, together with amino acids such as methionine, lysine, arginine, glutamic acid, and aspartate, can mitigate detrimental effects (Liu et al. 2020a). Moreover, AFB1 induced an inflammatory response at a dosage of 1 mg/kg by enhancing the gene expression of inflammatory markers (TLR4, NF-κB, and MyD88) and signaling molecules (TNF-α, iNOS, IL-6, and IL-1β). Furthermore, in broiler liver injury induced by AFB1, the signal crosstalk between TLR4 and TNF-α elicits inflammation, whereas RIPK1 and RIPK3 mediate necroptosis, as noted by Li et al. (2022).

10.3.2 Mycotoxin Mitigation

Ozone treatment of chicken feedstuffs eradicated various mycotoxins, including AF, ZEN, and FUM, achieving an efficacy of over 80% (Li et al. 2019). Ozone treatment significantly diminished the levels of AFB1 in chicken feed components (corn, sunflower, soybean meal, peanuts, and barley) without increasing lipid oxidation (Torlak et al. 2016). In recent decades, many adsorption binders have been investigated, demonstrating their efficacy in detoxifying mycotoxins (Liu et al. 2020b). As a result, the feed industry extensively uses the adsorbent decontamination method. Examples of possible absorbents include activated carbon, glucomannans, and aluminosilicates such as zeolite, bentonite, phyllosilicates, and hydrated sodium calcium aluminosilicate (HSCAS) (Vila-Donat et al. 2018). Certain mycotoxins, including aflatoxins, possess an ionic charge; hence, clay adsorbents, including bentonite, illite, kaolin, and zeolites, effectively reduce mycotoxin levels, achieving a clearance rate of 90% in feedstuffs (Hasheminejad et al. 2015; Elsherbeni et al. 2024).

The forthcoming generation of adsorbents is derived from the components of microbial cell walls (Attia et al. 2013). Glucomannan is characterized by animal pathogenic bacteria, mycotoxins, and highly adsorbent toxicants, which are not utilized by gastrointestinal microbes (Kolawole et al. 2019). The most effective treatment involved including 0.1% glucomannan-containing yeast in a diet supplemented with aflatoxin (250 ppb), which markedly mitigated the adverse impact of the aflatoxin mycotoxin on biochemical blood parameters (Azizpour and Moghadam 2015). Known mechanisms include the production of bacteriocin and pathogen exclusion, the synthesis of branched-chain and short-chain fatty acids, the generation of cell mucin adhesion, immune system modulation, regulation of neurological and endocrine functions, and interaction with the gut-brain axis (Panitsidis et al. 2023).

"*Lactobacillus*, *Bifidobacterium*, *Saccharomyces cerevisiae*, and some *Bacillus* species" are prominent probiotics studied for their role in detoxifying biological mycotoxins. Two main methods of mycotoxin elimination by these probiotics, particularly lactic acid bacteria and *Saccharomyces* yeasts, are surface adsorption and biodegradation (Wan et al. 2019; Afshar et al. 2020; Shruthi et al. 2025). Combining compound probiotics with enzymes presents an effective approach for degrading mycotoxins ZEA and AFB1. A blend of *Candida utilis* and *Lactobacillus casei* (1:1:1) achieved degradation rates of 45.49% for AFB1 and 44.90% for ZEA, surpassing *Bacillus subtilis*. Moreover, the integration of compound probiotics with *Aspergillus oryzae* mycotoxin-degradation enzymes (MDEs) resulted in enhanced degradation rates of 63.95% for AFB1 and 73.51% for ZEA (Huang et al. 2018). This method facilitates the biodegradation of toxins in broiler meals, reducing the risk of ZEA and AFB1 during chicken growth (Chang et al. 2020). Additionally, the cytotoxic effects of these mycotoxins on swine jejunal epithelial cells can be alleviated by combining MDEs with cell-free supernatants of compound probiotics. These results highlight the potential of microbial consortia in the biodegradation of mycotoxins (Fig. 10.3).

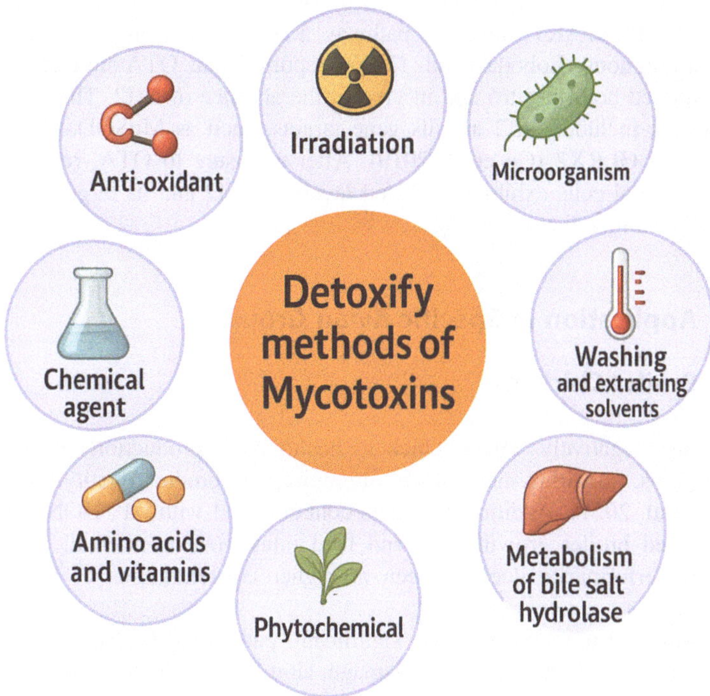

Fig. 10.3 Methods of mycotoxin detoxification. Overview of primary methods for mycotoxin detoxification, including physical (irradiation, washing, solvent extraction), chemical (antioxidants, chemical agents), biological (microorganisms, phytochemicals, bile salt hydrolase metabolism), and nutritional strategies (amino acids and vitamins), highlighting their diverse mechanisms for reducing mycotoxin toxicity in food and feed systems

10.3.3 Immunomodulatory Effects

The diminished mRNA synthesis of interferon-γ (IFN-γ) and interleukin-2 (IL-2), along with the degeneration of the *Bursa of Fabricius*, thymus, and spleen, may underlie the adverse impacts of mycotoxins on immune function (Tahir et al. 2022). From an alternative viewpoint, mycotoxins perturb the gut microbiota, resulting in a reduction of beneficial microbes and a proliferation of pathogenic bacteria. AFB1 diminished the prevalence of *Lactobacillus aviarius* in broilers while augmenting the levels of *E. coli*, *E. shigella*, and *Staphylococcus xylosus* (Guo et al. 2023).

Moreover, OTA-induced nephrotoxicity may be influenced by oxidative stress, as evidenced by Marin-Kuan et al. (2006). Furthermore, oxidative stress can modify Nrf2 activation and diminish overall Nrf2 protein levels by disrupting the signal transduction of the PI3K/AKT pathway (Deng et al. 2013). This PI3K/AKT pathway is essential for regulating cell proliferation, differentiation, migration, and apoptosis. Oxidative stress often serves as a mediator for the effects of OTA. Cavin et al. (2009)

elucidated the role of Nrf2 in renal disorders. Nonetheless, the mechanism by which the kidneys' PI3K/AKT signaling pathway activates Nrf2 in response to OTA remains ambiguous. Loboda et al. (2017) reported that OTA-simulated toxicity was heightened both in vitro and in vivo in the absence of Nrf2. The endogenous redox system includes Nrf2 and its gene targets, such as MnSOD, HO-1, CAT, GSH-px, and GLRX2 (Ge et al. 2019). After exposure to OTA, renal proximal tubular epithelial cells exhibited reduced levels of Nrf2 and its target gene, HO-1 (Boesch-Saadatmandi et al. 2008).

10.4 Application in Specific Avian Groups

10.4.1 Broiler Chickens

Mycotoxins negatively affect chicken health and production even at low concentrations, impairing metabolism, immunity, and product quality (Hernandez-Valdivia et al. 2021). Additionally, diets contaminated with AF, DON, ZEN, and OTA lowered broiler growth rates and feed intake (Metayer et al. 2019), with significant performance declines seen at higher concentrations of FUM, ZEN, and DON.

Aflatoxin consumption induces significant pathological changes in several organs, resulting in liver and kidney damage, along with reductions in feed conversion ratio (FCR), body weight gain (BWG), mortality rate, and feed intake (FI) (Zou et al. 2023). Moreover, T-2 toxin enhances oxidative damage in broiler chickens, leading to increased apoptosis, membrane destruction, and DNA damage (Chen et al. 2019). Reports indicate that AF induces immunosuppression and oxidative stress in broilers (Ali et al. 2021). The *Lactobacillus* species administered to hens was observed to mitigate the detrimental effects of "AFB1, ZEN, and DON" (Chang et al. 2020). Recent tests demonstrated that *Bacillus pumilus* enhanced FCR and increased weight gain and ultimate weight in broilers. Furthermore, it increased the populations of *Lactobacillus* and *Bacillus* spp. in the ileum and ceca (Bonos et al. 2021).

10.4.2 Laying Hens

The introduction of *Lactobacillus* to layer chickens' diet increased egg weight (Pambuka et al. 2014). Furthermore, it has been shown that *Bacillus* enhances the weight, thickness, calcium content, and egg production of laying hens (Wang et al. 2021). Both broiler and layer systems are equally vulnerable to the adverse effects of co-occurring mycotoxins. Co-exposure to mycotoxins results in substantial reductions in hatchability and egg production in laying hens (Shanmugasundaram et al. 2023).

Furthermore, Verma et al. (2003) found that laying hens subjected to simultaneous exposure to OTA (1–4 ppm) and AF (0.5–2 ppm) exhibited reduced egg production and feed efficiency. The interactions varied from synergistic to somewhat less than additive, contingent upon the dosage. Ogido et al. (2004) discovered that FB1 (10 mg/kg) and AFB1 (50–200 μg/kg) exhibited an antagonistic interaction in quails, resulting in a less pronounced reduction in egg production when both toxins were combined compared to their individual application. The research consistently indicates the detrimental effect of exposure to various mycotoxin types on laying performance. Chowdhury et al. (2005a) reported that the consumption of cereals naturally contaminated with *Fusarium* (DON) mycotoxins led to a notable rise in unmarketable egg and eggshell weights. This elicited significant apprehensions over food safety and productivity. Hens administered a diet contaminated with T-2 mycotoxins (4 ppm) and FB1 (20 ppm) exhibited a similar reduction in egg mass (Dazuk et al. 2020). Furthermore, the adverse effects of AFB1 on laying hens were alleviated by *Bacillus subtilis* ANSB060 (Ma et al. 2012).

10.4.3 Turkeys and Ducks

Aflatoxin concentrations of 200 μg/kg result in reduced weight gain and enhanced feed conversion (Rauber et al. 2007). Elevated dosages of aflatoxin may result in death. The aflatoxin in the feed influenced the expression of more than 200 genes associated with immune functions in the spleen and decreased the expression of IFNγ mRNA in blood leukocytes (Monson et al. 2015; Umar et al. 2015). According to Benlashehr et al. (2011), turkeys are more vulnerable to fumonisins than chickens, but less vulnerable than ducks. Turkeys' growth and feed consumption were unaffected by being fed feed tainted with this amount of fumonisin (Tardieu et al. 2007). Zeolite, smectite clays, bentonite clay, sodium bentonite, and kaolinite were found to lower AFB1 residues in liver, meat, and eggs by around 41–87% in broilers and ducks fed AFB1-contaminated diets, according to several studies conducted on poultry species (Bhatti et al. 2018). However, glucomannan did not stop the negative effects of a *Fusarium*-contaminated diet on hematocrit and hemoglobin levels in duck and turkey, according to Chowdhury et al. (2005b). In domestic turkeys, the inclusion of dietary probiotics during AFB1 exposure mitigated the expression alterations of genes such as serine/arginine repetitive matrix protein 1 (SRRM1), 28S ribosomal RNA (RNA28S), and ISG12-2 protein-related mRNA (Monson et al. 2015). Immune genes, including GZMA, THP2, and PRF1, had a more pronounced enhancement in the spleen (Monson et al. 2015).

Additionally, Sun et al. (2022a, b) found that compound probiotics, specifically *Bacillus subtilis* and *Bacillus Licheniformis*, improve intestinal health, antioxidant capacity, growth performance, and serum biochemical and immunological indices in Shaoxing ducks. Additionally, Hassan and Mosaad (2015) reported that supplementing broiler ducklings' diets with *Saccharomyces cerevisiae* significantly increases blood total protein levels.

Li et al. (2024) have shown that fermented feedstuff (FF) effectively enhances the developmental performance of Muscovy ducks by improving intestinal shape, microbial composition, and metabolic processes. These changes subsequently enhanced the uptake and digestion of feed. The suggested study offers empirical evidence that FF may serve as an effective strategy for improving the growth performance of Muscovy ducks. Finally, research has shown that probiotic strains of *Bifidobacterium* elevated immunoglobulin levels IgA, IgG, and IgM in turkeys and broilers, hence enhancing their disease resistance (Abdel-Moneim et al. 2019). The growth characteristics of turkeys at market age did not exhibit enhancements in absorption surface area due to probiotic treatment. Likewise, probiotic therapy enhanced intestinal microbial communities and gut architecture in turkeys, but without increasing ultimate body weight (Agboola et al. 2014). According to Omidiwura et al. (2018), adding bacterial culture (*Bacillus subtilis*) for 8 weeks, either alone or alongside prebiotic preparation, resulted in BWG and FCR rates comparable to those of untreated control pigeons.

10.4.4 Quail and Ornamental Birds (Pigeons, Parrots)

Furthermore, the incorporation of Se nanoparticles into the food (0.5 mg/kg) enhanced serum concentrations of GSH-Px, reductase, thioredoxin, and antibody titers against sheep red blood cells in Japanese quails, while mitigating the toxicity of AFB1 (Khazraei et al. 2022). Microorganisms can mitigate mycotoxicosis by biotransforming mycotoxins into less hazardous metabolites (Attia et al. 2025). An efficient approach to mitigate the detrimental effects of microbial and non-adsorbable mycotoxins is the utilization of microorganisms for their detoxification. The capacity of *Bacillus, Lactobacillus, Pseudomonas, Bifidobacterium, Lactococcus,* and *Propionibacterium* to bind aflatoxins can suppress mold growth in chicken feed ingredients.

Kasmani and Mehri (2015) have shown that dietary multi-strain probiotics may improve immunocompetence and growth performance in growing Japanese quails affected by aflatoxicosis. Aflatoxin exposure disrupted the microbial balance in the gut microbiota, reducing beneficial lactic acid bacteria (LAB) and increasing pathogenic coliforms in the PC group. Biochar supplementation facilitated the restoration of microbial equilibrium, particularly when paired with *Lactobacillus fermentum. Lactobacillus fermentum* diminishes dangerous bacteria by competitive exclusion and antibiotic production, while biochar acts as a microbial carrier; this combination likely exerted a synergistic effect (Raz et al. 2025). *Ligilactobacillus salivarius* S10 may serve as a probiotic in pigeons, potentially enhancing their growth performance by modifying intestinal flora and regulating the expression of differentially expressed genes linked to gut development (Zhao et al. 2025).

10.5 Notable Probiotic Strains and Their Antifungal Efficacy

The most prevalent probiotics in animal feed belong to the following genera: *Bacillus* (*subtilis*), *Lactobacillus* (*bulgaricus, acidophilus*, and *plantarum*), *Lactococcus* (*lactis*), *Enterococcus* (*faecalis, faecium*), *Bifidobacterium* (*thermophilum, pseudolongum*, and *longum*) (Mattarelli and Biavati 2018), and *Saccharomyces* (*boulardii* and *cerevisiae*) (Muzaffar et al. 2021; Deepthi et al. 2017). A specific strain of probiotics, a multi-spice blend, or a supplementary dietary supplement may be incorporated into feed or water. Probiotics containing many strains are often considered more effective than those with a single strain due to their synergistic advantages (Bhogoju and Nahashon 2022). A summary of the advantages of probiotic strains in different poultry species (Table 10.1).

Mohaghegh et al. (2017) assert that including 0.1% esterified glucomannan into broiler chicken diets is a safe and efficacious method for mycotoxin detoxification. It also improved FI, BWG, FCR, liver percentage, bursa of Fabricius percentage, and antibody titer through infectious bursal disease viruses and Newcastle disease virus. In broilers infected with Eimeria maxima, a dietary inclusion of 0.2% polymeric

Table 10.1 Positive effects of probiotic strains in poultry

Poultry strain	Probiotic species	References	Effect
Broilers	*Lactobacillus* spp.		Improve intestinal health, nutrient absorption, and modulate immunity.
	Saccharomyces Cerevisiae		Enhance nutrient utilization, improve intestinal integrity, and modulate immunity.
	Bifidobacterium spp.		Improve gut health, modulate the immune response, and inhibit pathogen growth of pathogens
Layer hens	*Enterococcus* spp.		Improve immunity, gut health, and inhibit pathogen colonization
	Streptococcus spp.		Improve gut health, enhance nutrient utilization, and modulate immunity.
Shaoxing ducks	*Bacillus subtilis* and *Bacillus licheniformis*	Sun et al. (2022b)	Enhance the microbiome and metabolome of the cecum
	Bacillus subtilis and *Bacillus licheniformis*	Sun et al. (2022a)	Enhance the growth traits, improve serum biochemical and immunity signals, and raise antioxidant capacity
Muscovy duck	*Bacillus* spp., *Bifidobacterium, Lactobacillus, Enterococcus faecium*	Li et al. (2024)	Increased the growth performance
Quail	*Lactobacillus Fermentum*	Raz et al. (2025)	Modulating immunity and oxidative response.
Pigeons	*Ligilactobacillus Salivarius*	Zhao et al. (2025)	Improve the growth traits and development of the intestinal
Turkeys	*Bacillus subtilis*	Omidiwura et al. (2018)	Increase BWG and FCR

glucomannan enhanced immunological function and decreased feed-associated *Fusarium* mycotoxins (Girgis et al. 2010). He et al. (2021) found that the modified yeast cell wall mannan-oligosaccharide (MOS), containing glucomannan isolated from the cell wall of *Saccharomyces cerevisiae*, might enhance immunity. Moreover, the adverse effects of OTA in broiler chicks were mitigated by the yeast extract cell wall (Mycosorb A+®) (Vartiainen et al. 2020). Incorporating Mycosorb A+® (a natural product comprising 10% mannan oligosaccharide extract from *Saccharomyces cerevisiae*) into diets contaminated with 200 ppb aflatoxin at concentrations of 0.1% and 0.2% enhanced growth, feed efficiency, and serum total protein levels, while not significantly affecting gamma-glutamyl transferase activity, bursa of Fabricius weight, or the weights of the liver, gizzard, and spleen (Vartiainen et al. 2020). In poultry gastrointestinal medium, yeast cell walls comprising β-glucans and mannan-oligosaccharides could sequester up to 90% of aflatoxin B1 in vitro (Attia et al. 2013).

10.6 Experimental and Field Evidence

The incorporation of *Lactobacillus rhamnosus* GG (LGG) and *Propionibacterium freudenreichii* (spp. *shermanii* JS) resulted in a reduction of AFB1 absorption by 74% and 63%, respectively, after an ex vivo infusion of AFB1 into the chicken duodenal cavity (El-Nezami et al. 2000). Furthermore, during the study's incubation durations, *Lactobacillus rhamnosus* TMU094, *Lactobacillus fermentum*, *P. pentosaceus*, and *Lactobacillus rhamnosus* PTCC1637 exhibited binding capacities of 25.64–75.06%, 38.63–72.15%, 24.86–63.21%, and 19.41–35% for AFB1, respectively (Bagherza deh Kasmani et al. 2012). The antioxidative and humoral-mediated immune responses in broilers were enhanced by the fermentation of *Bacillus subtilis* extract, which also mitigated the harmful effects of OTA (Elhady et al. 2020). Incorporating 2.0 kg of Eubacterium BBSH 797 per ton of meal alleviated the adverse effects of 2 mg T-2 toxin per kg of feed-on-feed efficiency and growth (Diaz et al. 2005). In domestic turkeys, the inclusion of dietary probiotics during AFB1 contamination mitigated AFB1-induced gene expression alterations in serine/arginine repetitive matrix protein 1 (SRRM1), ISG12-2 protein-related mRNA, and 28S ribosomal RNA (RNA28S) (Monson et al. 2015). Immune genes, including GZMA, THP2, and PRF1, showed significantly enhanced ameliorative effects in the spleen (Monson et al. 2015). Combination therapy elevated apolipoprotein A-IV (APOA4) in the liver and ISG12-2 and RNA28S in the spleen. This indicated that whereas oral probiotics modified liver and splenic expression, they failed to restore their normal contours (Attia et al. 2013).

10.7 Practical Considerations and Limitations

Yeast strains, including *Mastotermes darwiniensis, Trichosporon mycotoxinivorans,* and *Saccharomyces cerevisiae*, demonstrated the capacity to diminish the accumulation of AFB1, OTA, DON, T-2, and ZEN toxins in tissues and mitigate their detrimental effects (Chlebicz and Śliżewska 2020). *Saccharomyces cerevisiae* RC016 (1 g/kg feed) improved the intestinal histomorphology of broilers, exhibiting no malabsorption villi, no hyperemia, a reduced number of goblet cells, and atrophy, while alleviating the adverse effects of AFB1 and promoting normal hepatocyte function (Poloni et al. 2020). Çelýk et al. (2003) reported that broilers administered 2% *Saccharomyces cerevisiae* alongside 200 μg AF/kg of feed exhibited enhanced feed intake, growth, and improved levels of proteins, AST, albumin, and ALT, in addition to increased weight of the bursa of Fabricius. Moreover, broiler development, feed utilization, albumin, serum protein, and enzyme levels were adversely affected by the incorporation of AF at 1 and 2 ppm/kg diet; however, these effects were alleviated by *Saccharomyces cerevisiae* (Ellakany et al. 2011). The incorporation of the novel yeast strain *Pichia kudriazevii* (1 g/kg food) into the diet of broiler chickens decreased aflatoxin contamination and enhanced both humoral and cellular immunological responses, along with antioxidative properties (Ali et al. 2021). Conversely, the inclusion of 2% distillery yeast sludge in the food can inhibit the absorption of AFB1 and OTA in the gastrointestinal system and mitigate the immunotoxic effects on laying hens, specifically White Leghorns (Khan et al. 2017).

Effective anti-aflatoxin agents comprise *Lactobacillus acidophilus* and *Saccharomyces cerevisiae* (Attia et al. 2013). Saleemi et al. (2020) assert that the appropriate ratio of fungi and yeasts in poultry feed reduces the bioavailability of mycotoxins in the gastrointestinal tract, facilitating their excretion via feces. The European Commission provides guidelines about the maximum permissible quantities of OTA in animal feed. The concentration of OTA in incomplete feed for broiler chickens must not exceed 0.1 mg/kg, while the concentration of OTA in individual grain ingredients used for feed formulation should not surpass 0.25 mg/kg.

10.8 Integration into Commercial Poultry Systems

Of the more than 280 plant species assessed for their ability to suppress the growth of toxic *Aspergillus* or toxin production, over 100 showed varying levels of efficacy (Abd El-Ghany 2024). Garlic (*Allium sativum L.*) possesses extensive antifungal characteristics, capable of diminishing the growth of *Penicillium* and *Aspergillus* fungi transmitted by seeds by 60–82% (Abd El Ghany 2024). Neem (*Azadirachta indica*) seed oil inhibits the formation of aflatoxins (G and B) in mycelial fungi (Mossini et al. 2009). Crude extracts of ground red pepper, sage, bay, mint, and anise inhibit the proliferation of *Aspergillus parasiticus*. Flora that effectively obstruct the non-sorbic acid precursor of the aflatoxin biosynthetic pathway comprises *Ocimum gratissimum*, *Xylopia aethiopica*, *Cymbopogon citratus*, *Monodora myristica*, *Cinnamomum verum*, *Syzygium aromaticum*, and *Piper nigrum* (Tagoe et al. 2011).

Findings from another study indicate that the *Lactococcus* strain may serve as a carrier vaccine to enhance T-cell counts, cytokine production, and fungal immunological responses, while potentially protecting several *Vibrio* species (Lee et al. 2021). Treatment with *Bifidobacterium* augmented antibody production against the Newcastle disease virus in layers, showed an anti-*Campylobacter* effect, and elevated IFN levels in chickens afflicted with *Clostridium* infections (Hatab et al. 2016). Moreover, *Bifidobacterium* and lactic acid bacteria enhanced local antibody synthesis and systemic immunity in broilers (Haghighi et al. 2006).

To enhance gut health and microbial equilibrium, incorporate probiotics (beneficial bacteria) and prebiotics (certain substances that promote the proliferation of advantageous microflora) into the diet. This may mitigate the adverse effects and bioavailability of concealed mycotoxins. Mycotoxin modifiers, comprising bacteria and fungi, are employed to degrade mycotoxins into less toxic compounds, thereby mitigating their detrimental effects. Research indicates that including *Lactobacillus* species in broiler chicken feeds can alleviate the adverse effects of AFB1, masked AFB1, ZEN, and DON (Chang et al. 2020; de Souza et al. 2020). The findings indicated that *Bacillus subtilis* fermentation extract markedly decreased nephrotoxicity and immunological toxicity in broilers. The chemical probiotics and some proteases found in the *Bacillus subtilis* extract are responsible for the protective effects. OTA can be enzymatically degraded into OTα, a non-toxic variant (Elhady et al. 2020). The study asserts that bacteria, such as *Devosia* sp. ANSB714 and *Bacillus subtilis* ASAG 216, can markedly diminish the levels of DON in blood, liver, and kidney, while also exacerbating the adverse effects of DON on swine productivity (Jia et al. 2021).

10.9 Future Directions

Biofeed refers to feed products developed by biotechnological methods, including fermentation engineering, protein engineering, enzyme engineering, and genetic engineering, as described by the National Engineering Research Center of Biological Feed (China, January 1, 2018). Additionally, it is widely recognized that mycotoxin-degrading enzymes can be expressed in mature microbial hosts by genetic engineering or directly isolated and purified from biomaterials (Wang et al. 2017; Liu et al. 2018).

The cloning and heterologous expression of enzymes by genetic engineering facilitates efficient production at reduced economic and labor costs, making mycotoxin degradation by a recombinant enzyme particularly attractive. The fungus *Streptomyces cyaneus* CECT 3335 exhibits slow growth and a narrow optimum pH range, complicating the extraction of natural laccase. Nonetheless, heterologous expression in *E. coli* facilitated the large-scale production of laccase, achieving a yield of up to 104 mg L^{-1} (Ece et al. 2017). A unique strategy for addressing the co-contamination of various mycotoxins involves the recombinant genetic alteration of a minimum of two enzymes to facilitate the breakdown of distinct mycotoxins. Employing genetic modification techniques, Azam et al. (2019) identified a novel bifunctional recombinant fusion enzyme (ZHDCP) that integrates zearalenone hydrolase (ZHD) with carboxypeptidase. While the practical application of recombinant mycotoxin-degrading enzymes remains distant, the research has advanced the use of genetic engineering techniques for mycotoxin degradation. Furthermore, it has established a foundation for the industrial utilization of enzymes that decompose recombinant mycotoxins.

Phytogenics possess antibacterial capabilities that inhibit the development of detrimental bacteria and fungi in the stomach, thereby reducing the risk of infections and promoting a healthy gut microbiota (Austin and Ojebiyi 2025). Recent studies have investigated the influence of gut microbiota on several diseases, demonstrating that probiotics, synbiotics, prebiotics, and postbiotics (PSPP) confer beneficial effects by altering the microbiome (Lázaro et al. 2024). The strain type, dosage, and application method significantly influence the efficacy of probiotics (Gorzelanna et al. 2025).

10.10 Conclusion

Mycotoxins, including "aflatoxins (AF), zearalenone (ZEN), fumonisins (FUM), trichothecenes (TCN), ochratoxin A (OTA), T-2 toxin, and deoxynivalenol (DON)," adversely affected performance, antioxidative properties, product quality, immune activity, health status, economic profit, and production continuity. *Lactobacillus acidophilus*, *Bacillus* spp., *Lactobacillus casei*, *Bifidobacterium bifidum*, *Saccharomyces cerevisiae*, and *Lactococcus lactis* have been employed as probiotics. Probiotics exhibit several mechanisms, including pathogen elimination, bacteriocin production, modulation of enzymatic activity, synthesis of branched-chain and short-chain fatty acids, cell adhesion, mucin production, immune system modulation, and interaction with the gut-brain axis through the regulation of neurological and endocrine functions. Probiotics are regarded as safe and advantageous feed additions in chicken production. The integration of compound probiotics (CP) with mycotoxin-degrading enzymes for the concurrent detoxification of AFB1 and ZEA is an innovative and promising approach. A recent study has focused on the role of gut bacteria in many disorders. In light of this tendency, researchers must do further investigations to determine the optimal probiotic strain for each species and innovate new methodologies. Widespread implementation in avian health initiatives is urgently required.

References

Abd El-Ghany WA (2024) Potential effects of garlic (*Allium sativum* L.) on the performance, immunity, gut health, anti-oxidant status, blood parameters, and intestinal microbiota of poultry: an updated comprehensive review. Animals 14:498

Abd El-Hack ME, Kamal M, Alazragi RS, Alreemi RM, Qadhi A, Ghafouri K, Azhar W, Shakoori AM, Alsaffar N, Naffadi HM, Taha AE (2023a) Impacts of chitosan and its nanoformulations on the metabolic syndromes: a review. Braz J Biol 83:e276530

Abd El-Hack ME, Kamal M, Alqhtani AH, Alreemi RM, Alazragi RS, Khojah H, Swelum AA, Khafaga AF, Świątkiewicz S (2023b) Detoxification impacts of dietary probiotic and prebiotic supplements against aflatoxins: an updated knowledge–a review. Ann Anim Sci 23(4): 1049–1060

Abdel-Azeem AM (2010) The history, fungal biodiversity, conservation, and future perspectives for mycology in Egypt. IMA Fungus 1:123–142

Abdel-Moneim AE, Elbaz AM, Khidr RE, Badri FB (2019) Effect of *in vivo* inoculation of *Bifidobacterium* spp. on growth performance, thyroid activity, ileum histomorphometry and microbial enumeration of broilers. Probiotics Antimicrob Proteins 12(3):873–882

Abo Nouh FA, Gezaf SA, Abdel-Azeem AM (2020) Aspergillus mycotoxins: potential as biocontrol agents. In: Agriculturally important fungi for sustainable agriculture: volume 2: functional annotation for crop protection. Springer International Publishing, Cham, pp 217–237

Afshar P, Shokrzadeh M, Raeisi SN et al (2020) Aflatoxins biodetoxification strategies based on probiotic bacteria. Toxicon 178:50–58

Agboola AF, Aroniyo I, Suberu S et al (2014) Dietary supplementation of probiotics and synbiotics on intestinal microbial populations and gut morphology of Turkey poults. Afr J Livest Ext 14: 13–20

Aikore MOS, Ortega-Beltran A, Eruvbetine D (2019) Performance of broilers fed with maize colonized by either toxigenic or atoxigenic strains of *Aspergillus flavus* with and without an aflatoxin-sequestering agent. Toxins 11:565

Akinmusire OO, El-Yuguda AD, Musa JA et al (2018) Mycotoxins in poultry feed and feed ingredients in Nigeria. Mycotoxin Res 35:149–155

Alagawany M, El-Hack MEA, Farag MR et al (2018) The use of probiotics as eco-friendly alternatives for antibiotics in poultry nutrition. Environ Sci Pollut Res Int 25:10611–10618

Ali A, Khatoon A, Saleemi MK, Abbas RZ (2021) Aflatoxins associated oxidative stress and immunological alterations are mitigated by dietary supplementation of *Pichia kudriavzevii* in broiler chicks. Microb Pathog 161:105279

Antonissen G, De Baere S, Novak B et al (2020) Toxicokinetics of hydrolyzed Fumonisin B1after single oral or intravenous bolus to broiler chickens fed a control or a Fumonisins-contaminated diet. Toxins 12:413

Argañaraz-Martínez E, Apella MC, Chaia AP, Babot JD (2025) Probiotic applications of bifidobacteria in poultry: administration methods and microencapsulation techniques. Microb Res Rep 4(3):32

Asfaw M, Dawit D (2017) Review on major fungal disease of poultry. Br J Poult Sci 6(1):16–25

Attia Y, Allakany H, Abd Al-Hamid A et al (2013) Capability of different non-nutritive feed additives on improving productive and physiological traits of broiler chicks fed diets with or without aflatoxin during the first 3 weeks of life. J Anim Physiol Anim Nutr 97:754–772

Attia YA, Ebeid TA, Shafi ME et al (2025) Strategies for decontamination and alleviation of mycotoxins for sustainable poultry farming–a review. Ann Anim Sci 25(1):139–157

Austin K, Ojebiyi P (2025) Immune system development in poultry: role of early nutrition in enhancing immune responses in broilers

Ayana GU, Kamutambuko R (2024) Probiotics in disease management for sustainable poultry production. Adv Gut Microbiome Res 2024(1):4326438

Azam MS, Yu D, Liu N, Wu A (2019) Degrading ochratoxin A and zearalenone mycotoxins using a multifunctional recombinant enzyme. Toxins 11:301

Azizpour A, Moghadam N (2015) Assessment of serum biochemical parameters and pathological changes in broilers with chronic aflatoxicosis fed glucomannan-containing yeast product (Mycosorb) and sodium bentonite. J Vet Res 59:205–211

Bagherzadeh Kasmani F, Karimi Torshizi M, Allameh A et al (2012) A novel aflatoxin-binding *Bacillus* probiotic: performance, serum biochemistry, and immunological parameters in Japanese quail. Poult Sci 91:1846

Benlashehr I, Repussard C, Jouglar JY et al (2011) Toxicokinetics of fumonisin B2 in ducks and turkeys. Poult Sci 90:1671–1675. https://doi.org/10.3382/ps.2011-01434

Bermudez-Brito M, Plaza-Diaz J, Munoz-Quezada S et al (2012) Probiotic mechanisms of action. Ann Nutr Metab 61:160–174. https://doi.org/10.1159/000342079

Bhatti SA, Khan MZ, Hassan ZU et al (2018) Comparative efficacy of bentonite clay, activated charcoal and *Trichosporon mycotoxinivorans* in regulating the feed-to-tissue transfer of mycotoxins. J Sci Food Agric 98:884–890

Bhatti SA, Khan MZ, Saleemi MK et al (2019) Impact of dietary *Trichosporon mycotoxinivorans* on ochratoxin A induced immunotoxicity; *In vivo* study. Food Chem Toxicol 132:110696

Bhogoju S, Nahashon S (2022) Recent advances in probiotic application in animal health and nutrition: a review. Agriculture 12(2):304. https://doi.org/10.3390/agriculture12020304

Boesch-Saadatmandi C, Loboda A, Józkowicz A et al (2008) Effect of ochra toxin A on redox-regulated transcription factors, antioxidant enzymes and glutathione-S-transferase in cultured kidney tubulus cells. Food Chem Toxicol 46:2665–2671

Bonos E, Giannenas I, Sidiropoulou E et al (2021) Effect of *Bacillus pumilus* supplementation on performance, intestinal morphology, gut microflora and meat quality of broilers fed different energy concentrations. Anim Feed Sci Technol 274. https://doi.org/10.1016/J.ANIFEEDSCI.2021.114859

Cavanagh D, Gelb JJR (2008) Infectious bronchitis. In: Diseases of poultry, 12th edn. Iowa State University Press, Iowa, pp 117–135

Cavin C, Delatour T, Marin-Kuan M et al (2009) Ochratoxin a–mediated DNA and protein damage: roles of nitrosative and oxidative stresses. Toxicol Sci 110:84–94

Çelýk K, Denlý M, Savas T (2003) Redução dos efeitos tóxicos da aflatoxina B1, utilizando-se levedura de panificação (*Saccharomyces cerevisiae*), nadieta de pintos de corte em crescimento. Rev Bras Zootec 32:615–619

Chang J, Wang T, Wang P et al (2020) Compound probiotics alleviating aflatoxin B(1) and zearalenone toxic effects on broiler production performance and gut microbiota. Ecotoxicol Environ Saf 194:110420

Charlton BR, Chin RP, Barnes HJ (2008) In: Saif YM et al (eds) Diseases of poultry: fungal infections. Blackwell Publishing, Ames, pp 989–1001

Chen SS, Li YH, Lin MF (2017) Chronic exposure to the Fusarium Mycotoxin Deoxynivalenol: impact on performance, immune organ, and intestinal integrity of slow-growing chickens. Toxins 9:334

Chen Y, Han S, Wang Y et al (2019) Oxidative stress and apoptotic changes in broiler chicken splenocytes exposed to T-2 toxin. BioMed Res Int 2019:5493870

Chlebicz A, Śliżewska K (2020) In vitro detoxification of aflatoxin B1, deoxynivalenol, fumonisins, T-2 toxin and zearalenone by probiotic bacteria from genus *Lactobacillus* and *Saccharomyces cerevisiae* yeast. Prob Antimicrob Prot 12:289–301

Chowdhury SR, Smith TK, Boermans HJ et al (2005a) Effects of feed-borne Fusarium mycotoxins on hematology and immunology in laying hens. Poult Sci 84:1698–1706

Chowdhury S, Smith T, Boermans H et al (2005b) Effects of feeding blends of grains naturally con taminated with Fusarium mycotoxins on performance, metabo lism, hematology, and immunocompetence of ducklings. Poult Sci 84:1179–1185

Collado MC, Gueimonde M, Salminem S (2010) Probiotics in adhesion of pathogens: mechanisms of action. In: Watson RR, Preedy VR (eds) Bioactive foods in promoting health: probiotics and prebiotics, 1st edn. Academic Press Elsevier, London, pp 353–370

Cravens RL, Goss GR, Chi F et al (2013) The effects of necrotic enteritis, aflatoxin B1, and virginiamycin on growth performance, necrotic enteritis lesion scores, and mortality in young broilers. Poult Sci 92:1997–2004

D'Amelio P, Sassi F (2017) Gut microbiota, immune system, and bone. Calcif Tissue Int 102(4): 415–425. https://doi.org/10.1007/s00223-017-0331-y

Dass RS, Sreenivasa MY, Janardhana GR (2007) High incidence of *Fusarium verticillioides* in animal and poultry feed mixtures produced in Karnataka, India. Plant Pathol J 6(2):174–178

Dazuk V, Boiago MM, Rolim G (2020) Laying hens fed mycotoxin-contaminated feed produced by *Fusarium* fungi (T-2 toxin and fumonisin B1) and *Saccharomyces cerevisiae* lysate: impacts on poultry health, productive efficiency, and egg quality. Microb Pathog 149:104517

Deepthi BV, Rakesh S, Poornachandra Rao K, Deepa N, Dharanesha NK, Girish KS, Sreenivasa MY (2017) *Lactobacillus plantarum* MYS6 ameliorates fumonisin B1-induced hepatorenal damage in broilers. Front Microbiol. https://doi.org/10.3389/fmicb.2017.02317

Deng X, Rui W, Zhang F, Ding W (2013) PM 2.5 induces Nrf2-me diated defense mechanisms against oxidative stress by activating PIK3/AKT signaling pathway in human lung alveolar epithelial A549 cells. Cell Biol Toxicol 29:143–157

Devegowda G, Ravikiran D (2008) Mycotoxins and eggshell quality: cracking the problem. World Mycotoxin J 1:203–208

Diaz G, Cortes A, Roldan L (2005) Evaluation of the efficacy of four feed additives against the adverse effects of T-2 toxin in growing broiler chickens. J Appl Poult Res 14:226–231

Ece S, Lambertz C, Fischer R et al (2017) Heterologous expression of a *Streptomyces cyaneus* laccase for biomass modification applications. AMB Express 7:86

Elaroussi M, Mohamed F, El Barkouky E (et al (2006) Experimental ochratoxicosis in broiler chickens. Avian Pathol 35: 263–269

Elfadaly HA, Hassanain MA, Shaapan RM et al (2017) Detection of *Toxoplasma gondii* from wastage nourished small ruminant and poultry: zoonotic significance. Int J Zool Res 13(1):6–11

Elhady MA, Khalaf AAA, Ibrahim MA et al (2020) Protective effects of fermentation extract against ochratoxin A-induced nephrotoxicity and immuno toxicity in broiler chickens. J Vet Res 66:167–177

Ellakany H, Awad AM, Abdel-Baky K (2011) Comparison of the effect of three different anti-mycotoxin substrates on resistance to aflatoxins in broilers. Alex J Vet Sci 32:119–135

El-Nezami H, Mykkanen H, Kankaanpaa P et al (2000) Ability of a mixture of *Lactobacillus* and *Propionibacterium* to influence the faecal aflatoxin content in healthy Egyptian volunteers: a pilot clinical study. Biosci Microfl 19:41–45

Elsherbeni AI, Youssef IM, Kamal M, Youssif MA, El-Gendi GM, El-Garhi OH, Alfassam HE, Rudayni HA, Allam AA, Moustafa M, Alshaharni MO (2024) Impact of adding zeolite to broilers' diet and litter on growth, blood parameters, immunity, and ammonia emission. Poult Sci 103:103981

Ge J, Zhang C, Sun YC et al (2019) Cadmium exposure triggers mitochondrial dysfunction and oxidative stress in chicken (Gallus gallus) kidney via mito chondrial UPR inhibition and Nrf2-mediated antioxidant defense activation. Sci Total Environ 689:1160–1171

Girgis GN, Barta JR, Girish CK et al (2010) Effects of feed-borne *Fusarium* mycotoxins and an organic mycotoxin adsorbent on immune cell dynamics in the jejunum of chickens infected with *Eimeria maxima*. Vet Immunol Immunopathol 138(3):218–223

Gorzelanna Z, Mamrot A, B˛edkowska D et al (2025) Exploring the potential of novel animal-origin probiotics as key players in one health: opportunities and challenges. Int J Mol Sci 26: 5143. https://doi.org/10.3390/ijms26115143

Guo H, Wang P, Liu C et al (2023) Compound mycotoxin detoxifier alleviating aflatoxin B1 toxic effects on broiler growth performance, organ damage and gut microbiota. Poultry Sci 102: 102434

Haghighi HR, Gong J, Gyles CL (2006) Probiotics stimulate production of natural antibodies in chickens. Clin Vaccine Immunol 13:975–980

Hamid IS, Mahendra I, Kurniawan A, Febrian MB, Saptiama I, Marlina M, Solfaine R, Fikri F (2025) Recent updates on encapsulated probiotics in poultry: a review. Polish J Vet Sci 28(2): 345–353

Haque MA, Wang Y, Shen Z et al (2020) Mycotoxin contamination and control strategy in human, domestic animal and poultry: a review. Microb Pathog 142:104095

Hasheminejad SA, Makki OF, Nik HA et al (2015) The effects of aflatoxin B1 and silymarin-containing milk thistle seeds on ileal morphology and digestibility in broiler chickens. Vet Sci Develop 5:36–45

Hassan RI, Mosaad GM (2015) Effect of feeding probiotic on the performance of broiler ducks fed different protein levels. J Adv Vet Res 5:136–142

Hatab MH, Elsayed MA, Ibrahim NS (2016) Effect of some biological supplementation on productive performance, physiological and immunological response of layer chicks. J Radiat Res Appl Sci 9:185–192

He J, Evans NM, Liu H et al (2021) UV treat ment for degradation of chemical contaminants in food: a review. Compr Rev Food Sci Food Saf 20:1857–1886

Hernandez-Patlan D, Solis-Cruz B, Hargis BM et al (2019) The use of probiotics in poultry production for the control of bacterial infections and aflatoxins. In: Prebiotics and probiotics-potential benefits in human nutrition and health. Intech Open, London

Hernández-Ramírez JO, Nava-Ramírez MJ, Merino-Guzmán R et al (2020) The effect of moderate-dose aflatoxin B1 and *Salmonella enteritidis* infection on intestinal permeability in broiler chickens. Mycotoxin Res 36:31–39

Hernandez-Valdivia E, Valdivia-Flores A, Cruz-Vazquez C et al (2021) Diagnosis of subclinical aflatoxicosis by biochemical changes in dairy cows under field conditions. Pak Vet J 41:21–28

Higgins J, Higgins S, Vicente J et al (2007) Temporal effects of lactic acid bacteria probiotic culture on *Salmonella* in neonatal broilers. Poult Sci 86:1662–1666

Huang W, Chang J, Wan P et al (2018) Effect of the combined compound probiotics with mycotoxin degradation enzyme on detoxifying aflatoxin B1, and zearalenone. J Toxicol Sci 43:377–385

Idowu PA, Mpofu TJ, Magoro AM, Modiba MC, Nephawe KA, Mtileni B (2025) Impact of probiotics on chicken gut microbiota, immunity, behavior, and productive performance—a systematic review. Front Anim Sci 6:1562527

Jia R, Sadiq FA, Liu W et al (2021) Protective effects of *Bacillus subtilis* ASAG 216 on growth performance, antioxidant capacity, gut microbiota and tissues residues of weaned piglets fed deoxynivalenol contaminated diets. Food Chem Toxicol 148:111962

Kabir S (2009) The role of probiotics in the poultry industry. Int J Mol Sci 10:3531–3546

Kamal M, Sindi RA, El-Azzazi FE, Kishk WH, Khalil HA, Abdel-Khalek AM, Ayoub MA, Tufarelli V, Abd El-Hack ME (2023) Sexual behaviour response, testicular development and semen quality of New Zealand white rabbit bucks as influenced by dietary chitosan. Reprod Domest Anim 58(3):405–413

Kamal M, Linlin K, Gao J, Xinrui Z, Xinming C, Haibo W, Lulu D, Abd El-Hack ME, Mahrose K, Cheng Y (2025) Effects of *Saccharomyces cerevisiae* and *Bacillus subtilis* on in vitro fermentation in the rumen of Hu sheep. J Sci Food Agric 105(1):498–506

Karlovsky P, Suman M, Berthiller F et al (2016) Impact of food processing and detoxification treatments on mycotoxin contamination. Mycotoxin Res 32:179–205

Kasmani FB, Mehri M (2015) Effects of a multi-strain probiotics against aflatoxicosis in growing Japanese quails. Livestock Sci 177:110–116

Kemboi DC, Ochieng PE, Antonissen G (2020) Multi-mycotoxin occurrence in dairy cattle and poultry feeds and feed ingredients from Machakos town Kenya. Toxins 12:762

Khalifa E, Mohesien MT, Mossa MI et al (2022) Diversity of toxigenic fungi in livestock and poultry feedstuffs. Int J Environ Res Public Health 19(12):7250

Khan A, Aalim MM, Khan MZ et al (2017) Amelioration of immunosuppressive effects of aflatoxin and ochratoxin A in White Leghorn layers with distillery yeast sludge. Tox Rev 36:275–281

Khan S, Chand N, Hafeez A, Ali M, Qattan SY, Alam M, Siddiqui SA, Nass NM, AlMalki F, Abd El-Hack ME, Al-Gheffari HK (2025) Impact of *Bacillus subtilis* probiotic on growth, visceral and lymphatic organ weights, intestinal histomorphology, and pathogenic bacteria of broilers. Ann Anim Sci 25(2):661–668

Khazraei SK, Tabeidian SA, Habibian M (2022) Selenium nanoparticles are more efficient than sodium selenite in reducing the toxicity of aflatoxin B1 in Japanese quail. Vet Med Sci 8:254–266

Kolawole O, Meneely J, Greer B et al (2019) Comparative in vitro assessment of a range of commercial feed additives with multiple mycotoxin binding claims. Toxins 11:659

Kolawole O, Graham A, Donaldson C et al (2020) Low doses of mycotoxin mixtures below EU regulatory limits can negatively affect the performance of broiler chickens: a longitudinal study. Toxins 12:433

Kumar S, Chen C, Indugu N et al (2018) Effect of antibiotic withdrawal in feed on chicken gut microbial dynamics, immunity, growth performance and prevalence of foodborne pathogens. PLoS One 13:e0192450

Laxminarayan R, Van Boeckel T, Teillant A (2015) The economic costs of withdrawing antimicrobial growth promoters from the livestock sector. OECD Food Agriculture and Fisheries Papers: OECD Publishing

Lázaro Á, Vila-Donat P, Manyes L (2024) Emerging mycotoxins and preventive strategies related to gut microbiota changes: probiotics, prebiotics, and postbiotics–a systematic review. Food Funct 15:8998

Lee SH, Beck BR, Hwang S-H, Song SK (2021) Feeding olive flounder (*Paralichthys olivaceus*) with *Lactococcus lactis* BFE920 expressing the fusion antigen of *Vibrio* OmpK, and FlaB provides protection against multiple vibrio pathogens: a universal vaccine effect. Fish Shellfish Immunol 114:253–262

Legan JD (2000) Cereals and cereal products. In: The microbiological safety and quality of food. Springer, Boston, pp 759–783

Li M, Guan E, Bian K (2019) Structure elucidation and toxicity analysis of the degradation products of deoxynivalenol by gaseous ozone. Toxins 11:474

Li S, Liu R, Xia S, Wei G et al (2022) Protective role of curcumin on aflatoxin B1-induced TLR4/RIPK pathway mediated-necroptosis and inflammation in chicken liver. Ecotoxicol Environ Saf 233:113319

Li Z, Li C, Lin F et al (2024) Duck compound probiotics fermented diet alters the growth performance by shaping the gut morphology, microbiota and metabolism. Poult Sci 103(6): 103647

Liu B, Peng X, Meng X (2018) Effective biodegradation of mycotoxin patulin by porcine pancreatic lipase. Front Microbiol 9:615

Liu Y, Galani Yamdeu JH, Gong YY et al (2020a) A review of postharvest approaches to reduce fungal and mycotoxin contami nation of foods. Compr Rev Food Sci Food Saf 19:1521–1560

Liu J, Doupovec B, Schatzmayr D et al (2020b) The impact of deoxynivalenol, fumonisins, and their combination on performance, nutrient, and energy digestibility in broiler chickens. Poult Sci 99:272–279

Loboda A, Stachurska A, Sobczak M et al (2017) Nrf2 deficiency exacerbates ochra toxin A-induced toxicity in vitro and in vivo. Toxicol 389:42–52

Ma Q, Gao X, Zhou T et al (2012) Protective effect of *Bacillus subtilis* ANSB060 on egg quality, biochemical and histopathological changes in layers ex posed to aflatoxin B1. Poult Sci 91: 2852–2857

Ma R, Zhang L, Liu M et al (2018) Individual and combined occurrence of mycotoxins in feed ingredients and complete feeds in China. Toxins 10:113

Maciorowski KG, Herrera P, Jones FT et al (2007) Effects on poultry and livestock of feed contamination with bacteria and fungi. Anim Feed Sci Technol 133:109–136

Madheswaran R, Balachandran C, Murali-Manohar B (2005) Pathological effects of feeding aflatoxin and T-2 toxin in Japanese Quail. Indian J Vet Pathol 29(1):23–26

Manafi M, Mohan K, Ali MN (2011) Effect of ochratoxin A on coccidiosis-challenged broiler chicks. World Mycotoxin J 4:177–181

Marin-Kuan M, Nestler S, Verguet C et al (2006) A toxicogenomics approach to identify new plausible epigenetic mechanisms of ochratoxin A carcinogenicity in rat. Toxicol Sci 89:120–134

Mattarelli P, Biavati B (2018) Chapter 2- species in the genus *Bifidobacterium*. In: Mattarelli P, Biavati B, Holzapfel WH, Wood BJB (eds) The Bifidobacteria and related organisms. Academic Press, pp 9–48. https://doi.org/10.1016/B978-0-12-805060-6.00002-8

Metayer JP, Travel A, Mika A et al (2019) Lack of toxic interaction between Fusariotoxins in broiler chickens fed throughout their life at the highest level tolerated in the European Union. Toxins 11:455

Mikulski D, Kozłowski K, Jankowski J et al (2008) Efficacy of yeast extract *Saccharomyces cerevisiae* in Turkey feeding. Med Weter 64:1331–1334

Mngadi PT, Govinden R, Odhav B (2008) Co-occurring mycotoxins in animal feeds. Afr J Biotechnol 7:5

Mohaghegh A, Chamani M, Shivazad M, Sadeghi AA, Afzali N (2017) Effect of esterified glucomannan on broilers exposed to natural mycotoxin-contaminated diets. J Appl Anim Res 45:285–291. https://doi.org/10.1080/09712119.2016.1174122

Monson MS, Settlage RE, Mendoza KM et al (2015) Modulation of the spleen transcrip tome in domestic Turkey (*Meleagris gallopavo*) in response to aflatoxin B1 and probiotics. Immunogenetics 67:163–178

Mossini SA, Arrotéia CC, Kemmelmeier C (2009) Effect of neem leaf extract and neem oil on *Penicillium* growth, sporulation, morphology and ochratoxin A production. Toxins 1:3–13

Motbaynor A, Kassaye D, Keffale M et al (2021) Magnitude of Aflatoxigenic *Aspergillus* species, level of aflatoxin B1, and associated factors in stored feed at poultry farms in Dire Dawa, Ethiopia. Vet Med Int 2021:6638083

Murugesan G, Ledoux D, Naehrer K et al (2015) Prevalence and effects of mycotoxins on poultry health and performance, and recent development in mycotoxin counter acting strategies. Poult Sci 94(6):1298–1315. https://doi.org/10.3382/ps/pev075

Muzaffar K, Jan R, Ahmad Bhat N et al (2021) Chapter 25- commercially available probiotics and prebiotics used in human and animal nutrition. In: Dhanasekaran D, Sankaranarayanan A (eds) Advances in probiotics. Academic Press, pp 417–435

Naseem MN, Saleemi MK, Khan A et al (2018) Pathological effects of concurrent administration of aflatoxin B1 and fowl adenovirus-4 in broiler chicks. Microb Pathog 121:147–154

Ochieng PE, Scippo ML, Kemboi DC et al (2021) Mycotoxins in poultry feed and feed ingredients from Sub-Saharan Africa and their impact on the production of broiler and layer chickens: a review. Toxins 13:633

Ogido R, Oliveira CAF, Ledoux DR et al (2004) Effects of prolonged administration of aflatoxin B1 and fumoni-sin B1 in laying Japanese quail. Poult Sci 83:1953–1958

Omidiwura BRO, Agboola AF, Olaleye OO et al (2018) Influence of prebiotics, probiotics or synbiotics on performance, intestinal mucosal integrity and gut microbiota of Turkey poults. Niger J Anim Sci 20(2):64–75

Opazo RJ, Salinas CC, Villasante A (2025) Formulation strategies of probiotics in broilers: systematic review and meta-analysis of their effects on production performance. Front Anim Sci 6:1679614

Ouwehand AC, Salminen S, Isolauri E (2002) Probiotics: an overview of beneficial effects. In: lactic acid bacteria: genetics, metabolism and applications. Springer, Heidelberg/Berlin Germany, pp 279–289

Pambuka SR, Sjofjan O, Radiati LE (2014) Effect of liquid probiotics mixed culture supplements through drinking water on laying hens performance and yolk cholesterol. J World's Poult Res 4: 5–9

Panitsidis I, Barbe F, Chevaux E et al (2023) Probiotics, prebiotics, paraprobiotics, postbiotics. In: Sustainable use of feed additives in livestock: novel ways for animal production. Springer International Publishing, Cham, pp 173–227

Pavlovic N, Stankov K, Mikov M (2012) Probiotics—interactions with bile acids and impact on cholesterol metabolism. Appl Biochem Biotechnol 168:1880–1895

Pewan SB, Kabantiyok D, Emennaa PE, Dawurung JS, Dawurung CJ, Duwil RK, Olorundare OO, Ngukat HY, Umaru MG, Ugwuoke GM, Ezema C (2025) Advancing Nigerian indigenous poultry health and production, use of probiotics as viable alternatives to antibiotics: a review. Antibiotics 14(8):846

Poloni V, Magnoli A, Fochesato A, Cristofolini A, Caverzan M, Merkis C, Montenegro M, Cavaglieri L (2020) A *Saccharomyces cerevisiae* RC016-based feed additive reduces liver toxicity, residual aflatoxin B1 levels and positively influences intestinal morphology in broiler chickens fed chronic aflatoxin B1-contaminated diets. Anim Nutr 6:31–38

Popova T (2017) Effect of probiotics in poultry for improving meat quality. Curr Opin Food Sci 14: 72–77

Rauber RH, Dilkin P, Giacomini LZ et al (2007) Performance of Turkey poults fed different doses of aflatoxins in the diet. Poult Sci 86:1620–1624. https://doi.org/10.1093/ps/86.8.1620

Raz M, Bagherzadeh-Kasmani F, Karimi-Torshizi MA et al (2025) Prosopis farcta biochar neutralizes aflatoxin B1 and enhances health and productivity in quails. Discov Anim 2(1):12

Salamoura C, Kontogianni A, Katsipi D et al (2014) Probiotic fermented milks made of cowsmilk, goats milk and their mixture. J Biotechnol 185:125

Saleemi MK, Ashraf K, Gul ST et al (2020) Toxicopathological ef fects of feeding aflatoxins B1 in broilers and its amelioration with indigenous mycotoxin binder. Ecotoxicol Environ Saf 187: 109712

Salwa AA, Anwer W (2009) Effect of naturally contaminated feed with aflatoxins on performance of laying hens and the carryover of aflatoxin B1 residues in table eggs. Pak J Nutr 8:181–186

Shaapan RM, Girh ZM (2024) Overview of aspergillosis in poultry-a review. Egypt J Vet Sci 55(2): 407–419

Shanmugasundaram R, Lourenco J, Hakeem WA et al (2023) Subclinical doses of dietary fumonisins and deoxynivalenol cause cecal microbiota dysbiosis in broiler chickens challenged with Clostridium perfringens. Front Microbiol 14:1106604

Shoukat S, Wani H, Jeelani R et al (2018) An overview of avian Aspergillosis. Int J Avian Wildlife Biol 3(3):222–223. https://doi.org/10.15406/ijawb.2018.03.00088

Shruthi B, Deepthi BV, Adithi G, Vanitha PR, Sreenivasa MY (2025) *Saccharomyces cerevisiae* MYSY1 as a promising probiotic isolate with antifungal properties against foodborne *Aspergillus* species. Biocatal Agric Biotechnol G7:10374U. https://doi.org/10.101G/j.bcab.2025.103G4 U

Singh S, Fatima Z, Hameed S (2015) Predisposing factors endorsing *Candida* infections. Infez Med 23:211–223

Sokół I, Bobrek K, Tokarzewski S et al (2015) Candidiasis in poultry. Med Weter 71:731–735

Sokół I, Tokarzewski S, Bobrek K et al (2017) The effect of the administration of different antimicrobial formulations on the fungal infestation of the gastrointestinal tract in turkeys. Pak Vet J 37:475–479

de Souza M, Baptista AAS, Valdiviezo MJ et al (2020) *Lactobacillus* spp. reduces morphological changes and oxidative stress induced by deoxynivalenol on the intestine and liver of broilers. Toxicon 185:203–212

Sun H, Du X, Zeng T et al (2022a) Effects of compound probiotics on cecal microbiome and metabolome of Shaoxing duck. Front Microbiol 12:813598

Sun H, Gu T, Li G et al (2022b) Effects of compound probiotics on GrowthPerformance, serum biochemical and immune indices, antioxidant capacity, and intestinal tissue morphology of shaoxing duck. Animals 12:3219. https://doi.org/10.3390/ani12223219

Tagoe D, Nyarko H, Akpaka R (2011) A comparison of the antifungal properties of onion (*Allium cepa*), ginger (*Zingiber officinale*) and garlic (*Allium sativum*) against *Aspergillus flavus, Aspergil lusniger* and *Cladosporium herbarum*. Res J Med Plants 5:281–287

Tahir MA, Abbas A, Muneeb M et al (2022) Ochratoxicosis in poultry: occurrence, environmental fac tors, pathological alterations and amelioration strategies. Worlds Poult Sci J 78:727–749

Tardieu D, Bailly JD, Skiba F et al (2007) Chronic toxicity of fumonisins in turkeys. Poult Sci 86: 1887–1893. https://doi.org/10.1093/ps/86.9.1887

Torlak E, Akata I, Erci F, Uncu AT (2016) Use of gaseous ozone to reduce aflatoxin B1 and microorganisms in poultry feed. J Stored Prod Res 68:44–49

Umar S, Younus M, Rehman MU et al (2015) Role of aflatoxin toxicity on transmissibility and pathogenicity of H9N2 avian influenza virus in turkeys. Avian Pathol 44:305–310

Vartiainen S, Yiannikouris A, Apajalahti J, Moran CA (2020) Comprehensive evaluation of the efficiency of yeast cell wall extract to adsorb ochratoxin A and mitigate accumulation of the toxin in broiler chickens. Toxins 12:37

Verma J, Johri TS, Swain BK (2003) Effect of varying levels of aflatoxin, ochratoxin and their combinations on the performance and egg quality characteristics in laying hens. Asian Australas J Anim Sci 16:1015–1019

Vieco-Saiz N, Belguesmia Y, Raspoet R et al (2019) Benefits and inputs from lactic acid bacteria and their bacteriocins as alternatives to antibiotic growth promoters during food-animal production. Front Microbiol 10:57

Vila-Donat P, Marín S, Sanchis V, Ramos A (2018) A review of the mycotoxin adsorbing agents, with an emphasis on their multi binding capacity, for animal feed decontamination. Food Chem Toxicol 114:246–259

Wan MLY, Forsythe SJ, El-Nezami H (2019) Probiotics interaction with foodborne pathogens: a potential alternative to antibiotics and future challenges. Crit Rev Food Sci Nutr 59:3320–3333

Wang Y, Sun J, Zhong H et al (2017) Effect of probiotics on the meat flavour and gut microbiota of chicken. Sci Rep 7:1–13

Wang J, Wang WW, Qi GH et al (2021) Effects of dietary *Bacillus subtilis* supplementation and calcium levels on performance and eggshell quality of laying hens in the late phase of production. Poult Sci 100:100970

Wang Y, Wang X, Li Q (2023) Aflatoxin B1 in poultry liver: toxic mechanism. Toxicon 233: 107262

Witaszak N, Stepien L, Bocianowski J et al (2019) *Fusarium* species and mycotoxins contaminating veterinary diets for dogs and cats. Microorganisms 7:26

Yadav S, Jha R (2019) Strategies to modulate the intestinal microbiota and their effects on nutrient utilization, performance, and health of poultry. J Anim Sci Biotechnol 10:1–11

Yohannes T, Sharma AK, Sumi V et al (2012) Clinical manifestation and growth performance of broiler chickens fed with T-2 toxin and co-infected with infectious bronchitis virus. Food Agric Sci 2:30–35

Yu YH, Hsiao FSH, Proskura WS et al (2018) An impact of Deoxynivalenol produced by *Fusarium graminearum* on broiler chickens. J Anim Physiol Anim Nutr 102:1012–1019

Yushanaji B, Kamal M, Baishanbieke D, Zhu L, Zhang P, He Q, Li F (2025) Effects of raffinose supplementation on growth performance, carcass quality, blood chemistry, gut microbiota, and intestinal morphology in yellow-feathered broilers. Poult Sci 23:105888

Zhao P, Li Y, Yang Y et al (2025) Probiotic efficacy and mechanism of a pigeon derived *Ligilactobacillus salivarius* strain in promoting growth and intestinal development of pigeons. Front Microbiol 16:1584380. https://doi.org/10.3389/fmicb.2025.1584380

Ziółkowska G, Tokarzewski S (2007) Occurrence of moulds in reproductive goose flocks in southern-eastern Poland. Bull Vet Inst Pulawy 51:553–556

Zou Y, Liu SB, Zhang Q et al (2023) Effects of aflatoxin B1 on growth performance, carcass traits, organ index, blood bio chemistry and oxidative status in Chinese yellow chickens. J Vet Med Sci 85:1015–1022

Fungal Diseases and Mycotoxicosis in Aquaculture: Probiotic Measures

11

B. Shruthi, G. R. Supritha, Mohammed Kareem Abbood, and M. Y. Sreenivasa

Abstract

Fungal pathogens and mycotoxin-associated disorders pose significant challenges to aquaculture, causing notable economic losses and adversely affecting the health of cultured aquatic organisms. Opportunistic fungal infections can weaken fish's immune systems, increasing their susceptibility to secondary infections and elevating mortality rates. Moreover, mycotoxins produced by certain fungi may contaminate aquatic environments and commercial feeds, intensify disease incidence, and disrupt normal growth and physiological development in fish. Probiotic-based interventions have gained attention as effective strategies for managing fungal diseases and reducing mycotoxicosis in aquaculture systems. These beneficial microorganisms enhance host immune responses, improve water quality, and inhibit the growth of pathogenic fungi and other harmful microbes. Their use offers a sustainable alternative to antibiotics and chemical treatments, promoting healthier fish stocks while reducing environmental impact. This chapter highlights the role of probiotics in preventing fungal infections and mycotoxin contamination, focusing on their mechanisms of action, benefits, and potential integration into environmentally sustainable and economically viable aquaculture practices.

B. Shruthi · G. R. Supritha · M. K. Abbood · M. Y. Sreenivasa (✉)
Applied Mycology Laboratory, Department of Studies in Microbiology, University of Mysore, Mysuru, Karnataka, India
e-mail: mys@microbiology.uni-mysore.ac.in

M. Y. Sreenivasa et al. (eds.), *Probiotics as Antifungals*,
https://doi.org/10.1007/978-981-92-0002-3_11

Graphical Abstract

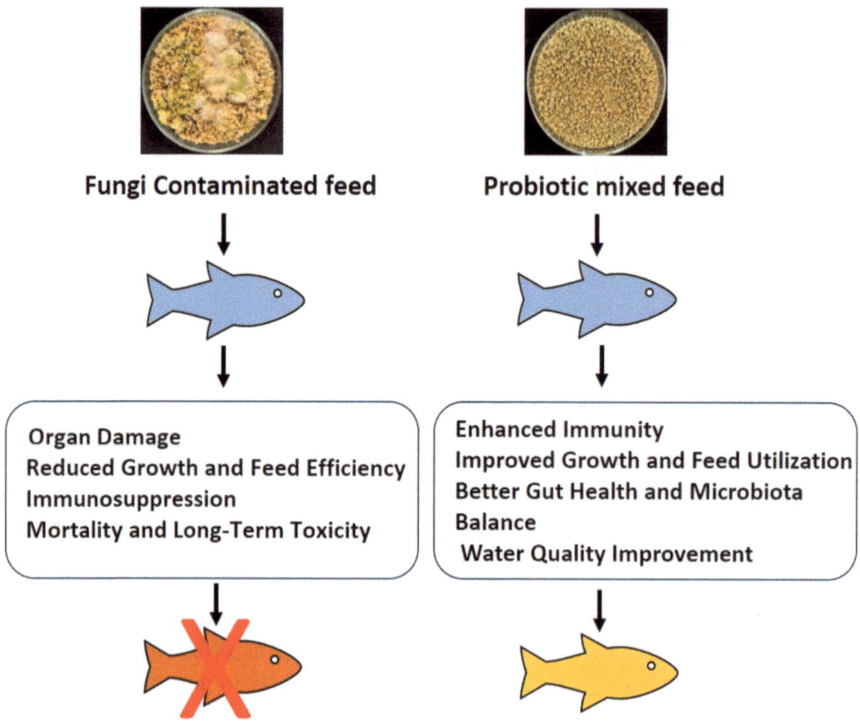

Keywords

Aquatic fungi · Disease management · Fish health · Mycotoxin · Probiotic efficacy

Learning Objectives

- Understand fungal diseases and mycotoxicosis in aquaculture species.
- Common fungal pathogens and mycotoxins affecting fish.
- Probiotics help prevent and control fungal infections and mycotoxicosis.
- Probiotic-based strategies to reduce fungal outbreaks and mycotoxin risks in aquaculture.
- Mode of action of probiotics on aquatic fungal pathogens

11.1 Introduction

Aquaculture has emerged as the most rapidly increasing sector of global food production, undergoing significant growth, strengthening, and geographic spread across most regions of the world. Growing human population numbers have driven a parallel rise in the demand for aquatic-based foods, positioning aquaculture as an important constituent of global food systems (Subasinghe et al. 2009). To meet this call, the sector now plays a main role in supporting food and dietary security, supplying roughly 43% of the world's aquatic animal protein with fish, crustaceans, and mollusks, while excluding aquatic plants, reptiles, and mammals (Bostock et al. 2010). Current trends specify that aquaculture will continue to inflate in the coming decades. Reflecting this growth, global aquaculture output reached an extraordinary level of 136.16 million tons in 2023 (Tacon et al. 2025).

In spite of the speedy expansion of aquaculture, the sector faces mounting challenges, particularly from the increasing occurrence and recurrence of infectious diseases that cause considerable financial damage. Among these, fungal diseases are some of the most frequently reported health problems in farmed fish, especially in temperate climates. Such infections often arise following physical injury, erstwhile infection by other pathogens, or exposure to adverse environmental conditions, including poor water quality and reduced temperatures. Aquatic fungi, which naturally multiply on decaying organic material, are often regarded as opportunistic pathogens; however, their effects on cultured fish populations can be severe. Economic assessments suggest that fungal diseases may contribute to approximately 10–15% of total production losses in aquaculture, mainly due to increased mortality and reduced fish health (Iqbal et al. 2023).

Disease epidemics caused by viral, bacterial, and fungal pathogens have resulted in severe monetary losses within aquaculture systems worldwide. The increasing incidence of these outbreaks represents one of the most pressing constraints on the sustainability of the aquaculture industry. The traditional disease management has often depended on the use of costly and ecologically harmful antibiotics, which have led to adverse impacts on aquatic ecosystems. As a result, there is growing interest in the application of probiotics as an alternative, environmentally friendly method to disease prevention and the improvement of fish growth and health (Mohammed et al. 2025). Besides, a range of environmental and biological stressors- together with inadequate farm management practices, imbalanced nutrition, toxin buildup, and genetic limitations- have been recognized as key contributors to large-scale stock losses (Kautsky et al. 2000; Dass et al. 2007).

Traditionally, disease management in aquaculture has hinged largely on the application of chemical agents and veterinary medicines, with antibiotics being the most widely used. While effective in the short term, this dependence has raised significant public health issues, predominantly due to the development and spread of antibiotic-resistant bacteria (Nomoto 2005). In retort, there has been increasing prominence on adopting aquaculture practices that are both sustainable and environmentally responsible (Onomu and Okuthe 2024). The development and execution of substitute approaches such as sustainable feed formulations and improved disease control strategies are therefore vital for maintaining the long-term sustainability of aquaculture production systems.

One of the most promising alternatives to conservative disease control methods in aquaculture is the application of probiotics, which are live microbial supplements intended to support and enhance host health. First introduced into aquaculture practices in the mid-1980s to enhance the growth of aquatic organisms, probiotics have since been widely adopted for additional purposes, including water quality improvement, pathogen suppression, and increased resistance to diseases. Their benefits are now understood to be wide ranging, encompassing enhanced nutrient utilization, greater tolerance to environmental stress, strengthened immune function, and improved reproductive outcomes. Commercial probiotics products used in aquaculture commonly contain beneficial microorganisms such as species of *Bacillus*, *Lactobacillus*, *Enterococcus*, *Carnobacterium* species, as well as the yeast *Saccharomyces cerevisiae*, with their effectiveness closely linked to appropriate application and management strategies (Cruz et al. 2012).

Recent research has emphasized the efficiency of probiotics beyond their roles in growth enhancement and general health maintenance, representing their potential in the prevention and control of fungal diseases in aquatic organisms. By strengthening immune defence systems and improving water quality, probiotics offer a practical and environmentally sustainable approach to managing fungal infections in aquaculture systems (Tuan et al. 2013; Deepa et al. 2016; Elsegeny et al. 2025). Consequently, this chapter scrutinizes the use of probiotic-based strategies for fungal disease management in aquaculture, supporting the extensive objective of promoting sustainable and resilient aquaculture practices.

11.2 Fungal Diseases in Aquaculture

Aquaculture has grown substantially over the last two decades and provides the most affordable protein source to the rapidly growing population around the world. Many fungi, such as *Saprolegnia, Achlya, Aphanomyces, Pythium, Aspergillus, Penicillium, and Fusarium* species, are common in the aquatic environment because they feed on decomposing organic waste. Fungal infection also occurs in shellfish like crab, lobster, and shrimp. The disease is thought to contribute 10–50% of the cost of production. The growth of fish disease is posing new obstacles and causing economic losses, particularly as fish mortality rises (Iqbal et al. 2023). Some of the fungal diseases in the case of fish are listed below.

11.2.1 Saprolegniasis

Saprolegniasis is the most common mold infection in freshwater aquariums, caused by the opportunistic fungal pathogen *Saprolegnia*. It typically presents as thin, cotton-like growths on the skin or gills of fish. These lesions often begin as small, localized infections but can rapidly spread across the body surface. Initially white in appearance, the affected areas may later turn red, brown, or green. The severity of the disease depends on the extent of skin and gill damage. Saprolegniasis usually arises as a secondary infection, developing after the fish's outer protective layer has been compromised (Mhaisen 1993; Shin et al. 2017).

11.2.2 Branchiomycosis

Gill rot disease is caused by two fungal species: *Branchiomyces sanguinis*, which primarily affects carp, and *Branchiomyces demigrans,* which targets pike and tench. These fungi characteristically infect fish that are stressed due to deprived ecological conditions, such as low pH levels (between 5.8 and 6.5), reduced dissolved oxygen, or dense algal blooms. The infection originates from fungal spores existing in the water and organic debris on the pond bottom. Once acquainted with, the fungi occupy and damage the gill tissue, damaging the fish's capability to respire and leading to serious health problems (Muraosa et al. 2006).

11.2.3 Icthyophoniasis

This ailment is caused by *Ichthyophonus hoferi*, a mycological pathogen that flourishes in cool temperatures ranging from 36 °F to 68 °F. It can endure in both freshwater and saltwater environments and disturbs both wild and farmed fish. The disease spreads primarily through the consumption of infected fish (cannibalism) and the release of fungal cysts through excreta. In advanced stages, the infection can spread beneath the skin and into the muscle tissue, giving the skin a rough sandpaper-like texture. Affected fish may also develop curved spines, and internal organs often become enlarged and covered with white to greyish lesions (Ramaiah 2006; Willoughby and Roberts 1992).

11.2.4 Exophialosis

This fungal infection is caused by *Exophiala salmonis* and *Exophiala psychrophila*, both of which produce septate, asymmetrical, and branching hyphae. These molds can taint a diversity of fish species. Diseased fish often show darkened coloration, lethargy, and abnormal or unreliable swimming behavior (Armwood et al. 2021).

11.2.5 Larval Mycosis

This fungal contamination is caused by *Exophiala salmonis* and *Exophiala psychrophila,* both of which produce septate, asymmetrical, and branching hyphae. These molds can blight a variety of fish species. Infected fish repeatedly revealed darkened coloration, lethargy, and uncharacteristic or erratic swimming behavior (Hatai 2011).

11.2.6 Black Gill Disease

Black gill disease is a mycological infection affecting decapods, chiefly caused by members of the phylum Ascomycota, with *Fusarium solani* identified as a key pathogen in *Penaeus* species. The disease is categorized by the occurrence of distinctive black spots on the gills, which often resulting in mortality in juvenile pond-raised prawns. Infections characteristically begin in previously damaged tissues, such as those affected by physical injuries, chemical exposure, pollutants or preexisting disease related lesions (Frischer Marc et al. 2017).

11.2.7 Aflatoxicosis (Red Disease)

Red disease is caused by aflatoxins produced by *Aspergillus flavus* and other *Aspergillus* species, commonly found in improperly stored or expired feed. This condition affects marine crustaceans such as *Penaeus monodon* and other *Penaeus* species. In juvenile pond-cultured prawns, the disease is marked by yellowish discoloration of the body and appendages, which progresses to a reddish hue. Infected prawns show lethargic swimming behavior, often gathering near pond dikes, and express signs of lethargy. Soft shelling is also normally observed in affected individuals (Landers 2021).

11.3 Mycotoxicosis in Aquaculture

Mycotoxins are noxious metabolites produced by certain molds, particularly *Aspergillus* and *Penicillium* species, often under poor storage conditions (Hussein and Brasel 2001). In aquaculture, these compounds pose a serious hazard to animal health, performance, and overall product quality. The industry's increasing dependence on plant-based constituents in aquafeeds has heightened worries around mycotoxin contamination (Tacon et al. 2011). While aflatoxins (AF) and ochratoxin A (OTA) are commonly connected with poor storage, other mycotoxins—especially those produced by *Fusarium* species, such as trichothecenes (Type A and B) and fumonisins are mostly introduced during the crop preharvest phase (Cheli et al. 2013).

These toxins are particularly stable during feed processing, permitting them to continue in finished feeds and then influence aquatic species. The strictness of their effects depends on the attentiveness and blend of mycotoxins present. Consequences include direct impacts such as mortality and reduced growth, and indirect impacts like increased susceptibility, all of which can result in substantial economic losses in aquaculture.

Major mycotoxins of concern in aquaculture include: Aflatoxins (e.g., Aflatoxin B1), Ochratoxin A (OTA), Fumonisins (e.g., Fumonisin B1), Deoxynivalenol (DON), T-2 toxin, Zearalenone (ZEN), Citrinin, Patulin, Ergot alkaloids, and Sterigmatocystin.

Globally, mycotoxin contamination in fish feed is widespread and varies by region. For instance, aflatoxins are more commonly detected in Southern Europe, Africa, South Asia, and Southeast Asia. Deoxynivalenol contamination is more prevalent in North America, Northern and Central Europe, Africa, and North Asia. Zearalenone frequently occurs in North and South America, Central Europe, Africa, and parts of Asia, while fumonisins are widely reported in South America, Southern Europe, Africa, and across Asia. Ochratoxin A is most often found in South Asia and Africa (Pinotti et al. 2016).

Ingestion of the contaminated feed not only compromises fish health, leading to higher mortality, reduced growth, and reproductive issues, but also poses risks to human consumers. Even low concentrations of mycotoxins can accumulate in fish muscle tissue, and when food safety regulations are less stringent, this accumulation adds to the overall burden of mycotoxin exposure. Long-term, low-level human consumption of mycotoxins has been linked to chronic health issues such as cancer and weakened immune function (Oliveira and Vasconcelos 2020).

11.4 Impact of Fungal Infections and Mycotoxicosis on Aquatic Species

Fish fungal infections are an emerging threat with significant implications for both animal and human health. Several fungal species that primarily infect fish also possess zoonotic potential, highlighting the global public health relevance of fish mycoses. These fungal pathogens are widespread in aquatic environments and can affect both wild and farmed fish populations. Infections can range from superficial to systemic, contributing substantially to illness and mortality in aquatic stocks (Alam et al. 2023).

While many reported cases involve opportunistic fungi that exploit weakened immune systems in fish, certain fungal species are capable of acting as primary pathogens, infecting even healthy individuals. The most commonly implicated genera in fish fungal diseases include *Aphanomyces*, *Aspergillus*, *Candida*, *Chrysosporium*, *Exophiala*, *Fusarium*, *Ichthyophonus*, *Paecilomyces*, *Penicillium*, *Phoma*, *Saprolegnia*, *Trichophyton*, and *Trichosporon*. Among these, zoonotic threats are primarily associated with genera such as *Chrysosporium*, *Histoplasma*, *Paracoccidioides*, *Sporothrix*, and *Cryptococcus* (Alam et al. 2023).

Aquatic fungi are capable of causing serious diseases in nearly all freshwater fish species, as well as some marine fish and their incubating eggs. Each year, the number of identified fungal and fungal-like pathogens affecting fish continues to rise globally. The most prevalent fungal diseases in fish include Saprolegniasis, Branchiomycosis, epizootic ulcerative syndrome (EUS), ichthyophoniasis, and exophialasis.

Saprolegniasis is one of the most economically significant fungal infections, primarily affecting freshwater and some estuarine fish species in warm and tropical regions. Another common fungal agent, *Achlya* spp., also infects fish; however, clinical diagnosis of *Achlya*-related infections remains inconsistent.

Branchiomycosis, also known as "gill rot," is a well-known fungal infection targeting the gill tissues of fish. EUS is a critical seasonal epizootic disease impacting both wild and farmed freshwater and estuarine fish populations. Ichthyophoniasis is a chronic, systemic, granulomatous disease caused by internal fungal infection. Opportunistic fungal species such as *Exophiala* can infect a wide variety of warm- and cold-blooded hosts (Qadir et al. 2024).

In addition to fungal infections, mycotoxins, secondary metabolites produced by fungi, pose a significant toxic threat to aquatic species. The severity of mycotoxin toxicity in fish depends not only on the concentration in feed but also on the duration of exposure, fish species, age, and sex (Pleadin et al. 2017). Among the various mycotoxins, aflatoxin B1 (AFB1) is the most extensively studied due to its widespread occurrence in tropical regions, potent hepatotoxicity, and classification as a human carcinogen (Marijani et al. 2019). Notably, channel catfish (*Ictalurus punctatus*) show high resistance to AFB1, whereas rainbow trout are highly susceptible; even exposure to as little as 0.4 μg/kg AFB1 can result in a 14% tumor incidence (Jantrarotai et al. 1990).

Fusarium-derived mycotoxins can cause both acute and chronic toxicity. Diets contaminated with deoxynivalenol (DON) have been associated with reduced growth, feed intake, and feed conversion efficiency in fish (Tola et al. 2015). Zearalenone (ZEN), known to disrupt reproduction in farm animals, has also shown adverse effects in aquatic species. In zebrafish (*Danio rerio*), ZEN exposure has been linked to reduced spawning frequency, altered fecundity across generations (Schwartz et al. 2010), and developmental abnormalities, including cardiac and ocular defects and spinal curvature at higher concentrations (Bakos et al. 2013).

Ochratoxin A (OTA) has been shown to negatively impact channel catfish, causing reduced weight gain, poor feed efficiency, lower survival rates, and decreased hematocrit levels (Manning et al. 2003). Another mycotoxin of concern sterigmatocystin, a precursor of aflatoxin biosynthesis with genotoxic and carcinogenic potential, has been reported to cause pathological effects in Nile tilapia. The effects include decreased body weight, increased frequencies of micro-nucleated red blood cells (MN RCBs), and chromosomal aberrations in kidney cells (Mahrous et al. 2006) (Fig. 11.1).

11.5 Probiotics in Aquaculture

Aquaculture plays a vital role in ensuring global food security for the growing human population and has expanded rapidly through the adoption of intensive farming practices. However, the widespread and often indiscriminate use of chemical additives and veterinary drugs as preventive and therapeutic measures has led to the emergence of antimicrobial-resistant pathogens and the deterioration of aquatic environments (Bachère 2000). As a result, disease outbreaks have become more frequent, causing substantial losses and posing a major challenge to aquaculture productivity and trade. This, in turn, hampers financial evolution in many provinces. In response, numerous substitute methods have been explored to improve the

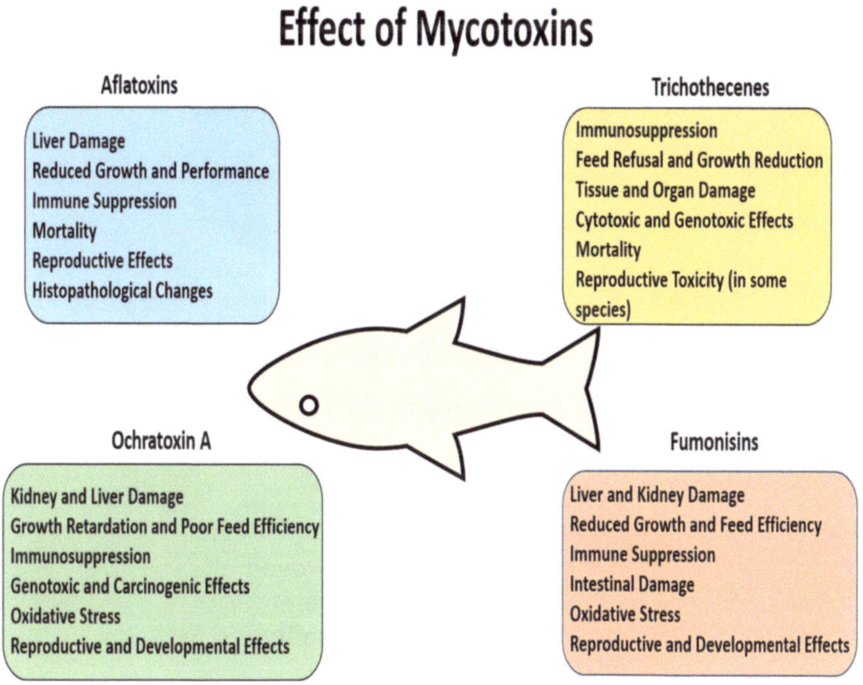

Fig. 11.1 Effect of mycotoxins in aquacultures

sustainability and excellence of aquaculture systems (Li et al. 2006). Probiotics have been extensively documented for their health benefits, and their role in aquaculture is gaining status as a sustainable alternative to conservative practices. This highlights the need for vigilant selection of host-specific probiotics to safeguard safety, effectiveness, and environmental compatibility in aquaculture systems. It discusses key challenges in probiotic application, including existence under variable conditions, storage, and delivery technologies, as well as potential risks. Overall, the effective use of probiotics offers a promising approach for cleaner, sustainable aquafarming and boosted protein production (Todorov et al. 2024).

To address the challenges, a variety of approaches have been developed to guard farmed aquatic animals from pathogenic threats. Among these, probiotics have arisen as a promising solution, representing significant benefits in disease management and overall farm health (Skjermo and Vadstein 1999) (Fig. 11.2). The concept of probiotics was initially proposed in 1974, when they were defined as organisms and substances that help maintain intestinal microbial balance. As scientific understanding grew, the definition of probiotics has been refined. Probiotics are described as cultured products or live microbial feed supplements that offer health benefits to the host by enhancing the microbial balance in the intestine (Hai 2015).

The intestinal microbiota of aquatic animals is in constant interaction with both the host's physiological functions and the surrounding environment. In this context,

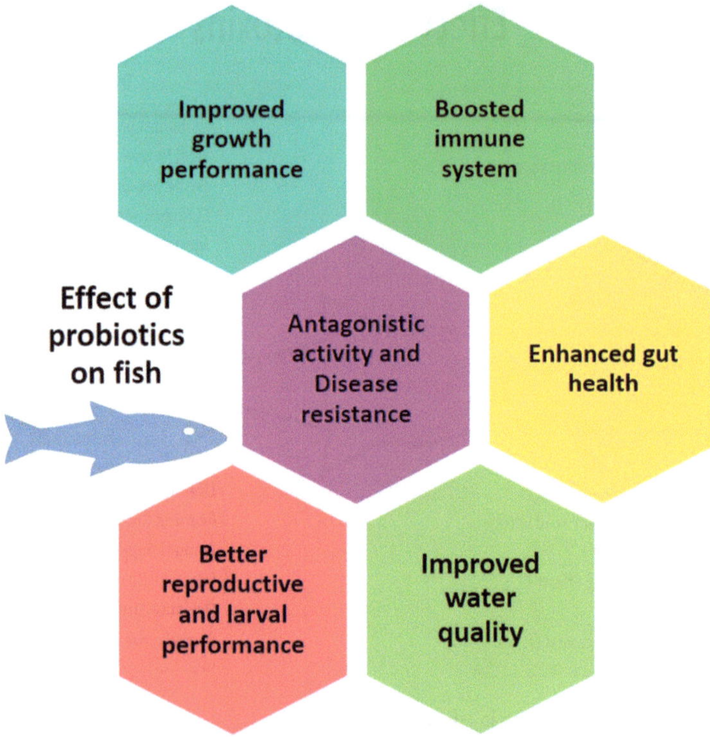

Fig. 11.2 Effect of probiotics on aquaculture

probiotics are defined as live microbial adjuncts that provide beneficial effects through various mechanisms, including (i) modulating the host associated or environmental microbial communities, (ii) enhancing feed utilization or improving its nutritional value, and (iii) boosting host immunity or improving the quality of the ambient environment (Verschuere et al. 2000). According to FAO/WHO (2001), probiotics are "live microorganisms which, when administered in adequate amounts, confer a health benefit on the host."

Expanding on this, Merrifield et al. (2010) described probiotics as live, dead, microbial components administered through feed or directly into rearing water, which enhance the host's health by improving disease resistance, growth performance, stress response, feed efficiency, and overall vitality—either by restoring the host's microbial balance or by modifying the microbiota of surrounding environment.

Probiotics have a long history of use in human and veterinary medicine, primarily consisting of lactic acid bacteria and yeast such as *Saccharomyces boulardii* in aquaculture. Their use has broadened to include both bacterial and nonbacterial strains, applied either through feed supplementation or water treatment. For instance, beneficial bacteria that promote fish growth include species such as *Streptococcus faecium* and *Lactobacillus acidophilus*, along with *Lactobacillus fermentum* and the yeast *Saccharomyces cerevisiae* and other *Bacillus* spp. (Elsegeny et al. 2025). These probiotics offer multiple benefits: promoting growth, reducing the incidence of disease, minimizing the reliance on chemical therapeutics, and performing effectively across a variety of aquatic environments, including fresh water, brackish water, and marine systems (Hai 2015).

11.6 Probiotics as a Strategy for Managing Fungal and Mycotoxin Contamination

Probiotics in aquaculture act by competitively excluding pathogens through adhesion to gut surfaces, producing antimicrobial and inhibitory compounds, and modulating the host immune system. They enhance phagocytosis, antioxidant defense, and immune gene expression while improving gut microbial balance. Probiotics also contribute to better water quality, stress tolerance, and protection against bacterial and fungal pathogens, leading to improved health and disease resistance in aquatic species (Gadhiya et al. 2025).

Aquatic probiotics promote aquaculture production and augment economic yield without compromising the safety of cultured organisms, human, and the environment. It is a promising alternative to antibiotics and chemotherapeutics, which are unsafe and have long-term negative consequences (Angahar 2016). The study isolated *Candida* spp., mainly *C. albicans*, from Nile tilapia, fish wash water, and fish sellers in Sharika, Egypt. Zinc nanocomposites (ZnNCs), *Lactobacillus salivarius*, and Zn nanoparticles showed antifungal effects (El-Gazzar et al. 2024). The study suggests that the administration of probiotics, herbal, and azadirachtin supplementation diets for 30 days protects the hematological and biochemical parameters in *Cirrhina mrigala* from *Aphanomyces invadans* (Harikrishnan et al. 2010). A bacteriocin-like inhibitory substance (BLIS) produced by *Aeromonas media* strain A199 inhibited the growth of *Saprolegnia* sp. in vitro, an opportunistic pathogen isolated from affected eels, *Anguilla australis* (Richardson) (Lategan and Gibson 2003). A study showed that dietary Galactooligosaccharide (GOS) (2.5 g/kg) improved growth, gut health, and immunity in *C. punctatus*, outperforming *S. cerevisiae*. GOS enhanced resistance to *A. invadans* by supporting beneficial gut microbes and immune responses. Further studies are needed to confirm these effects in other fish species before recommending GOS or *S. cerevisiae* as aquaculture feed additives (Devi et al. 2019). The study evaluated three probiotics, *Enterococcus gallinarum* L1, *Vagococcus fluvialis* L21, and *Lactobacillus plantarum* CLFP3 for controlling saprolegniosis in rainbow trout. While all strains inhibited *Saprolegnia parasitica* and reduced cyst adhesion in vitro, none provided in vivo protection. Results highlight the need to match probiotics and mechanisms to specific fish species and diseases (Fregeneda-Grandes et al. 2023). Dietary *Lactobacillus acidophilus* ATCC 4356 relieves the impacts of aflatoxin B_1 toxicity on the growth performance, hepatorenal functions, and antioxidative capacity of Thinlip Grey Mullet (*Liza ramada*) (Shukry et al. 2021).

11.7 Implementation of Probiotics in Aquaculture Systems

Probiotics are beneficial microorganisms, commonly including bacteria. For instance, lactic acid-producing bacteria are widely used in terrestrial animals (Lauzon et al. 2008). In aquaculture, a wider variety of microorganisms is utilized, including both Gram-positive and Gram-negative bacteria, which have shown effective results. Additionally, non-bacterial candidates such as bacteriophages, microalgae, and yeasts are also frequently explored for probiotic applications in aquaculture. Among these, Gram-positive bacteria are extensively employed worldwide, with particular emphasis on endospore-forming species from the *Bacillus* genus (Hong et al. 2005). *Bacillus subtilis,* in particular, is a commonly used probiotic in aquaculture systems due to its resilience and effectiveness. Yeasts have also demonstrated notable probiotic benefits in aquatic animals. *Saccharomyces cerevisiae*, for example, has shown potential as an alternative to live feed in fish production (Gunasundari et al. 2013). Probiotics can be administered to aquatic organisms through injection, bathing, artificial diets, live food, or direct addition to culture water. Injection and artificial diets are unsuitable for larvae, while bathing and water addition allow passive uptake. Live food is an effective method, providing nutrition while delivering probiotics and suppressing pathogen growth (Gadhiya et al. 2025).

Probiotics in aquaculture can be administered through various methods, including direct application to water; water-based administration is considered the most practical approach for prawns (Huang et al. 2006). However, the most notable benefits, for instance, are improving the fecundity of zebrafish (*Danio rerio*) (Gioacchini et al. 2010). Oral administration offers particular advantages in prawn farming, as it allows treatment at any stage of the culture period, regardless of the prawn's size (Sakai 1999).

Probiotics can be administered either as single strains or in combination. While most research to date has focused on single-strain applications, the potential benefits of multi-strain or multi-species combinations remain largely speculative (Hai 2015). However, some studies suggest that combining probiotic strains may offer enhanced protection against pathogenic infections (Timmermans et al. 2004). For example, a blend of *Bacillus subtilis* and *Lactobacillus acidophilus* was found to improve hematocrit levels and serum bactericidal activity in Nile tilapia more effectively than individual strains (Aly et al. 2008).

Recent strategies have also focused on using probiotics alongside prebiotics, immune-boosting compounds, or plant-derived additives to strengthen and improve disease resistance in farmed aquatic species. One particularly promising approach involves fortifying live feeds with probiotics through encapsulation techniques. This allows the microorganisms to remain active and, in some cases, multiply directly on the feed. As a result, probiotics can be delivered more efficiently to the target organisms, improving their uptake and potential benefits (Hai 2015). Live feeds such as *Artemia*, rotifers, and copepods have been effectively enhanced with probiotics and are considered appropriate vectors for management. Precisely, the copepod *Pseudodiaptomus annandalei* has been identified as an effective carrier of *Bacillus* spp. to grouper *(Epinephelus coioides)* larvae (Sun et al. 2013).

The effectiveness of live versus inactivated probiotics in aquaculture remains a topic of debate. Both viable cells and cellular components of *Bacillus licheniformis* and *Bacillus pumilus* have been shown to enhance lysozyme activity and respiratory burst responses in rohu *(Labeo rohita)* (Ramesh et al. 2015). Similarly, dietary supplementation with heat-inactivated probiotics has been found to stimulate innate immune parameters in fish (Irianto and Austin 2003). These findings suggest that non-viable probiotic preparations may serve as effective alternatives to live probiotics—especially in open aquatic systems, where the introduction of live microorganisms could pose ecological or safety concerns (Salinas et al. 2006).

11.8 Challenges and Limitations of Probiotic Use

The incorporation of feed additives, primarily probiotics, into the aquaculture diets became a subject of significant research interest in the latter part of the twentieth century. Since the early 2000s, numerous reviews have been published highlighting their potential benefits. However, despite their promising outcomes, the successful applications of these additives in aquaculture present several challenges. A key issue for the aquaculture sector is the financial burden associated with adopting probiotic technologies. Before their incorporation into feeds, newly developed strains must be thoroughly assessed to confirm their safety, effectiveness, and compliance with animal standards. In addition, manufacturers are required to implement up-to-date and standardized production and testing procedures to maintain consistent quality and safety. Together, these necessary measures lead to higher development and manufacturing expenses (Ayisi et al. 2017).

The global diversity of farmed aquatic species presents a major obstacle to the universal use of probiotics as feed supplements. Differences in feeding habits, physiological and habitat conditions, along with the chemical and physical variability of culture systems, make it difficult to apply uniform dosing and formulations. Consequently, the assumption that a single probiotic product can deliver reliable benefits across all species and production environments is increasingly viewed as unrealistic. Research conducted over recent decades highlights the need for precise selection and targeted application of these additives tailored to specific species and farming conditions. Despite notable advances, important gaps in knowledge remain, underscoring the need for continued investigation and refinement of existing strategies. In particular, the molecular processes involved in maintaining gut integrity and regulating immune responses are not fully understood. Moreover, the ways in which postbiotics interact with the immune cells, especially under conditions of chronic inflammation or immune dysregulation, warrant further detailed study. The long-term impacts of postbiotics on microbiome stability are unclear. Comprehensive omics-based research, well-designed clinical trials, and robust regulatory frameworks are essential for their safe and effective application (Amenyogbe et al. 2020; Thakur et al. 2025).

Overdosage administrations of probiotics can induce immune suppression of continuous responses of nonspecific immune systems. A probiotic dosage may bring positive and negative results to different receivers, whose responses to different dietary probiotic levels have been observed. The period of administration is also considered an important factor in using probiotics. Prolonged administrations of probiotics can induce immune suppression of continuous responses of nonspecific immune systems (Hai 2015).

11.9 Future Directions in Probiotic Research for Aquaculture

Although probiotics hold great promise in aquaculture as alternatives to antibiotics, much remains to be understood before their widespread application. Field trials have shown improved survival rates in fish and crustaceans when probiotics are routinely added to water, yet their exact modes of action remain unclear. Critical questions include whether probiotics act as nutrients, outcompete pathogens, or influence host immunity, especially under stress. Some bacteria, like *Vibrio alginolyticus,* have shown probiotic potential but also include pathogenic strains, raising safety concerns. The practical use of probiotics in aquaculture is hindered by the difficulty in clearly distinguishing beneficial microorganisms from potentially harmful ones. A deeper understanding of how these microbes function within living hosts is essential for ensuring both safety and effectiveness; however, only a limited number of studies have successfully connected laboratory results with real biological outcomes. While field-based evaluations provide valuable insights, controlled infection experiments are necessary to generate reproducible and comparable data. Given the high biological variability inherent in aquatic systems, advanced molecular tools, rigorous statistical methods, and well-structured experimental designs are critical. Probiotic performance often differs across host–pathogen combinations, highlighting the need for customized applications rather than generalized solutions. Methods of delivery, such as water treatment, formulated feeds, or live feed carriers, may influence probiotic stability and activity, yet their specific effects are not well understood. Sustained application appears to be required for reliable results, though ideal dosages, timing, and environmental parameters remain unclear. Regulatory uncertainty also persists, particularly regarding whether probiotics should be classified as therapeutic agents or environmental management tools. Although initial implementation costs may be higher, probiotics have the potential to provide long-term economic gains by improving system resilience and productivity, especially as restrictions on antibiotic use increase. Postbiotics further expand this potential by offering improved consistency, shelf life, and suitability for large-scale application. To fully realize these benefits, future studies should emphasize species-targeted evaluations, omics-driven validation, and assessments of ecological safety. Together, these advances could significantly enhance sustainable aquaculture practices, particularly in resource-limited regions.

11.10 Conclusion

Probiotics represent a viable and environmentally friendly approach to controlling fungal infections and mycotoxin-related problems in aquaculture. Their use supports fish health by strengthening immune functions, improving resistance to diseases, and decreasing reliance on antibiotics. Although existing studies demonstrate positive effects on water quality and pathogen control, several limitations still need to be addressed. These include variability in effectiveness among different aquatic living organisms. Achieving reliable results also depends on refining application strategies such as delivery systems, appropriate dosages, and frequency of use. As research advances and evaluation methods become more standardized, probiotics have the potential to play a central role in sustainable, cost-effective aquaculture, supporting long-term productivity and global food security.

Conflict of Interest The authors declare that they have no conflict of interest.

Author Contributions BS: Conceptualization, Visualization, Formal analysis, Software, and Writing- original draft. MYS: Conceptualization, Supervision, Review, and Editing.

Study Questions

1. What are fungal diseases in fish?
2. What is mycotoxicosis in aquaculture?
3. What are mycotoxins, and where do they come from?
4. How do fungal diseases harm fish?
5. How do mycotoxins affect fish health?
6. What are probiotics?

References

Alam S, Afzal G, Iqbal Z, Hussain R, Rizwan M, Afzal M, Mahmood Y, Yamin A, Raza GA, Farooq U, Iqbal S, Mustafa G (2023) Fungal zoonotic infections in fish: an emerging threat to aquatic and terrestrial life. In: Altaf S, Khan A, Abbas RZ (eds) Zoonosis, vol 4. Unique Scientific Publishers, Faisalabad, pp 611–624

Aly SM, Abdel-Galil AY, Abdel-Aziz GA, Mohamed MF (2008) Studies on *Bacillus subtilis* and *Lactobacillus acidophilus* as potential probiotics on the immune response and resistance of Oreochromis niloticus to challenge infections. Fish Shellfish Immunol 25:128–136

Amenyogbe E, Chen G, Wang Z, Huang J, Huang B, Li H (2020) The exploitation of probiotics, prebiotics and synbiotics in aquaculture: present study, limitations and future directions. Aquac Int 28(3):1017–1041

Angahar LT (2016) Applications of probiotics in aquaculture. Am J Curr Microbiol 4(1):66–79

Armwood AR et al (2021) Retrospective study of phaeohyphomycosis in aquarium-housed fish, with first descriptions of *Exophiala lecanii-corni* and *Neodevriesia cladophorae* in fish. J Fish Dis 44(10):1563–1577

Ayisi CL, Apraku A, Afriyie G (2017) A review of probiotics, prebiotics and synbiotics in crab: present research, problems and future perspectives. J Shellfish Res 36(3):799–806

Bachère E (2000) Introduction: shrimp immunity and disease control. Aquaculture 191:3–11

Bakos K, Kovács R, Staszny Á et al (2013) Developmental toxicity and estrogenic potency of zearalenone in zebrafish (Danio rerio). Aquat Toxicol 136–137:13–21

Bostock J, McAndrew B, Richards R et al (2010) Aquaculture: global status and trends. Philos Trans R Soc B 365(1554):2897–2912

Cheli F, Pinotti L, Rossi L, Dell'Orto V (2013) Effect of milling procedures on mycotoxin distribution in wheat fractions: a review. LWT Food Sci Technol 54:307–314

Dass RS, Sreenivasa MY, Janardhana GR (2007) High incidence of Fusarium verticillioides in animal and poultry feed mixtures produced in Karnataka, India. Plant Pathol J 6(2):174–178

Deepa N, Charith Raj AP, Sreenivasa MY (2016) Nested PCR method for the early detection of fumonisin-producing Fusarium verticillioides in pure cultures, cereal samples and plant parts. Food Biotechnol 30(1):18–29

Devi G, Harikrishnan R, Paray BA et al (2019) Comparative immunostimulatory effect of probiotics and prebiotics in *Channa punctatus* against *Aphanomyces invadans*. Fish Shellfish Immunol 86:965–973

El-Gazzar N et al (2024) Antifungal and antibiofilm effects of probiotic *Lactobacillus salivarius*, zinc nanoparticles, and zinc nanocomposites against *Candida albicans* from Nile tilapia (*Oreochromis niloticus*). Front Cell Infect Microbiol 14:1358270

Elsegeny SR et al (2025) A comprehensive overview of probiotics in aquaculture: from efficacy evaluation to diverse applications. Ann Microbiol 75(1):35

FAO/WHO (2001) Health and nutritional properties of probiotics in food including powder milk with live lactic acid bacteria. Joint FAO/WHO Expert Consultation Report. FAO/WHO, Rome

Fregeneda-Grandes JM et al (2023) Limited probiotic effect of *Enterococcus gallinarum* L1, *Vagococcus fluvialis* L21 and *Lactobacillus plantarum* CLFP3 to protect rainbow trout against saprolegniosis. Animals 13(5):954

Frischer ME et al (2017) Causes, diagnostics, and distribution of an ongoing penaeid shrimp black gill epidemic in the U.S. South Atlantic Bight. J Shellfish Res 36(2):487–500

Gadhiya A, Katariya S, Chhatrodiya D (2025) Probiotics as a sustainable alternative to antibiotics in aquaculture: a review of the current state of knowledge. The Microbe:100426

Gioacchini G et al (2010) Increase of fecundity by probiotic administration in zebrafish (Danio rerio). Reproduction 140:953–959

Gunasundari V et al (2013) An ex vivo model to evaluate the brewer's yeast *Saccharomyces cerevisiae* in clownfish aquaculture. Turk J Fish Aquat Sci 13:389–395

Hai NV (2015) The use of probiotics in aquaculture. J Appl Microbiol 119(4):917–935

Harikrishnan R et al (2010) Dietary supplementation with probiotics, herbal and azadirachtin on hematological and biochemical changes in *Cirrhina mrigala* and *Aphanomyces invadans*. Fish Aquac J 4:1–11

Hatai K (2011) Diseases of fish and shellfish caused by marine fungi. In: Biology of marine fungi. Springer, Berlin, Heidelberg, pp 15–52

Hong HA, Duc LH, Cutting SM (2005) The use of bacterial spore formers as probiotics. FEMS Microbiol Rev 29:813–835

Huang X, Zhou H, Zhang H (2006) Effect of *Sargassum fusiforme* polysaccharide extracts on vibriosis resistance in shrimp *Fenneropenaeus chinensis*. Fish Shellfish Immunol 20:750–757

Hussein HS, Brasel JM (2001) Toxicity, metabolism and impact of mycotoxins on humans and animals. Toxicol Appl Pharmacol 167:101–134

Iqbal G et al (2023) Fungal diseases in aquaculture: a review. Pharma Innov

Irianto A, Austin B (2003) Use of dead probiotic cells to control furunculosis in rainbow trout (Oncorhynchus mykiss). J Fish Dis 26:59–62

Jantrarotai W, Lovell RT (1990) Sub-chronic toxicity of dietary aflatoxin B1 to channel catfish. J Aquat Anim Health 2(4):248–254

Kautsky N et al (2000) Ecosystem perspectives on management of disease in shrimp pond farming. Aquaculture 191(1–3):145–161

Landers M (2021) Research connects black gill in shrimp to warming climate. The Current, 14 September

Lategan MJ, Gibson LF (2003) Antagonistic activity of *Aeromonas media* strain A199 against *Saprolegnia* sp. J Fish Dis 26(3):147–153

Lauzon HL et al (2008) Isolation of putative probionts from cod rearing environment. Vet Microbiol 132:328–339

Li J et al (2006) Comparative study between probiotic bacterium *Arthrobacter* XE-7 and chloramphenicol. Aquaculture 253:140–147

Mahrous KF et al (2006) Assessment of toxicity and clastogenicity of sterigmatocystin in Nile tilapia. Afr J Biotechnol 5:1180–1189

Manning BB et al (2003) Ochratoxin A fed to channel catfish causes reduced growth. Aquaculture 219(1–4):739–750

Marijani E et al (2019) Occurrence of fungi and mycotoxins in fish feeds and their impact on fish health. Int J Microbiol 2019(1):6743065

Martínez Cruz P et al (2012) Use of probiotics in aquaculture. Int Sch Res Notices 2012:916845

Merrifield DL et al (2010) Review: probiotic and prebiotic applications for salmonids. Aquaculture 302:1–18

Mhaisen FT (1993) A review on the parasites and disease in fish ponds and farms of Iraq. Iraqi J Vet Sci 6(2):20–28

Mohammed EAH et al (2025) Application of *Saccharomyces cerevisiae* in aquaculture. Antibiotics 14(8):736

Muraosa Y et al (2006) *Lagenidium thermophilum* isolated from shrimp eggs and larvae. Fish Pathol 41(1):35–40

Nomoto K (2005) Prevention of infections by probiotics. J Biosci Bioeng 100(6):583–592

Oliveira M, Vasconcelos V (2020) Occurrence of mycotoxins in fish feed. Toxins 12(3):160

Onomu AJ, Okuthe GE (2024) Application of fungi and secondary metabolites in aquaculture. J Fungi 10(10):711

Pinotti L et al (2016) Mycotoxin contamination in the EU feed supply chain. Toxins 8:45

Pleadin J et al (2017) Deoxynivalenol and zearalenone in cereals from Croatia. Food Addit Contam B 10:1–7

Qadir MS et al (2024) Aquatic fungi and fungal diseases of fish. Proc Zool Soc 77(4):430–442

Ramaiah N (2006) Review on fungal diseases of algae and marine organisms. J Mar Biol

Ramesh D et al (2015) Isolation of probiotic *Bacillus* spp. in Labeo rohita. Fish Shellfish Immunol 45:268–276

Sakai M (1999) Current research status of fish immunostimulants. Aquaculture 172:63–92

Salinas I et al (2006) Effect of heat-inactivated probiotics in *Sparus aurata*. Vet Immunol Immunopathol 111:279–286

Schwartz-Zimmermann HE et al (2010) Zearalenone impairs reproduction in fish. Sci Total Environ 409(2):326–333

Shin S et al (2017) *Saprolegnia parasitica* characterization and host–pathogen interaction. Mycobiology 45(4):297–311

Shukry M et al (2021) Dietary *Aspergillus oryzae* modulates immune responses in tilapia. Animals 11(6):1621

Skjermo J, Vadstein O (1999) Techniques for microbial control in marine larvae. Aquaculture 177: 333–343

Subasinghe R, Soto D, Jia J (2009) Global aquaculture and sustainable development. Rev Aquac 1(1):2–9

Sun Y-Z et al (2013) Application of autochthonous *Bacillus* in grouper larvae. Aquaculture 392–395:44–50

Tacon AGJ, Hasan MR, Metian M (2011) Demand and supply of feed ingredients for farmed fish and crustaceans. FAO Fish Aquac Tech Pap 564. FAO, Rome

Tacon AG et al (2025) Global aquaculture and aquafeed production update. Rev Fish Sci Aquac:1–10

Thakur K et al (2025) Potential of probiotics and postbiotics in aquaculture. The Microbe 8:100431
Timmermans HM et al (2004) Monostrain, multistrain and multispecies probiotics. Int J Food Microbiol 96:219–233
Todorov SD et al (2024) Probiotics for aquaculture: hope, truth and reality. Probiotics Antimicrob Proteins 16(6):2007–2020
Tola S et al (2015) Effects of *Fusarium* mycotoxins in red tilapia. Toxins 7(6):1929–1944
Tuan TN et al (2013) Overview of probiotics in aquaculture. Int J Res Fish Aquac 3(3):89–97
Verschuere L et al (2000) Probiotic bacteria as biological control agents in aquaculture. Microbiol Mol Biol Rev 64:655–667
Willoughby LG, Roberts RJ (1992) Strategic use of fungicides against *Saprolegnia parasitica*. J Fish Dis 15(1):1–10

Further Reading

Iqbal G, Mushtaq S, Sushil Singh L, Kumar Ganpatbhai AV (2023) Fungal diseases in aquaculture: a review. Pharma Innov
Tacon AG, Metian M, Parsons GJ, Shumway SE (2025) A global aquaculture and aquafeed production update: 2010 to 2023. Rev Fish Sci Aquacult:1–10

Part III

Guardians of the Gut: Probiotics Against Human Fungal Infections

Beyond Basics: Human Microbiome and Fungal Balance

12

Oves Siddiqui, Aadithya B. Urs, Aswini Y. Balappanavar, and Aarushi Garg

Abstract

Fungal organisms, though historically understudied in microbiome research, play a pivotal role in human health and disease. While many fungi exist as commensals, dysbiosis—disruption of the microbial ecosystem—can enable their transformation into opportunistic pathogens, particularly under immunocompromised conditions. This chapter explores the multifaceted dynamics of fungal colonization, host immune responses, and interkingdom interactions with bacterial populations. It highlights key environmental, dietary, and clinical factors contributing to fungal overgrowth, such as antibiotic misuse, poor dietary habits, and compromised barrier integrity. Special focus is given to the potential of probiotics, prebiotics, synbiotics, and metabolic modulators in restoring microbial balance and suppressing fungal virulence. Mechanisms of probiotic action, ranging from competitive exclusion to immunomodulation and SCFA production are examined, alongside emerging interventions such as fecal microbiota transplantation. By integrating host, microbial, and ecological perspectives, this chapter provides a comprehensive overview of the dynamics, mechanisms, and strategies to manage fungal dysbiosis and prevent infection. These insights are

O. Siddiqui
Department of Microbiology, Maulana Azad Medical College, New Delhi, India

A. B. Urs
Department of Oral Pathology, Maulana Azad Institute of Dental Sciences, New Delhi, India

A. Y. Balappanavar (✉)
Department of Public Health Dentistry, Maulana Azad Institute of Dental Sciences, New Delhi, India

A. Garg
Department of Oral Pathology, Maulana Azad Institute of Dental Sciences, New Delhi, India

M. Y. Sreenivasa et al. (eds.), *Probiotics as Antifungals*,
https://doi.org/10.1007/978-981-92-0002-3_12

increasingly relevant for translational applications in personalized medicine and microbiome-based therapeutics.

Graphical Abstract

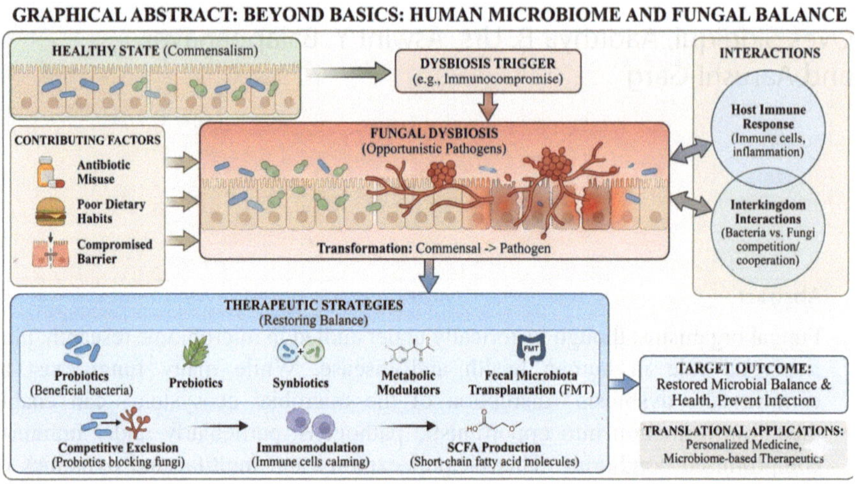

Keywords

Bacterial–Fungal Interactions · Fungal Balance · Fungal Dysbiosis · Human Microbiome · Antifungals · Probiotics

Learning Objectives

After reading this chapter, the reader will be able to:

- Understand the composition and functional role of the human mycobiome in health and disease.
- Identify key host, environmental, and clinical factors that contribute to fungal dysbiosis and the development of opportunistic infections.
- Explore the mechanisms by which probiotics, prebiotics, synbiotics, and dietary interventions modulate fungal balance.
- To understand emerging microbiome-based therapeutic strategies, such as metabolic modulation and fecal microbiota transplantation, and their public health significance.

12.1 Introduction

The human body is host to a vast and complex ecosystem of microorganisms collectively referred to as the human microbiota. These microbial communities, comprising bacteria, archaea, viruses, and fungi, are essential to the maintenance of health and homeostasis. They participate in a wide array of physiological processes, including nutrient metabolism, immune regulation, and protection against invading pathogens. Among these, microbial homeostasis, the balanced coexistence of diverse microbial species, is increasingly recognized as a cornerstone of human health. Disruptions to this balance, known as dysbiosis, have been implicated in the pathogenesis of various diseases, including metabolic disorders like diabetes, chronic inflammatory conditions such as colitis, and even malignancies like colorectal cancer. While much of the research to date has focused on the bacterial component of the microbiota, a growing body of evidence points to the important yet often overlooked role of the mycobiome, the fungal fraction of the human microbiota. Fungi, though present in lower abundance compared to bacteria, can exert a significant influence on microbial dynamics and host physiology. Moreover, under certain conditions, both commensal and environmental fungi can act as opportunistic pathogens, contributing to localized or systemic infections.

A key determinant of the transition from a healthy to a diseased state is the nature of microbe–microbe interactions within the host. These interactions, particularly those between bacteria and fungi, are central to shaping microbial community structure, influencing pathogenicity, and determining infection outcomes. However, the mechanistic underpinnings of such cross-kingdom interactions remain poorly understood. Another important determinant of fungal dysbiosis is the interaction between fungal pathogens and the human immune system. Fungi are ubiquitous in nature; approximately 155,000 species of fungi have been formally described and accepted, and even conservative estimates suggest there could be three million species (Hibbett et al. 2025). These fungi inhabit diverse environments such as air, soil, water, and living hosts. As a result, humans are exposed daily to various fungi, including *Penicillium*, *Mucor*, *Rhizopus*, *Cryptococcus*, *Aspergillus*, and *Candida* species. Among these, *Candida species like C. albicans, C. glabrata*, *C. parapsilosis*, and *C. tropica* commonly colonize the skin and mucous membranes without causing harm under normal conditions. Human immunity—particularly innate immunity—plays a central role in controlling fungal infections. The innate immune system relies on the recognition of conserved microbial structures known as pathogen-associated molecular patterns (PAMPs), such as fungal β-glucans, mannans, and chitin. These PAMPs are detected by pattern recognition receptors (PRRs), including Toll-like receptors and C-type lectin receptors. PRRs are expressed on various innate immune cells, such as macrophages, dendritic cells, and neutrophils, enabling the host to mount a rapid and effective defense against fungal invaders. These innate immune mechanisms form the first line of defense against fungal infection. However, in immunocompromised individuals, these normally harmless fungi can become opportunistic pathogens. When the innate immune system fails to contain the fungal pathogen, particularly in immunocompromised individuals, the—infection may progress and require medical intervention. This involves not only targeted antifungal therapy but also addressing the underlying causes of the illness. Key measures include de-escalation of broad-spectrum antibiotics, correction of metabolic imbalances, and restoration of the healthy microbiota using probiotics and prebiotics. These interventions help reverse the dysbiosis and immune disturbances that may have contributed to the infection.

This chapter aims to provide a foundational overview of the human microbiome, with an emphasis on the fungal component and its interactions with bacterial counterparts. We explore how bacteria can modulate fungal growth, morphology, and virulence, and conversely, how fungi can influence bacterial behavior, biofilm formation, and toxin production. Special attention is given to *Candida albicans, Cryptococcus neoformans,* and *Aspergillus fumigatus*, which together account for the majority of invasive fungal infections in humans. Recent advances in the field have begun to uncover chemical, physical, and protein-mediated pathways that govern bacterial–fungal interactions. These findings have significant implications for understanding not only the etiology of polymicrobial infections like dental caries, otitis media, and gingivitis, but also the potential for microbiota-derived therapeutics, such as novel antimicrobials and anti-virulence agents. By highlighting the complex interplay between microbial species within the human host, this chapter aims to deepen our understanding of fungal pathogenesis and inspire novel microbiome-based therapeutic strategies.

12.2 Human Microbiome

The human microbiome, encompassing diverse microorganisms (eukaryotes, archaea, bacteria, fungi, and viruses), inhabits various bodily sites, including the skin, oral cavity, respiratory, gastrointestinal, urinary, and reproductive tracts (Shreiner et al. 2015). While "microbiota" refers to a community of microorganisms in a specific environment, "microbiome" denotes the complete genomic elements associated with it (Dekaboruah et al. 2020). Functioning as a "secondary genome" with 50–100 times more genes than the host, the microbiome maintains a symbiotic relationship, profoundly influencing human health through advantageous, detrimental, or neutral interactions. Its enzymatic proteins regulate host physiology and metabolism via metabolite production.

Microbial composition is highly individualized and site specific. Although a precise definition of a "healthy microbiota" remains elusive, interventions like probiotics, prebiotics, and synbiotics are recognized for fostering a healthy microbial ecosystem. Distinguishing the stable "core microbiota" from fluctuating "transient microbiota" is crucial for effective experimental design and therapeutic applications in microbiome studies (Aggarwal et al. 2022).

12.2.1 Niche-Specific Microbiota Overview

Oral Microbiota The oral cavity harbors a diverse microflora (bacteria, archaea, fungi, viruses) forming a biofilm crucial for oral homeostasis and defense against pathogens (Deo and Deshmukh 2019; Rajasekaran et al. 2024). Key bacterial phyla include *Firmicutes, Bacteroidetes, Proteobacteria, Actinobacteria, Spirochaetes,* and *Fusobacteria*, with prevalent genera like *Streptococcus, Gemella,* and *Neisseria* (Dewhirst et al. 2010). Fungal flora is dominated by *Candida*, alongside others like *Cladosporium* and *Aspergillus* (Mark Welch et al. 2016). Methanogenic archaea, such as *Methanobrevibacter oralis*, are linked to periodontitis progression. Oral microbes can disseminate, influencing systemic health; for instance, nitrate metabolism by oral bacteria contributes to nitric oxide production, important for vascular health and antimicrobial defense (Wade 2013; Wescombe et al. 2009).

Vaginal Microbiota Characterized by low diversity, the vaginal microbiome is dominated by *Lactobacillus species*, which maintain an acidic pH (<4.5) via lactic acid production, acting as a chemical barrier against infections. *Lactobacillus* also produces bacteriocins, H_2O_2, and reactive oxygen species, inhibiting pathogen colonization (Valenti et al. 2018; Felten et al. 1999). Hormonal fluctuations across the reproductive cycle (puberty, menstruation, pregnancy, and menopause) influence microbial equilibrium by altering glycogen availability, which *Lactobacillus* utilizes for lactic acid production (Kaur et al. 2020).

Skin Microbiota The skin, the body's largest organ, hosts a remarkably stable yet diverse microbiota despite environmental exposure. Its composition varies individually and by site (oily, moist, or dry), with the feet often considered a distinct category due to unique microbial signatures (Aggarwal et al. 2022). Commensal bacteria, fungi, viruses, archaea, and mites inhabit the skin. External factors like antibiotic use (Chien et al. 2019), burns (Sanjar et al. 2020), hygiene products (Bouslimani et al. 2019), and lifestyle (Blaser et al. 2013) can alter its composition. Predominant genera include *Staphylococcus, Propionibacterium, Corynebacterium,* and *Streptococcus* (Bay et al. 2020). *Malassezia* is a prevalent lipophilic yeast, while other fungi like *Aspergillus* and *Cryptococcus* are site-specific (Gupta and Kohli 2004; Reynoso-García et al. 2022).

Gut Microbiota The gastrointestinal tract harbors the largest and most diverse microbial reservoir, with distinct environments along its length. Predominant phyla include *Firmicutes, Bacteroidetes, Proteobacteria,* and *Actinobacteria*. The esophagus mirrors oral composition (*Streptococcus, Veillonella, Prevotella*) (Reynoso-García et al. 2022). Infant gut microbiota, initially dominated by *Bifidobacterium*, diversifies to include *Clostridium* and *Bacteroides* by age one, stabilizing around age three with *Firmicutes* and *Bacteroidetes* as primary constituents (Ma et al. 2024). The stomach primarily contains *Proteobacteria* and *Firmicutes*. The small intestine, with higher oxygen and rapid transit, favors facultative anaerobes. The colon, the most diverse niche, is anaerobic, dominated by

Bacteroidetes. Non-bacterial components include bacteriophages, fungi (*Ascomycota, Basidiomycota*), and archaea (*Methanobrevibacter smithii*) (Reynoso-García et al. 2022).

Eye Microbiota The eye possesses a unique, less dense microbiome across its structures (ocular surface, meibomian glands, intraocular fluids). The ocular surface microbiome (OSM) is dominated by *Lactobacillus, Bacillus, Corynebacterium, Staphylococcus,* and *Cutibacterium* (Reynoso-García et al. 2022). Alterations in microbial diversity and dominant genera are noted in ocular diseases like dry eye. The aqueous humor, previously considered sterile, contains a unique microbial signature, including *Cutibacterium acnes, Enterococcus faecalis,* and *Staphylococcus epidermidis*, highlighting the ocular microbiome's role in eye health and disease (Kaur et al. 2025).

12.3 The Human Mycobiome: An Introduction

Fungi are omnipresent in the environment and integral to all of Earth's ecosystems (Peay et al. 2016). While bacterial communities of the human microbiota have been extensively studied, the mycobiome, the aggregate genome of all fungi (mycobiota) residing within the body, including unicellular yeasts and filamentous micromycetes, has received significantly less attention (Seed 2014). Despite this, the human mycobiome is recognized as a dynamic community responsive to environmental factors, playing a crucial role in host metabolism and immune homeostasis (Matijašić et al. 2020). The human mycobiome has been documented in five primary locations within the healthy human body: the oral cavity, intestinal tract, respiratory tract, genitourinary tract, and skin. The intestinal mycobiome is particularly emphasized for its significant contributions to human health, immune system development, and the maintenance of homeostasis (Belvoncikova et al. 2022).

The establishment of the mycobiome begins at birth, with fungal microorganisms acquired via vertical (maternal) or horizontal (environmental) transmission. Its early development is influenced by gestational age, birth weight, delivery method, and subsequent feeding practices (Wampach et al. 2017). During early life, newborns primarily acquire their gut mycobiome through vegetative fungal cells rather than spores. Fungi such as *Candida* spp., commonly present in the maternal vaginal microbiota, are transferred to the newborn during vaginal delivery (Turunen et al. 2023). Nevertheless, postnatal environmental exposure to fungal spores may also contribute to the shaping of the infant gut mycobiome.

In adulthood, key determinants of mycobiome composition include dietary intake and body mass. Fungi contribute to nutrient extraction and digestion through enzyme and vitamin production. Additional influencing factors include sex, age, lifestyle, hygiene, geography, and the use of antibiotics and antifungals (Belvoncikova et al. 2022; Borges et al. 2018).

12.3.1 Anatomic Distribution of the Human Mycobiota

12.3.1.1 Gut Mycobiome

The gut is the most extensively studied human body site for fungi and is inhabited by both yeasts and filamentous micromycetes. These colonizers contribute to the digestion of otherwise indigestible micro and macronutrients. The transition of certain fungi from benign yeast forms to pathogenic hyphal morphologies can be influenced by microbial metabolites within the gut ecosystem (Sam et al. 2017). The human gut mycobiome is primarily composed of fungi belonging to the phyla Ascomycota, Basidiomycota, and Zygomycota. Predominant genera include *Candida* (particularly *C. albicans*), *Saccharomyces (S. cerevisiae), Penicillium, Aspergillus, Cryptococcus,* and *Malassezia* (*M. restricta*) (Matijašić et al. 2020).

Based on fungal community composition and relative abundance profiling, two major gut mycotypes have been described. Mycotype 1 is characterized by the dominance of *Saccharomyces* and other unclassified fungal taxa, whereas Mycotype 2 is enriched in genera such as *Penicillium, Malassezia, and Mucor* (Belvoncikova et al. 2022). These mycotypes are identified through comparative analyses of fungal community structure and represent distinct ecological patterns rather than fixed taxonomic classifications.

12.3.1.2 Oral Mycobiome

The oral cavity represents a complex microbial ecosystem influenced by foodborne fungi. Its primary constituents are the phyla *Ascomycota, Basidiomycota, Glomeromycota,* and *Mucoromycota*. The predominant genus is the yeast like *Candida*, with lower abundances of *Aureobasidium, Saccharomyces,* and *Pichia*. Filamentous micromycetes such as *Penicillium, Cladosporium,* and *Aspergillus* are also identified (Monteiro-da-Silva et al. 2014). Similar to observations in gut mycobiome studies, two distinct mycotypes have also been reported in the oral cavity (Hong et al. 2020). These include a *Candida* mycotype and a *Malassezia* mycotype, each characterized by distinct fungal and bacterial associations.

The *Candida* mycotype is dominated by the genus *Candida*, accounting for approximately 59% to 99% of the relative fungal abundance. This mycotype is characterized by lower overall fungal diversity compared to the *Malassezia* mycotype. It shows positive associations with oral and systemic conditions such as oral cancer, steroid use, smoking, dental caries, removable prosthesis use, and higher plaque index. Additionally, the *Candida* mycotype is linked to aciduric bacterial species, including *Lactobacillus salivarius* and *Propionibacterium acidifaciens*, which preferentially thrive in acidic microenvironments. In contrast, the *Malassezia* mycotype is dominated by the genus *Malassezia*, with relative abundance ranging from approximately 1.5% to 98%. The most prevalent species within this mycotype are *Malassezia restricta* and *Malassezia globosa*. This mycotype exhibits greater fungal diversity than the *Candida*-dominated community and is associated with inflammation-associated bacterial taxa such as *Fusobacterium, Porphyromonas*, and *Treponema*. These bacteria favor mildly alkaline conditions and are commonly linked to gingival inflammation and periodontal disease.

12.3.1.3 Skin Mycobiome

The skin, a primary barrier, hosts a diverse commensal mycobiome that contributes to homeostasis and defense by secreting enzymes, inhibiting pathogens, and modulating immune responses (Jo et al. 2017). Composed of at least 168 genera, primarily from *Basidiomycota* and *Ascomycota*, the skin mycobiome varies significantly by body site. *Malassezia* is the most prevalent yeast, dominating many areas. *Cryptococcus* is also common, while *Aspergillus* and *Penicillium* are less frequently detected (Belvoncikova et al. 2022).

12.4 Bacterial and Fungal Interactions

Microbial communities within the human body are structured by anatomical niche, with each site oral cavity, vaginal tract, gastrointestinal tract, lungs, and bloodstream, hosting distinct consortia of bacteria and fungi. These microbes interact dynamically, with bacteria often modulating fungal behavior through nutrient competition, metabolite secretion, and quorum sensing. Depending on the local environment, these interactions can inhibit or promote fungal growth, morphogenesis, virulence, and immune evasion.

12.4.1 Oral Cavity

The oral microbiome harbors commensal bacteria such as *Lactobacillus*, *Actinomyces israelii*, *Prevotella nigrescens*, and *Porphyromonas gingivalis*, which inhibit *Candida albicans* proliferation and filamentation. These inhibitory effects are often mediated through acidic pH, metabolic byproducts, or nutrient competition. However, synergistic interactions also occur *Staphylococcus aureus*, for instance, forms biofilms with *C. albicans*, enhancing both organisms' virulence and resistance.

12.4.2 Vaginal Tract

In the vaginal niche, *Lactobacillus* species are dominant and protect against fungal overgrowth by secreting lactic and acetic acids, lowering pH, and competitively excluding *Candida* adhesion. Disruption of *Lactobacillus* populations due to antibiotics, hormonal changes, or hygiene practices can lead to fungal overgrowth and vulvovaginal candidiasis. While lactic acid inhibits fungal filamentation and iron acquisition, under certain conditions, it also remodels fungal cell walls, potentially influencing immune evasion.

12.4.3 Gastrointestinal Tract

In the gut, short-chain fatty acid (SCFA) producing bacteria like *Faecalibacterium* and *Roseburia* play critical roles in limiting fungal virulence through epigenetic modulation. SCFAs such as butyrate impair fungal growth, adhesion, and biofilm formation. Dysbiosis induced by antibiotics reduces SCFA levels and promotes *C. albicans* colonization. Meanwhile, *Escherichia coli* may either inhibit fungi through antifungal metabolites or support them indirectly via siderophore cross talk or amino acid exchange. Ethanol produced by yeasts like *Saccharomyces cerevisiae* fuels the growth and virulence of *Acinetobacter* species, highlighting bidirectional metabolic dependency.

12.4.4 Lungs and Respiratory Tract

In patients with cystic fibrosis (CF), co-infection with *Pseudomonas aeruginosa* and fungi such as *Candida albicans* or *Aspergillus fumigatus* is common. Here, interactions are typically antagonistic. *P. aeruginosa* suppresses fungal filamentation and growth using phenazines, siderophores, and quorum-sensing molecules, while fungal ethanol triggers bacterial biofilm formation and virulence. These interactions are tightly regulated by nutrient availability, especially phosphate and iron, influencing the balance between competition and coexistence.

12.4.5 Systemic and Clinical Infections

In bloodstream and deep-tissue infections, bacteria can influence fungal virulence indirectly by altering vitamin availability, modulating immune responses, or enhancing fungal resistance to antifungals. For example, *Klebsiella pneumoniae* inhibits *Cryptococcus neoformans'* virulence traits, such as capsule and melanin formation. Vitamin B competition (e.g., biotin, thiamine) has been shown to affect fungal survival in macrophages and system-wide infections. These interactions may determine clinical outcomes in immunocompromised hosts (Table 12.1).

12.4.6 Common Mechanisms of Bacterial-Fungal Interactions

12.4.6.1 Iron Uptake

Iron, along with other metals, is essential for microbial metabolism, playing a key role in respiration and the synthesis of biomolecules (Gao and Span 2025). Under aerobic conditions, iron is commonly available in its ferric (Fe^{3+}) form. However, most bacteria and fungi can utilize this iron only after either reducing it to the ferrous (Fe^{2+}) form via iron reductases or binding the ferric form to specialized iron-chelating molecules known as siderophores. Siderophores differ in their chemical structure, such as hydroxamate, catecholate, phenolate, or carboxylate, and each

Table 12.1 Summary of microbial and mycobiome community interaction in different parts of the body

Body site	State	Microbiome—dominant taxa (% composition)	Mycobiome—dominant taxa (% composition)	pH/environmental conditions	Functional characteristics	Key microbiome–mycobiome interactions	References
Oral cavity	*Healthy*	*Streptococcus* (44–54%), *Hemophilus* (8–12%), *Veillonella* (6–10%), *Rothia* (4–8%), *Neisseria* (3–7%)	*Candida* (5–15%), *Cladosporium* (3–8%), *Aureobasidium* (2–5%), *Aspergillus* (1–4%), *Fusarium* (1–3%)	pH 6.8–7.4; aerobic–anaerobic gradient	Biofilm stability, AMP production, immune modulation	Cross-kingdom biofilms; bacterial suppression of fungal overgrowth	Zangl et al. (2019a, b), Lamont et al. (2018), Perera et al. (2024)
Oral cavity	*Dysbiotic*	↑ *P. gingivalis* (15–25%), ↑ *T. forsythia* (10–20%), ↑ *S. mutans* (20–40%), ↑ *F. nucleatum* (10–15%)	↑ *C. albicans* (40–60%), ↑ *C. glabrata* (10–25%), ↑ *Aspergillus* spp. (5–15%)	Acidic (5.5–6.5) or alkaline (7.5–8.5)	Virulence expression, biofilm disruption, inflammation	Synergistic pathogenicity (e.g., *S. mutans* + *C. albicans*)	Zangl et al. (2019a, b), Richard and Sokol (2021)
Vagina	*Healthy*	*Lactobacillus* spp. (>70%): *L. crispatus* (40–60%), *L. iners* (20–30%), *L. jensenii* (5–15%)	Low fungal abundance (<5%); *Candida* spp. (1–3%)	Acidic pH 3.8–4.5	Lactic acid + H_2O_2 production; pathogen exclusion	Lactobacilli suppress *Candida* by lowering pH and producing antimicrobials	Ravel et al. (2016), Bradford and Ravel (2023)
Vagina	*Dysbiotic*	↑ *Gardnerella* (20–40%), ↑ *Prevotella* (15–30%), ↑ *Fannyhessea* (10–20%), ↑ *Atopobium* (5–15%)	↑ *C. albicans* (60–80%), ↑ *C. glabrata* (10–20%), *C. tropicalis* (5–15%)	pH >4.5–7.0; loss of acidic environment	Loss of protection; hyphal growth; immune evasion	Dysbiosis permits fungal expansion and virulence	Bradford and Ravel (2023), Richard and Sokol (2021)
Respiratory tract	*Healthy*	*Streptococcus* (20–30%), *Haemophilus*	*Malassezia* (30–40%), *Eremothecium*	pH ~7.0; oxygen gradient from	Colonization resistance;	Co-regulation by inhaled environmental	Dickson et al. (2023),

		(15–25%), *Neisseria* (10–15%), *Prevotella* (5–10%)	(10–15%), *Systenostrema* (8–12%), *Candida* (5–10%)	trachea to alveoli	immune homeostasis	fungi and commensal bacteria	Richard and Sokol (2021)
Respiratory tract	*Dysbiotic*	↑ pathogenic bacteria (20–50%); reduced diversity	↑ *Aspergillus* (15–30%), ↑ *Candida* spp. (20–40%); altered *Malassezia* composition	Altered pH and oxygen due to inflammation	Heightened inflammation; PAMP activation	Fungal dominance increases when bacterial diversity collapses	Dickson et al. (2023)
Gut	*Healthy*	Firmicutes (50–70%), Bacteroidetes (20–40%), Actinobacteria (3–8%), Proteobacteria (<5%)	*Saccharomyces* (40–60%), *Candida* (10–20%), *Malassezia* (5–15%), *Cladosporium* (3–8%)	Stomach pH 1.5–3.5 → SI 8.0–8.5 → Colon 5.5–7.0; anaerobic	SCFA production, vitamin metabolism, bile acid conversion	Bacterial SCFAs suppress *Candida*; fungi modulate immune tone	Richard and Sokol (2021)
Gut	*Dysbiotic*	↑ Proteobacteria (15–30%); ↓ Firmicutes (30–50%); reduced diversity	↑ *Candida* spp. (30–70%), ↑ *Aspergillus* (10–25%), ↓ *Saccharomyces* (20–40%)	Altered pH and reduced SCFA levels	Increased LPS, compromised barrier, metabolic disruption	Opportunistic fungi fill ecological gaps left by bacterial loss	Richard and Sokol (2021)
Skin	*Healthy*	*Staphylococcus epidermidis* (20–40%), *Cutibacterium acnes* (20–30%), *Corynebacterium* spp. (10–20%)	*Malassezia* spp. (70–90%), *Candida* spp. (2–5%)	pH 4.5–5.5; dry–moist zones	Barrier support, lipid metabolism	*S. epidermidis* inhibits *Malassezia* overgrowth; balanced lipid use	Byrd et al. (2018), Findley et al. (2013), Oh et al. (2016)
Skin	*Dysbiotic*	↑ *Staphylococcus aureus* (20–60%); ↓ *S. epidermidis*	↑ *Malassezia restricta* (40–60%), ↑ *Candida* spp. (10–20%)	Elevated pH (>5.5)	Inflammation, barrier disruption	*S. aureus* and *Malassezia* synergize in atopic dermatitis	Byrd et al. (2018), Oh et al. (2016)

class requires specific iron transporters known as SITs (Siderophore Iron Transporters) (Albelda-Berenguer et al. 2019).

Not all microbes produce their siderophores. For example, *Saccharomyces cerevisiae*, *Candida albicans*, and *Cryptococcus neoformans* lack siderophore biosynthesis genes (Bairwa et al. 2017). Similarly, certain mutant strains of *Escherichia coli* deficient in enterobactin production cannot synthesize siderophores but are capable of utilizing those produced by other microbes, referred to as xenosiderophores (Lesuisse et al. 2001). These interactions create both dependency and competition within microbial communities. For instance, when *Pseudomonas aeruginosa* and *Aspergillus fumigatus* occupy the same ecological niche, they compete for iron. Pyoverdine, a siderophore produced by *P. aeruginosa*, binds iron in a form that can be utilized only by *Pseudomonas* but not by *Aspergillus* or *Candida*, thereby suppressing the growth of these fungi (Lopez-Medina et al. 2015).

12.4.6.2 Microbe Interactions

Microbes interact within and across species when grown in close environments with specific metabolic relevance. One such interaction mechanism is known as quorum sensing (QS). Through QS, microbes regulate the expression of certain genes in response to population density, enabling coordinated behaviors such as biofilm formation, virulence regulation, antibiotic production, and resource optimization (Gray 1997; Mukherjee and Bassler 2019; Abisado et al. 2018). Quorum sensing has also been observed in fungi. For example, the QS molecule farnesol inhibits the yeast-to-hyphal transition in *Candida albicans*, a key virulence trait associated with its pathogenicity. Hyphal morphogenesis in *C. albicans* is considered a major factor contributing to tissue invasion and immune evasion. Moreover, QS signals can mediate interkingdom interactions. For instance, *Pseudomonas aeruginosa* produces 3-oxo-C12-HSL, a QS molecule that inhibits *C. albicans* hyphal formation and disrupts *Aspergillus fumigatus* biofilm formation and spore germination, acting through the same Ras1-cAMP-Efg1 pathway influenced by farnesol (Davis-Hanna et al. 2008; Mowat et al. 2010).

Other bacteria, such as *Burkholderia cenocepacia* and *Streptococcus mutans*, also inhibit *C. albicans* filamentation via their respective QS molecules BDSF and CSP, respectively (Boon et al. 2008; Hall et al. 2011; Jarosz et al. 2009). Notably, *S. mutans* is a common oral commensal, and such interkingdom signaling may play a role in maintaining microbial balance in the human oral microbiome.

12.4.6.3 Organic Acid Production by Bacteria

Lactic acid bacteria, such as *Lactobacillus rhamnosus*, *L. plantarum*, and *L. acidophilus*, secrete a variety of metabolites, including hydrogen peroxide, organic acids (such as lactic and acetic acid), and antimicrobial compounds. These substances play a crucial role in maintaining the homeostasis of the vaginal environment. The vaginal microbiota primarily consists of lactic acid bacteria, particularly *Lactobacillus* species, along with fungal species such as *Candida* (Mendling 2016; Bradford and Ravel 2017).

Lactobacilli help inhibit the overgrowth of *Candida* by acidifying the vaginal environment. However, dysbiosis—often triggered by antibiotics, hygiene practices, or contraceptive use—can disrupt this microbial balance. Such disruption may lead to *Candida* infections, including vulvovaginal candidiasis, which is commonly caused by *C. albicans*, *C. glabrata*, or *C. tropicalis* (Gonçalves et al. 2016; Sobel 2007). Restoring *Lactobacillus* populations has shown therapeutic potential in managing these infections (De Gregorio et al. 2019; Rose Jørgensen et al. 2020; Jang et al. 2019). These beneficial bacteria secrete antimicrobial metabolites such as lactic acid and acetic acid, which further acidify the vaginal environment and inhibit the growth of *Candida* species (Zangl et al. 2019a, b; Hickey et al. 2012). Notably, the most effective *Lactobacillus* strains for antifungal activity are those that produce high levels of these organic acids (Rose Jørgensen et al. 2020).

12.4.6.4 Short-Chain Fatty Acids (SCFAs)

Bacterial fermentation of dietary fibers produces short-chain fatty acids (SCFAs)—primarily acetate, butyrate, and propionate in the gastrointestinal tract (Wong et al. 2006). Fecal SCFA levels are widely recognized as indicators of gut health and microbial balance (Blaak et al. 2020; Tan et al. 2014). These metabolites also play a significant role in shaping bacterial–fungal interactions.

The use of antibiotics reduces the population of SCFA-producing bacteria, which may lead to fungal overgrowth in the gastrointestinal tract. For example, cefoperazone-treated mice showed significantly decreased SCFA levels accompanied by increased *Candida albicans* colonization (Guinan et al. 2019).

Although there is no conclusive evidence that fungi naturally produce SCFAs, certain species like *Saccharomyces boulardii* have been shown to increase butyrate levels, likely by stimulating SCFA-producing bacterial populations (Moens et al. 2019). For instance, *S. cerevisiae* promotes the growth of beneficial SCFA-producing bacteria such as *Faecalibacterium*, *Roseburia*, and *Bifidobacterium* and has been found to reduce the virulence of enterotoxigenic *Escherichia coli* (ETEC) (Roussel et al. 2021) (Fig. 12.1).

12.5 Immune System and Fungal Interactions

There are an estimated 1.5 to 5 million species of fungi, which are ubiquitously present in diverse environments, including air, water, soil, plants, and animals. Some fungi, such as *Aspergillus* species, produce airborne spores that we inhale on a daily basis, while others, like *Candida* species, exist as commensal organisms in the human body. Certain fungi, such as dermatophytes, are acquired through direct contact with infected animals or contaminated environments. Despite frequent exposure, these fungi usually do not cause disease in immunocompetent individuals, or they may lead only to mild, self-limiting superficial infections. Most fungal infections in humans are opportunistic, meaning they occur primarily when the immune system is weakened, for example, in individuals with HIV, those undergoing immunosuppressive therapy, or patients receiving cytotoxic

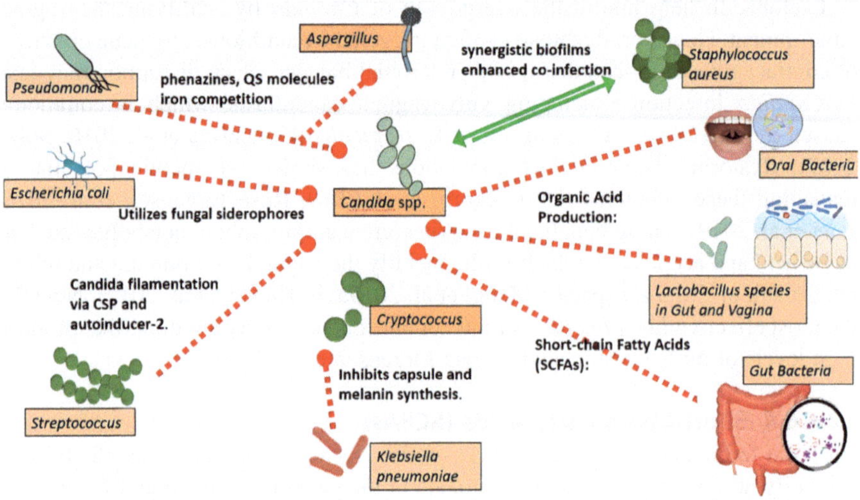

Fig. 12.1 Overview of myco-bacterial interactions

chemotherapy. The immune system typically provides robust protection against fungal pathogens, as demonstrated by the high susceptibility to fungal infections seen in individuals with inborn errors of immunity. Temperature plays a critical role in fungal growth and pathogenicity. Most fungi thrive at temperatures below 30 °C. Only a minority of fungi capable of surviving as commensals or pathogens in the human body can grow at 37 °C. This contrasts with lower invertebrate animals, where certain chytrid fungi have caused extinctions of amphibian species (Bellard et al. 2016).

The primary immune response against fungal infections is mediated by the innate immune system. Unlike the adaptive immune system, which relies on the precise recognition of specific antigens, often minor molecular variations in proteins through antibodies and T cell receptors, the innate immune system detects pathogens based on conserved molecular structures known as pathogen-associated molecular patterns (PAMPs). These PAMPs are unique to microbes and absent in mammalian cells, allowing the immune system to distinguish self from non-self.

In fungi, PAMPs include components of the cell wall such as β-glucans, mannans, and chitin. In bacteria, examples include lipopolysaccharides (LPS) from Gram-negative bacteria, lipoteichoic acids from Gram-positive bacteria, flagellin proteins, and unmethylated CpG DNA motifs. These PAMPs are recognized by specialized receptors on immune cells known as pattern recognition receptors (PRRs).PRRs include several major families, such as

- Toll-like receptors (TLRs): e.g., TLR2 and TLR4 recognize fungal cell wall components like phospholipomannan and mannans.
- C-type lectin receptors (CLRs): e.g., Dectin-1 recognizes β-glucans, and Dectin-2 and Mincle detect α-mannans and other fungal glycans.

- NOD-like receptors (NLRs): involved in intracellular sensing and inflammasome activation.
- RIG-I-like receptors (RLRs) and AIM2-like receptors (ALRs): primarily for viral and intracellular DNA sensing, but may contribute to the overall immune context.
- These PRRs are expressed on various innate immune cells, including macrophages, dendritic cells, and neutrophils, enabling the host to mount a rapid and effective defense against fungal invaders.

12.6 Factors Responsible for Dysbiosis and Fungal Imbalance or Infection

Dysbiosis refers to an imbalance or disruption in the composition of the gut microbiota. This disturbance can lead to fungal overgrowth or infections, such as candidiasis. Several host-related, environmental, and clinical factors contribute to this imbalance:

- Antibiotic Use: Broad-spectrum antibiotics can decrease or eliminate beneficial bacterial populations, thereby allowing opportunistic fungi like *Candida* to proliferate and potentially cause disease (Catanzaro et al. 2019; Levy et al. 2017). (Fig. 12.2).
- Dietary Changes: High-sugar, low-fiber diets can disrupt the intestinal barrier, trigger inflammation, and impair host metabolism. These effects often depend on the presence of gut microbiota; germ-free animals typically do not exhibit them, and microbial transplants can transfer disease phenotypes (Schulz et al. 2014; Collins et al. 2013; Shahinozzaman et al. 2021; Vrieze et al. 2021).

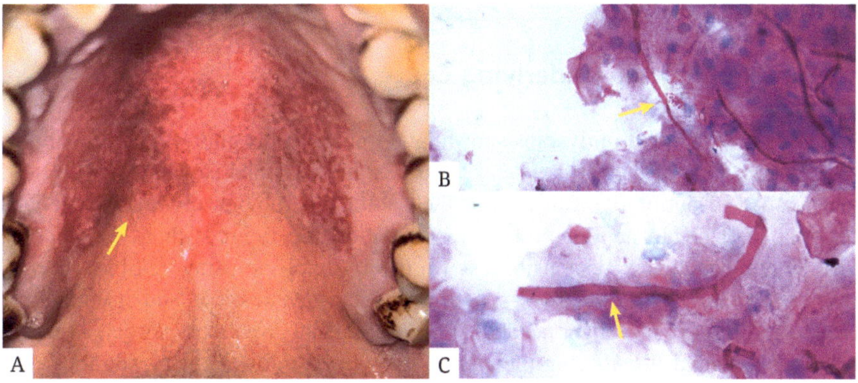

Fig. 12.2 A case of oral candidiasis in a 42-year-old male patient with a history of chronic smoking and antibiotic use. (**a**) Clinical photograph showing a non-homogenous red lesion over the entire hard palate, suggesting erythematous candidiasis to be the provisional diagnosis (yellow arrow); (**b**, **c**) photomicrographs of PAS-stained cytosmears showing ribbon-like, pseudoseptate fungal hyphae, suggestive of Candida species (10× and 40× respectively) (yellow arrows).

- Food Additives: Preservatives can harm sensitive gut microbes and promote proteobacteria overgrowth (Hrncirova et al. 2019). Emulsifiers alter microbiota composition and induce intestinal inflammation (Chassaing et al. 2017). Artificial sweeteners like sucralose may cause dysbiosis, worsen ileitis, and impair glucose tolerance (Rodriguez-Palacios et al. 2018; Suez et al. 2014).
- Host-Derived Factors: Genetic background, infections, chronic inflammation, and lifestyle choices can influence microbiota composition (Levy et al. 2017; Catanzaro et al. 2019). Stress, poor sleep, and sedentary habits also weaken microbial balance.
- Immunocompromised States: Conditions such as hematologic or oncologic diseases, corticosteroid therapy, neutropenia (<500/μL), HIV, or severe viral infections (e.g., COVID-19) increase susceptibility to fungal imbalance and infection.
- Infections: Viral infections like SARS-CoV-2 disrupt gut microbiota through ACE2 receptor-mediated entry into enterocytes, cytokine-driven inflammation, and increased intestinal permeability—factors linked to worse clinical outcomes (Yeoh et al. 2021; Hussain et al. 2021).
- Hospitalization and Medical Interventions: Intensive care unit admission, major surgeries, and broad-spectrum antibiotic use are known to promote *Candida* colonization. Use of total parenteral nutrition, indwelling catheters, prosthetic devices (e.g., shunts, LVADs), and urinary tract instrumentation can compromise anatomical barriers, facilitating fungal translocation.
- Barrier Disruption: Physical injuries such as burns, gastrointestinal perforations, pancreatitis, and mucositis increase the risk of fungal entry into the bloodstream.
- Genetic Susceptibility: Mutations in immune-related genes can predispose individuals to invasive *Candida* infections.

12.7 Management

12.7.1 Treating the Underlying Cause

Effective management of dysbiosis, or fungal imbalance, begins with addressing its root causes. One of the most critical factors is antimicrobial stewardship. Overuse and misuse of antibiotics can disrupt the natural microbiota, eliminate beneficial bacteria, and create an opportunity for opportunistic fungi such as *Candida* to proliferate. Practicing responsible stewardship involves using antibiotics only when necessary, selecting narrow-spectrum agents when possible, and discontinuing them once they are no longer needed. When treatment is necessary, targeted therapy becomes essential. Identifying the causative microbe through diagnostic tools like stool culture or sequencing allows for precision therapy that minimizes harm to commensal microbes. This targeted approach also helps prevent the collateral fungal overgrowth that is often triggered by broad-spectrum drugs.

There are cases where direct targeting is required, and selective antibiotic or antifungal use may be employed. Specific agents such as rifaximin or fluconazole can be used to reduce the overgrowth of particular pathogens. However, non-specific agents carry the risk of worsening dysbiosis by eliminating beneficial microbes. Therefore, they are often co-administered with recovery-supporting interventions like probiotics or fecal microbiota transplantation (FMT).

12.7.2 Probiotic Supplementation

Probiotics, defined as "live microorganisms [that] confer a health benefit on the host when administered in sufficient quantities" (Hill et al. 2014), can be a powerful tool to rebalance gut microbiota. These beneficial microbes help inhibit fungal overgrowth through multiple mechanisms: they produce organic acids such as lactic acid, compete for adhesion sites and nutrients, and modulate immune responses that prevent fungal colonization.

Among the commonly used probiotic strains, *Lactobacillus rhamnosus* is known to enhance mucosal immunity and reduce fungal adhesion and germ tube formation (Leão et al. 2015). *Bifidobacterium infantis* contributes by increasing the expression of antimicrobial peptides through IL-22 signaling, which inhibits *Candida* colonization. Additionally, certain food-derived yeasts such as *Saccharomyces cerevisiae* have demonstrated the ability to inhibit *Candida* virulence traits like adhesion, filamentation, and biofilm formation, making them promising candidates for alternative antifungal therapy (Kunyeit et al. 2019). They reduce the risk of developing oral candidiasis, with combined data showing about a 60% lower incidence among those using probiotics compared to controls. Health benefits vary widely depending on the strain. Probiotics can be taken as tablets, lozenges, yogurts, and dairy products. They can also be used locally as mouth rinses and chewing gums, with toothpastes in development. Synbiotics and postbiotics are useful for targeted disease prevention and restoring the microbiome (Wu et al. 2022).

12.7.3 Metabolic Modulation

This approach focuses not on adding microbes but on altering the activity and output of existing microbial communities. The goal is to promote the production of beneficial metabolites such as short-chain fatty acids (SCFAs), especially butyrate, which strengthen gut barrier integrity and reduce inflammation. At the same time, it seeks to suppress harmful microbial products like ammonia or hydrogen sulfide.

For example, *Lactobacillus paracasei* produces several metabolites, including lactic acid, SCFAs, indoleacetic acid, and 2-hydroxy-3-methylbutyric acid, all of which show antifungal activity (Honoré et al. 2016). Strategies to promote such metabolic activity often involve dietary changes that increase fiber intake, supplementation with SCFAs or their precursors, and encouraging microbial cross-feeding to support ecosystem cooperation.

12.7.4 Prebiotics

Prebiotics are substrates selectively utilized by host microorganisms to confer a health benefit, as redefined by the International Scientific Association for Probiotics and Prebiotics in 2016 (Gibson et al. 2017). Examples of prebiotics include inulin, fructooligosaccharides (FOS), galactooligosaccharides (GOS), and resistant starch, which are commonly incorporated into diets and supplements. These compounds help to enrich the population of beneficial bacteria that produce SCFAs like butyrate. Their intake has been associated with reduced populations of pro-inflammatory bacteria, improved stool consistency, lower intestinal pH (which inhibits pathogenic growth), and enhanced gut lining integrity (Z.C. Holmes et al. 2022).

12.7.5 Synbiotics

Combining probiotics and prebiotics can amplify their respective benefits through synergistic interactions. A synbiotic is defined as a product that contains both a probiotic and a prebiotic, formulated to enhance the survival and colonization of the probiotic strain (Swanson et al. 2020). For example, pairing *Lactobacillus rhamnosus* with inulin provides not only the organism but also its preferred nutrient source. This combination improves the engraftment of probiotics, accelerates microbial rebalancing, and supports long-term restoration of gut function.

12.7.6 Diet-Based Therapies

Diet remains one of the most impactful tools for reshaping gut microbiota. Tailoring dietary patterns can support beneficial microbes and inhibit the growth of pathobionts or fungi. For instance, high-fiber diets enhance SCFA production, promoting anti-inflammatory effects. In contrast, the low-FODMAP diet, often used short-term in individuals with irritable bowel syndrome (IBS), reduces fermentable substrates and is later followed by gradual food reintroduction to prevent long-term dysbiosis. The Mediterranean diet, rich in plant-based fibers and polyphenols, is also known to foster microbial diversity. These dietary strategies can shift the microbial balance rapidly, reduce inflammation, and alleviate symptoms of gut-related disorders.

12.7.7 Lifestyle Interventions

A holistic view of dysbiosis management includes lifestyle factors that influence the microbiome. Stress management techniques such as yoga and meditation help lower cortisol levels, which in turn improve gut permeability and microbial balance. Regular physical activity has been shown to increase microbial diversity and

SCFA production, while good sleep hygiene supports circadian regulation of both gut and immune functions. Together, these practices enhance the resilience of the gut microbiota and reduce the likelihood of relapse into dysbiosis.

12.7.8 Additional Therapeutic Options

Beyond these strategies, the gut microbiota can also be therapeutically modulated through fecal microbiota transplantation (FMT). This involves the transfer of stool from a healthy donor to restore the entire microbial ecosystem of a recipient. FMT has demonstrated over 90% success in treating recurrent *Clostridioides difficile* infections (CDI) (Smits et al. 2013), and its use is also being explored for conditions such as ulcerative colitis, constipation, IBS, liver diseases, and even neurological disorders like multiple sclerosis and Parkinson's disease.

12.8 Public Health Significance

Disturbance of the delicate balance is a public health issue, affecting everything from chronic diseases to common infections. These contribute to 1.5 million deaths each year. Dysbiosis is connected to diabetes, obesity, heart diseases, autoimmune disorders, allergies, inflammatory bowel disease, cancer, dental decay, gum diseases, and infections. Furthermore, an imbalance in the oral microbiome is associated with heart disease, diabetes, problematic pregnancy outcomes, and respiratory infections. This amplifies the public health impact (Turnbaugh et al. 2007). Dysbiosis also influences maternal and child health. It affects neonatal immunity, increases allergy risk, and heightens the risk of dental problems with early colonizers. Fungal dysbiosis presents a public health challenge in patients with weakened immune systems, such as those with HIV/AIDS, organ transplant recipients, cancer patients, and the elderly. There is also the threat of antibiotic resistance, fungal overgrowth syndromes, and opportunistic infections (Rautemaa and Ramage 2011). Policies that aim to maintain and restore microbiome diversity through dietary supplementation and nutrition, antibiotic stewardship, and environmental exposure are important for preventing diseases across the population. Public health strategies recommend diet and supplementation with probiotics. Using probiotic lozenges and chewing gums can help reduce dental issues. Targeted therapies can benefit immunocompromised individuals, older adults, and denture wearers (Brown et al. 2012).

Recommendations include surveillance and research on community microbiome health, its diversity, and its interactions. There should be nationwide initiatives for microbiome mapping. Health promotion and Literacy are essential (Aggarwal et al. 2022). Educating the public about the importance of modulating oral and gut microbiomes, including probiotics, should be part of public health, preventive medicine, and other health programs. Embracing microbiome awareness in public health strategies can help prevent a wide range of illnesses and improve the quality of life for communities. Integrating microbiome and probiotic science into public health practice may lead to significant benefits for both individuals and communities (Ma et al. 2024).

12.9 Future Directions and Challenges

The future of research on the human microbiome and fungal balance focuses on understanding complex interactions between fungi and bacteria. This includes mechanisms like competitive exclusion, metabolic cross-feeding, biofilm dynamics, and immune modulation. Current challenges include detecting low fungal biomass, dealing with contamination issues, and determining causality versus correlation in multi-species interactions.

Revolutionary prospects will involve single-cell fungal sequencing, real-time monitoring of the mycobiome, CRISPR edited bacteria that produce targeted antifungal compounds, and AI-guided precision therapies. Key mechanisms include TLR/CLR signaling pathways, networks for metabolite exchange, and modulation of trained immunity. Emerging therapeutic strategies feature engineering multi-kingdom microbiomes, personalized mycobiome transplants, and synthetic biology platforms for custom antimicrobial production. Critical research priorities focus on establishing causal mechanisms, understanding temporal dynamics, identifying therapeutic targets, and preventing antifungal resistance.

From advanced diagnostics to personalized therapies and precision mycobiome medicine becoming mainstream in the future, will fundamentally transform how we manage fungal infections and treat immune disorders. The human microbiome supports health and disease, influencing everything from mental health to cancer. It drives breakthroughs in precision medicine. Next-generation sequencing and AI enable real-time, personalized microbiome profiling that goes beyond traditional methods. Therapies are moving toward engineered probiotics, bacteriophages, and precise microbial mixtures. AI and quantum computing will analyze large datasets, addressing unknown species and their causes to change diagnostics and treatments. The prospects and challenges are present across sectors as enumerated (Table 12.2).

12.10 Conclusion

This chapter highlights the important but often overlooked role of the human mycobiome in supporting health and preventing disease. Fungi interact with bacteria and the immune system, affecting microbial balance, disease risk, and the body's defense mechanisms. Dysbiosis, caused by antibiotics, changes in diet, and immune suppression, can result in fungal overgrowth and infections. Understanding how bacteria and fungi interact, including quorum sensing, iron competition, and metabolite changes, could lead to targeted treatments. Probiotics, prebiotics, synbiotics, and metabolic adjustments are promising ways to restore fungal balance. Public health initiatives that focus on preserving the microbiome, responsible use of

Table 12.2 Future prospects of human microbiome and fungal balance applications

Domain	Current state	Future prospects	Revolutionary impact
Microbiome and mycobiome recognition	Health shaped by bacterial–fungal ecosystem; fungal dysbiosis linked to oral disease, cancer, IBD	Routine assessment of bacterial + fungal balance	Integrated precision medicine targeting both microbiome and mycobiome
Diagnostic technologies	NGS + AI, including fungal ITS profiling; detection of cross-kingdom shifts	Rapid diagnostics/ wearables measuring fungal load and interactions	Early detection and personalized monitoring of fungal–bacterial imbalance
Therapeutic interventions	Engineered probiotics and early mycobiome-focused therapies	Personalized bacterial–fungal consortia; targeted antifungal modulation	Custom therapeutics for resistant fungal infections and chronic inflammatory states
AI and computing	Cross-kingdom datasets integrated into AI models	Digital microbial–fungal twins, predictive modeling	Automated, optimized treatments addressing bacteria–fungi interactions
Research challenges	Sparse fungal data; unclear causal fungal mechanisms	Full microbial + fungal mapping and targeted interventions	Breakthroughs in understanding bacterial–fungal crosstalk and novel therapies

antibiotics, and dietary changes can benefit populations. Recent research on the microbiome is leading to personalized medicine that combines next-generation sequencing, data analysis, and designed probiotics. Ethical considerations, fair access, and teamwork across disciplines will be crucial for applying these findings in clinical settings. Ultimately, integrating microbiome science into healthcare policy and practice has the potential to change how we approach prevention and treatment. The future involves using microbial ecosystems to improve human health.

12.11 Summary

This chapter examines the human microbiome with a focus on fungal balance, moving beyond a bacteria-centric view. It surveys microbiota diversity across niches such as the oral cavity, gut, skin, vagina, and eyes, emphasizing the mycobiome's role in homeostasis and immune regulation. Commensal fungi like *Candida*, *Malassezia*, and *Saccharomyces* interact with bacteria via competitive exclusion, cross-feeding, and chemical signaling, shaping microbial communities and pathogenic potential. Key interaction mechanisms include quorum sensing, siderophore-mediated iron acquisition, and metabolite influence. The immune system regulates fungi through receptors like Dectin-1 and TLRs. Dysbiosis drivers antibiotics,

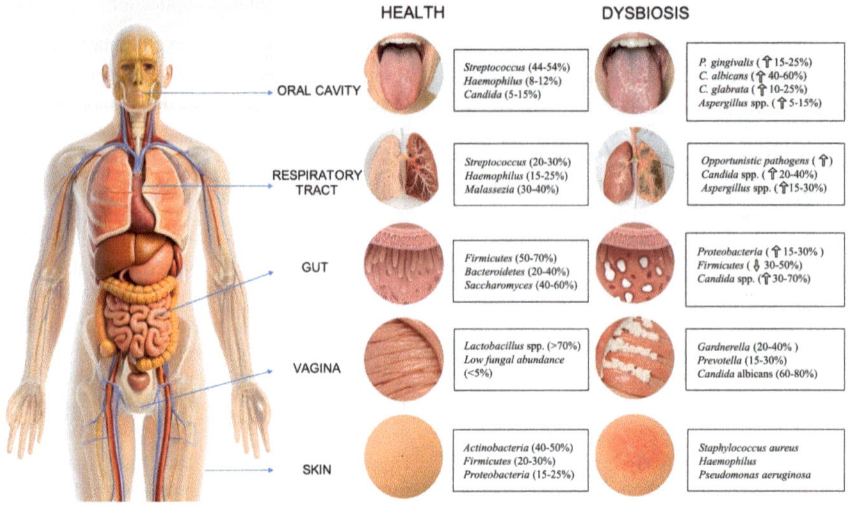

Fig. 12.3 Overview of microbiome, fungal balance, and dysbiosis

immune suppression, and diet shifts are linked to candidiasis, colitis, and systemic infections (Fig. 12.3). Probiotics emerge as promising tools for restoring balance and managing fungal disease, offering integrated, cross domain strategies for antifungal therapy.

Acknowledgement The AI tool ChatGPT version 5 is used for generating images and tables and refining text.

Conflict of Interest Nil

Author Contribution All the authors have contributed to conceptualizing, writing, and editing the chapter.

Further Reading

Procop GW, Church DL, Hall GS, Janda WM, Koneman EW (2020) Koneman's color atlas and textbook of diagnostic microbiology, 7th edn. Jones and Bartlett Learning, Burlington

Topley WWC, Wilson GS (eds) (2010) Topley and Wilson's microbiology and microbial infections, 10th edn. John Wiley and Sons, Chichester

Carroll KC, Pfaller MA (eds) (2023) Manual of clinical microbiology, 13th edn. ASM Press (John Wiley and Sons), Washington (DC)

Chibuye M, Mende DR, Spijker R, Simuyandi M, Luchen CC, Bosomprah S et al (2024) Systematic review of associations between gut microbiome composition and stunting in under-five children under five. NPJ Biofilms Microbiomes 10:46. https://doi.org/10.1038/s41522-024-00517-5

References

Abisado RG, Benomar S, Klaus JR, Dandekar AA, Chandler JR (2018) Bacterial quorum sensing and microbial community interactions. MBio 9(3):10–1128

Aggarwal N, Kitano S, Puah GRY, Kittelmann S, Hwang IY, Chang MW (2022) Microbiome and human health: current understanding, engineering, and enabling technologies. Chem Rev 123(1):31–72

Albelda-Berenguer M, Monachon M, Joseph E (2019) Siderophores: from natural roles to potential applications. Adv Appl Microbiol 106:193–225

Bairwa G, Hee Jung W, Kronstad JW (2017) Iron acquisition in fungal pathogens of humans. Metallomics 9(3):215–227. https://doi.org/10.1039/C6MT00301J

Bay L, Barnes CJ, Fritz BG, Thorsen J, Restrup MEM, Rasmussen L et al (2020) Universal dermal microbiome in human skin. MBio 11(1):10–1128

Bellard C, Genovesi P, Jeschke JM (2016) Global patterns in threats to vertebrates by biological invasions. Proc R Soc B 283(1823):20152454

Belvoncikova P, Splichalova P, Videnska P, Gardlik R (2022) The human mycobiome: colonization, composition and the role in health and disease. J Fungi 8(10):1046

Blaak EE, Canfora EE, Theis S, Frost G, Groen AK, Mithieux G et al (2020) Short chain fatty acids in human gut and metabolic health. Benef Microbes 11(5):411–455

Blaser MJ, Dominguez-Bello MG, Contreras M, Magris M, Hidalgo G, Estrada I et al (2013) Distinct cutaneous bacterial assemblages in a sampling of South American Amerindians and US residents. ISME J 7(1):85–95

Boon C, Deng Y, Wang LH, He Y, Xu JL, Fan Y et al (2008) A novel DSF-like signal from *Burkholderia cenocepacia* interferes with *Candida albicans* morphological transition. ISME J 2(1):27–36

Borges FM, de Paula TO, Sarmiento MR, de Oliveira MG, Pereira ML, Toledo IV et al (2018) Fungal diversity of human gut microbiota among eutrophic, overweight, and obese individuals based on aerobic culture-dependent approach. Curr Microbiol 75(6):726–735

Bouslimani A, Da Silva R, Kosciolek T, Janssen S, Callewaert C, Amir A, Dorrestein K, Melnik AV, Zaramela LS, Kim JN, Humphrey G, Schwartz T, Sanders K, Brennan C, Luzzatto-Knaan T, Ackermann G, McDonald D, Zengler K, Knight R, Dorrestein PC (2019) The impact of skin care products on skin chemistry and microbiome dynamics. BMC Biol 17(1). https://doi.org/10.1186/s12915-019-0660-6

Bradford LL, Ravel J (2017) The vaginal mycobiome: a contemporary perspective on fungi in women's health and diseases. Virulence 8(3):342–351

Bradford LL, Ravel J (2023) Vaginal microbiome: functional dynamics, health associations, and future directions. Nat Rev Microbiol 21(5):271–286

Brown GD, Denning DW, Levitz SM (2012) Hidden killers: human fungal infections. Sci Transl Med 4(165):165rv13

Byrd AL, Belkaid Y, Segre JA (2018) The human skin microbiome. Nat Rev Microbiol 16(3): 143–155

Catanzaro JR, Strauss JD, Bielecka A, Porto AF, Lobo FM, Urban A et al (2019) IgA-deficient humans exhibit gut microbiota dysbiosis despite secretion of compensatory IgM. Sci Rep 9: 13574

Chassaing B, Van de Wiele T, De Bodt J, Marzorati M, Gewirtz AT (2017) Dietary emulsifiers directly alter human microbiota composition and gene expression *ex vivo* potentiating intestinal inflammation. Gut 66(8):1414–1427

Chien AL, Tsai J, Leung S, Mongodin EF, Nelson AM, Kang S, Garza LA (2019) Association of systemic antibiotic treatment of acne with skin microbiota characteristics. JAMA Dermatol 155(4):425–434

Collins SM, Kassam Z, Bercik P (2013) The adoptive transfer of behavioural phenotype via the intestinal microbiota: experimental evidence and clinical implications. Curr Opin Microbiol 16(3):240–245

Davis-Hanna A, Piispanen AE, Stateva LI, Hogan DA (2008) Farnesol and dodecanol effects on the *Candida albicans* Ras1-cAMP signalling pathway and the regulation of morphogenesis. Mol Microbiol 67(1):47–62

De Gregorio PR, Silva JA, Marchesi A, Nader-Macías MEF (2019) Anti-Candida activity of beneficial vaginal *Lactobacilli* in *in vitro* assays and in a murine experimental model. FEMS Yeast Res 19(2):foz008

Dekaboruah E, Suryavanshi MV, Chettri D, Verma AK (2020) Human microbiome: an academic update on human body site specific surveillance and its possible role. Arch Microbiol 202(8): 2147–2167

Deo PN, Deshmukh R (2019) Oral microbiome: unveiling the fundamentals. J Oral Maxillofac Pathol 23(1):122–128

Dewhirst FE, Chen T, Izard J, Paster BJ, Tanner ACR, Yu WH, Lakshmanan A, Wade WG (2010) The human oral microbiome. J Bacteriol 192(19):5002–5017. https://doi.org/10.1128/JB.00542-10

Dickson RP, Erb-Downward JR, Huffnagle GB (2023) The respiratory tract microbiome and lung inflammation. Annu Rev Physiol 85:271–291

Felten A, Barreau C, Bizet C, Lagrange PH, Philippon A (1999) *Lactobacillus* species identification, H_2O_2 production, and antibiotic resistance and correlation with human clinical status. J Clin Microbiol 37(3):729–733

Findley K, Oh J, Yang J, Conlan S, Deming C, Meyer JA et al (2013) Topographic diversity of fungal and bacterial communities in human skin. Nature 498(7454):367–370

Gao H, Span I (2025) The diversity and applications of microbial iron metabolism and iron-containing proteins. Commun Biol 8(1):177

Gibson GR, Hutkins R, Sanders ME, Prescott SL, Reimer RA, Salminen SJ et al (2017) Expert consensus document: the International Scientific Association for Probiotics and Prebiotics (ISAPP) consensus statement on the definition and scope of prebiotics. Nat Rev Gastroenterol Hepatol 14(8):491–502

Gonçalves B, Ferreira C, Alves CT, Henriques M, Azeredo J, Silva S (2016) Vulvovaginal candidiasis: epidemiology, microbiology and risk factors. Crit Rev Microbiol 42(6):905–927

Gray KM (1997) Intercellular communication and group behavior in bacteria. Trends Microbiol 5(5):184–188

Guinan J, Wang S, Hazbun TR, Yadav H, Thangamani S (2019) Antibiotic-induced decreases in the levels of microbial-derived shortchain fatty acids correlate with increased gastrointestinal colonization of Candida albicans. Sci Rep 9(1). https://doi.org/10.1038/s41598-019-45467-7

Gupta AK, Kohli Y (2004) Prevalence of *Malassezia* species on various body sites in clinically healthy subjects representing different age groups. Med Mycol 42(1):35–42

Hall RA, Turner KJ, Chaloupka J, Cottier F, De Sordi L, Sanglard D et al (2011) The quorum-sensing molecules farnesol/homoserine lactone and dodecanol operate via distinct modes of action in *Candida albicans*. Eukaryot Cell 10(8):1034–1042

Hibbett D, Nagy LG, Nilsson RH (2025) Fungal diversity evolution and classification. Curr Biol 35(11):R463–R469. https://doi.org/10.1016/j.cub.2025.01.053

Hickey RJ, Zhou X, Pierson JD, Ravel J, Forney LJ (2012) Understanding vaginal microbiome complexity from an ecological perspective. Transl Res 160(4):267–282

Hill C, Guarner F, Reid G, Gibson GR, Merenstein DJ, Pot B et al (2014) Expert consensus document: ISAPP consensus statement on the scope and appropriate use of the term probiotic. Nat Rev Gastroenterol Hepatol 11(8):506–514

Holmes ZC, Villa MM, Durand HKR, Jiang S, Guillemin K, Obermiller L et al (2022) Microbiota responses to different prebiotics are conserved within individuals and associated with habitual fiber intake. Microbiome 10(1):114

Hong BY, Hoare A, Cardenas A, Dupuy AK, Choquette L, Salner AL et al (2020) The salivary mycobiome contains two ecologically distinct mycotypes. J Dent Res 99(6):730–738

Honoré AH, Aunsbjerg SD, Ebrahimi P, Thorsen M, Benfeldt C, Knøchel S et al (2016) Metabolic footprinting for investigation of antifungal properties of *Lactobacillus paracasei*. Anal Bioanal Chem 408(1):83–96

Hrncirova L, Hudcovic T, Sukova E, Machova V, Trckova E, Krejsek J, Hrncir T (2019) Human gut microbes are susceptible to antimicrobial food additives *in vitro*. Folia Microbiol 64(4): 497–508. https://doi.org/10.1007/s12223-018-00674-z

Hussain I, Cher GLY, Abid MA, Abid MB (2021) Role of gut microbiome in COVID-19: an insight into pathogenesis and therapeutic potential. Front Immunol 12:765965

Jang SJ, Lee K, Kwon B, You HJ, Ko G (2019) Vaginal *lactobacilli* inhibit growth and hyphae formation of *Candida albicans*. Sci Rep 9(1):8121

Jarosz LM, Deng DM, van der Mei HC, Crielaard W, Krom BP (2009) Streptococcus mutans competence-stimulating peptide inhibits *Candida albicans* hypha formation. Eukaryot Cell 8(11):1658–1664

Jo JH, Kennedy EA, Kong HH (2017) Topographical and physiological differences of the skin mycobiome in health and disease. Virulence 8(3):324–333

Kaur H, Merchant M, Haque MM, Mande SS (2020) Crosstalk between female gonadal hormones and vaginal microbiota across various phases of women's gynecological lifecycle. Front Microbiol 11:551

Kaur S, Patel BC, Collen A, Malhotra R (2025) The microbiome and the eye: a new era in ophthalmology. Eye 39(3):436–448

Kunyeit L, Kurrey NK, Anu-Appaiah KA, Rao RP (2019) Probiotic yeasts inhibit virulence of non-albicans *Candida* species. MBio 10(5):e02307–e02319

Lamont RJ, Koo H, Hajishengallis G (2018) The oral microbiota: dynamic communities and host interactions. Nat Rev Microbiol 16(12):745–759

Leão MV, Gonçalves e Silva CR, Santos SS, Leite PG (2015) *Lactobacillus rhamnosus* may change the virulence of *Candida albicans*. Rev Bras Ginecol Obstet 37(9):417–420

Lesuisse E, Blaiseau PL, Dancis A, Camadro JM (2001) Siderophore uptake and use by the yeast *Saccharomyces cerevisiae*. Microbiology 147(2):289–298

Levy M, Kolodziejczyk AA, Thaiss CA, Elinav E (2017) Dysbiosis and the immune system. Nat Rev Immunol 17(4):219–232

Lopez-Medina E, Fan D, Coughlin LA, Ho EX, Lamont IL, Reimmann C, Koh AY (2015) *Candida albicans* inhibits *Pseudomonas aeruginosa* virulence through suppression of pyochelin and pyoverdine biosynthesis. PLoS Pathog 11(8):e1005129

Ma Z, Zuo T, Frey N, Rangrez AY (2024) A systematic framework for understanding the microbiome in human health and disease: from basic principles to clinical translation. Signal Transduct Target Ther 9(1):237

Mark Welch JL, Rossetti BJ, Rieken CW, Dewhirst FE, Borisy GG (2016) Biogeography of a human oral microbiome at the micron scale. Proc Natl Acad Sci USA 113(6):E791–E800

Matijašić M, Meštrović T, Čipčić Paljetak H, Perić M, Barešić A, Verbanac D (2020) Gut microbiota beyond bacteria—mycobiome, virome, archaeome, and eukaryotic parasites in IBD. Int J Mol Sci 21(8):2668

Mendling W (2016) Vaginal microbiota. Adv Exp Med Biol 902:83–93. https://doi.org/10.1007/5584_2016_34

Moens F, Duysburgh C, van den Abbeele P, Morera M, Marzorati M (2019) *Lactobacillus rhamnosus* GG and *Saccharomyces cerevisiae boulardii* exert synergistic antipathogenic activity *in vitro* against enterotoxigenic *Escherichia coli*. Benef Microbes 10(8):923–936

Monteiro-da-Silva F, Araujo R, Sampaio-Maia B (2014) Interindividual variability and intraindividual stability of oral fungal microbiota over time. Med Mycol 52(5):498–505

Mowat E, Rajendran R, Williams C, McCulloch E, Jones B, Lang S, Ramage G (2010) *Pseudomonas aeruginosa* and their small diffusible extracellular molecules inhibit *Aspergillus fumigatus* biofilm formation. FEMS Microbiol Lett 313(2):96–102

Mukherjee S, Bassler BL (2019) Bacterial quorum sensing in complex and dynamically changing environments. Nat Rev Microbiol 17(6):371–382

Oh J, Byrd AL, Park M, Kong HH, Segre JA (2016) Temporal stability of the human skin microbiome. Cell 165(4):854–866. https://doi.org/10.1016/j.cell.2016.04.008

Peay KG, Kennedy PG, Talbot JM (2016) Dimensions of biodiversity in the Earth mycobiome. Nat Rev Microbiol 14(7):434–447

Perera M, Al-Hebshi NN, Johnson NW (2024) The changing landscape of the oral microbiome: new insights into health and disease. Oral Dis 30(2):215–229. https://doi.org/10.1111/odi.14230

Rajasekaran JJ, Krishnamurthy HK, Bosco J, Jayaraman V, Krishna K, Wang T, Bei K (2024) Oral microbiome: a review of its impact on oral and systemic health. Microorganisms 12(9):1797

Rautemaa R, Ramage G (2011) Oral candidosis—clinical challenges of a biofilm disease. Crit Rev Microbiol 37(4):328–336

Ravel J, Brotman RM, Gajer P, Ma B, Nandy M (2016) Translating the vaginal microbiome: gaps and challenges. Genome Med 8(1):35. https://doi.org/10.1186/s13073-016-0291-2

Reynoso-García J, Miranda-Santiago AE, Meléndez-Vázquez NM, Acosta-Pagán K, Sánchez-Rosado M, Díaz-Rivera J, Godoy-Vitorino F (2022) A complete guide to human microbiomes: body niches, transmission, development, dysbiosis, and restoration. Front Syst Biol 2:951403

Richard ML, Sokol H (2021) The gut mycobiota: insights into analysis, environmental interactions and role in gastrointestinal diseases. Nat Rev Gastroenterol Hepatol 18(6):333–345. https://doi.org/10.1038/s41575-020-00381-2

Rodriguez-Palacios A, Harding A, Menghini P, Himmelman C, Retuerto M, Nickerson KP et al (2018) The artificial sweetener Splenda promotes gut Proteobacteria, *Dsbiosis*, and *Myeloperoxidase* reactivity in Crohn's disease-like ileitis. Inflamm Bowel Dis 24(5):1005–1020

Rose Jørgensen M, Thestrup Rikvold P, Lichtenberg M, Østrup Jensen P, Kragelund C, Twetman S (2020) *Lactobacillus rhamnosus* strains of oral and vaginal origin show strong antifungal activity in vitro. J Oral Microbiol 12(1):1832832

Roussel C, De Paepe K, Galia W, De Bodt J, Chalancon S, Denis S, Van de Wiele T (2021) Multi-targeted properties of the probiotic *Saccharomyces cerevisiae* CNCM I-3856 against entero-toxigenic *Escherichia coli* (ETEC) H10407 pathogenesis across human gut models. Gut Microbes 13(1):1953246

Sam QH, Chang MW, Chai LYA (2017) The fungal mycobiome and its interaction with gut bacteria in the host. Int J Mol Sci 18(2):330

Sanjar F, Weaver AJ, Peacock TJ, Nguyen JQ, Brandenburg KS, Leung KP (2020) Identification of metagenomics structure and function associated with temporal changes in rat (Rattus norvegicus) skin microbiome during health and cutaneous burn. J Burn Care Res 41(2):347–358

Schulz MD, Atay C, Heringer J, Romrig FK, Schwitalla S, Aydin B et al (2014) High-fat-diet-mediated dysbiosis promotes intestinal carcinogenesis independently of obesity. Nature 514(7523):508–512

Seed PC (2014) The human mycobiome. Cold Spring Harb Perspect Med 5:a019810

Shahinozzaman M, Raychaudhuri S, Fan S, Obanda DN (2021) Kale attenuates inflammation and modulates gut microbial composition and function in C57BL/6J mice with diet-induced obesity. Microorganisms 9(2):238

Shreiner AB, Kao JY, Young VB (2015) The gut microbiome in health and in disease. Curr Opin Gastroenterol 31(1):69–75

Smits LP, Bouter KE, de Vos WM, Borody TJ, Nieuwdorp M (2013) Therapeutic potential of fecal microbiota transplantation. Gastroenterology 145(5):946–953

Sobel JD (2007) Vulvovaginal candidosis. Lancet 369(9577):1961–1971

Suez J, Korem T, Zeevi D, Zilberman-Schapira G, Thaiss CA, Maza O et al (2014) Artificial sweeteners induce glucose intolerance by altering the gut microbiota. Nature 514(7521): 181–186

Swanson KS, Gibson GR, Hutkins R, Reimer RA, Reid G, Verbeke K, Scott KP, Holscher HD, Azad MB, Delzenne NM, Sanders ME (2020) The international scientific association for probiotics and prebiotics (ISAPP) consensus statement on the definition and scope of synbiotics. Nat Rev Gastroenterol Hepatol 17(11):687–701. https://doi.org/10.1038/s41575-020-0344-2

Tan J, McKenzie C, Potamitis M, Thorburn AN, Mackay CR, Macia L (2014) The role of short-chain fatty acids in health and disease. Adv Immunol 121:91–119

Turnbaugh PJ, Ley RE, Hamady M, Fraser-Liggett CM, Knight R, Gordon JI (2007) The human microbiome project. Nature 449(7164):804–810

Turunen J, Paalanne N, Reunanen J et al (2023) Development of gut mycobiome in infants and young children: a prospective cohort study. Pediatr Res 94:486–494

Valenti P, Rosa L, Capobianco D, Lepanto MS, Schiavi E, Cutone A, Mastromarino P (2018) Role of *Lactobacilli* and *Lactoferrin* in the mucosal cervicovaginal defense. Front Immunol 9:376

Vrieze A, Van Nood E, Holleman F, Salojärvi J, Kootte RS, Bartelsman JF et al (2021) Transfer of intestinal microbiota from lean donors increases insulin sensitivity in individuals with metabolic syndrome. Gastroenterology 143(4):913–916

Wade WG (2013) The oral microbiome in health and disease. Pharmacol Res 69(1):137–143

Wampach L, Heintz-Buschart A, Hogan A, Muller EE, Narayanasamy S, Laczny CC, Wilmes P (2017) Colonization and succession within the human gut microbiome by archaea, bacteria, and microeukaryotes during the first year of life. Front Microbiol 8:738

Wescombe PA, Heng NC, Burton JP, Chilcott CN, Tagg JR (2009) *Streptococcal bacteriocins* and the case for *streptococcus salivarius* as model oral probiotics. Future Microbiol 4(7):819–835

Wong JM, De Souza R, Kendall CW, Emam A, Jenkins DJ (2006) Colonic health: fermentation and short chain fatty acids. J Clin Gastroenterol 40(3):235–243

Wu Y, Hu S, Wu C, Gu F, Yang Y (2022) Probiotics: potential novel therapeutics against fungal infections. Front Cell Infect Microbiol 11:793419

Yeoh YK, Zuo T, Lui GC, Zhang F, Liu Q, Li AY et al (2021) Gut microbiota composition reflects disease severity and dysfunctional immune responses in patients with COVID-19. Gut 70(4): 698–706

Zangl I, Pap IJ, Aspöck C, Schüller C (2019a) The role of Lactobacillus species in the control of *Candida* via biotrophic interactions. Microb Cell 7(1):1

Zangl I, Pap I, Aspöck C, Schicho R, Köstlbacher S (2019b) Oral microbiome shifts in health and disease: a comprehensive review. J Oral Microbiol 11(1):1609833

Exploring Probiotics for Managing Cutaneous Mycoses

13

P. R. Vanitha, J. Dhanuja, P. Monisha, C. R. Mohit Ram, and M. Y. Sreenivasa

Abstract

The increasing prevalence of cutaneous fungal infections, coupled with challenges of treating them, particularly the toxicity of antifungal drugs and the issue of antibiotic resistance, has prompted a search for alternative treatment strategies that are non-toxic and free from side effects. Probiotics have been investigated for various infections and have established a proven safety record, as they are part of the normal microflora found in the gut and are also present in different types of foods we consume. Different species of *Lactobacillus*, *Bifidobacterium*, *Streptococcus*, *Pediococcus*, *Bacillus*, *Saccharomyces*, etc., have been successfully tested against various fungal pathogens of skin, including dermatophytes, and both albicans and non-albicans *Candida* species. Positive outcomes from these studies have facilitated the development of formulations that incorporate probiotic whole cells or extracellular products. These formulations can be used for topical applications or as oral solutions to help control infections. Moreover, patents have been obtained for such products with probiotics. Extensive research should be conducted each time a new probiotic strain is isolated from a source and examined for its antipathogenic effect, as these effects are strain dependent. Furthermore, regulations and specific guidelines must be addressed when probiotics are marketed as drugs, medical devices, or dietary supplements.

P. R. Vanitha (✉)
Applied Mycology Laboratory, Department of Studies in Microbiology, University of Mysore, Mysore, Karnataka, India

Government Women's College, Chintamani, Karnataka, India

J. Dhanuja · P. Monisha · C. R. Mohit Ram · M. Y. Sreenivasa
Applied Mycology Laboratory, Department of Studies in Microbiology, University of Mysore, Mysore, Karnataka, India

M. Y. Sreenivasa et al. (eds.), *Probiotics as Antifungals*,
https://doi.org/10.1007/978-981-92-0002-3_13

Graphical Abstract

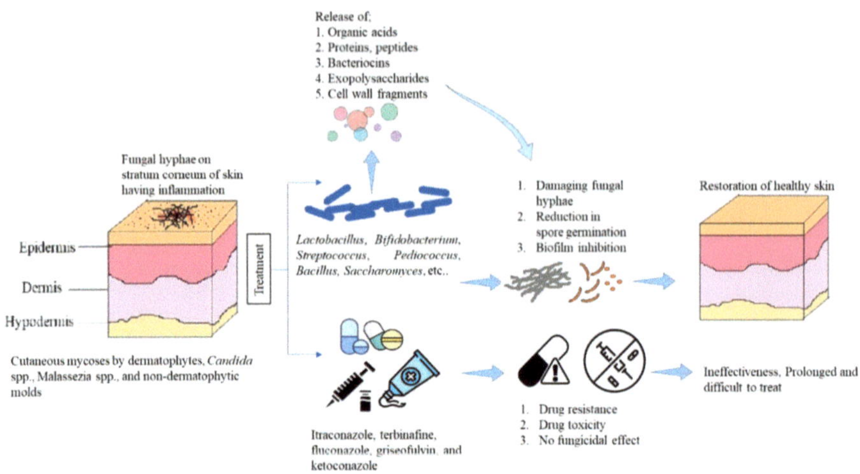

Keywords

Dermatomycoses · Candidiasis · *Lactobacillus* · Skin infections · Antifungals

Learning Objectives

By the end of the chapter, readers will be able

- To understand the growing concern about the prevalence of cutaneous mycoses and the drawbacks of continuous use of antifungal drugs.
- To learn about the different probiotic organisms involved in controlling skin pathogens and their mechanisms of action.
- To propose solutions for the challenges faced in the therapeutic use of probiotics.

13.1 Introduction

Cutaneous mycoses or superficial mycoses are infections of the keratinized tissues, which include skin, nails, or hair, caused by different species of fungi broadly grouped as dermatophytes and non-dermatophytes. The dermatophytes comprise nine genera, such as *Trichophyton, Epidermophyton, Lopophyton, Paraphyton,*

Microsporum, Arthroderma, Nannizia, Ctenomyces, and *Guarromyces* (de Hoog et al. 2017). Among these, the most commonly known genera include *Microsporum, Trichophyton*, and *Epidermophyton,* causing infection called tinea or ringworm. Based on the area of infection, it may be classified as tinea corporis (body), tinea capitis (scalp), tinea pedis (foot), tinea facie (face), tinea cruris (groin folds), and tinea unguium (nails) (Pei et al. 2022). Non-dermatophytes are any saprophytic fungi that cause infection in immunocompromised individuals. They include *Aspergillus* spp., *Fusarium* spp., *Alternaria* spp., *Acremonium* spp., *Cladosporium* spp., *Penicillium* spp., *Scopulariopsis brevicaulis*, and other moulds. Apart from moulds, many *Candida* spp. cause cutaneous candidiasis, specifically nail infections called onychomycosis affecting both fingernails and toenails. *Malassezia* spp. is another yeast that causes dermatologic conditions, which include Pityriasis versicolor (PV) or Tinea versicolor, Dandruff, Seborrheic dermatitis (SD), Atopic dermatitis, and Folliculitis.

13.2 Disease Incidence

According to a 2015 report from World Health Organization (WHO), dermatophyte infections mark approximately 25% of the global population, impacting individuals of all ages, regardless of their immunocompromised status, and leading to healthcare costs of at least half a billion dollars. While this issue is prevalent worldwide, its prevalence is notably higher in tropical regions due to warm climate and high humidity (Petrucelli et al. 2020). There has been a significant rise in recurrent and chronic cases of dermatophytosis in developing countries as India (Verma et al. 2021).

Tinea capitis, a scalp infection, is recognized as one of the most contagious forms of tinea diseases (Müller et al. 2021). The prevalence of tinea capitis varies globally, ranging from 0.4% to 91.2%. The primary causative agents for this condition, which include anthropophilic species *Trichophyton violaceum* and *T. tonsurans*, as well as zoophilic species *Microsporum canis* (Zhan and Liu 2017). Research indicates that *M. canis* accounts for approximately 37.1% to 98.7% of reported tinea capitis cases in regions such as North and South America, Southeast and Central Asia, and various European countries (Rodríguez-Cerdeira et al. 2021). This pathogen is widespread among animals, leading to dermatophytosis in both pets and their owners. Additionally, the incidence of the disease tinea capitis caused by this pathogen is currently growing, which raises concern. A study conducted in Germany over a four-year period from 2019 to 2022 documented a tenfold rise in *T. tonsurans* infections, escalating from 2.1% to 20.7% (Pilz et al. 2023). In the twenty-first century, *T. rubrum* has emerged as the major organism in places like Asia, Europe, Africa, and South America, causing tinea cruris, tinea corporis, onychomycosis, and tinea pedis (Dass et al. 2007; Chanyachailert et al. 2023). The incidence of infection caused by *T. rubrum* has increased from 45.8% in 2012 to 69.4% in 2016 (Sacheli et al. 2021). In India, there are endemic reports of chronic dermatophytosis, and the occurrence of this disease varies from 6.09 to 61.5% (Verma et al. 2021).

In addition to this, non-dermatophytic infections of moulds, yeasts, and yeast-like fungi have been associated with superficial mycoses. The genus *Candida* encompasses approximately 300 different species, yet around 90% of the infections are attributed to only five species, such as *C. albicans* (65.3%), *C. glabrata* (11.3%), *C. tropicalis* (7.2%), *C. parapsilosis* (6.0%), and *C. krusei* (2.4%). Over the past 15 to 20 years, *Candida* infections have become a significant concern, particularly among immunocompromised individuals. Those with underlying conditions that include advanced HIV infection, cancer, organ transplantation, granulocytopenia, severe burns, diabetes, malnutrition, trauma, and other forms of immunosuppression are at increased risk for these infections. A meta-analysis investigating *C. albicans* causing skin infections of individuals having type 2 diabetes reported an occurrence of 11.4% among a sample size of 1348 (Rasoulpoor et al. 2021).

Malassezia is another form of lipophilic yeast that contains lipases and hydrolyzes triglycerides to irritant unsaturated fatty acids on the scalp, back, face, and trunk. The organism is commonly present on the skin, ranging from 75% to 98% in healthy individuals (Dessinioti and Katsambas 2013). *Malassezia* comprises 18 species, and epidemiological prevalence of *Malassezia* showed that *M. furfur* is the most predominant in Pityriasis versicolor (PV) and Seborrheic dermatitis (SD) on the areas of scalp, whereas *M. sympodialis* is the most prevalent in atopic dermatitis. In healthy persons, *M. furfur, M. sympodialis*, and *M. globosa* occur in a low ratio (Hamdino et al. 2022). The factors like antibiotic usage, immunosuppression, and environmental changes that cause dysbiosis of the skin's normal microflora lead to a higher prevalence of *Malassezia* (Sasikumar et al. 2025).

SD is one of the chronic inflammatory skin disorders impacting 2–5% to the world population (Dessinioti and Katsambas 2013). It is clinically represented as red patches, flaking, and greasy skin on the parts of the ear, scalp, eyebrows, chest, and nasolabial folds. Dandruff is a condition characterized by loosely attached oily flakes that are restricted to the scalp and involve itching. It is a mild and non-inflammatory form of SD (Warner et al. 2001). The *Malassezia* species that are linked with SD and Dandruff are *M. furfur, M. globosa, M. restricta, M. sympodialis, M. obtuse,* and *M. sloofiae* (Saxena et al. 2018).

Pityriasis versicolor, commonly called Tinea versicolor, represents the hyper- or hypopigmented patches with translucent scales (Karray and McKinney 2023). *M. furfur, M. sympodialis,* and *M. globosa* have been isolated most often in PV cases (Saunte et al. 2020). It may occur in all age groups, more predominant in adolescents and in young adults, because of higher secretion of the sebum which creates a lipid-rich environment for the organism to grow. Prevalence of PV is more in tropical regions, i.e., 50%, whereas in moderate climates, around 1–4% and in cold climates, 1% (Karray and McKinney 2023). The predisposing factors for the infection are a hot and humid environment, use of corticosteroids and contraceptives, poor general health, malnutrition, and seborrhoea (Leung et al. 2022).

The occurrence of onychomycosis caused by non-dermatophytes varies from 0.8% to 65.8%, with significantly higher rates observed in countries such as Thailand, Sri Lanka, and Brazil (Ranawaka et al. 2016). The spectrum of pathogens also varies by region, including *Acremonium* spp., *Scopulariopsis brevicaulis,*

Aspergillus spp., and *Fusarium* spp. In tropical countries like Thailand, the *Neoscytalidium dimidiatum* is the main causative organism of onychomycosis with a reported occurrence rate of 13% (Phaitoonwattanakij et al. 2021).

13.3 Antifungal Medications and Their Side Effects

Dermatophytosis usually takes at least 2–4 weeks to be treated in all of its types and may take up to 6 months in cases of onychomycosis and tinea capitis (Al-Khikani 2020). Furthermore, tinea capitis and onychomycosis are not easy to treat and need systemic therapy (Kruithoff et al. 2024). The commonly used antifungal drugs come under the class azoles, allylamines, and polyenes used as both systemic (0.5–50 μg/ml) and topical medications (1–2%). Some of them are itraconazole, terbinafine, fluconazole, griseofulvin and ketoconazole. Most of the clinically used antifungals suffer from drawbacks, such as drug–drug interactions, toxicity, no fungicidal effect, and high cost.

Many of these antifungals are known to give side effects, which include liver damage and cause anaphylactic reactions. Long-term treatment with itraconazole had led to life-threatening adverse reactions like anaphylactic shock with acute lymphoblastic leukaemia (ALL). Further investigation indicated that immunosuppression-induced T-cell reduction and itraconazole accumulation were important factors for this delayed type hypersensitivity reaction in the patient (Chen et al. 2015). Amphotericin B (AmB), a polyene macrolide antimycotic agent, is the commonly used drug for most of the serious systemic and internal infections caused by *Aspergillus*, *Fusarium, Candida*, and others. The use of this drug is limited by dose-dependent nephrotoxicity and side effects that include fever, chills, nausea, vomiting, and headache. Approximately 80% of patients demonstrate acute side effects during the AmB treatment, and 30% show renal failure (Baginski and Czub 2009).

There are reports of several drug–drug interactions, which could be toxic, especially in immunocompromised individuals with conditions of diabetes, liver dysfunction, and renal failure (Al-Khikani 2020). Itraconazole and Ketoconazole are potential inhibitors of the drug-metabolizing CYP isoform enzyme in humans, CYP3A4. Coadministration of these antifungal drugs with CYP3A enzyme substrates such as cyclosporin, alprazolam, tacrolimus, midazolam, triazolam, felodipine, nifedipine, lovastatin, simvastatin, vincristine, terfenadine, or astemizole can lead to clinically significant drug interactions, some of which can be fatal (Venkatakrishnan et al. 2000).

Long-term usage of corticosteroids that are used for SD causes side effects, which include skin atrophy, folliculitis, telangiectasia, hypopigmentation, and hypertrichosis. The existing treatment is for the symptoms and not for the root causes, which creates recurrent or persistent infections (Jackson et al. 2024).

13.4 Drug Resistance

In addition, reports from different parts of the world indicate an alarming increase in the prevalence of resistance among clinically isolated strains, which is contributing to treatment failures in mycoses. This could be due to long-term drug exposure during therapy that favours the rise of resistant isolates (Saunte et al. 2019). In the study by Singh et al. (2019), a unique multidrug-resistant *Trichophyton* species was isolated in North India, having a high rate of resistance to the three commonly used oral antifungal drugs, namely, terbinafine, fluconazole, and griseofulvin. Similarly, terbinafine resistance was detected in the clinical isolates of *T. tonsurans* and *T. rubrum* at a rate of 2%, indicating the rise in the number of antifungal-resistant dermatophytes (Salehi et al. 2018). Additionally, cross-resistance among *Trichophyton* strains has emerged as a serious concern. For instance, terbinafine-resistant *T. rubrum* has been shown to exhibit reduced susceptibility to other allylamine and thiocarbamate antifungals, such as butenafine, tolnaftate, tolciclate, and naftifine (Sinha and Sardana 2018). Moreover, resistance to fluconazole and griseofulvin has also been reported in a limited number of *M. canis* isolates obtained from pet animals, further highlighting the expanding spectrum of antifungal resistance (Singh et al. 2021). It is also disturbing to know that dermatophytic fungi are becoming more virulent with an increased minimum inhibitory concentration (MIC) against the regularly used antifungal drugs (Pathania et al. 2018). The formation of biofilms within host tissues is also thought to play a major role in clinical resistance, which leads to poor therapeutic outcomes and persistent, hard-to-treat infections. In this regard, an increasing number of *Candida* strains have been reported to exhibit resistance to multiple antifungal drug classes, including fluconazole, caspofungin, and amphotericin B (Ademe and Girma 2020). Biofilms formed by *C. albicans* showed an inherent tolerance to most available antifungal agents, which makes infections associated with these biofilms difficult to manage therapeutically (Kaur and Nobile 2023). Terbinafine resistance was also observed among *Malassezia furfur* strains (Cafarchia et al. 2012). This increasing trend in resistance towards antifungal drugs demands the introduction of newer control agents which are more effective, less harmful, and cost-effective.

13.5 Probiotics as Therapeutic Agents

Human skin is home to different microflora and is known to carry out many functions, including antimicrobial, anti-inflammatory, and antineoplastic activity while also boosting the innate and acquired immune response (Nakatsuji and Gallo 2019). This is similar to the gut microbiome, which has significant antimicrobial and immune-enhancing activity. Several research studies have shown the importance of the gut–skin axis, which helps to maintain skin homeostasis. Disruptions in the gut microbiota balance have been associated with the development of various skin disorders, such as acne, atopic dermatitis, and psoriasis (Habeebuddin et al. 2022). Thus, strategies aimed at restoring microbial equilibrium both in the skin and the gut are increasingly being explored as therapeutic options. In this context, probiotics have gained attention for their ability to modulate microbial communities and are emerging as a promising approach in managing several skin conditions. Both oral and topical probiotic formulations have been tested with success for different skin conditions. Applying probiotic bacteria topically may enhance the skin's natural microbiota barrier function by direct site-specific action. Resident and probiotic microbiota produce various antimicrobial peptides that enhance immune functions and aid in eliminating pathogens (Al-Ghazzewi and Tester 2014). Similarly, oral probiotics help in restoring the gut microflora, which in turn will help in maintaining skin homeostasis (Szántó et al. 2019).

Numerous studies have shown that probiotics can inhibit the growth of a wide range of pathogenic fungi, including *Candida albicans*, *Aspergillus fumigatus*, *Trichophyton* spp., and *Microsporum* spp. (Rakesh et al. 2019; Shruthi et al. 2025; Vanitha et al. 2023). The antifungal potential of probiotics is attributed to their ability to inhibit filamentation of pathogenic fungi, inhibition of pathogen adhesion on the tissue surface, and biofilm inhibition (Wu et al. 2022; Adithi et al. 2025). The mechanism of probiotics action against fungal pathogens has been depicted in Fig. 13.1a, b.

Hence, employing probiotics either alone or along with traditional antifungals may act as potent therapeutic agents against fungal diseases.

13.6 Probiotics Against Dermatophytes

Probiotics are widely known for their ability to suppress the growth of dermatophytes and have shown potential in the prevention and treatment of dermatophytic infections. In particular, lactic acid bacteria (LAB), such as *L. plantarum*, *L. reuteri*, *L. acidophilus*, *L. fermentum*, *L. brevis*, *L. pentosus*, *L. casei*, *Enterococcus* spp., and *Pediococcus* spp., have been evaluated for their antifungal effect against the dermatophytes such as *T. tonsurans*, *T. rubrum*, *T. mentagrophytes*, *T. interdigitale*, *M. gypseum*, *M. canis*, and *E. floccosum*. These probiotic strains have been isolated from various sources, including fermented foods, cereals, human and animal intestine, forage silage, tropical fruits, and also from poultry (Guo et al. 2012; Arasu et al. 2013; Deepa et al. 2016; Divyashree S et al. 2023; Adeniyi et al. 2024; Vasundaradevi et al. 2025). Their anti-dermatophytic activity is observed to be strain-specific. For instance, Guo et al. (2011) reported that among five *Lactobacillus reuteri* strains that showed antidermatophytic activity, the *L. reuteri* R2 strain demonstrated the strongest inhibitory effect against *T. tonsurans*. Similarly, a study conducted in our laboratory (Fig. 13.2) detected that among 20 LAB isolates from a fermented food called Haria, the strain *L. plantarum Lp*MYSN7 was most effective against *T. tonsurans* (Vanitha et al. 2023), and the strain *L. fermentum* LfMYSY8 showed maximum antifungal activity against *M. canis* (Vanitha et al. 2024).

Collectively, many of these studies also confirmed the antifungal potential of cell-free supernatant (CFS) of the LAB isolates (Poornachandra Rao et al. 2017; Adithi

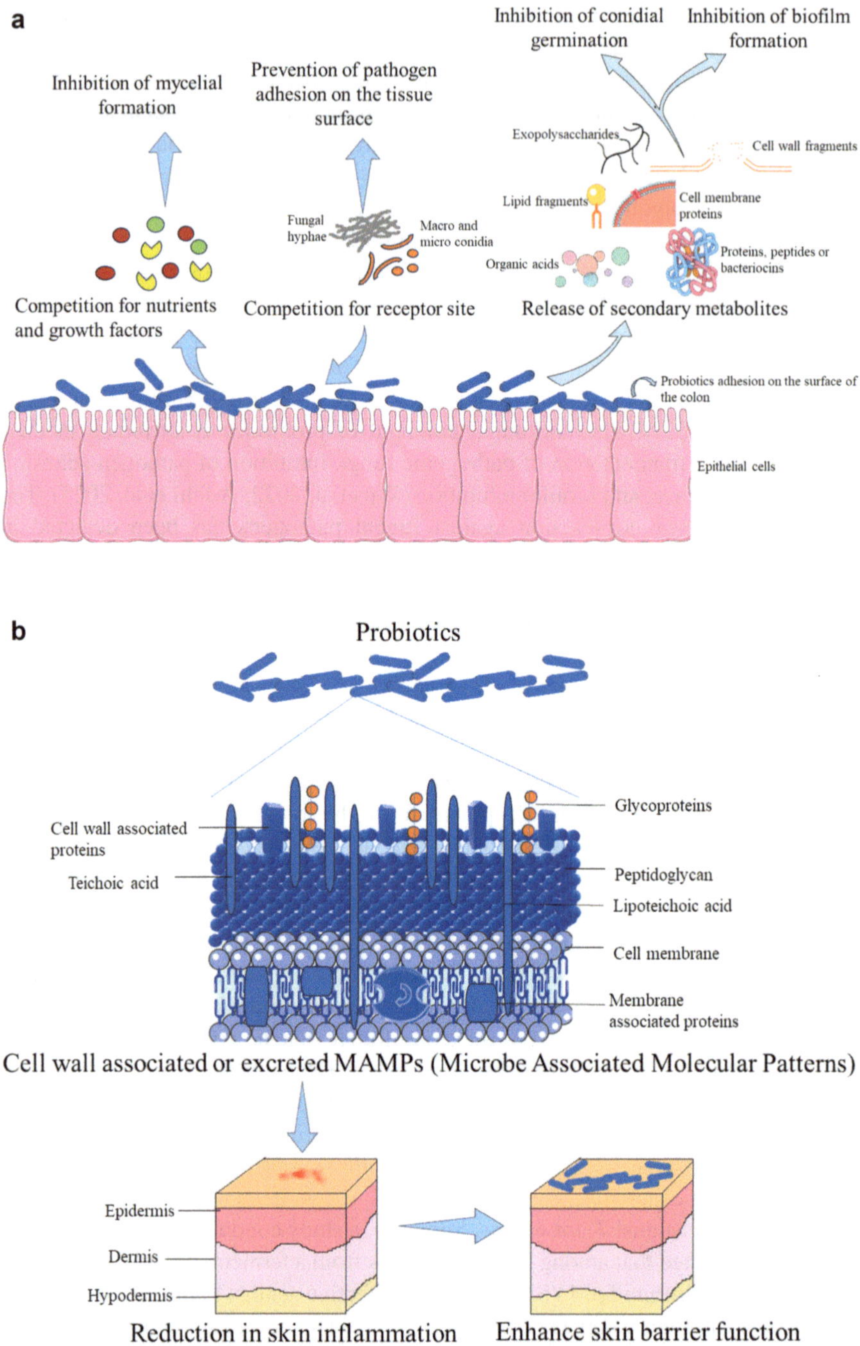

Fig. 13.1 (**a**) The mechanism of probiotics action against fungal pathogens: During metabolism, probiotics release secondary metabolites such as organic acids, proteins, peptides or bacteriocins, exopolysaccharides, cell membrane proteins, and lipid fragments, all of which affect the conidial germination and inhibit biofilm formation. Probiotics compete for adhesion to the tissue surfaces

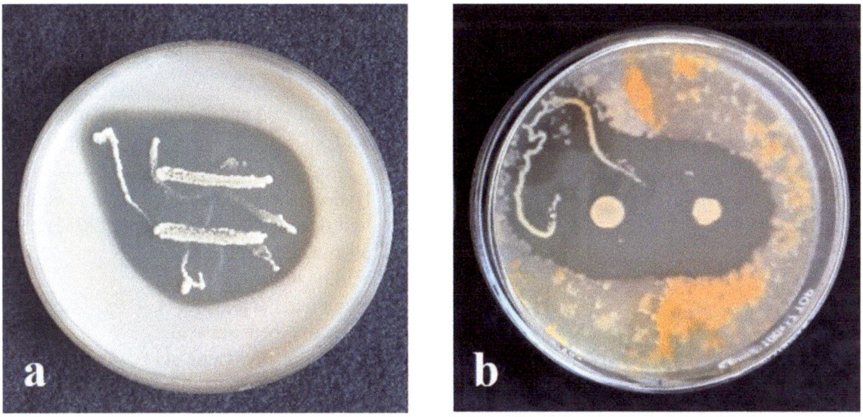

Fig. 13.2 MRS agar plates showing zone of inhibition of dermatophytic fungi in the presence of the Lactobacillus strains carried out by agar overlay procedure (**a**) *T. tonsurans* inhibited by *L plantarum* LpMYSN7, (**b**) *M. canis* inhibited by *L. fermentum* LfMYSY8

et al. 2022; Vasundaradevi et al. 2024; Vairagkar & Mirza 2021). Table 13.1 shows the selective probiotic isolates against specific dermatophyte cultures and the antifungal compounds detected in their CFS. It was also suggested that stabilised lyophilized extract of probiotics had additional antidermatophytic activity when compared to CFS (Mehdi-Alamdarloo et al. 2016), though this needs further confirmation. Another interesting observation was that when the identified antifungal compounds from CFS were purified and tested for efficacy, they seemed to be less effective when compared to the natural mixture in the supernatant (Danial et al. 2021). This affirms the preferable use of probiotic cultures and their crude CFS over the pure chemicals.

13.7 Probiotics Against *Candida* spp.

Cutaneous candidiasis is an opportunistic infection predominantly attributed to *Candida albicans*, although other species such as *C. tropicalis*, *C. parapsilosis*, *C. krusei*, *C. glabrata*, and *C. lusitaniae* have also been implicated with less frequency (Bilal et al. 2022). Several *Lactobacillus* species have been demonstrated to have antifungal activity against *Candida*, including *L. fermentum*, *L. plantarum*,

Fig. 13.1 (continued) and for the nutrients and growth factors that halt further growth and mycelia formation (Poornachandra Rao et al. 2021; Deepthi et al. 2016; Adithi et al. 2024). (**b**) Mechanism of probiotics action against fungal pathogens. Probiotics' cell wall components, such as peptidoglycan, lipoteichoic acid, teichoic acid, membrane-associated proteins, and cell wall-associated proteins or excreted components like CpG motifs as MAMPs (Microbe Associated Molecular Patterns), reduce skin inflammation and enhance the skin barrier function

Table 13.1 LAB strains and their antidermatophytic spectrum

Sl no	LAB isolate	Pathogen	Antifungal compounds detected in CFS	References
1.	*Enterococcus faecalis* TH10	*T. rubrum* *T. mentagrophytes*	Malonic acid and acetic acid	Kuwaki et al. (2002)
2.	*Lactobacillus reuteri* ee 1p	*M. gypseum*, *M. canis*, and *E. floccosum*	Hydroxyisocapric acid, hydrocinnamic acid, phenyllactic acid, azelaic acid, vanillic acid, p coumaric acid, 4-hydroxybenzoic acid, hydroxyphenyl lactic acid and 3-hydroxydecanoic acid	Guo et al. (2012)
3.	*Lactobacillus plantarum KCC 10*	*E. floccosum*, *T. mentagrophytes*, *T.roseum*	3-phenyllactic acid	Arasu et al. (2013)
4	*Enterococcus hirae* ATCC 9790 and *Enterococcus faecium* Aus0004	*M. canis*, *M. gypseum*, *M. nanum*, *T. mentagrophytes*, *A. niger*, *F. oxysporum*	Proteinceous substance (bacteriocins)	Allam et al. (2016)
5.	*L. plantarum*, *L. brevis*, *L. pentosus*	*T. rubrum*	Phenyl lactic acid	Maikhan and Researcher (2018)
6.	*L. plantarum* strain HT-W104-B1	*T. rubrum*	L-lactic acid and acetic acid. Minor compounds include Cyclo(L-Pro-L-Val), Tryptophol, Culmorin, Phenopyrrozin	Danial et al. (2021)
7.	*L. plantarum* *Lp*MYSN7	*T. tonsurans*	Mixture of 11 different acids. Succinic acid and lactic acid were predominant.	Vanitha et al. (2023)
8.	*L. casei*, *L. brevis*, *L. acidophilus*, *L. fermentum*, and *Pediococcus* spp.	*T. rubrum*, *C. albicans*, *C, tropicalis*, *C. krusie*, *E. rubrum*	Lactic acid, Diacetyl, and H_2O_2	Adeniyi et al. (2024)
9.	*L. fermentum* *Lf*MYSY8	*M. canis*	Mixture of organic acids: Lactic acid, maleic acid, malonic acid, fumaric acid, Tartaric acid, malic acid, succinic acid, shikimic acid, hydroxycitric acid, and citric acid	Vanitha et al. (2024)

L. rhamnosus, *L. reuteri*, *L. acidophilus*, *L. johnsonii*, *L. pentosus*, *L. paracasei*, *L. gasseri*, *L. crispatus*, and *L. vaginalis*. In addition to *Lactobacillus* species, several other probiotic microorganisms have been reported to show anticandidal activity. These include lactic acid bacteria such as *Pediococcus pentosaceus* and *Weissella confusa* (Bamidele et al. 2019), *Pediococcus acidilactici* (Bulgasem et al. 2016), and bifidobacteria, including *Bifidobacterium bifidum* (Denkova et al. 2013) and *Bifidobacterium breve*. Certain probiotic yeasts, namely *Saccharomyces boulardii* (Hager et al. 2019) and *S. cerevisiae* (Gabrielli et al. 2018), have also demonstrated inhibitory effects against *Candida*. The mechanisms underlying the anticandidal activity of selected probiotic isolates are summarized in Table 13.2.

13.7.1 Antibiofilm Activity

Biofilm formation in *Candida* species is a key virulence trait that contributes to their pathogenicity. Several studies have demonstrated that probiotic isolates and their culture filtrates possess a strong antibiofilm activity (Divyashree et al. 2022). In particular, the *Lactobacillus pentosus* LAP1 strain exhibited a marked inhibition of biofilm formation in *C. albicans*, *C. tropicalis*, and *C. krusei* (Aarti et al. 2018).

Table 13.2 Selective probiotic isolates and their mode of action against *Candida* spp.

S. no	*Candida* species	Probiotic used	Mode of action	Reference
1.	*C. albicans*	*Saccharomyces boulardii*	Reduction in inflammatory cytokines and colonization by yeast	Jawhara and Poulain (2007)
2.	*C. glabrata*	*L. reuteri RC-14*ATCC 55845 and *L. rhamnosus GR-1*ATCC 55826.	Strong coaggregation with the yeast, inhibition of colonization and filamentation	Chew et al. (2015)
3.	*C. albicans*	*L. acidophilus ATCC 4356*	Inhibition of filamentation and biofilm formation.	Vilela et al. (2015)
4.	*C. albicans*	*L. rhamnosus* GR-1, *L. reuteri RC*-14, *Lactobacillus johnsonii* PV016	Suppression of biofilm development	Kohler et al. (2012)
5.	*C. krusei*, *C. albicans*, *C. glabrata*, *C. tropicalis* *C. parapsilosis*	*L. plantarum*, *L. paracasei*, *L. rhamnosus*, *L. fermentum*, and *L. paracasei*	Production of antimicrobial compounds, i.e., Hydrogen peroxide	Verdenelli et al. (2014)
6.	*C. albicans*	*S. cerevisiae* CNCM I -3856	Effect on the expression of virulence characters of *C. albicans,* such as aspartyl proteinases	Gabrielli et al. (2018)

Mature biofilm formation by single culture and mixed culture of non-albicans *Candida,* including *C. tropicalis, C. krusie,* and *C. parapsilosis,* was disrupted by lactobacilli supernatants. These findings were validated by confocal laser scanning microscopy and scanning electron microscopy (Tan et al. 2018). In a similar study, Hager et al. (2019) demonstrated that culture filtrates derived from the probiotic strains *Saccharomyces boulardii, L. acidophilus, B. breve,* and *L. rhamnosus* effectively inhibited the polymicrobial biofilm formation involving *C. tropicalis* in association with *Escherichia coli* and *Serratia marcescens.*

13.7.2 In Vivo Studies

To gain deeper insight into the anticandidal effects of probiotic cultures, the human vaginal epithelial cell line VK2/E6E7 has been employed as an in vitro model. Using this system, *L. reuteri* RC-14, both alone and in combination with *L. rhamnosus* GR-1, was shown to suppress *Candida* growth. In addition to that, the corresponding cell-free supernatants were found to stimulate interleukin secretion by epithelial cells, suggesting a potential role in enhancing yeast clearance in vivo. The same research group later conducted a clinical trial in 2009, which demonstrated the effectiveness of these two strains as adjunctive therapy in the management of vulvovaginal candidiasis (Martinez et al. 2009).

A notable study published in 2017 demonstrated the protective role of *Lactobacillus rhamnosus* GG (LGG) against *Candida albicans*-induced damage in oral epithelial tissues (Mailänder-Sánchez et al. 2017). The authors employed the in vitro models comprising reconstructed human oral epithelial multilayers (RHOEs) and human keratinocyte monolayers (TR146). Preincubation with the LGG significantly reduced the epithelial damage produced by *C. albicans*, as reflected by the decreased lactate dehydrogenase (LDH) release and fewer histological lesions. Mechanistically, LGG was shown to deplete glucose at the infection site, thereby limiting the primary carbon source available to *C. albicans*, while also competitively occupying epithelial adhesion sites. The impact of LGG on fungal invasiveness was further assessed using immunofluorescence staining and microscopy, which revealed reduced fungal adhesion, invasion, and hyphal extension. Collectively, these findings not only highlight the protective effects of LGG but also suggest that probiotic-mediated nutrient depletion can induce metabolic reprogramming in *Candida*, thereby attenuating its virulence.

A study found that *L. paracasei* NCC2461 ST11 reduced the growth of *Malassezia* spp. the oral supplement of this improved the itching, redness of skin, and dandruff in a randomized double-blinded study. Anti-inflammatory cytokines (TGF-β and IL-10) were increased, and the proliferation and differentiation of keratinocytes increased with decreasing IFN-γ (Reygagne et al. 2017). Another study in the form of oily suspension of *L. crispatus* P17631 and *L. paracasei* 11688 reduced the severity of symptoms of erythema, pruritus, and desquamation in SD patients and also the surface adhesion and biofilm formation on the forehead and nasolabial region (Truglio et al. 2024). In the form of a topical lotion, the

Vitreoscilla filiformis lysate, when applied for 4 weeks to the affected area of skin, decreased the redness, itching, and scaling in SD (Guéniche et al. 2008).

Although there is a lot of literature on antagonistic activity of probiotic cultures on pathogenic moulds like *Rhizopus*, *Mucor*, *Fusarium*, *Aspergillus*, etc., they have primarily concentrated on plant pathogens and food spoilage causing species. Hence, though the antifungal potential of these probiotics has been well established, studies need to be carried to confirm their effect against human pathogens.

13.8 Probiotic Formulations for Skin Diseases

Formulations containing different probiotic species or their cell free extracts have been developed for the control of skin infections. One such formulation was developed by Mehdi-Alamdarloo et al. (2016), in which a topical gel incorporating the cell-free extract of *Lactobacillus casei* was tested against four species, such as *T. rubrum*, *T. verrucosum*, *M. canis*, and *M. gypseum*. Among the tested formulations, the gel containing 5% probiotic extract showed optimal antidermatophytic activity. Many such antimicrobial formulations have been developed, and patents have been obtained (Table 13.3). Commercially available skin protection products like topical facial serum, face cream, facial moisturiser, facial pad, face mask, toner, and body lotion have been prepared with either live cells or fermented lysate of probiotic cultures (Habeebuddin et al. 2022).

13.9 Prospects and Challenges

One of the common environments where immunocompromised individuals acquire skin infection is hospitals, where the clothes worn by patients, aprons of doctors and nurses, and other hospital staff can be contaminated with microbial pathogens. Use of topical probiotics in such clothing will be an innovative approach in controlling skin infections. As mentioned earlier, studies need to be done on the antifungal potential of probiotics against nondermatophytic moulds, which cause skin infection.

Probiotic skincare formulations are a recent development in dermatology with limited clinical evidence on their efficacy and safety. The clear distinctions between foods, drugs, and cosmetics also complicate developing and reporting probiotic cosmetic products. Few clinical trials explore probiotics for skin infections caused by *Candida* spp., though some reports suggest they help maintain skin balance and immune regulation (Guéniche et al. 2008).

Currently, there are no established regulations governing topical probiotics, and no topical probiotic products have been approved by the FDA, highlighting the need for improved regulation in this area.

Table 13.3 Probiotic formulations and the patents obtained

S. no	Year	Patent number	Product type	Probiotics involved	Activity against
1.	2009	EPO975227A1	Topical cream	*Bacillus* sp. spores or an extracellular product of *Bacillus coagulans*	Yeast, fungus, bacteria and Herpes simplex virus
2.	2019	US10238597 B2	Topical or oral solution	*Lactobacillus jensenii, Lactobacillus vaginalis*	Skin disorders of scalp (Dandruff/ seborrhoeic dermatitis)
3.	2019	US2019 / 0151382 A1	Oral ointment or chewing gum	*L. paracasei*	*C. albicans* infection
4.	2020	WO 2020/056359	Topical cream	*S. capitis* *S. epidermidis*	For topical management of scalp and skin dysbiosis–associated disorders.
5.	2018	BR112019017962A2	Topical cream	*S. epidermidis MO34*	For the treatment of skin infections, cancers, and other neoplastic conditions

13.10 Conclusion

A steady increase in the incidence of cutaneous mycoses has been observed. These infections are becoming difficult to treat because of drug toxicity and drug resistance. Certain lactic acid bacteria and yeast strains have demonstrated the ability to effectively suppress the growth of these fungal pathogens under both in vitro and in vivo conditions, indicating their potential application as oral and topical probiotics. Although many commercial formulations are available, further research is necessary to widen their application to related fields and to improve regulations in this area.

13.11 Summary

The chapter summarises that cutaneous mycoses are superficial skin infections caused by dermatophytes and non-dermatophytes. Their prevalence is high in tropical regions and has long-term treatment durations. Along with that, the side effects of drug toxicity, frequent relapse, and emergence of antifungal resistance limit the effectiveness of current therapies. Probiotic lactic acid bacteria and beneficial yeast have demonstrated antifungal, antibiofilm, antiadhesive, and immunomodulatory activities against these cutaneous fungal pathogens. The evidence from *in vitro* and *in vivo* studies suggests the use of probiotic cells, or CFS, for tropical formulations as well as for oral applications. However, the probiotic effects are strain-specific, and challenges remain for standardization and robust clinical validation.

Conflict of Interest The authors declare no conflicting interests.

Author Contribution VPR and MYS—conceptualization and idea of the contents, VPR and DJ—writing original draft, diagram, review and editing, MP and SGR -review and editing.

References

Aarti C, Khusro A, Varghese R, Arasu MV, Agastian P, Al-Dhabi NA et al (2018) *In vitro* investigation on probiotic, anti-*Candida*, and antibiofilm properties of *Lactobacillus pentosus* strain LAP1. Arch Oral Biol 89:99–106. https://doi.org/10.1016/j.archoralbio.2018.02.014

Ademe M, Girma F (2020) *Candida auris*: from multidrug resistance to pan-resistant strains. Infect Drug Resist 13:1287–1294. https://doi.org/10.2147/IDR.S249864

Adeniyi B, Adesuyi A, Ayeni F, Ogunbanwo T, Agidigbi T (2024) Poultry gastrointestinal-derived lactic acid bacteria (pGIT-d-LAB) inhibit multiple antibiotics resistance bacterial and fungal pathogens. Avicenna J Med Biotechnol 16(2):111–119. https://doi.org/10.18502/ajmb.v16i2.14862

Adithi G, Rakesh S, Divyashree S, Shruthi B, Sreenivasa MY (2022) Assessment of probiotic and antifungal activity of *Lactiplantibacillus plantarum* MYSAGT3 against mycotoxigenic *Aspergillus* species. Food Biosci:102118. https://doi.org/10.1016/j.fbio.2022.102118

Adithi G, Divyashree S, Shruthi B, Deepa N, Sreenivasa MY (2024) Evaluation of *Limosilactobacillus fermentum* MYSAGAM1 isolated from herbal amla juice as a probiotic candidate with antifungal characteristics against *Fusarium equiseti*. Food Biosci:103843. https://doi.org/10.1016/j.fbio.2024.103843

Adithi G, Rakesh S, Shruthi B, Divyashree S, Deepa N, Vanitha PR, Vasundaradevi R, Deepthi BV, Sreenivasa MY (2025) Biological decontamination of myco-toxigenic *Fusarium verticillioides* using *Lactiplantibacillus plantarum* MYSN128 isolated from traditional vellappam: implications for food safety and global mycotoxin management. Curr Microbiol 82:404. https://doi.org/10.1007/s00284-025-04401-z

Al-Ghazzewi FH, Tester RF (2014) Impact of prebiotics and probiotics on skin health. Benef Microbes 5:99–107

Al-Khikani F (2020) Dermatophytosis a worldwide contiguous fungal infection: growing challenge and few solutions. Biomed Biotechnol Res J 4(2):117–122. https://doi.org/10.4103/bbrj.bbrj_1_20

Allam NG, Abdel-mongy M, El-nahrawy NS, Mohammed A (2016) The potential role of lactic acid bacteriocins as an antifungal agent. J Vit Enol 31:66–105

Arasu MV, Jung MW, Ilavenil S, Jane M, Kim DH, Lee KD, Park HS, Hur TY, Choi GJ, Lim YC, Al-Dhabi NA, Choi KC (2013) Isolation and characterization of antifungal compound from *Lactobacillus plantarum* KCC-10 from forage silage with potential beneficial properties. J App Microbiol 115(5):1172–1185. https://doi.org/10.1111/jam.12319

Baginski M, Czub J (2009) Amphotericin B and its new derivatives—mode of action. Curr Drug Metab 10(5):459–469. https://doi.org/10.2174/138920009788898019

Bamidele TA, Adeniyi BA, Smith SI (2019) *In vitro*, acidic, non-proteinaceous antifungal activities of lactic acid bacteria isolated from salad vegetables against human pathogenic *Candida albicans*. Afr J Clin Exp Microbiol 20(2):137. https://doi.org/10.4314/ajcem.v20i2.7

Bilal H, Hou B, Shafiq M, Chen X, Shahid MA, Zeng Y (2022) Antifungal susceptibility pattern of *Candida* isolated from cutaneous candidiasis patients in eastern Guangdong region: a retrospective study of the past 10 years. Front Microbiol 13:981181. https://doi.org/10.3389/fmicb.2022.981181

Bulgasem BY, Lani MN, Hassan Z, Wan Yusoff WM, Fnaish SG (2016) Antifungal activity of lactic acid bacteria strains isolated from natural honey against Pathogenic *Candida* species. Mycobiology 44:302–309. https://doi.org/10.5941/MYCO.2016.44.4.302

Cafarchia C, Figueredo LA, Favuzzi V, Surico MR, Colao V, Iatta R, Montagna MT, Otranto D (2012) Assessment of the antifungal susceptibility of *Malassezia pachydermatis* in various media using a CLSI protocol. Vet Microbiol 159(3–4):536–540. https://doi.org/10.1016/j.vetmic.2012.04.034

Chanyachailert P, Leeyaphan C, Bunyaratavej S (2023) Cutaneous fungal infections caused by dermatophytes and non-dermatophytes: an updated comprehensive review of epidemiology, clinical presentations, and diagnostic testing. J Fungi 9:669. https://doi.org/10.3390/jof9060669

Chen J, Song X, Yang P, Wang J (2015) Appearance of anaphylactic shock after long-term intravenous Itraconazole treatment. Ann Pharmocother 43:537–541. https://doi.org/10.1345/aph.1L343

Chew SY, Cheah YK, Seow HF, Sandai D, Than LYL (2015) Probiotic *Lactobacillus rhamnosus* GR-1 and *Lactobacillus reuteri* RC-14 exhibit strong antifungal effects against vulvovaginal candidiasis-causing *Candida glabrata* isolates. J Appl Microbiol 118:1180–1190. https://doi.org/10.1111/jam.12772

Danial AM, Medina A, Magan N (2021) *Lactobacillus plantarum* strain HT-W104-B1: potential bacterium isolated from Malaysian fermented foods for control of the dermatophyte *Trichophyton rubrum*. World J Microbiol Biotechnol 37:1–11. https://doi.org/10.1007/s11274-021-03020-7

Dass RS, Sreenivasa MY, Janardhana GR (2007) High incidence of *Fusarium verticillioides* in animal and poultry feed mixtures produced in Karnataka, India. Plant Pathol J 6(2):174–178

de Hoog GS, Dukik K, Monod M, Packeu A, Stubbe D, Hendrickx M, Kupsch C, Stielow JB, Freeke J, Göker M et al (2017) Toward a novel multilocus phylogenetic taxonomy for the dermatophytes. Mycopathologia 182:5–31

Deepa N, Charith Raj AP, Sreenivasa MY (2016) Nested PCR method for the early detection of fumonisin-producing *Fusarium verticillioides* in pure cultures, cereal samples and plant parts. Food Biotechnol 30(1):18–29

Deepthi BV, Rao KP, Chennappa G, Naik MK, Chandrashekara KT, Sreenivasa MY (2016) Antifungal attributes of *Lactobacillus plantarum* MYSG against fumonisin-producing *Fusarium proliferatum* associated with poultry feeds. PLoS One 11(6):e0155122. https://doi.org/10.1371/journal.pone.0155122

Denkova R, Yanakieva V, Denkova Z, Nikolova V, Radeva V (2013) *In vitro* inhibitory activity of *Bifidobacterium* and *Lactobacillus* strains against *Candida albicans*. Bulg J Vet Med 16:186–197

Dessinioti C, Katsambas A (2013) Seborrheic dermatitis: etiology, risk factors, and treatments: facts and controversies. Clin in Derm 31(4):343–351. https://doi.org/10.1016/j.clindermatol.2013.01.001

Divyashree S, Ramith R, Sreenivasa MY (2023) Evaluation of new candidate probiotic *Lactobacillus* strains isolated from a traditional fermented multigrain-millet dosa batter. Food Biosci:103450. https://doi.org/10.1016/j.fbio.2023.103450

Divyashree S, Anjali PG, Deepthi BV, Rakesh S, Walid M, Hammami R, Sreenivasa MY (2022) Black cherry fruit as a source of probiotic candidates with antimicrobial and antibiofilm activities against *Salmonella*. S Afr J Bot 150:8G1–872. https://doi.org/10.1016/j.sajb.2022.08.045

Gabrielli E, Pericolini E, Ballet N, Roselletti E, Sabbatini S, Mosci P et al (2018) *Saccharomyces cerevisiae*-based probiotic as novel anti-fungal and anti-inflammatory agent for therapy of vaginal candidiasis. Benef Microbes 9:219–230. https://doi.org/10.3920/BM2017.0099

Guéniche A, Cathelineau AC, Bastien P, Esdaile J, Martin R, Queille Roussel C, Breton L (2008) *Vitreoscilla filiformis* biomass improves seborrheic dermatitis. J Eur Acad Dermatol Venereol 22(8):1014–1015. https://doi.org/10.1111/j.1468-3083.2007.02508.x

Guo J, Mauch A, Galle S, Murphy P, Arendt EK, Coffey A (2011) Inhibition of growth of *Trichophyton tonsurans* by *Lactobacillus reuteri*. J Appl Microbiol 111(2):474–483. https://doi.org/10.1111/j.1365-2672.2011.05032.x

Guo J, Brosnan B, Furey A, Arendt EK, Murphy P, Coffey A (2012) Antifungal activity of *Lactobacillus* against *Microsporum canis*, *Microsporum gypseum* and *Epidermophyton floccosum*. Bioeng Bugs 3:104–113. https://doi.org/10.4161/bbug.19624

Habeebuddin M, Karnati RK, Shiroorkar PN, Nagaraja S, Asdaq SMB, Khalid Anwer M, Fattepur S (2022) Topical probiotics: more than a skin deep. Pharmaceutics 14:557. https://doi.org/10.3390/pharmaceutics14030557

Hager CL, Isham N, Schrom KP, Chandra J, McCormick T, Miyagi M, Ghannoum MA (2019) Effects of a novel probiotic combination on pathogenic bacterial-fungal polymicrobial biofilms. MBio 10:e00338–e00319. https://doi.org/10.1128/mBio.00338-19

Hamdino M, Saudy AA, El-Shahed LH, Taha M (2022) Identification of *Malassezia* species isolated from some *Malassezia* associated skin diseases. J Med Mycol 32(4). https://doi.org/10.1016/j.mycmed.2022.101301

Jackson JM, Alexis A, Zirwas M, Taylor S (2024) Unmet needs for patients with seborrheic dermatitis. J Am Acad Dermatol 90(3):597–604. https://doi.org/10.1016/j.jaad.2022.12.017

Jawhara S, Poulain D (2007) *Saccharomyces boulardii* decreases inflammation and intestinal colonization by *Candida albicans* in a mouse model of chemically-induced colitis. Med Mycol 45:691–700

Karray M, McKinney WP (2023) Tinea versicolor. In: StatPearls. StatPearls Publishing. https://www.ncbi.nlm.nih.gov/books/NBK459156/

Kaur J, Nobile CJ (2023) Antifungal drug-resistance mechanisms in *Candida* biofilms. Curr Opin Microbiol 71:102237. https://doi.org/10.1016/j.mib.2022.102237

Kohler GA, Assefa S, Reid G (2012) Probiotic interference of *Lactobacillus rhamnosus* GR-1 and *Lactobacillus reuteri* RC-14 with the opportunistic fungal pathogen *Candida albicans*. Infect Dis Obstetr Gynecol. https://doi.org/10.1155/2012/636474

Kruithoff C, Gamal A, McCormick TS, Ghannoum MA (2024) Dermatophyte infections worldwide: increase in incidence and associated antifungal resistance. Life 14(1):1–13. https://doi.org/10.3390/life14010001

Kuwaki S, Ohhira I, Takahata M, Murata Y, Tada M (2002) Antifungal activity of the fermentation product of herbs by lactic acid bacteria against tinea. J Biosci Bioeng 94(5):401–405. https://doi.org/10.1016/S1389-1723(02)80216-X

Leung AKC, Barankin B, Lam JM, Leong KF, Hon KL (2022) Tinea versicolor: an updated review. Drugs Context 11:2022–9–2. https://doi.org/10.7573/dic.2022-9-2

Maikhan HK, Researcher I (2018) Study the effect of *Lactobacillus* spp. on the growth of *Trichophyton rubrum*. Online Int Interdiscip Res J VI:24–36

Mailänder-Sánchez D, Braunsdorf C, Grumaz C, Müller C, Lorenz S, Stevens P, Wagener J et al (2017) Antifungal defense of Probiotic *Lactobacillus rhamnosus* GG is mediated by blocking adhesion and nutrient depletion. PLoS One 12(10):1–19. https://doi.org/10.1371/journal.pone.0184438

Martinez RCR, Seney SL, Summers KL, Nomizo A, Martinis ECP De, Reid G (2009) Effect of *Lactobacillus rhamnosus* GR-1 and *Lactobacillus reuteri* RC-14 on the ability of *Candida albicans* to infect cells and induce inflammation. Microbiol Immunol (11):487–495. https://doi.org/10.1111/j.1348-0421.2009.00154.x

Mehdi-Alamdarloo S, Ameri A, Moghimipour E, Gholipour S, Saadatzadeh A (2016) Formulation development of a topical probiotic gel for antidermatophytosis effect. Jundishapur J Nat Pharm Prod 11. https://doi.org/10.17795/jjnpp-35893

Müller VL, Kappa-Markovi K, Hyun J, Georgas D, Silberfarb G, Paasch U et al (2021) Tinea capitis et barbae caused by *Trichophyton tonsurans*: a retrospective cohort study of an infection chain after shavings in barber shops. Mycoses 64:428–436. https://doi.org/10.1111/myc.13231

Nakatsuji T, Gallo RL (2019) The role of the skin microbiome in atopic dermatitis. Ann Allergy Asthma Immunol 122:263–269

Pathania S, Rudramurthy SM, Narang T, Saikia UN, Dogra S (2018) A prospective study of the epidemiological and clinical patterns of recurrent dermatophytosis at a tertiary care hospital. Indian J Dermatol Venereol Leprol 84(6):7

Pei Y, Zhang L, Liu ZH (2022) Tinea Faciei. J Pediatr 250:108–109

Petrucelli MF, de Abreu MH, Cantelli BAM, Segura G, Nishimura FG, Bitencourt TA et al (2020) Epidemiology and diagnostic perspectives of dermatophytoses. J Fungi 6:1–15. https://doi.org/10.3390/jof6040310

Phaitoonwattanakij S, Leeyaphan C, Lertrujiwanit K, Bunyaratavej S (2021) Predisposing factors, clinical features and treatment outcomes of *Fusarium* onychomycosis and comparison of its characteristics with *Neoscytalidium* onychomycosis. J Mycol Med 31:101165

Pilz JF, Köberle M, Kain A et al (2023) Increasing incidence of *Trichophyton tonsurans* in Munich—a single-centre observation. Mycoses 66:441–447. https://doi.org/10.1111/myc.13563

Poornachandra Rao K, Deepthi BV, Rakesh S, Ganesh T, Achar P, Sreenivasa MY (2017) Anti-aflatoxigenic potential of cell-free supernatant from *Lactobacillus plantarum* MYS44 against *Aspergillus parasiticus*. Probiotics Antimicrob Proteins. https://doi.org/10.1007/s12602-017-9338-y

Poornachandra Rao K, Hemanth Kumar NK, Rakesh S, Murali S, Jagannath S, Sreenivasa MY (2021) Probiotic attributes and inhibitory effects of *Lactobacillus plantarum* MYS84 against the growth and biofilm formation of *Pseudomonas aeruginosa*. Microbiology 90:361–369. https://doi.org/10.1134/S0026261721030103

Rakesh S, Shruthi S, Deepthi BV, Sreenivasa MY (2019) Probiotic properties of lactic acid bacteria isolated from neera, a naturally fermenting coconut palm nectar. Front Microbiol 10:1382. https://doi.org/10.3389/fmicb.2019.01382

Ranawaka RR, Nagahawatte A, Gunasekara TA, Weerakoon HS, de Silva SH (2016) Randomized, double-blind, comparative study on efficacy and safety of itraconazole pulse therapy and terbinafine pulse therapy on nondermatophyte mold onychomycosis: a study with 90 patients. J Dermatolog Treat 27:364–372

Rasoulpoor S, Shohaimi S, Salari N et al (2021) *Candida albicans* skin infection in patients with type 2 diabetes: a systematic review and meta-analysis. J Diabetes Metab Disord 20:665–672. https://doi.org/10.1007/s40200-021-00797-0

Reygagne P, Bastien P, Couavoux MP, Philippe D, Renouf M, Castiel-Higounenc I, Gueniche A (2017) The positive benefit of *Lactobacillus paracasei* NCC2461 ST11 in healthy volunteers with moderate to severe dandruff. Benefic Microbes 8(5):671–680. https://doi.org/10.3920/BM2016.0144

Rodríguez-Cerdeira C, Martínez-Herrera E, Szepietowski JC, Pinto-Almazán R, Frías-De-León MG, Espinosa-Hernández VM, Chávez-Gutiérrez E, García-Salazar E, Vega-Sánchez DC, Arenas R, Hay R, Saunte DM (2021) A systematic review of worldwide data on tinea capitis: analysis of the last 20 years. J Eur Acad Dermatol Venereol 35(4):844–883. https://doi.org/10.1111/jdv.16951

Sacheli R, Cuypers L, Seidel L, Darfouf R, Adjetey C, Lagrou K, Hayette MP (2021) Epidemiology of dermatophytes in Belgium: a 5-year' survey. Mycopathology 186:399–409

Salehi Z, Shams-Ghahfarokhi M, Razzaghi-Abyaneh M (2018) Antifungal drug susceptibility profile of clinically important dermatophytes and determination of point mutations in terbinafine-resistant isolates. Eur J Clin Microbiol Infect diseases:12–15

Sasikumar J, Ebrahim RA, Das SP (2025) Diverse colonisation and disease associations of the human commensal *Malassezia*: our body's secret tenant. Mycoses 68:e70014. https://doi.org/10.1111/myc.70014

Saunte DML, Hare RK, Jørgensen KM, Jørgensen R, Deleuran M, Zachariae CO, Thomsen SF, Bjørnskov-Halkier L, Kofoed K, Arendrup MC (2019) Emerging terbinafine resistance in *Trichophyton*: clinical characteristics, squalene epoxidase gene mutations, and a reliable EUCAST method for detection. Antimicrob Agents Chemother 63:e01126-19

Saunte DML, Gaitanis G, Hay RJ (2020) *Malassezia*-associated skin diseases, the use of diagnostics and treatment. Front Cell Infect Microbiol 10:112. https://doi.org/10.3389/fcimb.2020.00112

Saxena R, Mittal P, Clavaud C, Dhakan DB, Hegde P et al (2018) Comparison of healthy and dandruff scalp microbiome reveals the role of commensals in scalp health. Front Cell Infect Microbiol 8:346. https://doi.org/10.3389/fcimb.2018.00346

Shruthi B, Deepthi BV, Adithi G, Vanitha PR, Sreenivasa MY (2025) *Saccharomyces cerevisiae* MYSY1 as a promising probiotic isolate with antifungal properties against foodborne *Aspergillus* species. Biocatal Agric Biotechnol 67:103749. https://doi.org/10.1016/j.bcab.2025.103649

Singh A, Masih A, Monroy-nieto J, Kumar P, Bowers J, Travis J et al (2019) A unique multidrug-resistant clonal *Trichophyton* population distinct from *Trichophyton mentagrophytes* / *Trichophyton interdigitale* complex causing an ongoing alarming dermatophytosis outbreak in India: genomic insights and resistance profile. Fungal Genet Biol 133:103266. https://doi.org/10.1016/j.fgb.2019.103266

Singh AD, Debnath C, Banerjee A (2021) Epidemiological investigation, characterization and antifungal susceptibility profile of *Microsporum canis* isolated from pet animals. Vet Arh 91(4):339–347. https://doi.org/10.24099/vet.arhiv.0953

Sinha S, Sardana K (2018) Antifungal efficacy of amphotericin B against dermatophytes and its relevance in recalcitrant dermatophytoses: a commentary. Ind Dermatol Online J 9:120–122

Szántó M, Dózsa A, Antal D, Szabó K, Kemény L, Bai P (2019) Targeting the gut-skin axis-Probiotics as new tools for skin disorder management. Exp Dermatol 28:1210–1218

Tan Y, Leonhard M, Moser D, Ma S, Schneider-Stickler B (2018) Inhibitory effect of probiotic lactobacilli supernatants on single and mixed non-albicans *Candida* species biofilm. Arch Oral Biol 85:40–45. https://doi.org/10.1016/j.archoralbio.2017.10.002

Truglio M, Sivori F, Cavallo I, Abril E, Licursi V, Fabrizio G, Cardinali G, Pignatti M, Toma L, Valensise F, Cristaudo A, Pimpinelli F, Di Domenico EG (2024) Modulating the skin mycobiome-bacteriome and treating seborrheic dermatitis with a probiotic-enriched oily suspension. Sci Reports 14(1):2722. https://doi.org/10.1038/s41598-024-53016-0

Vairagkar U, Mirza Y (2021) Antagonistic activity of antimicrobial metabolites produced from seaweed-associated *Bacillus amyloliquefaciens* MTCC 10456 against *Malassezia* spp. Probiot Antimicrob Proteins 13:1228–1237. https://doi.org/10.1007/s12602-021-09742-2

Vanitha PR, Somashekaraiah R, Divyashree S, Pan I, Sreenivasa MY (2023) Antifungal activity of probiotic strain *Lactiplantibacillus plantarum* MYSN7 against *Trichophyton tonsurans*. Front Microbiol 14. https://doi.org/10.3389/fmicb.2023.1192449

Vanitha PR, Divyashree S, Shruthi B, Adithi G, Deepthi BV, Sreenivasa MY (2024) *Limosilactobacillus fermentum* MYSY8, a potential probiotic isolate from fermented rice beverage for the control of *Microsporum canis*. Food Biotechnol 38(3):314–342. https://doi.org/10.1080/08U0543G.2024.2383743

Vasundaradevi R, Sarvajith M, Somashekaraiah R, Adithi G, Sreenivasa MY (2024) Antagonistic properties of *Lactiplantibacillus plantarum* MYSVB1 against *Alternaria alternata*. Front Microbiol 15:1322758. https://doi.org/10.3389/fmicb.2024.1322758

Vasundaradevi R, Sarvajith M, Divyashree S, Deepa N, Achar PN, Sreenivasa MY (2025) Tropical fruit-derived *Lactiplantibacillus* as potential probiotic and antifungal agents against *Fusarium oxysporum*. Sci Rep 15:2144. https://doi.org/10.1038/s415U8-025-851U0-0

Venkatakrishnan K, Von Moltke LL, Greenblatt DJ (2000) Effects of the antifungal agents on oxidative drug metabolism. Clinical relevance. Clin Pharmacokinet 38(2):111–180. https://doi.org/10.2165/00003088-200038020-00002

Verdenelli MC, Coman M, Cecchini C, Silvi S, Orpianesi C, Cresci A (2014) Evaluation of antipathogenic activity and adherence properties of human *Lactobacillus* strains for vaginal formulations. J Appl Microbiol 116:1297–1307. https://doi.org/10.1111/jam.12459.42

Verma SB, Panda S, Nenoff P, Singal A, Rudramurthy SM, Uhrlass S et al (2021) The unprecedented epidemic-like scenario of dermatophytosis in India: I. Epidemiology, risk factors and clinical features. Indian J Dermatol Venereol Leprol 87:154–175

Vilela SFG, Barbosa JO, Rossoni RD, Santos JD, Prata MCA, Anbinder AL et al (2015) *Lactobacillus acidophilus* ATCC 4356 inhibits biofilm formation by *C. albicans* and attenuates the experimental candidiasis in *Galleria mellonella*. Virulence 6:29–39. https://doi.org/10.4161/21505594.2014.981486

Warner R, Schwartz JR, Boissy Y, Dawson TL Jr (2001) Dandruff has an altered stratum corneum ultrastructure that is improved with zinc pyrithione shampoo. J Am Acad Dermatol 45(6): 897–903. https://doi.org/10.1067/mjd.2001.117849

Wu Y, Hu S, Wu C, Gu F, Yang Y (2022) Probiotics: potential novel therapeutics against fungal infections. Front Cell Infect Microbiol 11:793419. https://doi.org/10.3389/fcimb.2021.793419

Zhan P, Liu W (2017) The changing face of dermatophytic infections worldwide. Mycopathologia 182(1–2):77–86. https://doi.org/10.1007/s11046-016-0082-8

Probiotic Efficiency in Fighting Subcutaneous Mycoses

14

Richard Kolade Omole, Timothy Bamgbose, Oluwafemi Bamidele Daramola, Modupe Stella Ayilara, Juliet Oluwayemisi Richard-Omole, and Bolanle Adenike Akinsanola

Abstract

Subcutaneous mycoses are chronic fungal infections that primarily affect the skin and subcutaneous tissues and may extend to deeper structures such as muscles and bones. These infections, prevalent in tropical and subtropical regions, pose significant clinical challenges, especially among immunocompromised individuals. Traditional antifungal treatments often encounter limitations, including toxicity, high cost, resistance development, and limited accessibility in resource-poor regions. Recent research highlights the potential of probiotics (live microorganisms with established health benefits) as alternative or adjunctive therapies for subcutaneous mycoses. Probiotics such as *Lactobacillus*, *Bifidobacterium*, and *Saccharomyces* species exhibit antifungal activity through mechanisms like biofilm inhibition, immune modulation, competitive exclusion, and secretion of antifungal metabolites, including organic acids and chitinases.

R. K. Omole · O. B. Daramola
Department of Microbiology, Osun State University, Osogbo, Nigeria

T. Bamgbose (✉)
Department of Biological Sciences, Microbiology Unit, Kings University, Odeomu, Nigeria

National Agency for Food and Drug Administration and Control (NAFDAC), Isolo, Nigeria
e-mail: timothy.bamgbose@nafdac.gov.ng; tm.bamgbose@kingsuniversity.edu.ng

M. S. Ayilara
Food Security and Safety Focus Area, Faculty of Natural and Agricultural Sciences, North-West University, Mmabatho, South Africa

J. O. Richard-Omole
Division of Astrobiology and Space Medicine, Centre for Space Life Sciences, National Space Research and Development Agency, University of Ibadan, Ibadan, Nigeria

B. A. Akinsanola
Department of Food Science and Microbiology, Landmark University, Omu-Aran, Nigeria

M. Y. Sreenivasa et al. (eds.), *Probiotics as Antifungals*,
https://doi.org/10.1007/978-981-92-0002-3_14

This chapter explores the pathophysiology, causative agents, and complications of subcutaneous mycoses while presenting a detailed discussion on how probiotics may contribute to their management. Specific case studies demonstrate how probiotics can modulate the immune system and fight fungal infections, suggesting new solutions to current treatment challenges. Furthermore, the chapter emphasises the need for innovative probiotic delivery systems and future human clinical trials to validate their efficacy in subcutaneous fungal infections.

Graphical Abstract

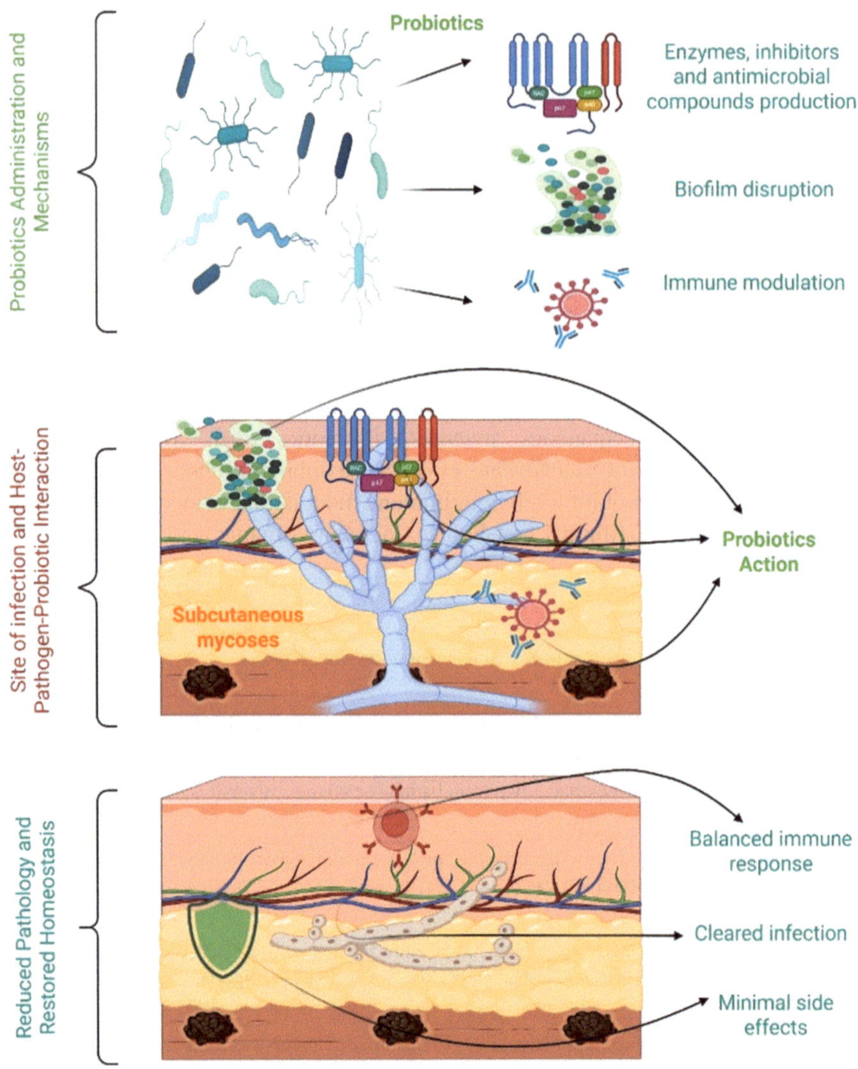

Keywords

Probiotics · Subcutaneous mycoses · Antifungal resistance · Immune modulation · *Lactobacillus* spp. · Alternative therapy

Learning Objectives

- Understand the epidemiology and pathogenesis of subcutaneous mycoses and their clinical implications.
- Identify common fungal pathogens responsible for subcutaneous mycoses and their modes of transmission and progression.
- Assess the limitations of current antifungal treatments, specifically addressing drug resistance and toxicity.
- Analyse the mechanisms of action of probiotics in the inhibition and modulation of fungal pathogens, and explore the therapeutic potential and prospects of probiotics as adjunctive or alternative treatments for subcutaneous fungal infections.

14.1 Introduction

Mycoses are infections caused by pathogenic fungi. These pathogenic fungi create infections that affect the human body's surfaces as well as deep tissues and could involve systemic organs (Thambugala et al. 2024). Medical professionals divide mycoses into three classification groups based on infection site, how hosts acquire pathogens, and fungal virulence traits (Queiroz-Telles and Santos 2023). Based on their infection site, mycoses are divided into superficial, cutaneous, subcutaneous, and deep/systemic. They are divided into exogenous and endogenous based on how hosts acquire pathogens. However, mycoses are grouped into primary and opportunistic depending on the fungal virulence traits (Queiroz-Telles and Santos 2023; Thambugala et al. 2024).

Medical settings now face broadened groups of individuals who become vulnerable to invasive fungal infections because treatment protocols have exposed them to such risks (Fang et al. 2023). This includes people whose immune systems have been compromised as a result of acquired immunodeficiency syndrome (AIDS), cancer therapies, or organ transplant operations, as well as those healing from extensive surgeries. Patients belonging to these groups encounter higher risks of developing such infections (Pisote et al. 2022). The increase in vulnerable patient numbers leads to more diverse opportunistic fungal pathogens that affect these groups. Medical experts usually face diagnostic obstacles with high drug resistance patterns during the early stages of invasive mycoses (Fang et al. 2023). Consequently, advancing novel strategies for diagnosing and treating these infections has become a critical

Table 14.1 Types of probiotics and their mechanisms of action

Probiotics	Mechanisms of action	References
Lactobacillus spp. and related genera. These include *L. casei, L. rhamnosus, L. acidophilus, L. johnsonii,* and *Limosilactobacillus fermentum.*	They act by lessening the pH in the intestine and preventing pathogens from thriving. They modulate the immune system and upregulate tight junction proteins	Bamgbose et al. (2024), Vazquez-Munoz et al. (2024)
Bifidobacterium spp. These include *B. longum, B. bifidum,* and *B. infantis.*	They compete for nutrients with pathogens. They release enzymes that could help to modulate the host's immune system	Latif et al. (2023), Gavzy et al. (2023)
Bacillus spp. These include *B. subtilis* and *B. coagulans*	They form spores that grow to produce antimicrobials and enzymes, which prevent pathogen proliferation	Cao et al. (2020), Zhu et al. (2023)
Streptococcus thermophilus	It works by assisting lactose-intolerant individuals in hydrolyzing lactose by synthesizing lactase	Martinović et al. (2020)
Saccharomyces boulardii	It works through anti-inflammatory effects and by neutralizing toxins produced by pathogens	Pais et al. (2020)

focus of ongoing medical research. Among existing treatment options for invasive mycoses, probiotics have shown efficiency against some, with sufficient evidence as a potential treatment for others (Divyashree et al. 2023).

Probiotics have been defined by the International Scientific Association for Probiotics and Prebiotics as living microorganisms that confer health benefits to their host when administered in suitably adequate amounts (Hill et al. 2014). Probiotics are generally regarded as safe (GRAS); hence, they are non-pathogenic/ non-toxic, acid-, and bile-stable, viable during storage/application, and possess numerous nutritional, physiological, and health benefits (Bamgbose et al. 2024; Lau and Quek 2024). Research has demonstrated that probiotics increase digestion, stimulate the production of essential nutrients (such as vitamins), and fortify the immune system against diverse forms of infections and diseases (Bamgbose and De La Fuente 2023; Lau and Quek 2024). Ideally, probiotics should not carry antibiotic resistance genes that can spread to other microbes (Das et al. 2020). Other vital features of probiotics include their ability to adhere to the intestinal mucosa, thereby preventing the colonization of pathogens and contributing to the metabolic and immunologic functions of their host (Bamgbose et al. 2021; Petrariu et al. 2024). Sometimes, probiotics are strain-specific in their actions (Bamgbose et al. 2022; Tegegne and Kebede 2022). As presented in Table 14.1, probiotics include *Lactobacillus* spp. and closely related genera, *Bifidobacterium* spp., *Bacillus* spp., *Streptococcus thermophilus*, and *Saccharomyces boulardii,* among others. In recent years, researchers have explored antifungal therapy innovations in combination with probiotic-based approaches to develop better subcutaneous mycosis treatment strategies, which have proved effective.

14.2 Subcutaneous Mycoses

Subcutaneous mycoses are fungal infections that affect human skin and subcutaneous tissue, and in rare cases, can extend to bones and muscles (Singh et al. 2024). These fungal infections develop when fungal elements (e.g., hyphae or conidia) penetrate skin tissue at cuts, abrasions, or puncture wounds to establish infections within the subcutaneous layers (Chemello et al. 2023). Subcutaneous mycoses penetrate deeper skin layers beyond the superficial areas than superficial mycoses because they generate continuous inflammation and granulomas and cause tissue destruction. Subcutaneous fungal infections progress gradually for long periods, which often makes them misdiagnosed or unattended until the late stages when notable tissue deterioration appears.

The tropical and subtropical geographical regions serve as the environment for these fungal infections because the fungi survive best in degraded material and earthy substances (Shankarnarayan and Shaw 2024). People who perform activities outside, such as agricultural, forestry, or gardening occupations, face an increased risk of subcutaneous mycoses infection. Fungi that cause these infections are saprophytic and usually cannot lead to diseases in people without health conditions. These fungi establish persistent subcutaneous infections following tissue penetration, complicating treatment. How severe subcutaneous mycoses become is affected by the patient's immune system condition and their surrounding environment. Some mycoses remain confined to the original penetration area, yet certain strains spread through lymphatic vessels and deep into the body tissue (Sanroman-Loza et al. 2025). The infections of subcutaneous mycoses often get complicated and more severe in immunocompromised patients and may lead to increased morbidity (Monika and Chandraprabha 2022). Due to the chronicity of these infections, extended antifungal therapy is necessary, and surgery may be required in complicated instances.

Clinical observation with tissue examination and laboratory identification (such as histopathological or molecular examinations) of the infecting fungi is important in diagnosis. This allows for the diagnosis of subcutaneous mycoses because fungal skin infections can appear as other skin disorders (Sanroman-Loza et al. 2025). To reduce complications, including secondary bacterial infections, fibrosis, and permanent disfigurement, patients need early diagnosis and appropriate treatment (Knobler et al. 2024).

14.2.1 Common Causative Agents of Subcutaneous Mycoses

A wide variety of pathogens from the fungi kingdom lead to subcutaneous mycoses; these organisms naturally inhabit the soil environment and decomposing organic materials (Oliveira et al. 2023). The fungi remain harmless until they invade subcutaneous tissue, where they turn into opportunistic pathogens. Majorly, subcutaneous mycoses are classified as phaeohyphomycosis, subcutaneous zygomycosis, sporotrichosis, mycotic mycetoma, lobomycosis, and chromoblastomycosis based on their causative fungal species (Queiroz-Telles and Santos 2023). For instance, *Sporothrix* sp. is the causative agent of sporotrichosis, which is acquired after touching contaminated plant materials like wood, rose thorns, and moss (Oliveira et al. 2023). When the infection starts, a small skin nodule usually forms before it ulcerates and spreads into multiple nodules along the lymphatic area. The most frequent manifestation of sporotrichosis is lymphocutaneous sporotrichosis. Similarly, chromoblastomycosis caused by different fungal species, namely, *Cladophialophora* sp., *Phialophora* sp., and *Fonsecaea* sp., exists as a persistent fungal infection (De Hoog et al. 2024). This infection features slow-growing warts that develop into rough, scaly growths with time. The disease takes time to develop until it causes extensive skin fibrosis and dangerous secondary bacterial infections when no treatment is made. Chromoblastomycosis is most commonly found in individuals exposed to agricultural materials such as soil and plant parts.

Another subcutaneous mycosis is phaeohyphomycosis, which develops from exposure to dematiaceous fungi such as *Exserohilum* sp., *Cladophialophora* sp., *Curvularia* sp., and *Exophiala* sp. (Kukhar and Kiyan 2022). The condition results in deep subcutaneous cysts, which may transform into invasive infections in immunocompromised individuals. Also, mycotic mycetoma (where fungal species such as *Acremonium*, *Scedosporium*, *Trematosphaeria,* and *Madurella* are implicated) is characterized by persistent swelling and sinus tracts (Van De Sande 2021; Borman and Johnson 2023). Also, subcutaneous zygomycosis, such as entomophthoromycosis (*Basidiobolus ranarum* and *Conidiobolus coronatus*) and mucormycosis (*Rhizopus and Rhizomucor*), is characterized by necrotic ulcers (White et al. 2020; Acosta-España and Voigt 2022). Lastly, lobomycosis, whose causative agent is *Lacazia loboi*, triggers keloid-like nodule growth, which mainly appears in South American patients (Talhari and Talhari 2012; Gonçalves et al. 2022).

14.2.2 Pathogenesis of Subcutaneous Mycoses

The disease process for subcutaneous mycoses starts after fungal spores or hyphae penetrate the skin as a result of traumatic injuries (Rodrigues Hoffmann et al. 2023). Following infection, the fungi penetrate the subcutaneous tissue and evade the host's immune response to establish localized infection. To survive within host tissues, the fungi produce melanin, maintain thermotolerance, and utilize enzymes to break down tissue components. These mechanisms allow the fungi to persist by circumventing the host's immune defences.

When an infection occurs, the host's immune response triggers a sequence where macrophages, neutrophils, and T-helper cells become activated to start the inflammatory response (Loh and Lam 2023). These persistent fungal infections exhibit chronic behaviour, overwhelming the immune response and leading to long-term inflammation and granuloma development. The granulomas formed contain fibroblasts, giant cells, and lymphocytes that aim to restrain the infection through their activity triggers, but, unfortunately, contribute to fibrosis and tissue scarring

(Hahn et al. 2022). When the infection advances, the fungal cells may either spread through the lymphatic channels (e.g., in sporotrichosis) or stay close to the site of infection to form nodular and verrucous lesions (e.g., in chromoblastomycosis) (Queiroz-Telles and Santos 2023). Extensive fungal infection can penetrate deeper tissue structures, such as bones and muscles, leading to severe health consequences. Fungal melanin enables these pathogens to resist oxidative stress, hindering their removal by the immune system.

The advanced stage of subcutaneous mycoses leads to systemic disease in patients living with human immunodeficiency virus or AIDS, individuals who have undergone organ transplants and patients with diabetes (Pisote et al. 2022). Such an infection spreads within the body, affects internal organs, and leads to severe and dangerous conditions. Hence, early medical intervention alongside proper antifungal treatments must be implemented because they act as prevention tools against these complications.

14.2.3 Clinical Manifestations and Complications of Subcutaneous Mycoses

The medical signs and symptoms of subcutaneous mycoses develop in various ways due to the fungal agents causing them, coupled with infection locations and the host's immune system functioning (Passero et al. 2021; Oliveira et al. 2023). Generally, infections present as slow-growing nodules, plaques, or ulcerative lesions at the site of fungal inoculation. These fungal diseases tend to develop symptoms gradually at a rate of months or years. Mycoses remain as chronic conditions until they receive proper, prompt diagnosis and treatment since they damage tissues extensively. Sporotrichosis reveals its initial nodule, which ulcerates before spreading through the lymphatic system to create lymphocutaneous sporotrichosis (Queiroz-Telles et al. 2022). Conversely, chromoblastomycosis develops verrucous scaly plaques, progressively becoming fibrotic and thickening over time (Queiroz-Telles et al. 2017; Damayanti et al. 2022).

Clinical features of Phaeohyphomycosis manifest as single cysts located under the surface of the skin, whereas mycotic mycetoma creates disfiguring tissue swelling and develops sinus tracts, fungal debris in draining pus (Tiwary et al. 2022). In cases of untreated subcutaneous mycoses, patients face severe complications that include long-lasting ulcers, secondary bacterial infections, and fibrosis, which affects body movement.

Advanced infections spread to inner body structures, where they produce multiple issues, including osteomyelitis with tissue death, as well as deformations (Akuri et al. 2024). The aggressive form of mucormycosis shows rapid blood vessel invasion, causing limb-threatening tissue necrosis, often leading to amputation (Patel et al. 2015; Tran and Schmit 2020). Healthcare providers should perform early assessments through clinical evaluations with histopathological studies and fungal cultures to manage these conditions effectively. Medical protocols for treating subcutaneous mycosis demand extended treatment with itraconazole, along with amphotericin B or terbinafine, for therapeutic success (Shirsat et al. 2021). When the prevailing condition becomes severe, medical professionals might need to perform surgical debridement to treat it. This chapter presents a succinct analysis of how probiotics might strengthen immune responses and improve patient outcomes in subcutaneous mycosis management. Thus, this establishes promising therapeutic possibilities for the future.

14.2.4 Challenges in Current Treatment Approaches for Subcutaneous Mycoses

The fact that subcutaneous mycoses affect the dermis and subcutaneous tissues poses significant treatment challenges (Lee et al. 2021). Their complex nature and the limitations of current antifungal therapies make offering lasting solutions difficult. The infections typically arise following traumatic inoculation of fungi and can lead to dissemination either locally or, although less frequently, systemically, especially in the immunocompromised host (Gupta ct al. 2021). The current treatment options for subcutaneous mycoses involve the administration of antifungal drugs and surgical removal. These two approaches are not without challenges, including drug resistance, toxicity, and treatment failures (Santos et al. 2018; Gupta et al. 2021).

There are few available antifungal drugs, most targeting either ergosterol in the fungal cell membrane or 1,3-β-D-glucan in the cell wall (Table 14.2). These therapies, such as polyenes, azoles, and echinocandins, are frequently linked with

Table 14.2 Current treatment approaches and their limitations

Treatment approach	Description	Limitations	References
Antifungal agents	These agents include polyenes, azoles and echinocandins, which target the fungal cell membrane/wall.	They could exhibit high levels of toxicity, poor interaction with other medications when co-administered and treatment ineffectiveness due to resistance by the fungi	Lee et al. (2021), Sahadevan (2021)
Surgical excision	This is the physical removal of infected tissues	This approach is often invasive with a risk of recurrence. It is not always feasible to conduct due to the location and the level and depth of the infection's spread	Carrasco-Zuber et al. (2016), Araújo et al. (2018), Tirado-Sánchez et al. (2020)
Combination therapy	This is the simultaneous administration of multiple antifungal therapies	This causes an increased risk of adverse effects and complex drug interactions	Bellmann and Smuszkiewicz (2017), Wambaugh et al. (2020)

toxicities and drug interactions, which restrict their utility (Lee et al. 2021). Furthermore, the development of antifungal resistance, especially in *Candida* species, compounds the problem, as resistance mechanisms like efflux pump overexpression and biofilm formation decrease drug effectiveness (Pereira 2021). The small number of distinct molecular targets in fungi, as a result of their eukaryotic nature, also makes it more complicated to develop novel antifungal drugs (Lee et al. 2021).

Early and accurate diagnosis is crucial for controlling subcutaneous mycoses, yet it presents a significant challenge. Many subcutaneous mycoses exhibit non-specific symptoms, often resembling bacterial or parasitic infections. Traditional diagnostic methods, such as histopathology and culture, are time-consuming and show inconsistent sensitivity. While molecular diagnostics like polymerase chain reaction-based tests have improved diagnostic accuracy, they remain inaccessible in resource-poor areas where these infections are most prevalent (Mendonça et al. 2022; Fang et al. 2023).

14.3 Probiotics to the Rescue of Subcutaneous Mycoses

The dire requirement for alternative treatments is evident, with probiotics offering a promising area for future research. This alternative strategy could potentially enhance treatment efficacy, reduce adverse effects, and address the growing problem of antifungal resistance (Andrade et al. 2022).

14.3.1 How Probiotics Interact with Fungal Pathogens as a Potential Treatment for Subcutaneous Mycoses with Case Studies

Probiotics protect against pathogens through several mechanisms (Fig. 14.1). These mechanisms include the production of inhibitory compounds (such as bacteriocins, hydrogen peroxide, reuterins, and organic acids), competition for nutrients and binding sites, reduction of their niche pH, and the production of enzymes (Charlet et al. 2020; Divyashree et al. 2023). Probiotics could also improve barrier functions to inhibit attachment, decrease pro-inflammatory mediators, enhance immune responses (stimulating phagocytosis, IgA, and anti-inflammatory cytokine production) and confer immunity (Mazziotta et al. 2023).

Among the many microbes that have been identified as probiotics, those of the genera *Lactobacillus*, *Pediococcus*, *Saccharomyces*, *Lactococcus*, and *Bifidobacterium* have received more acceptance and applications (Singh et al. 2022; Latif et al. 2023; Vanitha et al. 2024; Adithi et al. 2025). Despite the cross-site functionalities of probiotics, their impact on the gastrointestinal tract has been the focus of many studies (Kim et al. 2019; Rau et al. 2024). Interestingly, more studies have begun to highlight the contributory effects of probiotics in other body sites, such as the respiratory tract (Darbandi et al. 2021), urinary tract (Forster et al. 2021; Mishra et al. 2024), urogenital tract (Abad and Safdar 2009), and skin (Cong et al. 2023; Vanitha et al. 2023, 2024; Deng et al. 2024). Concerning skin health, the

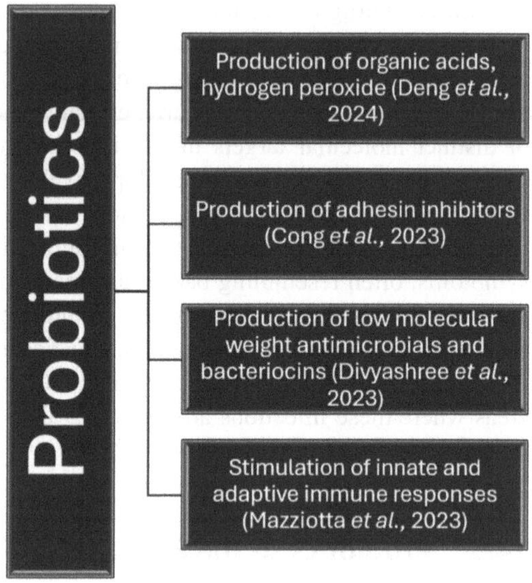

Fig. 14.1 Mechanisms of action of probiotics in the treatment of subcutaneous mycoses

oral route approach has promoted immune response with systemic effects on every part of the skin tissue (Gao et al. 2023; Deng et al. 2024). Some studies have explored this method to control skin inflammation and treat dermal infection (Hacini-Rachinel et al. 2009; Kober and Bowe 2015). This has broadened the scope of how probiotics might be used within the pharmaceutical industry.

Preclinical evidence has shown that the management of subcutaneous mycoses by probiotics can be achieved in many ways. One potential and practical property of probiotics in managing subcutaneous mycoses is their ability to stabilize or improve microbial homeostasis either by competitive exclusion or by producing antagonistic bacteriocins, reactive oxygen species, and other metabolites (Plaza-Diaz et al. 2019; Bermúdez-Humarán et al. 2024). For instance, a 29 kDa collagen-binding protein found in lactobacilli often binds to collagen in host tissue epithelial cells, which inhibits pathogen adhesion (Heinemann et al. 2000). A study demonstrated an expanded antagonistic attribute of *L. johnsonii* on *Candida albicans* in mice despite the availability of essential sugar (sucrose) required for their proliferation (Bertolini et al. 2021). This was attributed to the pH-dependent and pH-independent abilities of *L. johnsonii* to produce metabolites and inhibit biofilm formation in *Candida* (Vazquez-Munoz et al. 2022).

In addition, probiotics play a significant role in activating and modulating macrophages and IgA for local and systemic actions such as subcutaneous mycoses management (Fig. 14.1). Some studies have reported the ability of *Lactobacillus* to regulate the expression of some specific genes relevant to immune system regulation and also exert innate and adaptive immune responses when bound to infected tissues (Chon et al. 2009; Abedin-Do et al. 2015). Probiotics induce interleukins, with a critical function in immune system regulation and responses. *Lactobacillus rhamnosus* was documented in a study to have elicited the production of cytokines by macrophages as a defensive action (Jorjão et al. 2015). In a recent study, the chitinase activity of probiotics (*Lactobacillus johnsonii*) was reviewed as a potential compromising mechanism against fungal pathogens that could cause subcutaneous mycoses (Vazquez-Munoz et al. 2024). The oral inoculation of *L. johnsonii* in *Candida*-infected mice was reported to have damaged cell walls, inhibited metabolic activity and compromised the viability of *Candida* due to the production of chitinase by *L. johnsonii* (Vazquez-Munoz et al. 2024). The authors reported that chitinase cleaves chitin, while another study reported the chitinolytic action facilitated by chitinase secreted by *L. rhamnosus* against *C. albicans* (Allonsius et al. 2019).

A few clinical studies have discussed the applications of probiotics in managing fungal infections, such as cutaneous and subcutaneous mycoses (Xie et al. 2023; Al-Madhagi and Alramo 2024). A typical example is the worsening symptoms of atopic dermatitis (AD) or eczema, caused by fungi such as *Malassezia* and *Candida*. These fungi exacerbate AD by triggering allergic reactions, contributing to the established inflammation (Faergemann 2002). A study examined the effect of probiotics *Lactobacillus paracasei* and *Bifidobacterium lactis* on 133 early infants (3–6 months) with AD and its progression, with a follow-up until age 3 (Gore et al. 2012). The trial did not find any benefit from supplementing standard topical treatment with *Lactobacillus paracasei* and *Bifidobacterium lactis* in the treatment of AD, and no impact on the progression of allergic disease. Similarly, the supplementation of pregnant mothers (from 14–16 gestation weeks) with *L. rhamnosus* has no significant effect on preventing AD in infants, indicating that breast milk is less likely to act as an indirect supplement channel for infants (Wickens et al. 2018). Another trial established the potential of *Bifidobacterium bifidum* triple viable capsules in the treatment of AD in 40 infants (12–36 months) (Lin et al. 2015). The study showed that the short-term (4 weeks) administration of *Bifidobacterium bifidum* has a positive effect on the prevention and treatment of infantile AD (Lin et al. 2015). This was also validated for the infant age group (6–36 months) using *Lacticaseibacillus rhamnosus* (Carucci et al. 2022).

Similarly, a clinical study assessed the mixture of probiotics (*Lactobacillus paracasei* and *Lactobacillus fermentum*) for the treatment of AD in children (1–18 years) (Wang and Wang 2015). Their findings report clinical improvement in children with AD with a reduction in the serum eosinophil cationic protein levels, an AD inflammation marker (Wang and Wang 2015). This was similar to the clinical findings of Jeong et al. (2020). In another trial that involved 50 AD patients ($\geq$16 years) who received *L. acidophilus* for 24 weeks, it was observed that the long ingestion of *L. acidophilus* significantly decreased and increased the lactate dehydrogenase and transforming growth factor-β (TGF-β), respectively (Yamamoto et al. 2016). It was also reported to have suppressed the inflammatory response of Th2 cells by activating regulatory T (Treg) and Th1 cells (Yamamoto et al. 2016). A recent trial using *L. sakei* as a probiotic treatment for AD in ages 3–18 reported an improvement in the skin barrier function due to the significant increase in the skin sebum content (Rather et al. 2021). Evidently, the effectiveness of probiotics in treating cutaneous and subcutaneous mycoses may depend on the specificity of the strain, the target age group, and the duration of the treatment (Xie et al. 2023; Fijan et al. 2023).

14.3.2 Potential Advantages of Probiotics over Traditional Antifungal Therapies

Probiotics offer some advantages over traditional treatment options like antifungals because the likelihood of the fungi developing resistance to probiotics is minimal (Table 14.3 and Fig. 14.2).

Fungi like *Fonsecaea* (chromoblastomycosis) form stubborn biofilms. *Lactobacillus* spp., among other probiotics, produce lactic acid and proteolytic enzymes, which degrade biofilms, thereby enhancing deeper penetration beyond the surface of the infection site (Barzegari et al. 2020). Topical and local injection applications of

Table 14.3 Comparison of the efficacy, mechanisms, and clinical potential of probiotics versus traditional antifungal therapies in managing subcutaneous mycoses

Aspect	Traditional antifungal therapies	Probiotics	References
Mechanism of action	Targets ergosterol in fungal membrane or 1,3-β-D-glucan in cell wall (e.g., azoles, polyenes, echinocandins)	Produces organic acids, bacteriocins, reuterins, hydrogen peroxide, chitinase, and immunomodulatory metabolites	Lee et al. (2021), Divyashree et al. (2023), Vazquez-Munoz et al. (2024)
Risk of resistance	High due to overuse and limited drug targets (especially in *Candida* and *Fonsecaea* spp.)	Low, due to non-specific and synergistic action mechanisms	Pereira (2021), Barzegari et al. (2020)
Side effects	Hepatotoxicity, nephrotoxicity, and adverse drug interactions	Generally regarded as safe (GRAS); minimal side effects when used appropriately	Bellmann and Smuszkiewicz (2017), Hill et al. (2014)
Cost and accessibility	High cost, limited availability in low-resource settings	Cost-effective, especially in fermented foods and supplements	Andrade et al. (2022), Gao et al. (2023)
Effectiveness in immunocompromised hosts	Reduced efficacy; higher risks of toxicity	Stimulates immune responses (e.g., phagocytosis, cytokines); supports systemic immunity	Bamgbose et al. (2021), Jorjão et al. (2015)
Potential for combination therapy	Can increase toxicity and complexity	Synergizes with antifungals; reduces dose and toxicity	Allonsius et al. (2019), Saravanan et al. (2023)
Delivery routes	Oral, intravenous, and limited topical applications	Oral, topical, and emerging options like bioengineered formulations and microneedle patches	Deng et al. (2024), Bermúdez-Humarán et al. (2024), Vazquez-Munoz et al. (2024)

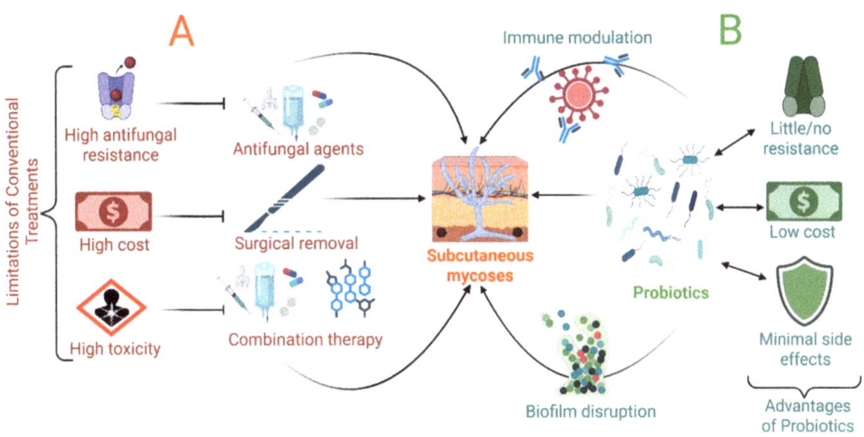

Fig. 14.2 Comparison of conventional treatments and probiotic-based strategies for managing subcutaneous mycoses

probiotics via bioengineered formulations allow direct inhibition of fungi at subcutaneous infection sites without dependence on systemic drug distribution (Allonsius et al. 2019). The healing of ulcerated chronic subcutaneous lesions potentially responds favourably to treatment with *Lactobacillus plantarum*, among other probiotics, since this bacterium accelerates tissue repair through the production of growth factors (TGF-β) that simultaneously control oxidative stress and inflammation at the infection area (Saravanan et al. 2023). Likewise, probiotics offer the benefit of causing fewer adverse side effects in their hosts compared to conventional antifungal treatments that could impair liver and kidney functions. Probiotics regularly work through different mechanisms that act simultaneously to combat fungal pathogens (Latif et al. 2023). Probiotics are usually cheaper compared to other antifungals. With this, probiotics offer a solution that is sustainable, particularly in low-resource environments like most countries in Africa and Asia. When probiotics are administered as adjunctive therapy in combination with traditional antifungals, they help to lower the likelihood of toxicity of the antifungal agent because the antifungals are administered in lower doses. This improves the treatment outcomes for patients.

14.4 Challenges and Future Directions

Technical challenges with topical probiotics exist because subcutaneous tissue remains hard to access; therefore, innovative delivery systems are currently being investigated. Effective subcutaneous fungal treatment depends on the selection of specific probiotic strains. As an example, an improved anti-*Candida* immunity in the gut due to the use of *Lactobacillus* strain produces no effect, or possibly an adverse effect on a subcutaneous infection of *Sporothrix schenckii*. The type of immune signals (cytokines, different metabolites such as short-chain fatty acids) generated by various bacterial strains used as probiotics may differ significantly. To treat a particular fungal pathogen, a therapy involves the identification of the precise strain to generate a particular immune modulation. The picture of what bacterial strains induce what systemic immune responses is yet to be fully drawn. In contrast to antibiotics, the choice of a probiotic in the case of subcutaneous mycosis is approximately the choice of a certain immune instructor. This needs thorough, mechanistic research that has yet to be fully conducted on such neglected tropical diseases. Although the ideal strain may be discovered, it has to go through the gastrointestinal system (when taken orally) to colonize the gut and have its action. This brings in such variables as acid and bile tolerance, which are also strain-specific.

There is a definite and evident risk of systemic infection among immunocompromised patients, which necessitates the selection of patients and constant monitoring. At the same time, horizontal gene transfer of antibiotic resistance is a more pernicious and population-wide risk factor that makes the safety-by-design approach of developing live biotherapeutics more necessary. These are the harsh truths that the promise of probiotics will have to counter. In addition, the regulatory environment for probiotics is complicated and country-dependent. This poses a significant obstacle to making probiotics legitimate treatment options. When a specific health claim is made by a company or researcher, regarding the treatment of a disease such as sporotrichosis, the probiotic will certainly be considered a drug (or a biologic) by the regulatory authorities (such as the Food and Drug Administration (FDA), USA, or the European Medicines Agency (EMA), Europe). This initiates a need to successfully pass the same stringent, costly, and lengthy approval process as a new chemical substance. A probiotic has to exhibit long-term safety in immunocompromised people (a major patient group in the case of mycoses) to be licensed as a drug. It should have a specified, viable amount of the given strain(s) and should be contamination-free, which may be hard with living microorganisms.

Research on subcutaneous fungal infections treated with probiotics is limited to studies using in vitro and a few animal models, as no human trials are currently ongoing, based on the information available to us. Therefore, research requires continuous energy and focus in this direction to confirm the effectiveness of probiotics in treating subcutaneous mycoses. Human efficacy studies are large, randomized, and placebo-controlled clinical trials, which are prohibitively costly. Moreover, pharmaceutical firms have inadequate financial incentive to invest hundreds of millions into drug-level development of a probiotic against a neglected tropical disease when they cannot obtain a solid patent on a naturally occurring bacterium. Therefore, there should be production of strong antifungal peptides through the genetic modification of specific probiotic strains that release defensins. Medical professionals should administer probiotics alongside antifungal agents and immunostimulants to treat chronic conditions. Injectable probiotics using microneedle patches should be developed for deep tissue delivery.

14.5 Conclusion

The research on the probiotic treatment of subcutaneous mycoses is a very interesting and promising branch of medical mycology. Probiotics present promise to work alongside antifungal treatments or substitute them in subcutaneous fungal infections through their ability to reduce toxicity, overcome resistance, and help the immune system clear such infections. The idea of using the immune system of the host through the microbiome is intellectually convincing. However, the journey between concept and clinic is full of challenges. The strain-dependent character of probiotic actions requires a degree of mechanistic accuracy that remains in its development. The regulatory system considers disease-targeting probiotics as new drugs, which is an expensive financial and logistical barrier. Lastly, the absence of large-scale clinical trials leaves us deeply wanting, as we do not have the strong evidence to support the recommendation of probiotics as a standard or adjuvant treatment of these infections. The future will require a concerted effort by incorporating academic research that will determine the appropriate strains and mechanisms, public-private collaboration to fund the required clinical trials and well-considered regulation that provides the safety and efficacy but does not hamper innovation in this potentially lucrative area. At present, probiotics treatment for subcutaneous mycoses is still a field of experimental investigation and not clinical practice. For preclinical research findings to translate into useful treatments, there is a need to verify them in practice while also developing better delivery methods.

14.6 Summary

The chapter gives a thorough analysis of the action of probiotics against subcutaneous mycoses. It describes the epidemiology and pathogenesis of such infections that determine their clinical importance. It highlights the difficulties associated with the latent onset, delayed diagnosis, and complex progression. The major pathogenic species of fungi, including *Sporothrix schenckii* and mycotic mycetoma, were identified, while their modes of transmission and pathological mechanisms were succinctly elucidated. A critical review of traditional antifungal treatment reveals the main shortcomings of this treatment, which include emerging drug resistance, systemic toxicity, and inconsistent patient response. To address such challenges, the chapter assigns substantial attention to probiotics as a novel form of therapy. It critically examines their complex mechanisms, including the release of antifungal metabolites, biofilm disruption, and the strong regulation of the host immune system. The evidence presented establishes the potential of probiotic interventions as complementary or alternative therapies for clinical management of subcutaneous fungal infections.

Conflicts of Interest The authors have no interests that could bias the presentation of information.

Authors Contributions RKO—Conceptualization, Data curation, Investigation, Project administration, Validation, Visualization, Writing—original draft, Writing—review and editing; TB—Conceptualization, Visualization, Writing—original draft, Writing—review and editing; OBD—Investigation, Visualization, Writing—original draft; MSA—Investigation, Writing—original draft; JOR—Investigation, Writing—original draft; BAA—Investigation, Writing—original draft. All authors approved the final manuscript.

References

Abad CL, Safdar N (2009) The role of *Lactobacillus* probiotics in the treatment or prevention of urogenital infections—a systematic review. J Chemother. https://doi.org/10.1179/joc.2009.21.3.243

Abedin-Do A, Taherian-Esfahani Z, Ghafouri-Fard S, Ghafouri-Fard S, Motevaseli E (2015) Immunomodulatory effects of *Lactobacillus* strains: emphasis on their effects on cancer cells. Immunotherapy 7(12):1307–1329. https://doi.org/10.2217/imt.15.92

Acosta-España JD, Voigt K (2022) An old confusion: Entomophthoromycosis versus mucormycosis and their main differences. Front Microbiol 13:1035100. https://doi.org/10.3389/fmicb.2022.1035100

Adithi G, Rakesh S, Shruthi B, Divyashree S, Deepa N, Vanitha PR et al (2025) Biological decontamination of myco-toxigenic *Fusarium verticillioides* using *Lactiplantibacillus plantarum* MYSN128 isolated from traditional vellappam: implications for food safety and global mycotoxin management. Curr Microbiol 82:404. https://doi.org/10.1007/s00284-025-04401-z

Akuri MC, Bencardino JT, Peixoto JB, Sato VN, Miyahara LK, Kase DT et al (2024) Fungal musculoskeletal infections: comprehensive approach to proper diagnosis. Radiographics 44(7):e230176. https://doi.org/10.1148/rg.230176

Allonsius CN, Vandenheuvel D, Oerlemans EFM, Petrova MI, Donders GGG, Cos P, Delputte P, Lebeer S (2019) Inhibition of *Candida albicans* morphogenesis by chitinase from *Lactobacillus rhamnosus* GG. Sci Rep 9(1):2900. https://doi.org/10.1038/s41598-019-39625-0

Al-Madhagi H, Alramo A (2024) Clinical trials of probiotics: current outlook. Food Sci Nutr 12(3):2192–2194. https://doi.org/10.1002/fsn3.3884

Andrade JC, Kumar S, Kumar A, Černáková L, Rodrigues CF (2022) Application of probiotics in candidiasis management. Crit Rev Food Sci Nutr 62(30):8249–8264. https://doi.org/10.1080/10408398.2021.1926905

Araújo MG, Cirilo NS, Santos SNMBD, Aguilar CR, Guedes ACM (2018) Lobomycosis: a therapeutic challenge. An Bras Dermatol 93(2):279–281. https://doi.org/10.1590/abd1806-4841.20187044

Bamgbose T, De La Fuente J (2023) Exploring the potential of probiotics in boosting the immune system's response to reduce the severity of malaria. J Cell Immunol 5(2):36–40. https://doi.org/10.33696/immunology.5.166

Bamgbose T, Anvikar AR, Alberdi P, Abdullahi IO, Inabo HI, Bello M et al (2021) Functional food for the stimulation of the immune system against malaria. Probiot Antimicro Prot 13(5):1254–1266. https://doi.org/10.1007/s12602-021-09780-w

Bamgbose T, Alberdi P, Abdullahi IO, Inabo HI, Bello M, Sinha S et al (2022) Functional characterization of α-Gal producing lactic acid bacteria with potential probiotic properties. Sci Rep 12(1):7484. https://doi.org/10.1038/s41598-022-11632-8

Bamgbose T, Quadri A, Abdullahi IO, Inabo HI, Bello M, Kori LD et al (2024) Antiplasmodial activity of probiotic *Limosilactobacillus fermentum* YZ01 in *Plasmodium berghei* ANKA infected BALB/c mice. J Trop Med 2024(1):6697859. https://doi.org/10.1155/jotm/6697859

Barzegari A, Kheyrolahzadeh K, Hosseiniyan Khatibi SM, Sharifi S, Memar MY, Zununi Vahed S (2020) The Battle of probiotics and their derivatives against biofilms. IDR 13:659–672. https://doi.org/10.2147/IDR.S232982

Bellmann R, Smuszkiewicz P (2017) Pharmacokinetics of antifungal drugs: practical implications for optimized treatment of patients. Infection 45(6):737–779. https://doi.org/10.1007/s15010-017-1042-z

Bermúdez-Humarán LG, Chassaing B, Langella P (2024) Exploring the interaction and impact of probiotic and commensal bacteria on vitamins, minerals and short chain fatty acids metabolism. Microb Cell Factories 23(1):172. https://doi.org/10.1186/s12934-024-02449-3

Bertolini M, Vazquez Munoz R, Archambault L, Shah S, Souza JGS, Costa RC, Thompson A, Zhou Y, Sobue T, Dongari-Bagtzoglou A (2021) Mucosal bacteria modulate Candida albicans virulence in oropharyngeal candidiasis. mBio 12(4):10.1128/mbio.01937-21

Borman AM, Johnson EM (2023) Changes in fungal taxonomy: mycological rationale and clinical implications. Clin Microbiol Rev 36(4):e00099–e00022. https://doi.org/10.1128/cmr.00099-22

Cao J, Yu Z, Liu W, Zhao J, Zhang H, Zhai Q, Chen W (2020) Probiotic characteristics of *Bacillus coagulans* and associated implications for human health and diseases. J Funct Foods 64:103643. https://doi.org/10.1016/j.jff.2019.103643

Carrasco-Zuber JE, Navarrete-Dechent C, Bonifaz A, Fich F, Vial-Letelier V, Berroeta-Mauriziano D (2016) Cutaneous involvement in the deep mycoses: a literature review. Part I—Subcutaneous Mycoses. Actas Dermo-Sifiliográficas (English Edition) 107(10):806–815. https://doi.org/10.1016/j.adengl.2016.05.026

Carucci L, Nocerino R, Paparo L, De Filippis F, Coppola S, Giglio V, Cozzolino T, Valentino V, Sequino G, Bedogni G, Russo R, Ercolini D, Berni Canani R (2022) Therapeutic effects elicited by the probiotic Lacticaseibacillus rhamnosus GG in children with atopic dermatitis. The results of the ProPAD trial. Pediatr Allergy Immunol 33(8):e13836. https://doi.org/10.1111/pai.13836

Charlet R, Bortolus C, Sendid B, Jawhara S (2020) Bacteroides thetaiotaomicron and Lactobacillus johnsonii modulate intestinal inflammation and eliminate fungi via enzymatic hydrolysis of the fungal cell wall. Sci Rep 10(1):11510. https://doi.org/10.1038/s41598-020-68214-9

Chemello RL, Castagna RD, Cappelletti T, Stramari JM, Da Silva CB, Comazzetto LR (2023) Fungal infections. In: Rangel Bonamigo R (ed) Dermatology in public health environments. Springer International Publishing, Cham, pp 293–338

Chon H, Choi B, Lee E, Lee S, Jeong G (2009) Immunomodulatory effects of specific bacterial components of *Lactobacillus plantarum* KFCC11389P on the murine macrophage cell line RAW 264.7. J Appl Microbiol 107(5):1588–1597. https://doi.org/10.1111/j.1365-2672.2009.04343.x

Cong L, Chen C, Mao S, Han Z, Zhu Z, Li Y (2023) Intestinal bacteria—a powerful weapon for fungal infections treatment. Front Cell Infect Microbiol 13:1187831. https://doi.org/10.3389/fcimb.2023.1187831

Damayanti N, Qomariah LN, Purwanto H, Siswati AS (2022) Diagnosis and therapy of chromoblastomycosis. J Pak Assoc Dermatol 32(2):443–448

Darbandi A, Asadi A, Ghanavati R, Afifirad R, Darb Emamie A, kakanj M, Talebi M (2021) The effect of probiotics on respiratory tract infection with special emphasis on COVID-19: systemic review 2010–20. Int J Infect Dis 105:91–104. https://doi.org/10.1016/j.ijid.2021.02.011

De Hoog S, Walsh TJ, Ahmed SA, Alastruey-Izquierdo A, Arendrup MC, Borman A et al (2024) Nomenclature for human and animal fungal pathogens and diseases: a proposal for standardized terminology. J Clin Microbiol 62(12):e00937-24. https://doi.org/10.1128/jcm.00937-24

Deng K, Fan X, Yuan Z, Li D (2024) Probiotic effects on skin health: comprehensive visual analysis and perspectives. Front Microbiol 15:1453755. https://doi.org/10.3389/fmicb.2024.1453755

Divyashree S, Shruthi B, Vanitha PR, Sreenivasa MY (2023) Probiotics and their postbiotics for the control of opportunistic fungal pathogens: a review. Biotechnol Reports 38:e00800. https://doi.org/10.1016/j.btre.2023.e00800

Faergemann J (2002) Atopic dermatitis and fungi. Clin Microbiol Rev 15(4):545–563. https://doi.org/10.1128/CMR.15.4.545-563.2002

Fang W, Wu J, Cheng M, Zhu X, Du M, Chen C, Liao W, Zhi K, Pan W (2023) Diagnosis of invasive fungal infections: challenges and recent developments. J Biomed Sci 30(1):42. https://doi.org/10.1186/s12929-023-00926-2

Fijan S, Kolč N, Hrašovec M, Jamtvedt G, Pogačar MŠ, Mičetić Turk D, Maver U (2023) Single-strain probiotic Lactobacilli for the treatment of atopic dermatitis in children: a systematic review and meta-analysis. Pharmaceutics 15(4):1256. https://doi.org/10.3390/pharmaceutics15041256

Forster CS, Hsieh MH, Cabana MD (2021) Perspectives from the Society for Pediatric Research: probiotic use in urinary tract infections, atopic dermatitis, and antibiotic-associated diarrhea: an overview. Pediatr Res 90(2):315–327. https://doi.org/10.1038/s41390-020-01298-1

Gao T, Wang X, Li Y, Ren F (2023) The role of probiotics in skin health and related gut–skin axis: a review. Nutrients 15(14):3123. https://doi.org/10.3390/nu15143123

Gavzy SJ, Kensiski A, Lee ZL, Mongodin EF, Ma B, Bromberg JS (2023) *Bifidobacterium* mechanisms of immune modulation and tolerance. Gut Microbes 15(2):2291164. https://doi.org/10.1080/19490976.2023.2291164

Gonçalves FG, Rosa PS, Belone ADFF, Carneiro LB, De Barros VLQ, Bispo RF et al (2022) Lobomycosis epidemiology and management: the quest for a cure for the most neglected of neglected tropical diseases. JoF 8(5):494. https://doi.org/10.3390/jof8050494

Gore C, Custovic A, Tannock GW, Munro K, Kerry G, Johnson K et al (2012) Treatment and secondary prevention effects of the probiotics Lactobacillus paracasei or Bifidobacterium lactis on early infant eczema: randomized controlled trial with follow-up until age 3 years. Clin Exp Allergy 42(1):112–122. https://doi.org/10.1111/j.1365-2222.2011.03885.x

Gupta AK, Renaud HJ, Quinlan EM, Shear NH, Piguet V (2021) The growing problem of antifungal resistance in onychomycosis and other superficial mycoses. Am J Clin Dermatol 22(2):149–157. https://doi.org/10.1007/s40257-020-00580-6

Hacini-Rachinel F, Gheit H, Luduec J-BL, Dif F, Nancey S, Kaiserlian D (2009) Oral probiotic control skin inflammation by acting on both effector and regulatory T cells. PLoS One 4(3): e4903. https://doi.org/10.1371/journal.pone.0004903

Hahn RC, Hagen F, Mendes RP, Burger E, Nery AF, Siqueira NP, Guevara A, Rodrigues AM, De Camargo ZP (2022) Paracoccidioidomycosis: current status and future trends. Clin Microbiol Rev 35(4):e00233–e00221. https://doi.org/10.1128/cmr.00233-21

Heinemann C, van Hylckama Vlieg JE, Janssen DB, Busscher HJ, van der Mei HC, Reid G (2000) Purification and characterization of a surface-binding protein from Lactobacillus fermentum RC-14 that inhibits adhesion of Enterococcus faecalis 1131. FEMS Microbiol Lett 190(1): 177–180. https://doi.org/10.1111/j.1574-6968.2000.tb09282.x

Hill C, Guarner F, Reid G, Gibson GR, Merenstein DJ, Pot B et al (2014) The International Scientific Association for Probiotics and Prebiotics consensus statement on the scope and appropriate use of the term probiotic. Nat Rev Gastroenterol Hepatol 11(8):506–514. https://doi.org/10.1038/nrgastro.2014.66

Jeong K, Kim M, Jeon SA, Kim Y-H, Lee S (2020) A randomized trial of *Lactobacillus rhamnosus* IDCC 3201 tyndallizate (RHT3201) for treating atopic dermatitis. Pediatr Allergy Immunol 31(7):783–792. https://doi.org/10.1111/pai.13269

Jorjão AL, de Oliveira FE, Leão MVP, Carvalho CAT, Jorge AOC, de Oliveira LD (2015) Live and heat-killed Lactobacillus rhamnosus ATCC 7469 may induce modulatory cytokines profiles on macrophages RAW 264.7. Sci World J 2015:716749. https://doi.org/10.1155/2015/716749

Kim S-K, Guevarra RB, Kim Y-T, Kwon J, Kim H, Cho JH, Kim HB, Lee J-H (2019) Role of probiotics in human gut microbiome-associated diseases. 29(9):1335–1340. https://doi.org/10.4014/jmb.1906.06064

Knobler R, Geroldinger-Simić M, Kreuter A, Hunzelmann N, Moinzadeh P, Rongioletti F et al (2024) Consensus statement on the diagnosis and treatment of sclerosing diseases of the skin, part 1: localized scleroderma, systemic sclerosis and overlap syndromes. Acad Dermatol Venereol 38(7):1251–1280. https://doi.org/10.1111/jdv.19912

Kober M-M, Bowe WP (2015) The effect of probiotics on immune regulation, acne, and photoaging. Int J Women's Dermatol 1(2):85–89. https://doi.org/10.1016/j.ijwd.2015.02.001

Kukhar YV, Kiyan VS (2022) Phoma as an infectious pathogen in medical practice. In: Rai M, Zimowska B, Kövics GJ (eds) Phoma: diversity, taxonomy, bioactivities, and nanotechnology. Springer International Publishing, Cham, pp 161–179

Latif A, Shehzad A, Niazi S, Zahid A, Ashraf W, Iqbal MW et al (2023) Probiotics: mechanism of action, health benefits and their application in food industries. Front Microbiol 14:1216674. https://doi.org/10.3389/fmicb.2023.1216674

Lau LYJ, Quek SY (2024) Probiotics: health benefits, food application, and colonization in the human gastrointestinal tract. Food Bioeng 3(1):41–64. https://doi.org/10.1002/fbe2.12078

Lee Y, Puumala E, Robbins N, Cowen LE (2021) Antifungal drug resistance: molecular mechanisms in *Candida albicans* and beyond. Chem Rev 121(6):3390–3411. https://doi.org/10.1021/ACS.CHEMREV.0C00199

Lin R-J, Qiu L-H, Guan R-Z, Hu S-J, Liu Y-Y, Wang G-J (2015) Protective effect of probiotics in the treatment of infantile eczema. Exp Ther Med 9(5):1593–1596. https://doi.org/10.3892/etm.2015.2299

Loh JT, Lam K-P (2023) Fungal infections: immune defense, immunotherapies and vaccines. Adv Drug Deliv Rev 196:114775. https://doi.org/10.1016/j.addr.2023.114775

Martinović A, Cocuzzi R, Arioli S, Mora D (2020) Streptococcus thermophilus: to survive, or not to survive the gastrointestinal tract, that is the question! Nutrients 12(8):2175. https://doi.org/10.3390/nu12082175

Mazziotta C, Tognon M, Martini F, Torreggiani E, Rotondo JC (2023) Probiotics mechanism of action on immune cells and beneficial effects on human health. Cells 12(1):184. https://doi.org/10.3390/cells12010184

Mendonça A, Santos H, Franco-Duarte R, Sampaio P (2022) Fungal infections diagnosis—past, present and future. Res Microbiol 173(3):103915. https://doi.org/10.1016/j.resmic.2021.103915

Mishra A, Jena SK, Srinivasan A, Tripathy A, Maiti R, Hota D (2024) Effect of an add-on multi-strain probiotic formulation in the prevention of recurrent urinary tract infections: a double-blind randomized placebo-controlled trial. Microbiol Res 15(4):2330–2340. https://doi.org/10.3390/microbiolres15040156

Monika P, Chandraprabha MN (2022) Risks of mucormycosis in the current Covid-19 pandemic: a clinical challenge in both immunocompromised and immunocompetent patients. Mol Biol Rep 49(6):4977–4988. https://doi.org/10.1007/s11033-022-07160-3

Oliveira M, Oliveira D, Lisboa C, Boechat J, Delgado L (2023) Clinical manifestations of human exposure to fungi. JoF 9(3):381. https://doi.org/10.3390/jof9030381

Pais P, Almeida V, Yılmaz M, Teixeira MC (2020) *Saccharomyces boulardii*: what makes it tick as successful probiotic? JoF 6(2):78. https://doi.org/10.3390/jof6020078

Passero LFD, Cavallone IN, Belda W (2021) Reviewing the etiologic agents, microbe-host relationship, immune response, diagnosis, and treatment in Chromoblastomycosis. J Immunol Res 2021:1–23. https://doi.org/10.1155/2021/9742832

Patel V, Squires S, Shaath T, Hinthorn DR, Fraga GR, Horvat R (2015) Explosive cutaneous Mucormycosis requiring limb amputation: case report and literature review. J Dermatol Res Ther 1(1). https://doi.org/10.23937/2469-5750/1510005

Pereira FDO (2021) A review of recent research on antifungal agents against dermatophyte biofilms. Med Mycol 59(4):313–326. https://doi.org/10.1093/mmy/myaa114

Petrariu O-A, Barbu IC, Niculescu A-G, Constantin M, Grigore GA, Cristian R-E, Mihaescu G, Vrancianu CO (2024) Role of probiotics in managing various human diseases, from oral pathology to cancer and gastrointestinal diseases. Front Microbiol 14:1296447. https://doi.org/10.3389/fmicb.2023.1296447

Pisote DK, Menon G, Choure RL (2022) A brief overview of invasive fungal infections in immunocompromised individuals. CBIOT 11(2):117–122. https://doi.org/10.2174/2211550111666220802162126

Plaza-Diaz J, Ruiz-Ojeda FJ, Gil-Campos M, Gil A (2019) Mechanisms of action of probiotics. Adv Nutr 10:S49–S66. https://doi.org/10.1093/advances/nmy063

Queiroz-Telles F, Santos DWDCL (2023) Fungal infections of implantation (Chromoblastomycosis, Mycetoma, Lobomycosis, and Entomophthoromycosis). In: Hospenthal DR, Rinaldi MG, Walsh TJ (eds) Diagnosis and treatment of fungal infections. Springer International Publishing, Cham, pp 369–389

Queiroz-Telles F, De Hoog S, Santos DWCL, Salgado CG, Vicente VA, Bonifaz A, Roilides E, Xi L, Azevedo CDMPES, Da Silva MB, Pana ZD, Colombo AL, Walsh TJ (2017) Chromoblastomycosis. Clin Microbiol Rev 30(1):233–276. https://doi.org/10.1128/CMR.00032-16

Queiroz-Telles F, Bonifaz A, Rossow J, Chindamporn A (2022) Sporothrix and Sporotrichosis. In: Encyclopedia of infection and immunity. Elsevier, pp 376–396

Rather IA, Kim B-C, Lew L-C, Cha S-K, Lee JH, Nam G-J, Majumder R, Lim J, Lim S-K, Seo Y-J, Park Y-H (2021) Oral administration of live and dead cells of *Lactobacillus sakei* proBio65 alleviated atopic dermatitis in children and adolescents: a randomized, double-blind, and placebo-controlled study. Probiot Antimicro Prot 13(2):315–326. https://doi.org/10.1007/s12602-020-09654-7

Rau S, Gregg A, Yaceczko S, Limketkai B (2024) Prebiotics and probiotics for gastrointestinal disorders. Nutrients 16(6):778. https://doi.org/10.3390/nu16060778

Rodrigues Hoffmann A, Ramos MG, Walker RT, Stranahan LW (2023) Hyphae, pseudohyphae, yeasts, spherules, spores, and more: a review on the morphology and pathology of fungal and oomycete infections in the skin of domestic animals. Vet Pathol 60(6):812–828. https://doi.org/10.1177/03009858231173715

Sahadevan NV (2021) Drug interactions of azole antifungals. JSSTD 5:50–54. https://doi.org/10.25259/JSSTD_61_2021

Sanroman-Loza EA, Medina-Guerrero EO, Holguín-Aguirre TE, Licona-Lasteros KC, Zermeño-Ruiz M, Gutiérrez-Sevilla JE, León-Hernández CY, Anaya-Ambriz EJ (2025) Decoding fungal skin infections: current and emerging trends in diagnosis and management. Curr Trop Med Rep 12(1):4. https://doi.org/10.1007/s40475-024-00334-y

Santos GC d O, Vasconcelos CC, Lopes AJO, De Sousa Cartágenes MDS, Filho AKDB, Do Nascimento FRF et al (2018) *Candida* infections and therapeutic strategies: mechanisms of action for traditional and alternative agents. Front Microbiol 9:1351. https://doi.org/10.3389/fmicb.2018.01351

Saravanan P, Pooja R, Balachander N, Singh KR, Silpa S, Rupachandra S (2023) Anti-inflammatory and wound healing properties of lactic acid bacteria and its peptides. Folia Microbiol 68(3):337–353. https://doi.org/10.1007/s12223-022-01030-y

Das DJ, Shankar A, Johnson JB, Thomas S (2020) Critical insights into antibiotic resistance transferability in probiotic lactobacillus. Nutrition 69:110567. https://doi.org/10.1016/j.nut.2019.110567

Shankarnarayan SA, Shaw D (2024) Epidemiology of fungal Zoonoses. In: Parija SC, Rudramurthy SM (eds) Textbook of fungal Zoonoses and Sapronoses. Springer Nature Singapore, Singapore, pp 69–79

Shirsat SP, Tambe KP, Dhakad GG, Patil PA, Jain Ritik S (2021) Review on antifungal agents. RJPPD:147–154. https://doi.org/10.52711/2321-5836.2021.00028

Singh RP, Shadan A, Ma Y (2022) Biotechnological applications of probiotics: a multifarious weapon to disease and metabolic abnormality. Probiot Antimicro Prot 14(6):1184–1210. https://doi.org/10.1007/s12602-022-09992-8

Singh N, Kumari V, Agrawal K, Kulshreshtha M (2024) Molecular pathway, epidemiological data and treatment strategies of Fungal infection (mycoses): a comprehensive review. CNSAMC 24(1):68–81. https://doi.org/10.2174/0118715249274215231205062701

Talhari S, Talhari C (2012) Lobomycosis. Clin Dermatol 30(4):420–424. https://doi.org/10.1016/j.clindermatol.2011.09.014

Tegegne BA, Kebede B (2022) Probiotics, their prophylactic and therapeutic applications in human health development: a review of the literature. Heliyon 8(6):e09725. https://doi.org/10.1016/j.heliyon.2022.e09725

Thambugala KM, Daranagama DA, Tennakoon DS, Jayatunga DPW, Hongsanan S, Xie N (2024) Humans vs. fungi: an overview of fungal pathogens against humans. Pathogens 13(5):426. https://doi.org/10.3390/pathogens13050426

Tirado-Sánchez A, Franco-Paredes C, Bonifaz A (2020) Subcutaneous mycoses in travelers. Curr Trop Med Rep 7(4):141–152. https://doi.org/10.1007/s40475-020-00216-z

Tiwary AK, Kumar P, Vinay S, Anand V, Barkat R, Fatima T (2022) Tropical diseases of the skin. In: Smoller B, Bagherani N (eds) Atlas of dermatology, dermatopathology and venereology. Springer International Publishing, Cham, pp 267–308

Tran D, Schmit B (2020) An aggressive case of Mucormycosis. Cureus 12(8):e9610. https://doi.org/10.7759/cureus.9610

Van De Sande WWJ (2021) *In vitro* susceptibility testing for black grain eumycetoma causative agents. Trans R Soc Trop Med Hyg 115(4):343–354. https://doi.org/10.1093/trstmh/traa184

Vanitha PR, Somashekaraiah R, Divyashree S, Pan I, Sreenivasa MY (2023) Antifungal activity of probiotic strain *Lactiplantibacillus plantarum* MYSN7 against *Trichophyton tonsurans*. Front Microbiol 14:11U244U. https://doi.org/10.3389/fmicb.2023.1192449

Vanitha PR, Divyashree S, Shruthi B, Adithi G, Deepthi BV, Sreenivasa MY (2024) *Limosilactobacillus fermentum* MYSY8, a potential probiotic isolate from fermented rice beverage for the control of *Microsporum canis*. Food Biotechnol 38(3):314–342. https://doi.org/10.1080/08U0543G.2024.2383743

Vazquez-Munoz R, Thompson A, Russell JT, Sobue T, Zhou Y, Dongari-Bagtzoglou A (2022) Insights from the *Lactobacillus johnsonii* genome suggest the production of metabolites with Antibiofilm activity against the Pathobiont *Candida albicans*. Front Microbiol 13. https://doi.org/10.3389/fmicb.2022.853762

Vazquez-Munoz R, Thompson A, Sobue T, Dongari-Bagtzoglou A (2024) *Lactobacillus johnsonii* is a dominant *Lactobacillus* in the murine oral mucosa and has chitinase activity that compromises fungal cell wall integrity. mBio 15(10):e02416-24. https://doi.org/10.1128/mbio.02416-24

Wambaugh MA, Denham ST, Ayala M, Brammer B, Stonhill MA, Brown JC (2020) Synergistic and antagonistic drug interactions in the treatment of systemic fungal infections. eLife 9:e54160. https://doi.org/10.7554/eLife.54160

Wang I-J, Wang J-Y (2015) Children with atopic dermatitis show clinical improvement after *Lactobacillus* exposure. Clin Exp Allergy 45(4):779–787. https://doi.org/10.1111/cea.12489

White SD, Théon AP, Angelos JA, Makhdoomi MM (2020) Diseases of the skin. In: large animal internal medicine. Elsevier, pp 1316–1351.e11

Wickens K, Barthow C, Mitchell EA, Stanley TV, Purdie G, Rowden J et al (2018) Maternal supplementation alone with *Lactobacillus rhamnosus* HN001 during pregnancy and breastfeeding does not reduce infant eczema. Pediatr Allergy Immunol 29(3):296–302. https://doi.org/10.1111/pai.12874

Xie A, Chen A, Chen Y, Luo Z, Jiang S, Chen D, Yu R (2023) *Lactobacillus* for the treatment and prevention of atopic dermatitis: clinical and experimental evidence. Front Cell Infect Microbiol 13. https://doi.org/10.3389/fcimb.2023.1137275

Yamamoto K, Yokoyama K, Matsukawa T, Kato S, Kato S, Yamada K, Hirota T (2016) Efficacy of prolonged ingestion of *Lactobacillus acidophilus* L-92 in adult patients with atopic dermatitis. J Dairy Sci 99(7):5039–5046. https://doi.org/10.3168/jds.2015-10605

Zhu J, Chen Y, Imre K, Arslan-Acaroz D, Istanbullugil FR, Fang Y, Ros G, Zhu K, Acaroz U (2023) Mechanisms of probiotic *Bacillus* against enteric bacterial infections. One Health Adv 1(1):21. https://doi.org/10.1186/s44280-023-00020-0

Navigating the Challenge of Opportunistic Fungal Infections: Probiotics in Action

15

Bhima Bhukya and Priyadarshini Boda

Abstract

Opportunistic fungal pathogens represent a persistent threat, causing significant morbidity and mortality in immunocompromised individuals. Host microbial disruption induced by immunosuppressive therapies or broad-spectrum antibiotics creates a vacant ecological niche within the host environment, allowing uncontrolled proliferation of opportunistic fungi, ultimately leading to invasion and infection. The challenge of these fungal infections is increased by a wide variety of fungal species, such as *Candida* spp. causing candidiasis, *Aspergillus* spp. resulting in aspergillosis, *Cryptococcus neoformans* leading to cryptococcosis, and the rapidly progressive zygomycetes causing zygomycotic infections. The effectiveness of conventional antifungal therapies is often hindered by toxicity and the growing issue of drug resistance. Moreover, most fungal pathogens are increasingly developing resistance to synthetic antifungal agents, underscoring the urgent need for alternative or complementary strategies. In this context, probiotics have emerged as promising adjuncts due to their multifaceted antifungal mechanisms. Probiotics modulate mucosal microbiota, enhancing competitive exclusion against fungi, stimulating innate and adaptive immunity, and reinforcing the epithelial barrier, reducing the likelihood of fungal translocation. This chapter provides an overview of key opportunistic fungal infections, caused by *Candida*, *Aspergillus,* and other potential opportunistic mycoses, and how probiotics offer a novel, supportive approach to managing these infections, paving the way for more comprehensive and integrative patient care.

B. Bhukya (✉) · P. Boda
Centre for Microbial and Fermentation Technology, Department of Microbiology, University College of Science, Osmania University, Hyderabad, Telangana State, India
e-mail: bhima.ou@osmania.ac.in

M. Y. Sreenivasa et al. (eds.), *Probiotics as Antifungals*,
https://doi.org/10.1007/978-981-92-0002-3_15

Graphical Abstract

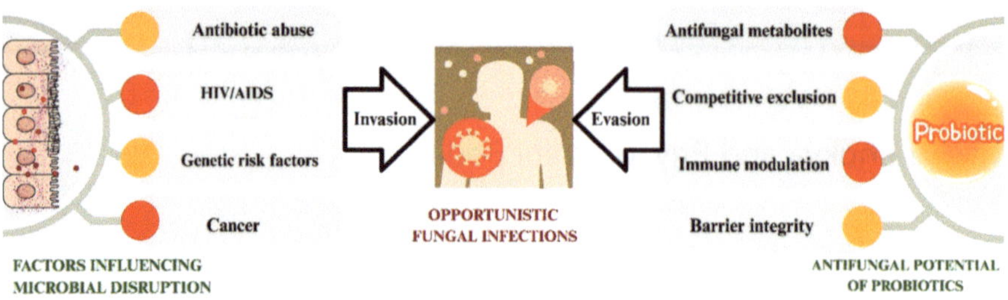

Keywords

Microbiome disruption · Opportunistic mycoses · Antifungal resistance · Probiotic therapy

Learning Objectives

After reading this chapter, learners will be able to:

- Evaluate how the indigenous microbiota contribute to maintaining fungal homeostasis and preventing dysbiosis.
- Assess the antifungal potential of probiotics and the mechanisms by which they inhibit fungal growth, adhesion, hyphal transition, and biofilm formation.
- Differentiate between various bioactive compounds produced by probiotic microorganisms and their modes of antifungal action.
- Summarize key findings from preclinical studies that demonstrate probiotic efficacy against fungal pathogens.
- Interpret the One Health perspective of probiotic-mediated approaches in controlling fungal infections across humans, animals, and the environment.
- Critically appraise the advantages, limitations, and future research directions of probiotic-based antifungal strategies.

15.1 Introduction

15.1.1 Understanding Host–Fungal Relationships Through the Influence of Microbiota

The microbial communities of humans show astounding diversity, with thousands of different bacterial species making up each person's unique microbial fingerprint that can be used for personal identification in some research contexts. Although some core microbes are common among people, the exact types and abundance of these species differ greatly from person to person. This variability between individuals is due to many factors, including lifestyle, genetics, diet, geography, and environment, as depicted in Fig. 15.1a (Heidrich et al. 2025).

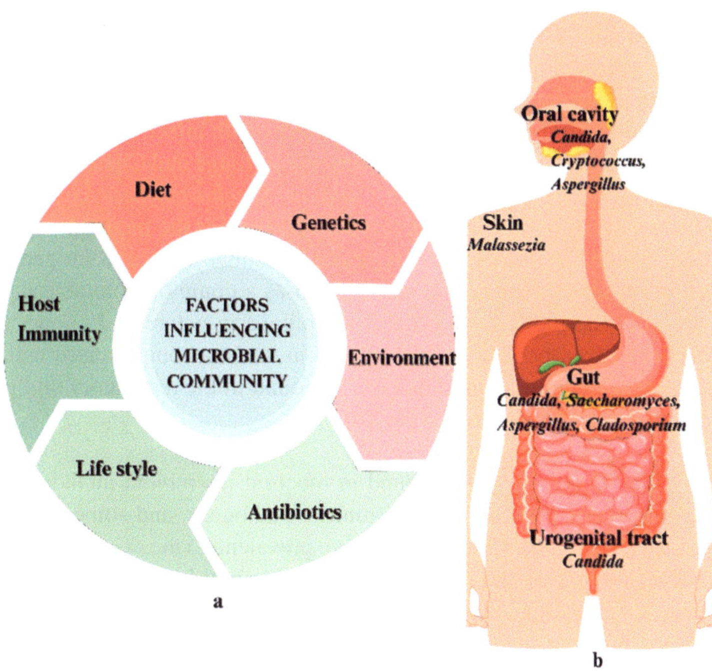

Fig. 15.1 Factors influencing human microbial community and distribution of prevalent fungal species across the human host. (**a**) The diversity of human microbiota results from a complex interplay of six major factors. These include intrinsic influence from the host, such as genetics, immunity, and lifestyle, and external influence such as diet, environment, and antibiotics; together these factors determine the health and balance of the entire microbial community. (**b**) Anatomical distribution of human Mycobiota. *Malassezia* is characteristic of the skin, while the gut harbors a diverse mix including *Candida*, *Saccharomyces,* and *Aspergillus*. The oral and urogenital tracts are also specific sites for colonization by genera like *Candida*, *Cryptococcus,* and *Aspergillus.*

15.1.1.1 Mucosal Microbiota: Composition and Function

The Mucosal microbiota is a complex and diverse group of microbial populations colonized on the mucus-lined surfaces of the human body, including the oral cavity, respiratory passages, gastrointestinal tract, and urogenital tract, which act as a crucial interface between the human body and external environment. Bacteria constitute one of the most predominant groups within these microbial communities (Ding et al. 2025). They colonize different sites, forming specific microbial makeups based on the local environment. The gut is home to many anaerobic bacteria, whereas the oral cavity contains a mix of aerobic and anaerobic species. Major bacterial phyla commonly found include *Firmicutes*, *Bacteroidetes*, *Proteobacteria*, and *Actinobacteria*. Being less abundant than bacteria, fungi are integral members of the mucosal microbiota. Fungi can exert a disproportionately significant impact due to their larger cell size and specialized metabolic capabilities (Nash et al. 2017). The fungal community distributed throughout the human body is largely dominated by two main phyla as described below (Fig. 15.1b).

Ascomycota Includes many common yeasts and molds. Yeasts like *Candida* and *Saccharomyces* are unicellular and often function as normal commensal flora with pathogenesis initiated by endogenous overgrowth and morphological switching when host immunity is compromised (Gu et al. 2020). Molds like *Aspergillus* are filamentous and are primarily exogenous pathogens with infection occurring through inhalation of airborne spores from the environment (Latgé and Chamilos 2019). Key genera found in mucosal sites include *Candida, Saccharomyces, Aspergillus, and Cladosporium*. Among all, the yeast *Candida albicans* is the most prevalent fungal commensal, colonizing the oral cavity, gastrointestinal tract, and urogenital tract. The other yeast *Saccharomyces,* is also known as a common commensal, particularly in the gut. The mold *Aspergillus* is ubiquitous in nature and known to cause acute infections like aspergillosis, particularly in immunocompromised individuals. *Cladosporium* is a common mold that can be found in the respiratory and gastrointestinal tracts (Hallen-Adams and Suhr 2017).

Basidiomycota Common genera found in mucosal microbiota include some species of Cryptococcus that can be opportunistic pathogens, and some others can be part of healthy mycobiota. *Malassezia* is more prevalent on the skin and is also found in the mucosal lining of the gut. The mycobiota contributes to human health in several ways, including aiding in metabolism by breaking down complex carbohydrates that are inaccessible to bacteria, modulating the immune system, and interacting with bacterial communities in both synergistic and competitive manners. A growing body of evidence suggests that fungal dysbiosis plays a major role in various health issues like inflammatory bowel disease, allergies, and an increased susceptibility to other infections (Huang et al. 2024).

15.2 Microbial Disruption Leading to Increasing Risk of Opportunistic Mycoses

Human microbiota is a complex system of bacteria, fungi, viruses, and other microorganisms that act as the primary line of defense by competing with harmful pathogens for nutrients and preventing their growth by producing antimicrobial substances. When this balance is disrupted, dysbiosis occurs, increasing the risk of infections. The human body encounters fungi everywhere, from inhaling *Aspergillus* spores to carrying *Candida* on the body naturally. Yet, the human body rarely gets sick because of the immune system that effectively defends. The body uses special receptors to recognize fungal components and kill the fungi. Immune cells like macrophages in the lungs quickly eliminate inhaled spores before they can grow (Garcia-Rubio et al. 2020). Furthermore physical barriers and adaptive T cells help in maintaining balance that keeps *Candida* controlled on mucosal surfaces. Fungi only transform into life-threatening opportunistic pathogens when these vital host defense mechanisms fail.

15.2.1 Incidence of Opportunistic Fungal Infections and Morbidity in Vulnerable Population

Historically fungal infections have been neglected but later in twenty-first century these have become a major health concern. This shift was largely due to certain factors like increase in immunocompromised individuals driven by chronic medical conditions such as diabetes, cancer, and autoimmune diseases that often requires immunosuppressive treatment regimes, the advent of COVID-19 pandemic, the rise of HIV/AIDS crisis, and anthropogenic ecological alterations (Casadevall 2018). Globally, every year over 6.5 million people are prone to life-threatening fungal infections, which lead to over 3.5 million deaths. Candidiasis is the most common mycotic infection with an annual incidence of more than 4 lakh cases, which may vary by country. It is about 1 to 12 per 1 lakh population in India, 21 and 21.8 per 1 lakh in Pakistan and Italy, respectively (Denning 2024). Invasive aspergillosis has been the most common infectious cause of death among patients with leukemia, those on long-term antibiotics and HIV/AIDS. But the incidence has decreased in recent years because of antifungal prophylaxis and proactive diagnostic approaches. Cryptococcosis is an opportunistic fungal infection that causes meningitis. It became a major cause of illness during the HIV/AIDS epidemic as the number of immunosuppressed individuals drastically increased worldwide. Approximately 15% of the AIDS related deaths are linked to cryptococcal meningitis (Rajasingham et al. 2017). Globally, about 2.2 lakh cases are estimated, of which the majority were reported in sub-Saharan Africa. The majority of mucormycosis cases are reported in India and neighboring countries and India witnessed a major increase in cases during the COVID-19 pandemic due to the use of systemic corticosteroids and a rise in diabetic cases (Rudramurthy et al. 2021).

15.2.1.1 Opportunistic Fungi and the Spectrum of Human Diseases

Opportunistic fungi reside as commensals, but can transform into pathogens and cause a wide array of infections when the host's immune system is compromised. *Candida* resides as a commensal on mucosal surfaces and skin of the host. When local or systemic immune defense is disrupted, it causes disease. The least severe cases include mucocutaneous infections, which are common and easily treatable, such as oral thrush, vaginal infections, cutaneous candidiasis, candidemia, and disseminated candidiasis (Pappas et al. 2018). The fungi *Aspergillus* is widely spread in the environment, from soil to air that we breathe, and cause diseases ranging from mild allergic reactions to life-threatening infections (Challa 2018). Mild allergic infections like Allergic Bronchopulmonary Aspergillosis (ABPA), Aspergilloma, Invasive Aspergillosis, and Pulmonary Aspergillosis. The soil contaminated by bird droppings harbors *Cryptococcus neoformans* that infects the central nervous system. In immunosuppressed individuals, the infection becomes serious and causes Pulmonary Cryptococcosis, Cryptococcal Meningitis (Johnson and Perfect 2010). Mucorales such as *Rhizopus* and *Mucor* species often cause fatal fungal infections in people with a weakened immune system, which particularly infects the sinus and lungs. Life-threatening disease of mucormycosis is known as rhino cerebral syndrome, which targets the sinuses and brain.

15.3 Unravelling the Pathogenesis of Opportunistic Fungi

15.3.1 Candidiasis

The yeast *C. albicans* is an opportunistic fungus that exists as a commensal inhabiting the oral, vaginal, and gastrointestinal tract. When it attaches to a host cell, yeast senses the contact and transforms into an invasive hyphal form. This ability of yeast to switch from harmless to invasive hyphal form is known as Dimorphism. The hyphal form has the ability to secrete enzymes and burrow deeper into the tissues. The yeast form is responsible for spreading throughout the body, and the hyphal form, however, is important in invading tissues and causing infection (Jacobsen and Hube 2017). For *Candida* to survive and thrive inside the host, it has to adapt to some environmental factors, including pH, nutrient availability, and temperature. Successfully adapting to these conditions directly increases *Candida's* ability to be virulent and pathogenic. The identification of *Candida* by epithelial receptors triggers many signaling pathways which in turn activate the genes that are important for its virulence specifically related to filamentous growth and biofilm formation.

15.3.2 Aspergillosis

Aspergillus is a common soil fungus; it releases large number of spores or conidia into the air. These spores don't cause any illness in healthy individuals as they have multiple layers of defense. The mucus layer and the cilia in airways deactivate these spores from germinating. By any chance if the spores cross the initial barrier, immune cells like neutrophils and macrophages destroy them (Challa 2018). On the other hand, immunocompromised individuals cannot effectively fight against fungus which leads to deadly invasive infections. Conidia or spores have an outer layer made up of Rod A protein and pigment called DHN-melanin that acts as a protective layer. Once conidia enter the lungs, they attach to laminin and fibrinogen proteins on the pulmonary epithelium (Hohl and Feldmesser 2007). The conidia swell and become metabolically active and transform into hyphae. This swelling causes shedding of the outer membrane, hyphae, then breaks out and invades the blood vessels and initiates infection by local angioinvasion and systemic angioinvasion. Aspergillosis affects multiple organs by taking several forms, like invasive aspergillosis, as illustrated in Fig. 15.2, sinusitis. The pathology ranges from localized pus-filled inflammation to granulomas that lead to tissue damage.

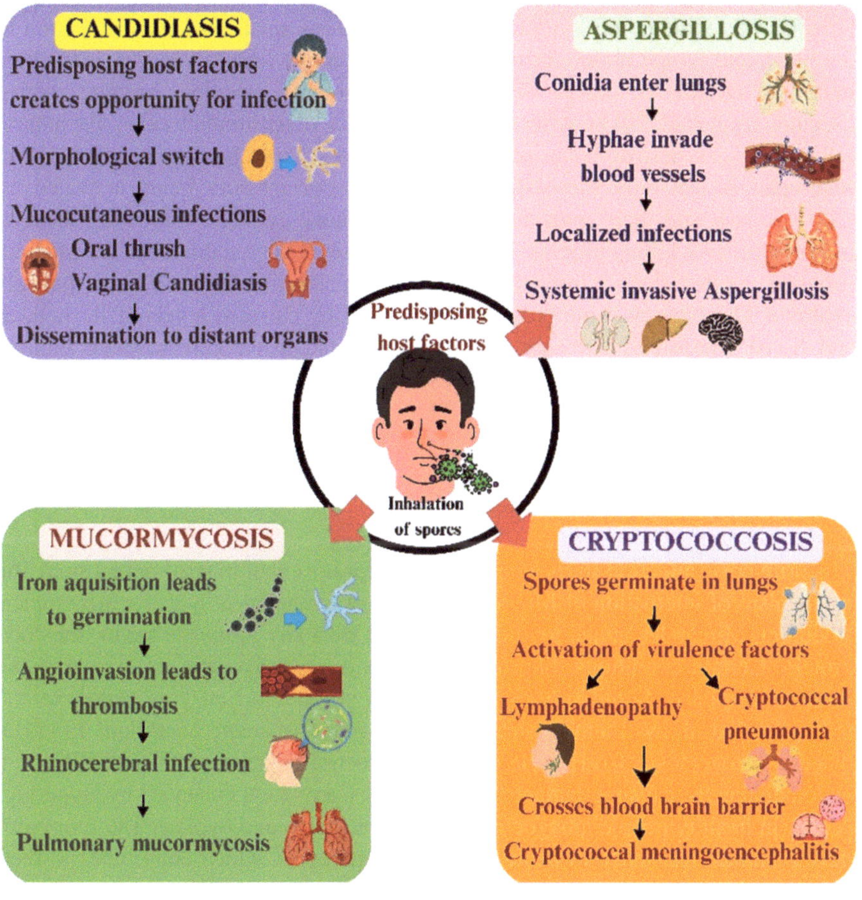

Fig. 15.2 Pathogenesis and organ-specific infections associated with opportunistic fungi in humans

Manifestation of this disease depends on the immune status of the patient. The more severe the patient's immunosuppression, the less inflammatory response is seen (Latgé and Chamilos 2019).

15.3.3 Cryptococcosis

Cryptococcosis is a systemic fungal infection caused by the encapsulated yeast *Cryptococcus*. The primary source of this infection is through the lungs. *Cryptococcus neoformans* and *C. gattii* are able to cause disease mainly due to their polysaccharide capsule composed of glucuronoxylomannan. This capsule protects the fungi from the host immune system, acting as a barrier against immune cells of the host. Other factors like their ability to survive in human body temperature, melanin production, secretion of protease, and phospholipase B help to invade tissues (de Sousa et al. 2022). In most of the cases, *Cryptococcus* infection in the lung remains mild without any symptoms. They remain as small pockets of infection in the lungs and lymph nodes surrounded by a protective fibrous capsule. If the host immune system becomes weak, these dormant capsules become active (Setianingrum et al. 2019). After this initial evasion, other immune cells, like natural killer cells, can destroy some of the fungal cells without having to engulf them. Macrophages can only successfully fight *Cryptococcus after* being activated by CD4+ T cells and their associated pro-inflammatory cytokines, allowing them to use the oxidative killing process. In patients with severe cryptococcosis, the immune response is often altered. Instead of mounting a protective response, these patients exhibit a Th2-type cytokine profile characterized by elevated levels of IL-4, IL-5, and IL-10. These cytokines are commonly detected in the cerebrospinal fluid of individuals with cryptococcal meningitis, where a robust immune defense is critically needed (Negroni 2012).

15.3.4 Mucormycosis

Mucormycosis infection starts when fungal spores are either inhaled or ingested. The mechanism of interaction of Mucorales with host epithelial cells is not fully understood yet, Mucorales attach to endothelial cells by a key virulence factor, the cotH protein. Certain conditions, like neutropenia and hyperglycemia, also weaken the immune response. Some species of *Rhizopus* produce an enzyme called ketone reductase that allows them to survive in high glucose, acidic conditions found in diabetic ketoacidosis. CotH proteins bind to glucose receptor proteins (GPR78) on endothelial cells, causing damage by endocytosis, a process enhanced by Hyperglycemia. In healthy people, phagocytes kill fungal spores and hyphae. The fungi also have a natural resistance to being killed by phagocytes, which contributes to their virulence. Iron metabolism also plays a critical role; individuals with excess iron or those undergoing certain chelation therapies are more susceptible to mucormycosis, as the fungus utilizes excess iron to support its growth. In acidic environments, the iron-binding capacity of transferrin is reduced, making more free iron available for fungal growth. Mucormycosis has the strong ability to invade blood vessels, which leads to thrombosis and tissue death, as illustrated in Fig. 15.2. The interaction of fungal spores with host epithelial cells helps in fungal spread and cell death facilitated by angioinvasion (Ibrahim et al. 2012).

15.3.5 Antifungal Resistance in Fungal Pathogens

The management of opportunistic fungal infections has become increasingly challenging due to the alarming rise in antifungal resistance (Hui et al. 2024). Unlike bacterial pathogens, fungi possess eukaryotic cell structures, making them inherently more difficult to target without affecting host cells. The prolonged and widespread use of azoles, echinocandins, and polyenes has led to the emergence of resistant strains, particularly in *Candida*, *Aspergillus*, and *Cryptococcus* species (Geddes-McAlister and Shapiro 2019). *Candida auris*, for instance, has gained global attention as a multidrug-resistant yeast with high mortality rates. Resistance mechanisms include alterations in drug targets, upregulation of efflux pumps, and biofilm formation, which contribute to reducing drug efficacy. In immunocompromised patients, these resistant strains can lead to persistent, hard-to-treat infections with limited therapeutic options (Lockhart 2019). The growing threat of antifungal resistance calls for the development of alternative strategies that can work synergistically or independently to prevent or mitigate fungal infections without contributing to resistance pressure.

15.3.6 Role of Indigenous Microbiota in Prevention of Fungal Dysbiosis

The body's natural microbiome maintains fungal homeostasis and prevents infections through a variety of methods. Primarily by competitive exclusion, causing colonization resistance by which the establishment of pathogens in the gut is prevented (Ricci et al. 2022). In the gut, these microbiota form interconnecting networks that fill most of the niches and make it difficult for the invading pathogens to find nutrients for their growth. The native microbes, which are already established in the gut, outcompete the pathogen for nutrients and space. Furthermore, they produce antifungal compounds and make the environment hostile for fungal growth. Ultimately, these microbes interfere with fungal virulence factors, preventing the transition of opportunistic fungi into pathogenic. The ability of indigenous microbes to inhibit fungal growth varies significantly from person to person (Donia and Fischbach 2015).

15.4 The Role of Probiotics in Combating Fungal Infections

Probiotics are defined by the Food and Agriculture Organization of the United Nations (FAO) and the World Health Organization (WHO) as "live microorganisms that, when administered in adequate amounts, confer health benefits on the host." Historically consumed as part of fermented foods, probiotics, primarily Lactic acid bacteria including *Lactobacillus* and *Bifidobacterium,* and certain yeasts like *Saccharomyces boulardii*, have gained considerable attention for their diverse health-promoting properties. These benefits extend beyond gut health to include immune modulation, allergy prevention, and protection against various bacterial and viral infections. In recent times, the potential of probiotics as an alternative strategy in the prevention and treatment of fungal infections has emerged as a promising area of research, offering a natural and potentially safer approach to manage these challenging pathogens.

15.4.1 Probiotics with Antifungal Potential

Probiotics are known for their ability to maintain microbial homeostasis and have become a promising alternative to conventional antifungal drugs, providing a wide range of approaches to combat fungal infections and boost host immunity without contributing to drug resistance (Peters et al. 2017). Their effectiveness is strain-specific, depending on the particular probiotic's mechanism of action and the target fungal pathogen. While many probiotic species offer general health benefits, certain strains have demonstrated significant and well-documented antifungal properties (Tiwari et al. 2020; Adithi et al. 2022, 2025). *Lactobacillus* species are more extensively studied probiotics for their antifungal properties, particularly against *Candida species*. Probiotic yeast, such as *Saccharomyces*, has been demonstrated to have antifungal potential against *Candida*. Due to its resilience to antibiotic, it can be used as an adjunct in antibiotic therapy against antibiotic-associated candidiasis. Species of *Bifidobacterium,* along with gut health benefits, also exhibit antifungal potential, contributing to a healthy gut environment, which indirectly inhibits fungal growth (Zeng et al. 2020).

15.4.2 Production of Fungicidal and Fungistatic Compounds by Probiotic Bacteria

The antifungal potential of probiotics lies in their ability to synthesize and secrete various bioactive compounds/postbiotics that inhibit fungal growth. These bioactive compounds show fungicidal or fungistatic effects through different mechanisms, targeting fungal cellular processes or structures (Table 15.1).

Table 15.1 Probiotic microorganisms along with bioactive compounds and their antagonistic effects against fungal pathogens

Probiotics	Target pathogen	Bioactive compound implicated	Mode of action	Antifungal effect	Reference
Lactobacillus reuteri	*Aspergillus niger, Aspergillus versicolor*	Reuterin	Inhibition activity against Yeast and filamentous fungi	Fungicidal	Vimont et al. (2019)
Lactobacillus pentosus, Pediococcus acidilactici	*Rhizopus* sp., *Mucor* sp.	Phenyl lactic acid	Inhibition of Mucormycosis infection	Fungicidal	Lunavath et al. (2023)
Lactobacillus rhamnosus	*Candida albicans*	Exo-metabolites	Inhibition of Biofilm formation	Candicidal	Matsubara et al. (2016)
Lactobacillus crispatus	*Candida* spp., *E. coli*	Antimicrobial metabolites	Antagonistic activity	Fungistatic	Hütt et al. (2016)
Saccharomyces boulardii	*C. albicans*	Caproic, capric, and caprylic acids	Inhibitory against filamentous growth, adhesion, and biofilm formation	Fungicidal	Bahuguna et al. (2024)
Pediococcus pentosaceus, Pediococcus acidilactici, Enterococcus faecium	*Staphylococcus aureus, Listeria monocytogenes Candida albicans*	Bioactive active metabolites	Antimicrobial activity	Fungistatic	Choi et al. (2025)
Lactobacillus acidophilus and Lactobacillus reuteri	*Candida glabrata*	Antifungal compounds	Inhibition of fungal growth and cell death	Fungicidal	Chew et al. (2015)
Saccharomyces cerevisiae, Issatchenkia occidentalis	Non albican *Candida* spp.	Antimicrobial metabolites	Inhibits cell adhesion and biofilm formation	Fungicidal	Kunyeit et al. (2019)

15.4.2.1 Organic Acids and H_2O_2

Most of the probiotic bacteria belonging to the genera *Lactobacillus* and *Bifidobacterium* are known to produce organic acids by carbohydrate fermentation. These organic acids include lactic acid and acetic acid, which drastically lowers the environmental pH, making the conditions unfavorable for the growth of fungus by inhibiting their metabolic pathways. These acids can diffuse passively into fungal cell membranes and cause internal acidification, which leads to disruption of cellular functions like ATP synthesis, nutrient transport, and finally leads to fungal cell death. Lactic acid bacterial strains showed antifungal activity against several fungi by producing organic acids and antifungal peptides that cause food spoilage. Among them, *Lactobacillus fermentum* CRL 251 has shown promising antifungal activity. Researchers are currently investigating the specific antifungal peptides it produces and exploring their mechanisms of action (Gerez et al. 2013). Hydrogen peroxide (H_2O_2) induces oxidative stress in fungal cells by damaging essential cellular components such as DNA, proteins, and lipids, ultimately leading to cellular dysfunction and death. While fungi have an antioxidant defense system, excess production of H_2O_2 can submerge these defenses. The antifungal activity of H_2O_2 and organic acids is synergistic, as the acidic environment enhances the ability of H_2O_2 to damage the fungal cell membrane. Certain probiotic strains, notably *Lactobacillus* species including *L. crispatus*, *L. jensenii*, *L. gasseri*, produce hydrogen peroxide and other bioactive substances by their metabolic activity, particularly through the action of NADH oxidase or flavin oxidase (Hütt et al. 2016). The probiotics *Lactobacillus pentosus* and *Pediococcus acidilactici* have been reported to produce phenyllactic acid, a compound with inhibitory effects against mucormycosis (Lunavath et al. 2023).

15.4.2.2 Bacteriocins and Antifungal Peptides

Bacteriocins are ribosomally synthesized antimicrobial peptides that act against other bacteria and fungi. Probiotic bacteria can synthesize these bacteriocins and other non-bacteriocin antifungal peptides that exhibit fungicidal or fungistatic effect. Most of the peptides interact with fungal cell membranes and disrupt them by forming pores, leakage of intracellular components, and finally leading to the death of cell. Few peptides cause damage to intracellular processes, including DNA replication, protein synthesis, and ultimately lead to growth arrest. Reuterin is a broad-spectrum antimicrobial compound that shows inhibitory activity against various fungi. Purified reuterin from *L. reuteri* ATCC 53608 inhibited various molds and yeast, and was found to have a fungicidal effect. Based on its chemical properties and its ability to inhibit both bacteria and fungi, it is used as a food preservative (Vimont et al. 2019).

15.5 Mechanism of Action Against Fungal Pathogens

Probiotic microorganisms use a variety of direct and indirect mechanisms to fight against fungal pathogens and improve host immunity. In direct antifungal mechanism, probiotics involve production of antimicrobial compounds including organic acids like lactic acid, acetic acid, and propionic acid that help in damaging fungal cell membrane, proteinaceous bacteriocins, other metabolites, fatty acids that contribute to antifungal activity. They also have the ability to improve host defense by indirect fungal mechanisms, like immune system modulation and restoration of microbial balance. They interact with the immune cells of the host and trigger antifungal responses, which eventually enhance innate immunity. They also promote antibody production to target fungal antigens. Probiotics help in the establishment of a healthy and diverse microbial population and reduce gut inflammation by suppressing fungal proliferation indirectly. The probiotic ability to fight fungal pathogens comes from a complex interplay of mechanisms that improve host defense, outcompete resources and inhibit fungal growth, illustrated in Fig. 15.3 and Table 15.2.

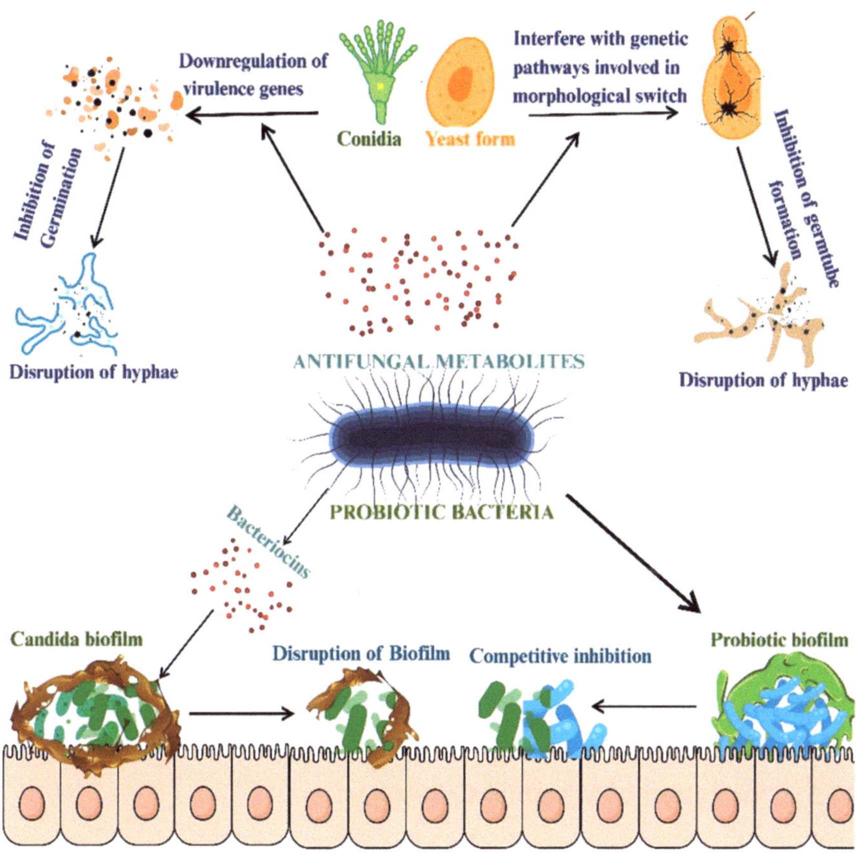

Fig. 15.3 Hypothetical illustration of the mechanism of action of probiotic bacteria inhibiting hyphae and biofilm formation

Table 15.2 Mechanism of action of probiotic microorganisms against fungal pathogens

Mechanism of action	Probiotics and target pathogen
Production of antifungal metabolites	*Lactobacillus* species produce organic acids and short-chain fatty acids that inhibit fungal growth and biofilm gene expression in *Candida* (Choi et al. 2025)
Disruption of quorum sensing	*Saccharomyces cerevisiae* interferes with signalling molecules in *Candida albicans* and disrupts them (Kunyeit et al. 2019)
Competitive exclusion	*Lactobacillus* species adheres to epithelial cells and blocks the adhesion sites of *Candida albicans* (Atanasov et al. 2023)
Enzymatic degradation of biofilm structure	Probiotic microorganisms produce proteolytic enzymes and biosurfactants that breakdown and degrade biofilm matrix (Rossoni et al. 2018)
Downregulation of genes associated with biofilm formation	*Lactobacillus acidophilus* supresses the genes EFG1, HWP1, ALS3 and BRC1 and CPH1 involved in filamentation and biofilm formation in *Candida* (Ribeiro et al. 2020)

15.5.1 Inhibiting the Formation of Invasive Hyphae

The mycelium of the pathogenic fungi is a key virulence factor that helps to form invasive hyphae. These hyphae attach to the host tissue and form biofilms. They also produce hydrolytic enzymes with which the hyphae can penetrate epithelium and endothelium layers, leading to tissue damage and enter the bloodstream. These invasive hyphae are difficult to detect by the immune system, and they also provoke a specific immune response from macrophages. *Lactobacillus acidophilus* lowers the pH of the mucosal environment, as filamentation in *Candida* is pH sensitive; it inhibits yeast—hyphal transition. This acid stress downregulates the EFG1, HWP1, and ALS3 genes involved in filamentation. These pathways are suppressed by short-chain fatty acids and probiotic conditioned media (Ribeiro et al. 2020). Pathogenic fungi use quorum-sensing molecules like farnesol and tyrosol to regulate filamentation. Probiotics can modulate these molecules or produce analogues that inhibit hyphal formation by blocking Ras1-cAMP-PKA, the signal transduction pathway crucial for initiation of hyphae. These results suggest that probiotics contain a non-protein substance that acts as an effective antifungal agent (Kunyeit et al. 2019).

15.5.2 Inhibition of Mycobial Cell Adhesion

Under normal conditions, opportunistic bacteria or fungi can live on the human body without causing harm. However, under favorable conditions, these microbes can overgrow and become hazardous to host health. Microbial adhesion to the host cells is one of the crucial steps that allows them to invade and spread infection. Therefore, inhibiting this adhesion is a key strategy for preventing disease. Probiotics attach to the epithelial cells by blocking the adhesion sites and preventing fungal colonization, which is an important prerequisite for hyphal formation (Atanasov et al. 2023). This physical barrier further helps in the reduction of fungal biofilm formation. Probiotics can stimulate the host to produce more mucin, which is a key component of the mucin layer that acts as a barrier and prevents fungal cell adhesion. Studies have reported that probiotics could prevent *C. albicans* from adhering to epithelial cells by testing three different scenarios (Kunyeit et al. 2019). When probiotics were introduced pre-inoculation, they were remarkably effective, reducing adhesion by 95% to 99%. Even when the probiotics were co-inoculated or administered post-inoculation, they still showed strong inhibitory effects, with adhesion rates dropping by 72% to 98%. These findings suggest that probiotics have adapted multiple mechanisms to prevent pathogenic adhesion. They may produce antimicrobial substances like bacteriocins that directly target pathogens (Choi et al. 2025). Additionally, they can act as a physical barrier, blocking pathogens from attaching to host cells, a process that is often mediated by lectin adhesion components.

15.5.3 Prevention of Biofilm Formation

Fungal biofilm formation is a three-step process. Initially, fungal cells adhere to a surface and begin to form microcolonies as illustrated in Fig. 15.3. In the middle stage, these colonies merge and start to release their protective matrix, forming a solid base. In the mature stage, the matrix is released in large quantities as the fungi form more complex structures like hyphae, which makes fungal cells resistant to treatment. Studies have shown that probiotics can be effective at preventing the formation of these biofilms by secreting proteases, DNases, and biosurfactants that prevent the synthesis of components of the extracellular matrix. This results in weakening of the biofilm structure, which in turn makes the immune cells and antifungal agents easily penetrate the fungal cells and disrupt them. Fully mature biofilms disperse both the yeast and hyphal forms of fungi to new locations. Probiotic metabolites, including lactate and hydrogen peroxide, inhibit the BRC1 and CPH1 gene activation of biofilm expansion, eventually disintegrating the biofilm and making it less virulent (Rossoni et al. 2018). Several research groups have found that probiotics, like *Lactobacillus* species, can inhibit biofilm development in the early stages. However, their effect is significantly weaker once the biofilm has already matured (Kean and Ramage 2019). Further research suggests that the inhibitory effects of probiotics are due to substances they release into their environment.

15.6 Insights into the Mechanisms of Probiotic Action Derived from Preclinical Studies

Since Probiotics are considered beneficial microorganisms, having the ability to alter the gut microbiota and protect against lung infections, their application in treating fungal infections has become a new area of research. Clinical studies have found that probiotics, particularly those from the Lactobacilli family, can help prevent and treat hospital-acquired lung infections (Du et al. 2022). These probiotics work by improving the immune system's ability to clear bacteria, lowering the bacterial load in the lungs, and reducing inflammation. In a recent study, researchers found that the probiotic strains *Lactobacillus fermentum*, *L. paracasei*, and *L. rhamnosus* are effective at preventing the formation of *Candida albicans* biofilms. They do this by reducing the number of *Candida* cells, stopping the growth of hyphae, and breaking down the structure of the biofilm itself. These effects were also observed when only the cell-free supernatant from the *Lactobacillus* cultures was used, suggesting that the probiotics produce acids or other substances that are responsible for inhibiting *Candida* growth. The antifungal effects of *Lactobacillus* strains are linked to the downregulation of key genes in *C. albicans* that are involved in biofilm formation (Rossoni et al. 2018). These findings indicate that the specific *Lactobacillus* strains have potential as a probiotic treatment to prevent oral candidiasis. A study conducted by Chew et al. (2015) showed that *L. rhamnosus* GR-1 and *L. reuteri* RC-14 probiotic strains not only inhibited the growth of *C. glabrata* but also completely halted its metabolic activity, leading researchers to believe they may be fungicidal. The strains of *Lactobacillus pentosus* and *Pediococcus acidilactici* isolated from human and buffalo milk (Lunavath et al. 2023) showed promising probiotic qualities and were effective at fighting off both bacterial and fungal pathogens, including those that cause mucormycosis, as illustrated in Fig. 15.4. The study highlighted that Phenyl lactic acid (PLA) produced by these bacteria is used in preventing infections from mucormycosis-causing fungi like *Rhizopus* and

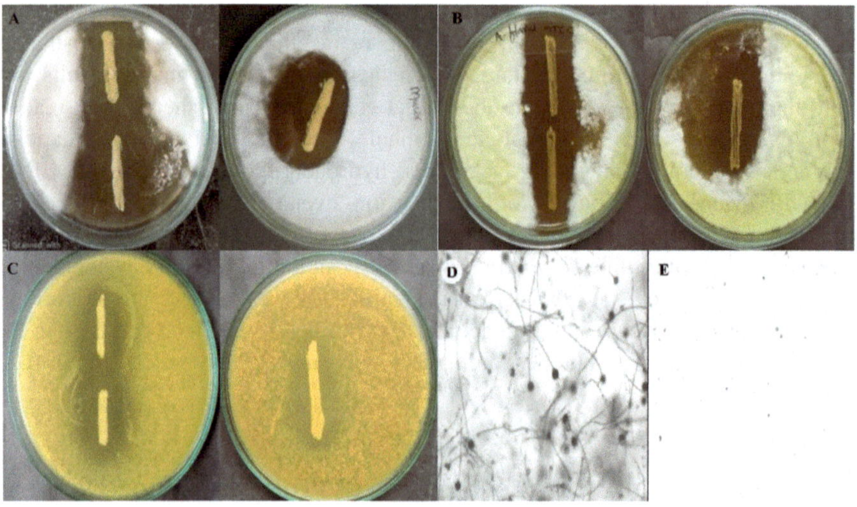

Fig. 15.4 Antifungal activity of *Lactobacillus pentosus* and *Pediococcus acidilactici against* (**a**) *Mucro* sp. (**b**) *Aspergillus* sp. and (**c**) *Candida albicans.* Mycelial inhibition of *Mucor* sp. using HPLC eluted PLA under EVOS live cell imaging microscopy resulted in a minimum inhibitory concentration (180 mg/mL). (**d**) *Mucor* sp. Untreated. (**e**) *Mucor* sp. treated with PLA

Mucor. The findings suggest that learning more about how PLA works against these specific fungi could lead to new ways to prevent mucormycosis.

Yeast belonging to the *Saccharomyces* species is emerging as a potential probiotic (Shruthi et al. 2024, 2025). They can survive the stomach's acidic environment, are not harmful and can outcompete pathogenic bacteria in the gut. Clinical trials using *Saccharomyces* species have shown encouraging results for treating conditions like diarrhoea, inflammatory bowel disease, and candidiasis. Yeast *Saccharomyces boulardii* was found to release a bioactive compound that stops *C. albicans* from forming filaments and growing into its invasive hyphal form (Bahuguna et al. 2024). This function of probiotic yeast is important as hyphae are key virulence factors for the fungi. Studies conducted by Nikaein et al. (2023) showed that *Saccharomyces boulardii* cell extracts could be a potential new treatment for allergies and asthma triggered by *Aspergillus*, especially those involving the cytokines IL-13 and IL-17. However, more research is needed to confirm its safety and effectiveness in living organisms, as well as its full range of effects on allergic reactions.

15.7 Probiotic-Mediated Approaches to Fight Fungal Infections: A One Health Perspective

Probiotic-based strategies to counter fungal infections represent a compelling example of the One Health approach, which recognizes the intrinsic link between human, animal, and environmental health (Fig. 15.5). By leveraging the multifaceted benefits of probiotics, we can develop integrated solutions that simultaneously benefit all three domains. This approach is particularly important because fungal pathogens such as *Candida* and *Aspergillus* are ubiquitous, present in water, food, soil, animals, and humans. Overuse of antifungal agents in any one sector can lead to the emergence of drug-resistant strains that may spread across sectors, posing a serious global threat (Wu et al. 2022).

In humans, probiotic strains like *Lactobacillus* and *Saccharomyces* are increasingly used to manage fungal infections (Gopalan et al. 2023; Vanitha et al. 2024). These probiotics help maintain microbial balance in the gut and mucosa, reducing reliance on antifungal drugs and thereby lowering the selective pressure that drives resistance development. Similarly, in veterinary settings, probiotics are employed to inhibit colonization by fungi such as *Aspergillus* and *Candida* in poultry, pigs, and aquaculture systems. Their use reduces the need for antifungal medications, which are significant contributors to the global reservoir of drug-resistant pathogens (Arsène et al. 2021). Incorporating probiotics into animal feed not only enhances animal health and productivity but also lowers the risk of fungal pathogens entering the human food chain. Probiotics are also applied in environmental and agricultural contexts as biopreservatives. They protect crops and stored grains from fungal contamination and help detoxify mycotoxins, particularly those produced by *Aspergillus* species. Acting as natural alternatives to chemical fungicides, probiotics reduce environmental pollution and curb the emergence of resistant fungal strains that could affect both crops and human health (de Souza Vandenberghe et al. 2017).

Fig. 15.5 Role of probiotics in One Health approach involving human, animal, and environment

The One Health approach fosters a positive feedback loop: using probiotics in agriculture reduces mycotoxin spread to animals and humans; their use in livestock improves animal health while reducing drug reliance; and maintaining a balanced microbiome in any one sector supports the stability of the entire ecosystem.

15.8 Merits and Demerits

Although probiotics have a long history of use, their mechanisms and efficacy in clinical antifungal applications remain not fully understood. Nevertheless, they are emerging as promising alternatives to conventional antifungal therapies. Numerous studies have demonstrated the antifungal potential of probiotics against pathogenic fungi, suggesting that probiotics, either alone or in combination with antifungal drugs, could open new avenues for managing fungal infections. However, the use of live microorganisms, particularly fungal probiotics, is not without risk. For instance, strains like *Candida albicans*, which are known to cause invasive infections, may pose a threat to immunocompromised individuals if used therapeutically. This highlights the need for caution in developing probiotic-based therapies. Researchers must carefully balance the potential benefits with the associated risks, especially when considering live fungal probiotics. A more prudent approach may involve the use of prebiotics or postbiotics, which can offer therapeutic benefits without the complications associated with live organisms.

15.9 Research Gaps and Future Prospects

Due to an increase in the number of fungal infections and resistance to current antifungal drugs, it has become critically important to find new treatments with a different mechanism of action. To limit drug resistance, better strategies to manage pathogens and create new disease control that reduce our reliance on medication. Although the exact way by which probiotic bacteria act against pathogens is not clearly understood, it's evident that they have a strong antifungal effect. This makes probiotics safe and affordable option for treating fungal infections. The potential of probiotics is not limited to live cells; even the cell-free supernatants possess antimicrobial effect, broadening their application. As a result, future research will focus on understanding these mechanisms and identifying more potential probiotic strains with antifungal potential for clinical applications.

15.10 Chapter Summary

Probiotics are beneficial microorganisms and work through different mechanisms to fight against pathogens. Initially, they engage in competitive exclusion followed by the production of various antimicrobial compounds, thereby modulating the host immune system. Although limited studies have explored the probiotics against opportunistic fungal pathogens, this chapter consolidates the available data and provides a comprehensive overview of the potential mechanisms through which probiotics exert their antifungal effects. The antifungal potential of the probiotic bacteria is highly valuable because they offer a safe and cost-effective alternative to conventional antifungal drugs. Probiotics could become an important part of a contemporary and integrated approach to combat fungal infections; therefore, more research is needed.

15.11 Conclusion

The emerging evidence strongly supports the significance of probiotics against fungal infections. They offer a multifaceted strategy to fight against pathogenic fungi, unlike traditional methods, to provide a sustainable approach. The antifungal properties of probiotics are rooted in a diverse range of mechanisms. Furthermore, they are capable of producing a diverse range of bioactive compounds that directly inhibit fungal growth. Probiotics act as immunomodulators, improve host immune responses and eliminate fungal invasion. Their ability to disrupt key virulence factors like biofilm formation makes them effective against some of the tough fungal pathogens. Despite these promising findings, significant research is still needed to unlock the potential of probiotics as antifungals.

Acknowledgments Authors are grateful to the Indian Council of Medical Research (ICMR), RUSA 2.0 program of the Ministry of Education, and Council of Scientific and Industrial Research (CSIR), Govt. of India, for financial support to carry out the research work on probiotics.

Conflict of Interest The authors declare that they have no conflict of interest.

Author Contributions BB conceived the idea, guided the structure and content of the manuscript, contributed to the writing, and critically revised and edited the final version. PB gathered relevant literature, drafted the initial manuscript, and contributed to the writing process. Both authors reviewed and approved the final manuscript.

Study Questions

1. Emphasize the role of the indigenous microbiota in maintaining fungal balance with the key factors and steps that drive a commensal fungus like *C. albicans* to cause opportunistic disease.
2. Explain the concept of dimorphism in *C. albicans*. What are the distinct roles of the yeast form and the hyphal form in the pathogenesis of candidiasis.
3. In a healthy host, multiple layers of defense prevent *Aspergillus* conidia from germinating. What specific fungal components (protein and pigment) form the protective outer layer of the conidia?
4. *Cryptococcus* infection often remains mild or asymptomatic in the lungs. Explain how this fungus can establish a state of dormancy.
5. What is the synergistic relationship between organic acids (like lactic acid) and hydrogen peroxide in combating fungal cells and the specific mechanism by which each compound damages the fungus.
6. Why is the effectiveness of a probiotic considered "strain-specific" in its antifungal action?
7. Fungal biofilms provide resistance to treatment. Explain how probiotics disrupt the mature stage of biofilm formation, referencing the specific compounds/ enzymes produced by probiotics.
8. Identify and explain the evidence for the use of *Lactobacillus* and *Pediococcus* strains against mucormycosis. What is the specific bioactive compound implicated in this inhibitory effect.
9. Define the One Health concept as it relates to managing fungal infections. Explain why this holistic approach is particularly relevant to fungal pathogens like *Candida* and *Aspergillus*.

References

Adithi G, Rakesh S, Divyashree S, Shruthi B, Sreenivasa MY (2022) Assessment of probiotic and antifungal activity of *Lactiplantibacillus plantarum* MYSAGT3 isolated from locally available herbal juice against mycotoxigenic *Aspergillus* species. Food Biosci:102118. https://doi.org/10.1016/j.fbio.2022.102118

Adithi G, Rakesh S, Shruthi B, Divyashree S, Deepa N, Vanitha PR, Vasundaradevi R, Deepthi BV, Sreenivasa MY (2025) Biological decontamination of mycotoxigenic *Fusarium verticillioides* using *Lactiplantibacillus plantarum* MYSN128 isolated from traditional vellappam: implications for food safety and global mycotoxin management. Curr Microbiol 82:404. https://doi.org/10.1007/s00284-025-04401-z

Arsène MM, Davares AK, Andreevna SL, Vladimirovich EA, Carime BZ, Marouf R, Khelifi I (2021) The use of probiotics in animal feeding for safe production and as potential alternatives to antibiotics. Vet World 14(2):319

Atanasov N, Evstatieva Y, Nikolova D (2023) Antagonistic interactions of lactic acid bacteria from human oral microbiome against *Streptococcus mutans* and *Candida albicans*. Microorganisms 11(6):1604

Bahuguna N, Venugopal D, Rai N (2024) Biotherapeutic potential of probiotic yeast *Saccharomyces boulardii* against *Candida albicans* biofilm. Indian J Microbiol. https://doi.org/10.1007/s12088-024-01124-3

Casadevall A (2018) Fungal diseases in the 21st century: the near and far horizons. Pathog Immun 3(2):183

Challa S (2018) Pathogenesis and pathology of invasive aspergillosis. Curr Fungal Infect Rep 12(1):23–32

Chew SY, Cheah YK, Seow HF, Sandai D, Than LTL (2015) Probiotic *Lactobacillus rhamnosus* GR-1 and *Lactobacillus reuteri* RC-14 exhibit strong antifungal effects against vulvovaginal candidiasis-*causing Candida glabrata*. J Appl Microbiol 118(5):1180–1190

Choi GH, Fugaban JII, Dioso CM, Bucheli JEV, Holzapfel WH, Todorov SD (2025) Antimicrobial peptides produced by *Lactococcus lactis* and *Pediococcus pentosaceus* with activity against clinical and food-borne pathogens. Probiotics Antimicrob Proteins 17(3):1171–1192

Denning DW (2024) Global incidence and mortality of severe fungal disease. Lancet Infect Dis 24(7):e428–e438

de Sousa HR, de Oliveira Jr GP, Frazão SDO, Gorgonha KCDM, Rosa CP, Garcez EM, Nicola AM (2022) Faster Cryptococcus melanization increases virulence in experimental and human cryptococcosis. J Fungi 8(4):393

de Souza Vandenberghe LP, Garcia LMB, Rodrigues C, Camara MC, de Melo Pereira GV, de Oliveira J, Soccol CR (2017) Potential applications of plant probiotic microorganisms in agriculture and forestry. AIMS Microbiol 3(3):629

Ding G, Yang X, Li Y, Wang Y, Du Y, Wang M et al (2025) Gut microbiota regulates gut homeostasis, mucosal immunity and influences immune-related diseases. Mol Cell Biochem 480(4):1969–1981

Donia MS, Fischbach MA (2015) Small molecules from the human microbiota. Science 349(6246):1254766

Du T, Lei A, Zhang N, Zhu C (2022) The beneficial role of probiotic *Lactobacillus* in respiratory diseases. Front Immunol 13:908010

Garcia-Rubio R, de Oliveira HC, Rivera J, Trevijano-Contador N (2020) The fungal cell wall: *Candida, Cryptococcus* and *Aspergillus* species. Front Microbiol 10:2993

Geddes-McAlister J, Shapiro RS (2019) New pathogens, new tricks: emerging drug-resistant fungal pathogens and future prospects for antifungal therapeutics. Ann N Y Acad Sci 1435(1):57–78

Gerez CL, Torres MJ, De Valdez GF, Rollán G (2013) Control of spoilage fungi by lactic acid bacteria. Biol Control 64(3):231–237

Gopalan S, Ganapathy S, Mitra M et al (2023) Unique properties of yeast probiotic *Saccharomyces boulardii* CNCM I-745: a narrative review. Cureus 15(10):e38147

Gu ZQ, Tseng KY, Tsai YH (2020) *Candida* gut commensalism and inflammatory disease. Med Microecol 3:100008

Hallen-Adams HE, Suhr MJ (2017) Fungi in the healthy human gastrointestinal tract. Virulence 8(3):352–358

Heidrich V, Valles-Colomer M, Segata N (2025) Human microbiome acquisition and transmission. Nat Rev Microbiol 23:568–584. https://doi.org/10.1038/s41579-025-00985-4

Hohl TM, Feldmesser M (2007) *Aspergillus fumigatus*: principles of pathogenesis and host defense. Eukaryot Cell 6(11):1953–1963

Huang H, Wang Q, Yang Y, Zhong W, He F, Li J (2024) The mycobiome as an integral part of the gut microbiome. Gut Microbes 16(1):2440111

Hui ST, Gifford H, Rhodes J (2024) Emerging antifungal resistance in fungal pathogens. Curr Clin Microbiol Rep 11(2):43–50

Hütt P, Lapp E, Štšepetova J et al (2016) Characterisation of probiotic properties in human vaginal lactobacilli strains. Microb Ecol Health Dis 27:30484

Ibrahim AS, Spellberg B, Walsh TJ, Kontoyiannis DP (2012) Pathogenesis of mucormycosis. Clin Infect Dis 54(Suppl 1):S16–S22

Jacobsen ID, Hube B (2017) *Candida albicans* morphology: still in focus. Expert Rev Anti-Infect Ther 15:327–330

Johnson MD, Perfect JR (2010) Use of antifungal combination therapy: agents, order, and timing. Curr Fungal Infect Rep 4(2):87–95

Kean R, Ramage G (2019) Combined antifungal resistance and biofilm tolerance. mSphere 4(4): e00458–e00419

Kunyeit L, Kurrey NK, Anu-Appaiah KA, Rao RP (2019) Probiotic yeasts inhibit virulence of non-albicans *Candida* species. mBio. https://doi.org/10.1128/mBio.02307-19

Latgé JP, Chamilos G (2019) *Aspergillus fumigatus* and aspergillosis. Clin Microbiol Rev 33(1): e00140–e00118

Lockhart SR (2019) *Candida auris* and multidrug resistance: defining the new normal. Fungal Genet Biol 131:103243

Lunavath R, Mohammad SH, Bhukya KK et al (2023) Antimycotic effect of phenyllactic acid against COVID-19-associated mucormycosis-causing fungi. PLoS One 18(3):e0279118

Matsubara VH, Wang Y, Bandara HMHN et al (2016) Probiotic lactobacilli inhibit early stages of *Candida albicans* biofilm development. Appl Microbiol Biotechnol 100(14):6415–6426

Nash AK, Auchtung TA, Wong MC et al (2017) The gut mycobiome of the Human Microbiome Project healthy cohort. Microbiome 5:153

Negroni R (2012) Cryptococcosis. Clin Dermatol 30(6):599–609

Nikaein D, Malekmadani H, Beikzadeh B, Mardanpour R, Khosravi A, Moghadami SM (2023) Effect of yeast probiotic *Saccharomyces boulardii* cell wall extract on *Aspergillus fumigatus* allergenicity in A549 cells. Curr Med Mycol 9(4):1

Pappas PG, Lionakis MS, Arendrup MC et al (2018) Invasive candidiasis. Nat Rev Dis Primers 4: 18026

Peters BM, Jabra-Rizk MA, O'May GA, Costerton JW, Shirtliff ME (2017) Polymicrobial interactions: impact on pathogenesis and human disease. Clin Microbiol Rev 25(1):193–213

Rajasingham R, Smith RM, Park BJ et al (2017) Global burden of HIV-associated cryptococcal meningitis. Lancet Infect Dis 17(8):873–881

Ribeiro FC, Rossoni RD, De Barros PP et al (2020) Action mechanisms of probiotics on *Candida* spp. J Appl Microbiol 129(2):175–185

Ricci L, Mackie J, Donachie GE et al (2022) Human gut bifidobacteria inhibit the growth of *Candida albicans*. FEMS Microbiol Ecol 98(10):fiac095

Rossoni RD, de Barros PP, de Alvarenga JA et al (2018) Antifungal activity of clinical *Lactobacillus* strains against *Candida albicans* biofilms. Biofouling 34(2):212–225

Rudramurthy SM, Hoenigl M, Meis JF et al (2021) ECMM/ISHAM recommendations for clinical management of COVID-19-associated mucormycosis. Mycoses 64(9):1028–1037

Setianingrum F, Rautemaa-Richardson R, Denning DW (2019) Pulmonary cryptococcosis: a review. Med Mycol 57(2):133–150

Shruthi B, Deepa N, Adithi G, Divyashree S, Sreenivasa MY (2024) Probiotic and functional attributes of yeasts isolated from different traditional fermented foods and products. Probiotics Antimicrob Proteins. https://doi.org/10.1007/s12602-024-10342-z

Shruthi B, Deepthi BV, Adithi G, Vanitha PR, Sreenivasa MY (2025) *Saccharomyces cerevisiae* MYSY1 as a promising probiotic isolate with antifungal properties against foodborne *Aspergillus* species. Biocatal Agric Biotechnol 67:103749. https://doi.org/10.1016/j.bcab.2025.103649

Tiwari SK, Dicks LMT, Popov IV et al (2020) Probiotics at the frontier of preventive and therapeutic microbiome interventions. Microorganisms 8(8):1237

Vanitha PR, Divyashree S, Shruthi B, Adithi G, Deepthi BV, Sreenivasa MY (2024) *Limosilactobacillus fermentum* MYSY8, a potential probiotic isolate from fermented rice beverage for the control of *Microsporum canis*. Food Biotechnol 38(3):314–342. https://doi.org/10.1080/08U0543G.2024.2383743. IF 1.8 (Taylor & Francis)

Vimont A, Fernandez B, Ahmed G et al (2019) Quantitative antifungal activity of reuterin. Int J Food Microbiol 289:182–188

Wu Y, Hu S, Wu C et al (2022) Probiotics: potential novel therapeutics against fungal infections. Front Cell Infect Microbiol 11:793419

Zeng Z, Luo J, Zeng Y et al (2020) *Bifidobacterium* species and their anti-inflammatory potential. Microorganisms 8(9):1296

Probiotics Tackling Other Fungal Infections and Intoxications

16

A. H. Preethu, Tushar Shaw, Supriya Dhoni, K. N. Hanumantharaju, and Chennappa Gurikar

Abstract

Probiotics have gained a lot of attention as a biological therapy for fungal infections and intoxications because of their safety, wide range of health benefits, and minimal side effects. *Aspergillus flavus, Candida albicans*, *Mucour racemosus*, *Rhizopus, Aspergillus, Fusarium*, and dermatophytes are the pathogens that continue to pose a serious risk to human health. Chemical-based antifungal treatments are still the norm, but their long-term usage has created drug-resistant fungus strains and major health risks. Probiotic strains such as *Lactobacillus* spp., *Leuconostoc* spp., and *Saccharomyces* spp. have shown antifungal efficacy by producing bioactive postbiotic compounds like organic acids, short-chain fatty acids, hydrogen peroxide, cyclic dipeptides, and bacteriocin-like substances that prevent fungal growth and biofilm formation. Furthermore, through binding and degradation pathways, probiotics have demonstrated potential in detoxifying harmful contaminants such as heavy metals, acrylamide, benzopyrene, and mycotoxins. Several fungal species produce mycotoxins, such as *Aspergillus flavus*, which produces aflatoxins, and *Fusarium oxysporum*, which produces fumonisins. Their potential as a treatment is further strengthened by their antimutagenic and anticarcinogenic properties. This chapter emphasizes the role of probiotics in preventing fungal infections and intoxications in humans as well as its potential as a sustainable, affordable, and adaptable solution for improving food safety and public health.

A. H. Preethu · S. Dhoni · K. N. Hanumantharaju · C. Gurikar (✉)
Department of Food Technology, Ramaiah University of Applied Sciences, Bangalore, Karnataka, India
e-mail: chennappa.ft.ls@msruas.ac.in

T. Shaw
Department of Allied Health Sciences, Ramaiah University of Applied Sciences, Bangalore, Karnataka, India

M. Y. Sreenivasa et al. (eds.), *Probiotics as Antifungals*,
https://doi.org/10.1007/978-981-92-0002-3_16

Graphical Abstract

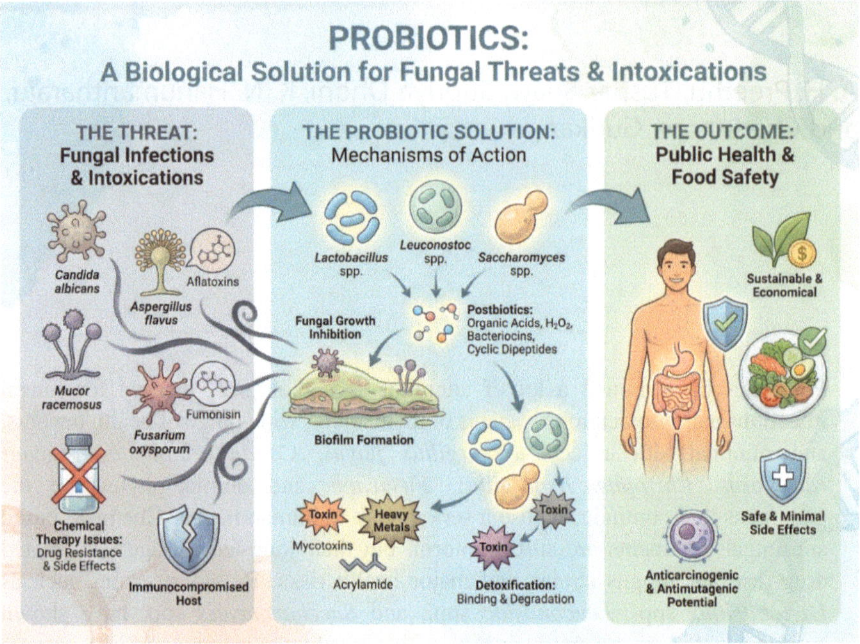

Keywords

Mycotoxins · Antifungal efficacy · *Lactobacillus* spp. · Drug-resistant · Therapeutic

Learning Objectives

By the end of this chapter, readers will be able

- To understand the common fungal infections and the risks they pose to human health.
- To learn about traditional methods used to control fungal infections and their limitations.
- To know how probiotics can help prevent and control fungal infections and harmful toxins.

16.1 Introduction

The concerns regarding the impact of fungal infections on global health are raised by the fact that their occurrence is rising rapidly along with the number of immunocompromised people (Martins et al. 2023). Treating fungal infections is challenging due to the limited availability of antifungal drugs, many of which are known to cause severe toxicity (Azole-based antifungal drugs, fluconazole, amphotericin B) (Mathew and Nath 2009). In addition, many infections show resistance to current treatments, emphasizing the growing demand for alternative medications (Fisher et al. 2022). In clinical practice, probiotics are gaining wider acceptance as a promising treatment for fungal infections. According to clinical research, probiotics have a number of health advantages beyond gut health, such as immunomodulation, lowering the risk of cancer, improving urogenital health, lowering cholesterol, preventing allergies and oral infections, boosting skin immunity, and having antifungal and dermatological qualities. As a result, they are crucial for controlling fungal infections (Cristofori et al. 2021).

The antibacterial properties of probiotics have been well studied, with promising results against various harmful infections (Raheem et al. 2021). High rates of morbidity and mortality are a result of opportunistic fungal infections, which are emerging as a major worldwide health concern among immunocompromised people. People with diseases including HIV, cancer, granulocytopenia, organ transplantation, severe burns, diabetes, trauma, and malnutrition are the main reasons for these infections (Denning 2024). In healthcare settings, opportunistic fungal pathogens such as *Candida*, *Cryptococcus, Aspergillus*, and *Zygomycetes* are more common. These organisms frequently cause invasive infections that are challenging to identify and treat (Bajpai et al. 2019). *Candida albicans* is a frequently observed pathogen that is one of the main causes of nosocomial bloodstream infections. *Aspergillus* species also have a major role in respiratory and systemic infections, especially in immunocompromised individuals (Suleyman and Alangaden 2021).

Traditional antifungal medications, including nystatin, amphotericin B, and fluconazole, have been used extensively to treat and prevent fungal infections. However, the effectiveness of antifungal medications has decreased due to an increase in antifungal drug resistance, which has prompted research into alternative treatment possibilities. Natural products, including peptides, phytochemicals, and essential oils, are being recognized as possible antifungal medicines despite issues with toxicity and bioavailability (Stan et al. 2021).

The Food and Agriculture Organization (FAO) and the World Health Organization (WHO) define probiotics as live bacteria that, when provided in sufficient quantities, confer health benefits to the host (Sengun and Guney 2021). The most studied probiotic strains belong to the genera *Lactobacillus* and *Bifidobacterium*, along with other beneficial microorganisms such as *Saccharomyces boulardii, Pediococcus acidilactici, Lactococcus lactis*, and *Bacillus subtilis* (Kwoji et al. 2021).

Probiotics combat fungal infections through various mechanisms, including immunomodulation, competitive exclusion, and the production of antimicrobial metabolites. Probiotics are a group of beneficial bacteria that produce a variety of secondary metabolites at various stages of growth, and which act as a biocontrol tool for various fungal infections (candidiasis, mucormycosis, aspergillosis, and fusariosis) (Chaudhari et al. 2022). Probiotic microorganisms inhibit fungal growth by producing organic acids such as lactic acid and acetic acid, hydrogen peroxide, short-chain fatty acids, and bacteriocins (Divyashree et al. 2023). Furthermore, probiotics influence gene expression and host signaling pathways, enhancing immunity against fungal pathogens. As a safe and natural therapy, probiotics offer a promising approach to managing opportunistic fungal infections (Cong et al. 2023).

16.2 Common Opportunistic Fungal Infections

Fungal infections are caused by a group of fungi that affect humans, animals, and plants at various stages. These infections can range from minor skin issues to severe systemic diseases, particularly in immunocompromised individuals (Oliveira et al. 2023). While many fungi are harmless and commonly found in the environment, some species can pose significant health risks (Table 16.1) (Ma et al. 2021). Fungal infections are categorized as superficial, cutaneous, subcutaneous, or systemic mycoses based on the location and severity of the infection (Table 16.1).

16.2.1 *Penicillium* Infections

Penicillium is a genus of filamentous fungi commonly found in the environment, particularly in soil, decaying matter, and in the air (Yadav et al. 2018). While most species are harmless and even beneficial (antibiotic and cheese production), some can act as opportunistic pathogens, primarily affecting individuals with compromised immune systems (Delgado et al. 2021). Among different species, *Penicillium marneffi* (dimorphic fungus) is the main fungal species responsible for infections in humans. Ingestion or inhalation of fungal conidia could be the mode of transmission (Fig. 16.1) (Boonsarngsuk et al. 2015), which mainly causes systemic mycosis (penicilliosis).

People living with human immunodeficiency virus (HIV) are the most affected. Population studies indicate that Penicilliosis is often found as a disseminated infection, with relatively non-specific signs and symptoms (Stone and Park 2011). Common manifestations of disseminated *Penicillium marneffi* infection in AIDS (acquired immunodeficiency syndrome) patients are fever, anemia, weight loss, lymphadenopathy, hepatosplenomegaly, respiratory signs, and skin lesions (Fig. 16.1) (Vanittanakom et al. 2006). Infection caused by other *Penicillium* species is rare, but some diseases are caused by various *Penicillium* spp. (*P. chrysogenum, P. citrinum, P. expansum*) are invasive diseases, namely, endophthalmitis, peritonitis, kidney infection, pericarditis, esophagitis, lung infection, and endocarditis. Superficial diseases include onychomycosis, dermatitis (skin infections), fungal keratitis (eye infection), and otomycosis (Lyratzopoulos et al. 2002).

Table 16.1 Overview of various fungal pathogens associated with human infections, their affected body parts, and routes of infection

Sl no.	Name of fungi	Fungal-associated infections	Affected body parts	Routes of infection	References
1	*Candida albicans*	Candidiasis	Skin, nails, and, in severe conditions, the gastrointestinal tract, genital tract, and lungs	Person to person, food, water, and soil	Rasheed (2025)
2	*Aspergillus fumigatus*	Aspergillosis	Primarily affects lungs and affects brain, kidney, and heart	Air, dust, soil, food, water, and decaying matter	Paulussen et al. (2017)
3	*Histoplasma capsulatum*	Histoplasmosis	Mainly affects the lungs and, in severe cases, the central nervous system, liver, spleen, and bone marrow	Soil with bird and bat droppings and decaying organic matter	Mittal et al. (2019)
4	*Coccidiodes immitis*	Coccidiomycosis	Lungs, skin, bones, joints, and brain	Dry, sandy soil and dusty environment	Crum (2022)
5	*Blastomyces dermatitidis*	Blastomycosis	Lungs, skin, bones, joints, and central nervous system	Moist soil and decaying organic matter	Capoor and Sharma (2024)
6	*Sporothrix schenckii*	Sporotrichosis	Primarily affects skin but in rare cases bones, joints and lungs	Soil, plants and decaying organic matter	Queiroz-Telles et al. (2022)
7	*Dermatophytes*	Dermatomycosis	Primarily affects skin and nails, foot, inner thigh and scalp	Soil, person to person and contaminated objects	Durdu and Ilkit (2021)
8	*Mucor racemosus*	Mucormycosis	Primarily affect lungs and in severe conditions sinuses, brain, skin gastrointestinal tract and multiple organs	Soil, air, dust and contaminated surfaces	Bharadwaj and Thilagavathy (2024)

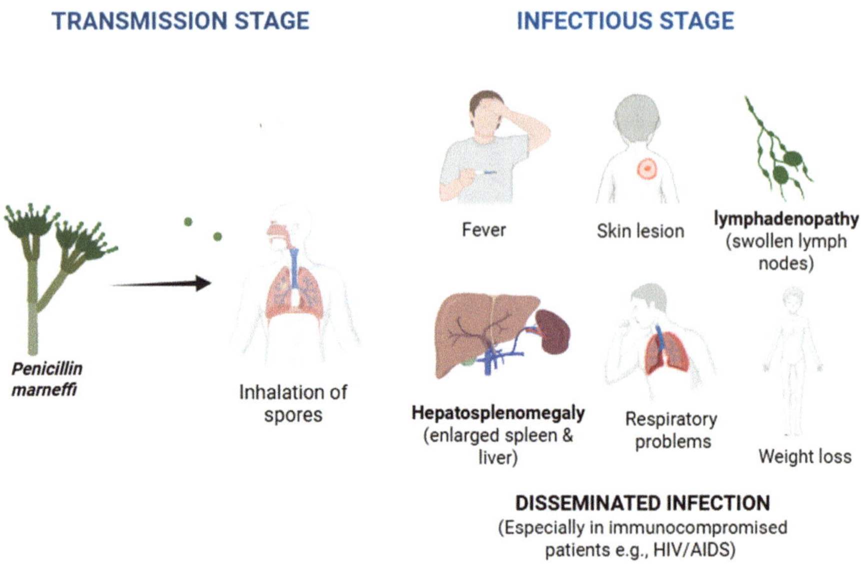

Fig. 16.1 Transmission stage and infectious stage of penicilliosis in humans

Penicillium infections are primarily treated with amphotericin B, and secondary prophylaxis with itraconazole is effective (Vanittanakom et al. 2006). The growing challenge of antifungal drug resistance has led researchers to explore alternative remedies, including antifungal compounds produced by probiotics as possible fungal infection control measures (Sharma et al. 2024).

Some of the studies reported that the probiotic species are active and efficient against most of the fungal infections in humans (Sullivan and Nord 2002). Recent studies reported that probiotics are not used as a primary treatment for *Penicillium* infections, mainly penicilliosis. But some of the probiotic bacteria showed inhibitory effects against other *Penicillium* fungal pathogens. However, *Lactobacillus casei* isolate showed maximum fungal inhibition against *Penicillium chrysogenum* (Abbaszadeh et al. 2015), and similarly, *Lacticaseibacillus rhamnosus* C1 effectively inhibited *Penicillium roqueforti.* The biocontrol mechanisms pathways include preventing and damaging mycelium, hyphae and microconidia growth (An et al. 2023). Four *Lactobacillus plantarum strains* (KCC-25, KCC-26, KCC-27, and KCC-28) showed efficient antifungal activity against *Penicillium chrysogenum* and *Penicillium roqueforti.* Among them, KCC-26 proved the maximum inhibitory effect against *Penicillium* species by producing organic acids (lactic acid, acetic acid, and succinic acid) and extracellular enzymes (β-galactosidase, α-glucosidase, β-glucosidase, acid phosphatase) (Ilavenil et al. 2015).

16.2.2 Fusariosis

Fusariosis is a disseminated fungal infection caused by *Fusarium* species, opportunistic molds that are found in soil, plants, and water. Although they are known for causing plant infections, the most dominant seed-borne fungi are commonly found in food grains (Dass et al. 2007; Deepa et al. 2016; Gurikar et al. 2023). *Fusarium* can also cause superficial, locally invasive, and disseminated infections in humans, especially in immunocompromised individuals (Muraosa et al. 2017). The most clinically relevant species are *Fusarium solani, Fusarium oxysporum,* and *Fusarium verticillioides*. Fusariosis occurs in several forms, including superficial infections like onychomycosis (fungal nail infection) and keratitis (fungal eye infection), locally invasive infections affecting the sinuses and lungs, and disseminated fusariosis, a life-threatening infection in which the infection spreads through the bloodstream, causing fever, skin lesions, pneumonia, and septic shock (Fig. 16.2). The infection is typically seen in patients with hematologic malignancies, prolonged neutropenia, organ transplants, severe burns, or prolonged immunosuppressive therapy (Sáenz et al. 2020). Diagnosis is difficult and involves microscopy, fungal culture, PCR, and histopathological examination (Nucci and Anaissie 2023). Fusariosis is extremely resistant to standard antifungal therapy, and making liposomal Amphotericin B and voriconazole the primary treatment options in severe conditions, with surgical debridement and immune restoration therapy may be the secondary treatment options (Al-Yazidi and Al-Hatmi 2021). Recent studies have indicated that probiotics, especially *Lactobacillus* and *Bacillus* species, can inhibit

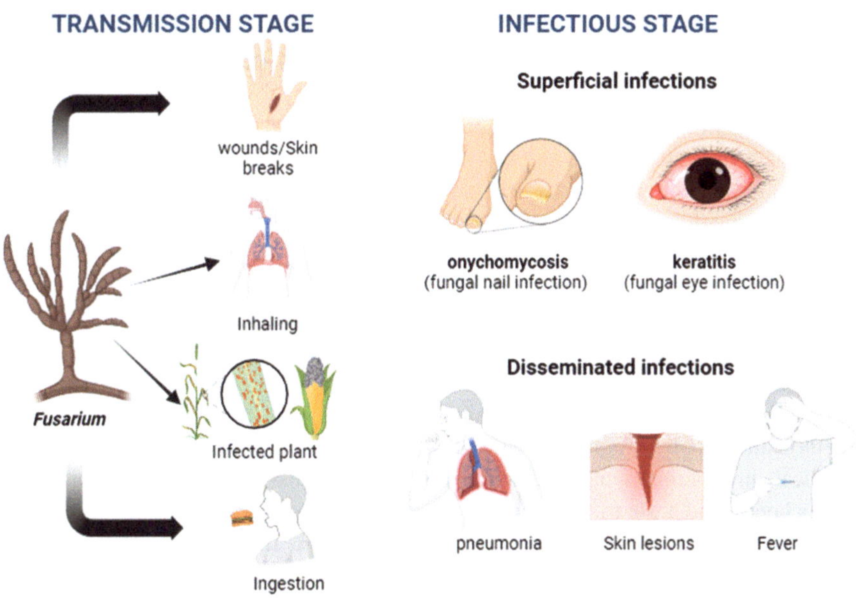

Fig. 16.2 Transmission stage and infectious stage of fusariosis in humans

Fusarium infections by releasing organic acids, augmenting immune responses, and inhibiting biofilm formation, which is the reason for fungal resistance (Table 16.2). Because of its high mortality rate and antifungal resistance, early detection and aggressive management are necessary, and new therapies, such as probiotic-based antifungal therapy, are being investigated for improved management of fusariosis (Byrne et al. 2022).

16.2.3 Blastomycosis

The fungi that cause Blastomycosis belong to the *Blastomyces* species, while *Blastomyces dermatitidis* is the primary species responsible for fungal infection. These fungi are frequently found in moist, acidic soil in wooded areas, commonly transmitted through inhalation (Capoor and Sharma 2024). Approximately half of individuals exposed remain asymptomatic, whereas symptomatic cases include lung infection, with extrapulmonary dissemination to the skin, bones, genitourinary system, and central nervous system in severe instances (Pullen et al. 2022). Treatment depends on infection severity, which includes mild to moderate infection stages that are managed with antifungal medication, i.e., oral itraconazole, while severe infection stages require initial medication with amphotericin B followed by itraconazole (Fan et al. 2023) and CNS infection stage are medicated with prolonged amphotericin B and azoles (voriconazole or fluconazole) therapy (McBride et al. 2017). Based on Recent research, there is limited work on treating blastomycosis using probiotics.

16.3 Conventional Approaches for Fungal Management

Fungal infections are a major concern in both clinical and agricultural settings requiring a diversity of management strategies. Conventional techniques for fungal infection treatment include physical and mechanical control strategies and chemical antifungal medicines (Garvey et al. 2022).

16.3.1 Physical and Mechanical Control Methods

Physical and mechanical methods are often utilized in agricultural, environmental, and medical settings to prevent fungal growth and disease. Heat treatment is a useful method that applies high temperature to sterilize soil, surgical instruments, and medical environments and thus prevents the transmission and growth of fungi. Autoclaving and pasteurization are the foremost methods used for this intent (Tucker and Featherstone 2021). Another significant method is ultraviolet (UV) light treatment, which is antifungal due to its ability to interfere with fungal DNA and is very useful in hospitals and food storage (Garvey and Rowan 2023; Deepthi et al. 2017, 2022). In case of invasive fungal infections, like mucormycosis, surgical

Table 16.2 Comprehensive overview of probiotic bacteria with antifungal properties

Probiotic bacteria	Fungal pathogen	Fungal infection caused	Inhibitory compounds	References
Lactobacillus plantarum	*Candida albicans*	Candidiasis	Bacteriocins, 3-Phenyllactic acid	Vazquez and Dongari (2021)
Leuconostoc citreum THY-F51	*Candida albicans, Candida tropicalis*	Vulvovaginal candidiasis	Bacteriocins, organic acid	Jin et al. (2025)
Lactobacillus acidophilus	*Candida albicans*	Mucosal candidiasis	Organic acids	Spaggiari et al. (2022)
Saccharomyces cerevisiae	*Candida albicans*	Mucosal candidiasis	Citric acid, propionic acid	Matsubara et al. (2016)
Bacillus subtilis R0179	*Candida albicans, Candida parapsilosis*	Oral candidiasis	Iturin A.	Zhao et al. (2016)
Lactiplantibacillus plantarum MYSVCF3, Lpb. argentoratensis MYSVCF5	*Fusarium oxysporum*	Fusariosis	Ciric acid, lactic acid and malic acid	Vasundaradevi et al. (2025)
Lactobacillus plantarum MYS6	*Fusarium proliferatum*	Fumonisin contamination in feeds	10-Octadecenoic acid, methyl ester; palmitic acid, methyl ester; heptadecanoic acid, 16-methyl ester; stearic acid; lauric acid	Deepthi et al. (2016)
Lactobacillus plantarum TR71	*F. verticillioides*	Fusariosis	Lactic acid, phenyl acetic acid, benzoic acid	Li et al. (2023)
Limosilactobacillus fermentum SD4 and *Lactiplantibacillus plantarum FCW4*	*Fusarium verticillioides, Fusarium oxysporum.*	Contamination in grain and nuts	Phenyl lactic acid, propionic acid, acetic acid	Krishnan et al. (2025)
Lactiplantibacillus pentosus	*Aspergillus flavus, Fusarium oxysporum*	Aspergillosis, Fusariosis	Phenyl lactic acid, hydroxy phenyl lactic acid	Divyashree et al. (2023)
Lactobacillus pentosus BMOBR013, Lactobacillus pentosus BMOBR061 and *Pediococcus acidilactici BMOBR041*	*Rhizopus* sp., *Mucor* sp.	Mucormycosis	3-Phenyllactic acid	Lunavath et al. (2023)

debridement is often needed to eliminate infected tissues and prevent the infections from spreading across the body (Mitchell et al. 2014). In addition, hygiene and environmental practices, like proper ventilation, dehumidification, and regular cleanliness, are important in preventing fungal growth, especially in immunocompromised individuals (Kolweni 2022).

16.3.2 Chemical Antifungal Drugs and Their Limitations

Chemical antifungal drugs are needed for the treatment of fungal infections through targeting a specific fungal structure and functions (Ivanov et al. 2022). Azoles like fluconazole, voriconazole, and itraconazole work by inhibiting the synthesis of ergosterol, a key building block of the fungal cell membrane. These drugs are frequently applied to treat superficial and systemic fungal infections (Maksimov et al. 2021). Polyenes like amphotericin B and nystatin exert antifungal activity through binding with ergosterol and the formation of pores within fungal membranes, leading to elevated permeability and cell death. Though these drugs exhibit broad-spectrum activity, their severe nephrotoxicity limits their application (Table 16.3) (Ngece et al. 2024).

Echinocandins like caspofungin, micafungin, and anidulafungin inhibit β-glucan synthesis, a vital mechanism for the sustenance of fungal cell walls. These drugs are very effective against *Aspergillus* and *Candida* infections but weakly active against *Cryptococcus* and *Mucorales* (Szymański et al. 2022). Allylamines, including terbinafine and naftifine, primarily target dermatophytes by blocking squalene epoxidase, which disrupts ergosterol production and causes fungal cell death (Hammoudi et al. 2022). Flucytosine, which is typically used in conjunction with other antifungals such as amphotericin B, inhibits fungal DNA and RNA synthesis, but its usage is limited due to severe toxicity and quick resistance development (Zhang et al. 2021). Current studies draw attention to the toxicity issues and new resistance patterns linked to these antifungal classes, highlighting the necessity of combination therapy, new antifungal drugs, and alternative therapeutic strategies to successfully treat chronic fungal infections (Fisher et al. 2022).

Table 16.3 Associated antifungal drugs and their side effects with risk factors

Associated drug	Side effect	Risk factor	References
Amphotericin B and azole antifungals	Hepatotoxicity and nephrotoxicity	Liver and kidney damage	Tragiannidis et al. (2021)
Azole-based antifungal therapy	Gastrointestinal disturbances	Nausea, vomiting, and diarrhea	Mathew and Nath (2009)
Fluconazole and ketoconazole	Hormonal imbalances	Endocrine dysfunction and hormonal imbalances	Bhattacharya et al. (2021)
Voriconazole	Neurological and cardiovascular effects	Visual problems, hallucinations, and cardiac arrhythmias	Levine and Chandrasekar (2016)

16.3.3 Side Effects of Prolonged Antifungal Drug Use

Prolonged usage of antifungal medications is associated with a range of adverse effects, affecting several organ systems and reducing the effectiveness of overall treatment (Table 16.3).

16.4 Biological Approaches in Fungal Infection Control

16.4.1 Natural Products as Antifungal Agents

Long-term usage of antifungal medications might result in toxicity issues, drug resistance, and adverse effects. So, alternative methods are required. Natural antifungal medicines, such as substances derived from plants and probiotics, provide a more sustainable, safer alternative method (Abd El-Hack et al. 2018). Fungal resistance to conventional antifungal medicines is becoming a significant threat to public health (Vitiello et al. 2023). Natural products (NPs) with a variety of bioactivities provide a viable route for creating novel antifungal therapies. NPs have been used to prevent fungal growth, some of which are in clinical use today. NPs exhibit beneficial antifungal activity, notably against drug-resistant fungal diseases (Roemer et al. 2011). The limited efficiency of current antifungal therapies can be attributed to a variety of resistance mechanisms, including drug target changes, multidrug transporter upregulation, stress response modulation, and genetic mutations (Perlin et al. 2015). Several NPs have been identified with cell wall-disrupting properties, such as plagiochin E from *Marchantia polymorpha* L., which damages the *Candida albicans* cell wall by affecting chitin synthetase genes (Wu et al. 2008), and nagilactone E from *Podocarpus nagi* inhibits β-(1,3)-D-glucan biosynthesis, weakening the fungal cell wall (Hayashi et al. 2018), while compounds like O-methyl pisiferic acid from *Chamaecyparis pisifera* inhibit chitin synthase, leading to structural instability (Chaudhary 2012). Other prominent NPs include poacic acid, which binds to β-(1,3)-D-glucan and disrupts cell wall integrity. Emodin lowers β-(1,3)-D-glucan synthase activity and alkamides from *Echinacea*, which damage fungal cell wall-membranes (García et al. 2021).

Another efficient antifungal method is to target the fungal cell membrane, which includes ergosterol, a crucial component that controls membrane fluidity, permeability, and stress tolerance (Choy et al. 2023). Pinoresinol from *Sambucus williamsii* and papyriflavonol A from *Broussonetia papyriferra* disrupt membrane integrity in *C. albicans*. Magnolol from *Magnolia officinalis* ruptures cell walls and membranes (Long and Li 2024), whereas dill seed essential oil (DSEO) breaks cytoplasmic membranes, lowers ergosterol levels, and causes oxidative stress (Taheri et al. 2023). These diverse processes highlight natural products' potential to be useful antifungal drugs, offering novel therapeutic strategies for invasive fungal infections. Besides these bioactive compounds, probiotics have also proven to be another feasible biological choice in antifungal therapy. Probiotics not only compete with fungal pathogens for nutrients and space, but also release antifungal metabolites, regulate gut microbiota and enhance the immune system, thus enhancing the body's resistance against fungal infection (Tegegne and Kebede 2022).

16.4.2 Probiotics: A Promising Biological Solution

The word "probiotic" was coined in the 1960s, and since then, research has grown exponentially. Probiotics are living, non-pathogenic microbes that, when taken in sufficient amounts, give positive health benefits. They are known by the term "health-friendly bacteria" and are often found in supplements and functional foods (Omakobia 2015). Probiotic strains like *Bifidobacterium*, *Lactobacillus*, and *Saccharomyces boulardii* are some of the most common isolates. Such microbes regulate gut microbiota, enhance digestion, immunomodulation, and infection resistance (Raheem et al. 2021). They also used to treat or prevent several conditions, such as antibiotic-associated diarrhea caused by *Candida albicans* overgrowth in the gastrointestinal tract, irritable bowel syndrome caused by *Candida tropicalis,* and assist in the management of metabolic disorders such as obesity and type 2 diabetes caused by *Candida albicans* and *Candida krusei* (He and Shi 2017). Their effects also extend beyond the gut, with new studies highlighting their potential for oral health, skin health, and even mental health across the gut-brain axis (Bermúdez et al. 2019).

As concerns over antibiotic resistance and synthetic medication side effects grow, probiotics offer a viable, natural alternative to infection prevention and treatment. Their applications are expanding, with research exploring their use in functional foods, medications, and even agricultural and environmental sustainability (Yang et al. 2024).

16.4.3 Common Probiotic Strains with Antifungal Properties

A probiotic strain with antifungal properties offers a promising natural alternative for preventing fungal growth and fungal infections (Table 16.2). Their ability to combat fungal infections by numerous mechanisms makes them beneficial for medicinal and food applications (Arena et al. 2018; Adithi et al. 2025; Vanitha et al. 2024). The consumption of potentially beneficial probiotics is associated with a variety of health advantages. As a result, the use of probiotics for both prevention and treatment of a wide range of diseases has gained acceptance during the last 10 years (Kerry et al. 2018). Recent studies highlight the importance of LAB's antibiofilm characteristics in controlling pathogens that form biofilms. The efficiency of LAB species in controlling biofilm development by several human diseases has been extensively documented (Poornachandra et al. 2015). Various probiotic strains have been explored for their antifungal activities and specific effects against fungal species (Table 16.2).

16.5 Mechanism of Antifungal Action by Probiotics

Probiotics, which are commonly accessible as GRAS (Generally Recognized as Safe)-status dietary supplements, help to maintain the host's microbiome without causing damage. They improve gut barrier function and keep away harmful pathogens from attaching to gut epithelial cells, limiting colonization and providing a complete defense against infections (O'Connell et al. 2024). Probiotics are potential alternatives to standard antibiotics that provide a variety of health advantages, including the prevention of diarrheal illness, cholesterol reduction, and immune system regulation. They also aid in regulating nosocomial and respiratory tract infections, and immune-inflammatory diseases (Khani et al. 2012). The increasing interest in probiotics is determined by the critical need for alternatives to traditional antifungals and antibiotics, especially in response to antimicrobial resistance (Habteweld and Asfaw 2023). By competing with pathogens for available space and resources in the gut, probiotics deprive pathogens of essential substrates and inhibit them from colonizing and growing. Additionally, they produce antimicrobial compounds such as organic acids and bacteriocins that directly hinder the development of harmful pathogens (Fig. 16.3) (Mokoena et al. 2021).

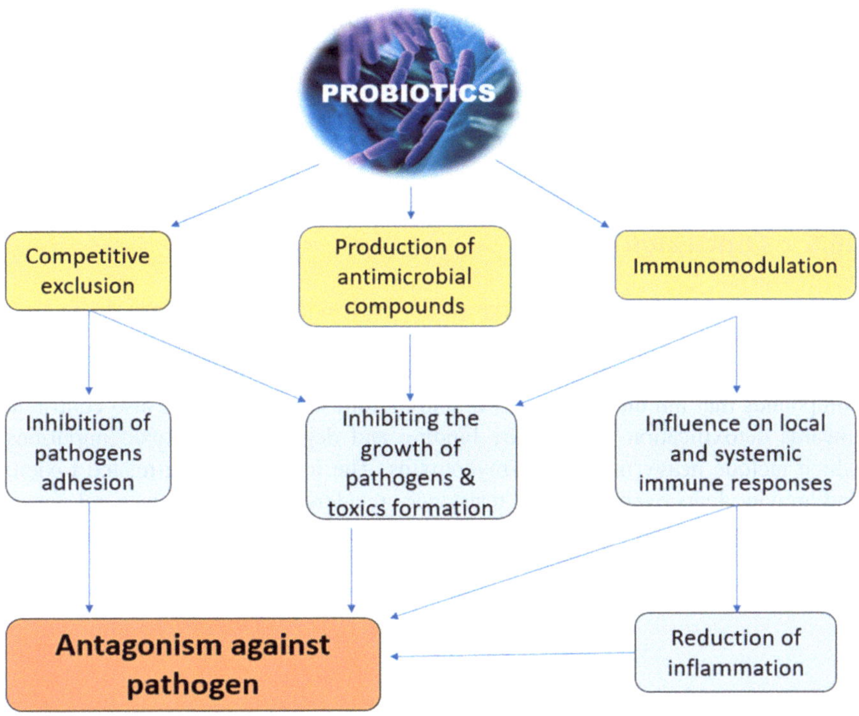

Fig. 16.3 Antagonistic action of probiotics against fungal pathogens

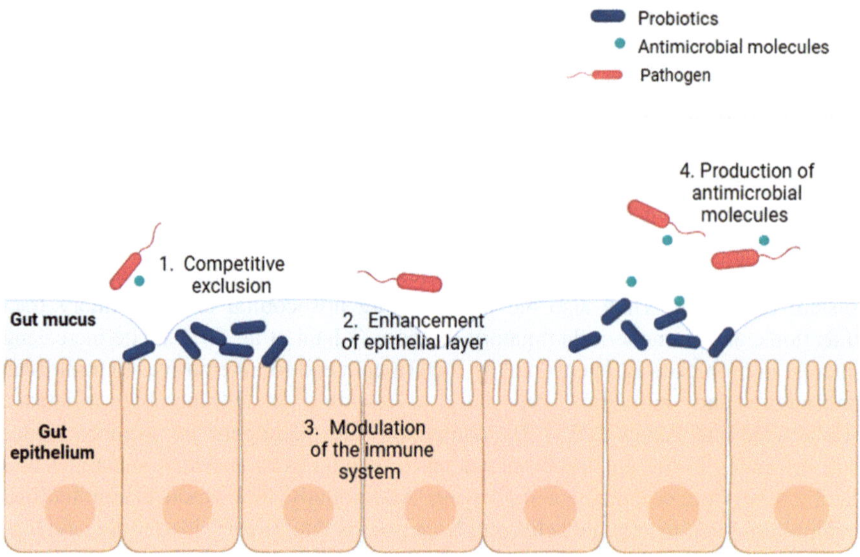

Fig. 16.4 Mechanism of action of probiotics against pathogens

Probiotics support gut integrity by reinforcing the gut barrier, which restricts pathogens' entry into host tissues. They also control immune responses by boosting innate and adaptive immunity, strengthening the body's ability to recognize and eliminate pathogens (Raheem et al. 2021). These mechanisms show how probiotics can treat infection, especially in the gut, and serve as natural substances to combat antimicrobial resistance (Fig. 16.4).

16.6 Conclusion

Probiotics have the potential to serve as a biological treatment for fungal infections and intoxications. Probiotics, especially the strains of *Lactobacillus*, *Leuconostoc*, and *Saccharomyces,* have antifungal activity through the production of antifungal compounds that inhibit fungal growth and biofilm formation. They also contribute towards detoxification through the binding and degradation of toxic impurities, which include heavy metals and mycotoxins. The long-term drug-related toxicity and growing fears over antifungal resistance are addressed through the employment of probiotics. Probiotics offer a possible method of preventing opportunistic fungal infections, which have a tendency to occur more in immunocompromised patients, by enhancing immune responses and ensuring healthy microbiomes.

16.7 Chapter Summary

The major opportunistic fungal infections are reviewed in this chapter along with the challenges with conventional antifungal therapies, such as medication resistance and adverse side effects. It emphasizes biological methods of fungal control, with a particular emphasis on probiotics as a sustainable and safe alternative. Probiotic strains such as *Lactobacillus, Leuconostoc,* and *Saccharomyces* show strong antifungal activity through the production of organic acids, hydrogen peroxide, cyclic dipeptides, and bacteriocin-like compounds, and their ability to inhibit fungal growth, disrupt biofilms, and compete for nutrients is highlighted. The chapter also discusses the detoxification potential of probiotics against mycotoxins like aflatoxins and fumonisins, as well as other toxic contaminants. Overall, probiotics offer a promising, economical, and multifunctional solution for managing fungal infections and improving food safety and public health.

Author Contributions PAH and CG conceptualized the chapter, designed the framework, and wrote the complete manuscript. SD assisted in the preparation of tables and figures. TS and HKN contributed to critical reviewing and editing and provided valuable intellectual comments to improve the quality of the chapter. All authors read and approved the final version of the chapter.

References

Abbaszadeh S, Tavakoli R, Sharifzadeh A, Shokri H (2015) Lactic acid bacteria as functional probiotic isolates for inhibiting the growth of *Aspergillus flavus, A parasiticus, A niger* and *Penicillium chrysogenum.* J Mycol Med 25(4):263–267. https://doi.org/10.1016/j.mycmed.2015.10.011. Epub 2015 Nov 17. PMID:26597145

Abd El-Hack ME, Samak DH, Noreldin AE, El-Naggar K, Abdo M (2018) Probiotics and plant-derived compounds as eco-friendly agents to inhibit microbial toxins in poultry feed: a comprehensive review. Environ Sci Pollut Res 25:31971–31986

Adithi G, Rakesh S, Shruthi B, Divyashree S, Deepa N, Vanitha PR, Vasundaradevi R, Deepthi BV, Sreenivasa MY (2025) Biological decontamination of myco-toxigenic *Fusarium verticillioides* using *Lactiplantibacillus plantarum* MYSN128 isolated from traditional vellappam: implications for food safety and global mycotoxin management. Curr Microbiol 82:404. https://doi.org/10.1007/s00284-025-04401-z

Al Yazidi LS, Al-Hatmi AM (2021) Fusariosis: an update on therapeutic options for management. Expert Opin Orphan Drugs 9(3):95–103

An P, Li L, Huang P, Zheng Y, Jin Z, Korma SA, Zhang N (2023) *Lacticaseibacillus rhamnosus* C1 effectively inhibits *Penicillium roqueforti*: effects of antimycotic culture supernatant on toxin synthesis and corresponding gene expression. Front Microbiol 13:1076511

Arena MP, Capozzi V, Russo P, Drider D, Spano G, Fiocco D (2018) Immunobiosis and probiosis: antimicrobial activity of lactic acid bacteria with a focus on their antiviral and antifungal properties. Appl Microbiol Biotechnol 102:9949–9958

Bajpai VK, Khan I, Shukla S, Kumar P, Rather IA, Park YH, Han YK (2019) Invasive fungal infections and their epidemiology: measures in the clinical scenario. Biotechnol Bioprocess Eng 24:436–444

Bermúdez-Humarán LG, Salinas E, Ortiz GG, Ramirez-Jirano LJ, Morales JA, Bitzer-Quintero OK (2019) From probiotics to psychobiotics: live beneficial bacteria which act on the brain–gut axis. Nutrients 11(4):890

Bharadwaj R, Thilagavathy S (2024) Mucormycosis in COVID-19: A clinico-microbiological dilemma. Kauverian Med J 1(11):13–27

Bhattacharya S, Kubiha S, Tyagi P (2021) Fungi and endocrine dysfunction. Endotext

Boonsarngsuk V, Eksombatchai D, Kanoksil W, Tantrakul V (2015) Airway obstruction caused by penicilliosis: a case report and review of the literature. Arch Bronconeumol (Engl Ed) 51(5): e25–e28

Byrne MB, Thapa G, Doohan FM, Burke JI (2022) Lactic acid bacteria as potential biocontrol agents for *Fusarium* head blight disease of spring barley. Front Microbiol 13:912632

Capoor MR, Sharma S (2024) Blastomycosis. In: Textbook of fungal zoonoses and sapronoses. Springer Nature, Singapore, pp 237–249

Chaudhari A, Bharti A, Dwivedi MK (2022) Role of probiotics in the management of fungal infections. In: Probiotics in the prevention and management of human diseases. Academic Press, pp 305–320

Chaudhary PM (2012) Development of novel antifungal agents using fungal cell wall as a target

Choy HL, Gaylord EA, Doering TL (2023) Ergosterol distribution controls surface structure formation and fungal pathogenicity. mBio 14(4):e01353–e01323

Cong L, Chen C, Mao S, Han Z, Zhu Z, Li Y (2023) Intestinal bacteria—a powerful weapon for fungal infections treatment. Front Cell Infect Microbiol 13:1187831

Cristofori F, Dargenio VN, Dargenio C, Miniello VL, Barone M, Francavilla R (2021) Anti-inflammatory and immunomodulatory effects of probiotics in gut inflammation: a door to the body. Front Immunol 12:578386

Crum NF (2022) Coccidioidomycosis: a contemporary review. Infect Dis Ther 11(2):713–742

Dass RS, Sreenivasa MY, Janardhana GR (2007) High incidence of *Fusarium verticillioides* in animal and poultry feed mixtures produced in Karnataka. India Plant Pathol J 6(2):174–178

Deepa N, Charith Raj AP, Sreenivasa MY (2016) Nested PCR method for the early detection of fumonisin-producing *Fusarium verticillioides* in pure cultures, cereal samples and plant parts. Food Biotechnol 30(1):18–29

Deepthi BV, Poornachandra Rao K, Chennapa G, Naik MK, Chandrashekara KT, Sreenivasa MY (2016) Antifungal attributes of *Lactobacillus plantarum* MYS6 against fumonisin producing *Fusarium proliferatum* associated with poultry feeds. PLoS One 11(6):e0155122

Deepthi BV, Gnanaprakash AP, Sreenivasa MY (2017) Effect of γ-irradiation on fumonisin-producing *Fusarium* associated with animal and poultry feed mixtures. 3 Biotech 7:57. https://doi.org/10.1007/s13205-017-0GU3-x

Deepthi BV, Deepa N, Vanitha PR (2022) Stress responses on the growth and mycotoxin biosynthesis of *Fusarium proliferatum* associated with stored poultry feeds. Appl Food Res. https://doi.org/10.1016/j.afres.2022.100091

Delgado J, Bermúdez E, Asensio MA, Núñez F (2021) Molecular identification and subtyping of toxigenic and pathogenic *Penicillium* and *Talaromyces*. In: Molecular food microbiology. CRC Press, pp 289–307

Denning DW (2024) Global incidence and mortality of severe fungal disease. Lancet Infect Dis 24(7):e428–e438

Divyashree S, Shruthi B, Vanitha PR, Sreenivasa MY (2023) Probiotics and their postbiotics for the control of opportunistic fungal pathogens: a review. Biotechnol Rep 38:e00800

Durdu M, Ilkit M (2021) Dermatophytic infections: a group of imitator diseases. In: Dermatomycology. Akademisyen Kitabevi, Ankara, pp 59–112

Fan D, Ahmad A, Frisby J, Darji K, Ojeaga A, Behshad R, Kunnath P (2023) Classical presentation of disseminated blastomycosis in a 44-year-old healthy man 3 months after diagnosis of COVID-19. Am J Case Rep 24:e938659–e938651

Fisher MC, Alastruey-Izquierdo A, Berman J, Bicanic T, Bignell EM, Bowyer P, Verweij PE (2022) Tackling the emerging threat of antifungal resistance to human health. Nat Rev Microbiol 20(9):557–571

García R, Itto-Nakama K, Rodríguez-Peña JM, Chen X, Sanz AB, de Lorenzo A, Arroyo J (2021) Poacicacid, a β-1,3-glucan–binding antifungal agent, inhibits cell-wall remodeling and activates

transcriptional responses regulated by the cell-wall integrity and high-osmolarity glycerol pathways in yeast. FASEB J 35(9):e21778
Garvey M, Rowan NJ (2023) Pathogenic drug-resistant fungi: a review of mitigation strategies. Int J Mol Sci 24(2):1584
Garvey M, Meade E, Rowan NJ (2022) Effectiveness of front line and emerging fungal disease prevention and control interventions and opportunities to address appropriate eco-sustainable solutions. Sci Total Environ 851:158284
Gurikar C, Shivaprasad DP, Sabillón L, Gowda NN, Siliveru K (2023) Impact of mycotoxins and their metabolites associated with food grains. Grain Oil Sci Technol 6(1):1–9
Habteweld HA, Asfaw T (2023) Novel dietary approach with probiotics, prebiotics, and synbiotics to mitigate antimicrobial resistance and subsequent out marketplace of antimicrobial agents: a review. Infect Drug Resist 16:3191–3211
Hammoudi Halat D, Younes S, Mourad N, Rahal M (2022) Allylamines, benzylamines, and fungal cell permeability: a review of mechanistic effects and usefulness against fungal pathogens. Membranes 12(12):1171
Hayashi K, Yamaguchi Y, Ogita A, Tanaka T, Kubo I, Fujita KI (2018) Effect of nagilactone E on cell morphology and glucan biosynthesis in budding yeast *Saccharomyces cerevisiae*. Fitoterapia 128:112–117
He M, Shi B (2017) Gut microbiota as a potential target of metabolic syndrome: the role of probiotics and prebiotics. Cell Biosci 7:1–14
Ilavenil S, Park HS, Vijayakumar M, Valan Arasu M, Kim DH, Ravikumar S, Choi KC (2015) Probiotic potential of *Lactobacillus* strains with antifungal activity isolated from animal manure. Sci World J 2015:802570
Ivanov M, Ćirić A, Stojković D (2022) Emerging antifungal targets and strategies. Int J Mol Sci 23(5):2756
Jin X, Zheng Q, Nguyen TTM, Yang SJ, Park SJ, Yi GS, Yi TH (2025) Lactic acid bacteria from kiwi: antifungal and biofilm-inhibitory activities against *Candida albicans*. Appl Sci 15(3):1647
Kerry RG, Patra JK, Gouda S, Park Y, Shin HS, Das G (2018) Benefaction of probiotics for human health: a review. J Food Drug Anal 26(3):927–939
Khani S, Hosseini HM, Taheri M, Nourani MR, Imani Fooladi A (2012) Probiotics as an alternative strategy for prevention and treatment of human diseases: a review. Inflamm Allergy Drug Targets 11(2):79–89
Kolweni ON (2022) Identification and evaluation of bacterial and fungal microbial load on surfaces in a health facility dispensary. University of Johannesburg, South Africa
Krishnan SV, Anaswara PA, Nampoothiri KM, Kovács S, Adácsi C, Szarvas P, Pusztahelyi T (2025) Biocontrol activity of new lactic acid bacteria isolates against *Fusaria* and *Fusarium* mycotoxins. Toxins 17(2):68
Kwoji ID, Aiyegoro OA, Okpeku M, Adeleke MA (2021) Multi-strain probiotics: synergy among isolates enhances biological activities. Biology 10(4):322
Levine MT, Chandrasekar PH (2016) Adverse effects of voriconazole: over a decade of use. Clin Transpl 30(11):1377–1386
Li Q, Zeng X, Fu H, Wang X, Guo X, Wang M (2023) *Lactiplantibacillus plantarum*: a comprehensive review of its antifungal and anti-mycotoxic effects. Trends Food Sci Technol 136:224–238
Long N, Li F (2024) Antifungal mechanism of natural products derived from plants: a review. Nat Prod Commun 19(8):1934578X241271747
Lunavath R, Mohammad SH, Bhukya KK, Barigela A, Banoth C, Banothu AK, Bhukya B (2023) Antimycotic effect of 3-phenyllactic acid produced by probiotic bacterial isolates against COVID-19 associated mucormycosis causing fungi. PLoS One 18(3):e0279118
Lyratzopoulos G, Ellis M, Nerringer R, Denning DW (2002) Invasive infection due to *Penicillium* species other than *P. marneffei*. J Infect 45(3):184–195

Ma Y, Wang X, Li R (2021) Cutaneous and subcutaneous fungal infections: recent developments on host–fungus interactions. Curr Opin Microbiol 62:93–102

Maksimov AY, Balandina SY, Topanov PA, Mashevskaya IV, Chaudhary S (2021) Organic antifungal drugs and targets of their action. Curr Top Med Chem 21(8):705–736

Martins-Santana L, Rezende CP, Rossi A, Martinez-Rossi NM, Almeida F (2023) Addressing microbial resistance worldwide: challenges over controlling life-threatening fungal infections. Pathogens 12(2):293

Mathew BP, Nath M (2009) Recent approaches to antifungal therapy for invasive mycoses. ChemMedChem 4(3):310–323

Matsubara VH, Bandara HMHN, Mayer MP, Samaranayake LP (2016) Probiotics as antifungals in mucosal candidiasis. Clin Infect Dis 62(9):1143–1153

McBride JA, Gauthier GM, Klein BS (2017) Clinical manifestations and treatment of blastomycosis. Clin Chest Med 38(3):435

Mitchell TA, Hardin MO, Murray CK, Ritchie JD, Cancio LC, Renz EM, White CE (2014) Mucormycosis attributed mortality: a seven-year review of surgical and medical management. Burns 40(8):1689–1695

Mittal J, Ponce MG, Gendlina I, Nosanchuk JD (2019) Histoplasma capsulatum: mechanisms for pathogenesis. Fungal Physiol Immunopathog 422:157–191

Mokoena MP, Omatola CA, Olaniran AO (2021) Applications of lactic acid bacteria and their bacteriocins against food spoilage microorganisms and foodborne pathogens. Molecules 26(22): 7055

Muraosa Y, Oguchi M, Yahiro M, Watanabe A, Yaguchi T, Kamei K (2017) Epidemiological study of *Fusarium* species causing invasive and superficial fusariosis in Japan. Med Mycol J 58(1): E5–E13

Ngece K, Ntondini TL, Khwaza V, Paca AM, Aderibigbe BA (2024) Polyene-based derivatives with antifungal activities. Pharmaceutics 16(8):1065

Nucci M, Anaissie E (2023) Invasive fusariosis. Clin Microbiol Rev 36(4):e00159–e00122

O'Connell Motherway M, O'Connor L, Smith S, Bolton D, Cormican M, De Waal T, Duffy G (2024) Assessment of the safety of "probiotics" in food supplements. 2(4):0050E

Oliveira M, Oliveira D, Lisboa C, Boechat JL, Delgado L (2023) Clinical manifestations of human exposure to fungi. J Fungi 9(3):381

Omakobia M (2015) The immunomodulatory effects of probiotics. Swansea University

Paulussen C, Hallsworth JE, Álvarez-Pérez S, Nierman WC, Hamill PG, Blain D, Lievens B (2017) Ecology of aspergillosis: insights into the pathogenic potency of *Aspergillus fumigatus* and some other *Aspergillus* species. Microb Biotechnol 10(2):296–322

Perlin DS, Shor E, Zhao Y (2015) Update on antifungal drug resistance. Curr Clin Microbiol Rep 2(2):84–95

Poornachandra Rao K, Chennappa G, Suraj U, Nagaraja H, Charith Raj AP, Sreenivasa MY (2015) Probiotic potential of *Lactobacillus* strains isolated from sorghum-based traditional fermented food. Probiotics Antimicrob Proteins 7:146–156

Pullen MF, Alpern JD, Bahr NC (2022) Blastomycosis—some progress but still much to learn. J Fungi 8(8):824

Queiroz-Telles F, Bonifaz A, Rossow J, Chindamporn A (2022) Sporothrix and sporotrichosis. Encycl Infect Immun 2:376–396

Raheem A, Liang L, Zhang G, Cui S (2021) Modulatory effects of probiotics during pathogenic infections with emphasis on immune regulation. Front Immunol 12:616713

Rasheed SO (2025) Fungal candidiasis and its role in superficial and systemic infections. Eur J Ecol Biol Agric 2(2):3–11

Roemer T, Xu D, Singh SB, Parish CA, Harris G, Wang H, Bills GF (2011) Confronting the challenges of natural product-based antifungal discovery. Chem Biol 18(2):148–164

Sáenz V, Alvarez-Moreno C, Pape PL, Restrepo S, Guarro J, Ramírez AMC (2020) A one-health perspective to recognize *Fusarium* as important in clinical practice. J Fungi 6(4):235

Şengün İY, Güney D (2021) Probiotic potential of fermented foods and their effects on immune system. Turk J Agric Food Sci Technol 9(10):1744–1750

Sharma K, Parmanu PK, Sharma M (2024) Mechanisms of antifungal resistance and developments in alternative strategies to combat *Candida albicans* infection. Arch Microbiol 206(3):95

Spaggiari L, Sala A, Ardizzoni A, De Seta F, Singh DK, Gácser A, Pericolini E (2022) *Lactobacillus acidophilus, L. plantarum, L. rhamnosus,* and *L. reuteri* cell-free supernatants inhibit *Candida parapsilosis* pathogenic potential upon infection of vaginal epithelial cells monolayer and in a transwell coculture system *in vitro*. Microbiol Spectr 10(3):e02696–e02621

Stan D, Enciu AM, Mateescu AL, Ion AC, Brezeanu AC, Stan D, Tanase C (2021) Natural compounds with antimicrobial and antiviral effect and nanocarriers used for their transportation. Front Pharmacol 12:723233

Stone A, Park BJ (2011) *Penicillium marneffei* infection: knowledge, gaps, and future directions. Curr Fungal Infect Rep 5:193–198

Suleyman G, Alangaden GJ (2021) Nosocomial fungal infections: epidemiology, infection control, and prevention. Infect Dis Clin N Am 35(4):1027–1053

Sullivan Å, Nord CE (2002) Probiotics in human infections. J Antimicrob Chemother 50(5): 625–627

Szymański M, Chmielewska S, Czyżewska U, Malinowska M, Tylicki A (2022) Echinocandins—structure, mechanism of action and use in antifungal therapy. J Enzyme Inhib Med Chem 37(1): 876–894

Taheri P, Soweizy M, Tarighi S (2023) Application of essential oils to control some important fungi and bacteria pathogenic on cereals. J Nat Pestic Res 6:100052

Tegegne BA, Kebede B (2022) Probiotics, their prophylactic and therapeutic applications in human health development: a review of the literature. Heliyon 8(6):e09725

Tragiannidis A, Gkampeta A, Vousvouki M, Vasileiou E, Groll AH (2021) Antifungal agents and the kidney: pharmacokinetics, clinical nephrotoxicity, and interactions. Expert Opin Drug Saf 20(9):1061–1074

Tucker G, Featherstone S (2021) Essentials of thermal processing. John Wiley and Sons

Vanitha PR, Divyashree S, Shruthi B, Adithi G, Deepthi BV, Sreenivasa MY (2024) *Limosilactobacillus fermentum* MYSY8, a potential probiotic isolate from fermented rice beverage for the control of *Microsporum canis*. Food Biotechnol 38(3):314–342. https://doi.org/10.1080/08U0543G.2024.2383743

Vanittanakom N, Cooper CR Jr, Fisher MC, Sirisanthana T (2006) *Penicillium marneffei* infection and recent advances in the epidemiology and molecular biology aspects. Clin Microbiol Rev 19(1):95–110

Vasundaradevi R, Sarvajith M, Divyashree S, Deepa N, Achar PN, Sreenivasa MY (2025) Tropical fruit-derived *Lactiplantibacillus* as potential probiotic and antifungal agents against *Fusarium oxysporum*. Sci Rep 15(1):2144

Vazquez-Munoz R, Dongari-Bagtzoglou A (2021) Anticandidal activities by *Lactobacillus* species: an update on mechanisms of action. Front Oral Health 2:689382

Vitiello A, Ferrara F, Boccellino M, Ponzo A, Cimmino C, Comberiati E, Sabbatucci M (2023) Antifungal drug resistance: an emergent health threat. Biomedicines 11(4):1063

Wu XZ, Cheng AX, Sun LM, Lou HX (2008) Effect of plagiochin E, an antifungal macrocyclic bis (bibenzyl), on cell wall chitin synthesis in *Candida albicans*. Acta Pharmacol Sin 29(12): 1478–1485

Yadav AN, Verma P, Kumar V, Sangwan P, Mishra S, Panjiar N, Saxena AK (2018) Biodiversity of the genus *Penicillium* in different habitats. In: New and future developments in microbial biotechnology and bioengineering. Elsevier, pp 3–18

Yang S, Qiao J, Zhang M, Kwok LY, Matijašić BB, Zhang H, Zhang W (2024) Prevention and treatment of antibiotics-associated adverse effects through the use of probiotics: a review. J Adv Res 71:209–226

Zhang Q, Liu F, Zeng M, Mao Y, Song Z (2021) Drug repurposing strategies in the development of potential antifungal agents. Appl Microbiol Biotechnol 105:5259–5279

Zhao C, Lv X, Fu J, He C, Hua H, Yan Z (2016) *In vitro* inhibitory activity of probiotic products against oral *Candida* species. J Appl Microbiol 121(1):254

Probiotics in Gastrointestinal Mycotoxin Detoxification: Innovative Approaches for Managing Fungal Toxicity

17

Subhasree Mal, Kalyanbrata Pal, Tridip Kumar Das, Shrabani Pradhan, and Kuntal Ghosh

Abstract

Mycotoxins are secondary metabolites which are produced by filamentous fungi that often contaminate food and animal feed on a global scale. Persistent exposure to compounds such as aflatoxins, deoxynivalenol, ochratoxins, and zearalenone is linked to serious health consequences, including hepatotoxicity, immunosuppression, and an increased carcinogenic risk. The existing conventional methods of decontamination are not effective; therefore, there is a need for alternative biological methods. The probiotic microorganisms are viable beneficial bacteria and yeast, which offer a promising approach to reducing mycotoxins. Probiotics like *Lactobacillus*, *Bifidobacterium,* and *Saccharomyces* strains have demonstrated the capacity to bind, enzymatically degrade, and biotransform mycotoxins and therefore reduce their bioavailability and toxicity. These help in the reconstruction of the gut microbiota and improve the mucosal damage and host defense. This chapter explored probiotic-based mycotoxin detoxification methods, with the main consideration being the strain-specific mechanism of action, namely physicochemical binding, enzymatic biotransformation, immune regulation, and strengthening of intestinal barrier functionality. Moreover,

S. Mal
Department of Biological Sciences, Midnapore City College, Midnapore, West Bengal, India

Biodiversity and Environmental Studies Research Centre affiliated to Vidyasagar University, Midnapore City College, Midnapore, West Bengal, India

K. Pal · S. Pradhan
Department of Paramedical and Allied Health Sciences, Midnapore City College, Midnapore, West Bengal, India

T. K. Das · K. Ghosh (✉)
Department of Biological Sciences, Midnapore City College, Midnapore, West Bengal, India
e-mail: kuntalghosh@mcconline.org.in

M. Y. Sreenivasa et al. (eds.), *Probiotics as Antifungals*,
https://doi.org/10.1007/978-981-92-0002-3_17

continuous attention and future prospects for incorporating probiotic technologies into food safety frameworks are also addressed. Overall, probiotic-based approaches are concluded to be a safe, sustainable, and eco-friendly method of addressing fungal toxins with significant potential to improve human and animal health.

Graphical Abstract

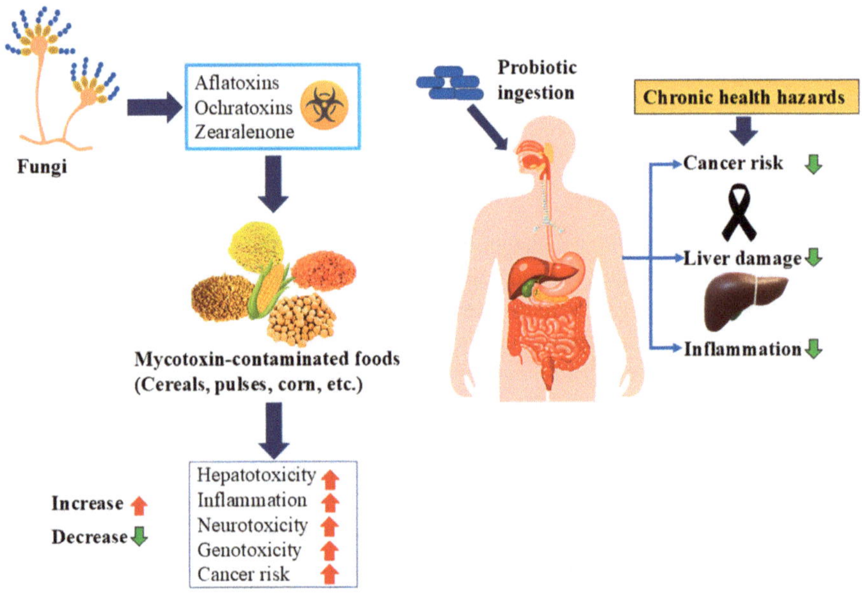

Keywords

Mycotoxins · Aflatoxins · Probiotic · Biotransformation · Immunomodulation

Learning Objectives

- Probiotics have the ability to break down toxins through their toxin-binding capabilities.
- Probiotics efflux the intestinal toxic substances and reduce the chances of adsorption in the gut.
- Probiotics maintain the gut microbial imbalance and alleviate the mucus secretion, which protects the gut.

17.1 Introduction

Mycotoxins are a group of harmful secondary metabolites, which are produced by filamentous fungi such as *Aspergillus*, *Penicillium*, and *Fusarium* (Greeff-Laubscher et al. 2020). These toxins are commonly found in a vast number of agricultural fields, especially cereal, nuts, and spices, thus presenting a severe risk to both human and veterinary health due to their powerful and diversified toxicological attributes (Yu and Pedroso 2023). The GI tract is the main pathway of mycotoxin exposure, leading to adverse effects of oxidative stress, inflammatory reactions, and loss of mucosal regulatory integrity. The dysregulations can reach the point of systemic toxicity and onset of chronic conditions, including hepatocellular carcinoma and immunosuppression (Kozieł et al. 2021). Among various mycotoxins, aflatoxins, ochratoxins, zearalenone, and deoxynivalenol can be distinguished by their high chemical stability and inertness to food processing procedures. As a result, these toxins are constantly present in the food, increasing the likelihood of continuous dietary exposure (Awuchi et al. 2021). The conventional strategies for managing mycotoxins, such as physical elimination, chemical adsorption, and thermal degradation, have limitations in their effectiveness, cost-effectiveness, and suitability in the dynamic environment of the gastrointestinal tract (Khan et al. 2024).

Over the past few years, biological detoxification using probiotics has emerged as a promising approach due to its safety, sustainability, and physiological compatibility. Probiotics are beneficial live microorganisms that can bind, degrade, and neutralize mycotoxins, reducing their bioavailability and adverse effects on health (Baralić et al. 2023). As an example, *Lactobacillus rhamnosus* and *Bifidobacterium longum* have been shown to adsorb aflatoxin in vitro other strains show enzyme activity that transforms mycotoxin to less toxic products (Baralić et al. 2023; Gratz et al. 2007). In addition, recent advances in microbial biotechnology have led to the development of the new approach, e.g., genetically modified probiotics and symbiotic formulations that could increase the quality of detoxification and offer better protection to the gut microbiome (Vinderola and Ritieni 2014).

This chapter discusses the role of probiotics in detoxifying gastrointestinal mycotoxins, with a specific focus on new measures that may be implemented to mitigate fungal toxicity. The proposed research aims to investigate the biotransformation mechanisms by which probiotic strains facilitate the degradation of mycotoxins, assessing the translational feasibility of these interventions and identifying the obstacles that must be overcome to translate them into clinical and industrial practice. Since the contamination of food with mycotoxins has remained a major challenge to food safety worldwide, probiotic designs are a paradigm that aligns with the growing trend of microbiome-based health promotion and disease prevention.

17.2 Effects of Mycotoxins on Gastrointestinal Health

Some of the mycotoxins that impact negatively on gastrointestinal health include AFB (mainly AFB_1), ochratoxins (in particular ochratoxin A), zearalenone, trichothecenes, and DON (Table 17.1). These toxins specifically differ in their

Table 17.1 The impact of major mycotoxins on gastrointestinal health.

Mycotoxin	Producing fungi	Common sources	Gastrointestinal effects	Health implications	References
Aflatoxin B_1 (AFB_1)	*Aspergillus flavus*, *A. parasiticus*	Peanuts, maize, spices	Oxidative stress, epithelial damage, leaky gut	Hepatotoxicity, cancer risk	Grenier and Applegate (2013)
Ochratoxin A (OTA)	*Penicillium verrucosum*, *A. ochraceus*	Cereals, coffee, and grapes	Disrupts tight junctions, inflammation	Nephrotoxicity, immune suppression	Lambert et al. (2007)
Zearalenone (ZEN/ZEA)	*Fusarium graminearum*, *F. culmorum*	Wheat, barley, maize	Microbial dysbiosis, endocrine disruption	Reproductive issues, gut inflammation	Zinedine et al. (2007)
Deoxynivalenol (DON)	*Fusarium graminearum*, *F. culmorum*	Wheat, oats, barley	Villus atrophy, protein synthesis inhibition, emesis	Reduced nutrient absorption, infection risk	Pinton et al. (2012)
Fumonisin B_1 (FB_1)	*Fusarium verticillioides*, *F. proliferatum*	Maize, sorghum	Sphingolipid disruption, epithelial apoptosis	Esophageal cancer, neural tube defects	Marin et al. (2013)
Patulin (PAT)	*P. expansum*, *A. clavatus*	Apples, pears, fruit juices	Oxidative damage, ulcer formation, cytotoxicity	Genotoxicity, immune modulation	Puel et al. (2010)
T-2 Toxin	*F. sporotrichioides*, *F. poae*	Oats, barley, wheat	Mucosal necrosis, immune cell depletion, emesis	Alimentary toxic aleukia, immunosuppression	Li et al. (2011)

chemical structures and toxicological properties (Zain 2011). Every year, nearly 25% of the world's food crops are infected by mycotoxins, highlighting a concerning problem and a pressing need to address this (Stoev 2024). To mitigate such issues, effective detoxification plans must be incorporated into the growth of any food crops that have been afflicted. There is a need to outline the impacts of these mycotoxins on gastrointestinal physiology, as the gastrointestinal tract is the primary route of absorption for contaminated toxins (Shekhar et al. 2025). This becomes an important issue, as the way these interactions occur is fundamental to the creation of effective detoxification and intervention measures.

As the mycotoxins pass through the digestive system, they exert a wide range of adverse effects, causing both local and systemic damage. Especially, aflatoxins are the most potent hepatotoxins that destabilize the gut condition by causing oxidative stress and damage the integrity of epithelial cells, and thus they are associated with increased intestinal barrier permeability (Grenier and Applegate 2013). This condition, similar to that of the so-called leaky gut condition, allows translocation of toxins and pathogens into the systemic circulation, resulting in increased toxicity in the system. Simultaneously, ochratoxin A, a type of nephrotoxic mycotoxin, has been shown to alter various tight junction proteins in intestinal cell lines, thereby weakening barrier functionality and inducing inflammation (Lambert et al. 2007). Another mycotoxin family, trichothecenes, for example, DON, sometimes shows its emetic effect; therefore, it is referred to as "vomitoxin". It reduces protein synthesis in rapidly dividing enterocytes; as a result, villus atrophy, diminished nutrient absorption, and increased infection susceptibility occur (Pinton et al. 2012). One estrogenic mycotoxin, Zearalenone, disrupts the balance of gut microbiota and leads to endocrine system-related dysfunctions, which contribute to the aggravation of gastrointestinal distress (Zinedine et al. 2007).

In the normal condition, the gastrointestinal microbiota play a pivotal role in maintaining gut health, which is significantly affected by mycotoxin contamination. During mycotoxin contamination, the gut microbiota becomes imbalanced, and beneficial bacteria, such as *Lactobacillus* and *Bifidobacterium*, become less abundant and increase the number of pathogenic microorganisms (Liew and Mohd-Redzwan 2018). The AFB_1 mycotoxin reduces the synthesis of SCFAs, which eventually disrupts mucosal integrity, as SCFAs-like metabolites are crucial for maintaining mucosal integrity (Wang et al. 2016). Mycotoxin-like PAT can disturb the abundance of tryptophan-degrading bacteria like Bacteroides and *Lactobacillus* to disrupt the tryptophan metabolism and their dysbiosis (Gasperini et al. 2025). The mycotoxins not only spread their toxicity but also enhance the mitochondrial ROS production to accelerate the cell apoptosis (Gao et al. 2020).

The inadequacy of traditional mitigation plans, including physical or chemical deprivation, is highlighted by exposing the body to mycotoxins, either acutely or chronically. The solutions often do not adapt to the diverse and dynamic interactions with the gastrointestinal microenvironment (Shekhar et al. 2025). Therefore, it necessitates a shift towards biologically based solutions, especially probiotics, which are the most viable candidate in solving these mycotoxin problems in the intestine. This probiotic-based strategy not only helps neutralize mycotoxin toxicity but also supports the restoration of intestinal homeostasis, making it a promising and holistic approach to health improvement (Mal et al. 2025a).

17.3 The Action Mechanism of Probiotics to Detoxify the Mycotoxin

Probiotics primarily comprise LAB, such as *Lactobacillus* and *Bifidobacterium*, as well as yeast (*Saccharomyces* sp., *Pichia kudriavzevii*), which are well-characterized live microorganisms that, when ingested in adequate amounts, confer health benefits (Das et al. 2022; Mal et al. 2025b; Adithi et al. 2025; Shruthi et al. 2025). They are effectively compatible for detoxifying fungal toxicity in the gastrointestinal tract and maintaining gut health, while also regulating toxicity related to fungal metabolites. Unlike conventional physical or chemical mitigation strategies, which are often limited by cost, scalability, and unexpected side effects, probiotics take advantage of their natural microbial process, by which they bind, degrade, and neutralize the mycotoxins' activity, minimising mycotoxin availability and alleviating gut damage. Few in vitro studies in animal models and a limited number of human trials have demonstrated the potential therapeutic importance of probiotics in managing mycotoxin activity (Table 17.2).

17.3.1 Adhesion and Adsorption

The physical adhesion or binding of mycotoxins to the cell surface of probiotics is one of the well-known mechanisms by which probiotics prevent them from adsorbing across the intestinal epithelium. The probiotics chelate the mycotoxin through non-covalent binding, facilitated by hydrogen bonding or hydrophobic forces, due to the presence of peptidoglycan and teichoic acids in their cell walls (Javanshir et al. 2021). Eiri et al. (2024) demonstrated that *L. rhamnosus* GG and *L. plantarum* exhibit a high affinity towards AFB_1, resulting in a decrease in bioavailability of up to 50–80% in simulated gastrointestinal environments. Similarly, the *B. bifidum* also demonstrated binding with ochratoxin A, where the binding capacity varies depending on the pH of the environment (Massoud and Zoghi 2022). Ragoubi et al. (2021) reported the mechanism of OTA and AFB_1 removal by using viable and heat-inactivated cells of *L. acidophilus* CIP: 76.13T and *L. delbrueckii* subsp. bulgaricus CIP: 101027T and confirmed that the concentration of the mycotoxin did not vary in both cases; therefore, they were removed by the binding and adsorbing mechanism. This strain-specific reversible binding strategy enables the toxin to be excreted with feces without absorption, ensuring the health benefits of the probiotics through pathogen inhibition and immune-modulatory effects.

Table 17.2 Probiotic mechanisms for mycotoxin detoxification

Probiotic strains	Target mycotoxin	Primary mechanisms	Effects	References
Lactobacillus rhamnosus GG	Aflatoxin B1	Binding/ adsorption	Reduces bioavailability by 50–80%	El-Nezami et al. (1998)
Bifidobacterium longum	Aflatoxin B1	Biodegradation	Transforms to less toxic aflatoxicol	Oatley et al. (2000)
Lactobacillus + *Propionibacterium*	Aflatoxin B1	Synergistic binding	Enhanced removal in multi-strain mix	Gratz et al. (2005)
L. plantarum	Ochratoxin A	Binding/ adsorption	Traps toxins for fecal excretion	Fuchs et al. (2008)
B. animalis	T-2 toxin	Antioxidant activity	Mitigates oxidative stress via SCFAs	Wang et al. (2016)
L. reuteri	Deoxynivalenol (DON)	Gut protection	Upregulates tight junctions, reduces leakiness	Akbari et al. (2017)
L. brevis	Zearalenone	Biodegradation	Converts to non-estrogenic α-zearalenol	Piotrowska and Żakowska (2005)

17.3.2 Bio-regeneration

The probiotics detoxified the mycotoxin contamination not only through physical attachment but also through enzymatic activation, which degrades the mycotoxin into less toxic compounds. This is a highly effective approach for mycotoxin detoxification. Moreover, probiotics have the ability to degrade the mycotoxins by some special biological actions, such as esterases, oxidases, and reductases that change the mycotoxins' structures. Thereby, the mycotoxins become weak or less toxic for the body. One study reported that *L. brevis* and *S. cerevisiae* can break down zearalenone into α-zearalenol, a non-estrogenic metabolite that exhibited minimal endocrine-disruptive activity (Keller et al. 2015). An in vitro model study showed 72.26 ± 7.54% reductions of AFB_1 for *Pediococcus oentosaceus*, and 69.67 ± 9.70% for *L. curvatus* due to enzymatic activity (Rafai et al. 2025). Likewise, Oatley et al. (2000) found that *B. longum* could produce reductase enzymes, which convert AFB_1 into less mutagenic, aflatoxicol. Another study also reported that *L. acidophilus* CIP: 76.13T and *L. delbrueckii* subsp. *bulgaricus* CIP: 101027T reduced the ZEA and DON in PBS by 57% and 30%, respectively (Ragoubi et al. 2021). These biotransformation activities are direct and specific to the metabolic potentialities of the probiotics, which can also be enhanced by genetic engineering or by co-aggregating with other microorganisms that have detoxifying properties. Despite the potential of biodegradation, it is not a well-established method compared to the binding and adhesion method. Therefore, it needs further safety and stability assessment of the resulting metabolites.

17.3.3 Gut Integrity Protection by Probiotics

The toxicity of mycotoxins is also diminished by enhancing the functionality of the gastrointestinal barrier and regulating the gut microenvironment, which acts indirectly to reduce toxin-induced damage. The mycotoxins, such as DON and T-2, cause disruption of tight junction proteins (including claudins and occludin), leading to increased gut permeability and inflammation (Pinton et al. 2012). Akbari et al. (2017) reported that probiotics, such as *L. reuteri* and *B. animalis*, enhanced the expression of tight junction proteins and anti-inflammatory metabolites, including short-chain fatty acids (SCFAs), thereby protecting the intestine from toxic damage. Additionally, probiotics can mitigate oxidative stress, a hallmark of mycotoxin exposure, by enhancing antioxidant enzymes, such as superoxide dismutase (SOD) and catalase, and neutralize reactive oxygen species (ROS) production to minimize oxidative damage (Das et al. 2024; Das et al. 2025; Mal et al. 2025a, 2025b). The probiotic establishment not only prevents the translocation of the mycotoxin but also restores the microbial balance (Gratz et al. 2007).

17.3.4 Compound Effect and Gut Circumstantial Elements

The effectiveness of these mechanisms is notably influenced by synergistic interactions and specific microenvironment within the gastrointestinal environment. Elaborately, the probiotic-based detoxification is enhanced when multiple strains of probiotics are used in combination with mycotoxin-degrading enzymes, which together increase effectiveness by restoring the gut environment and counteracting the mycotoxin, making the unfavourable condition for the mycotoxin-producing fungi (Markowiak and Śliżewska 2017). Huang et al. (2018) reported that *Bacillus subtilis*, *L casein*, and *Candida utilis*, together with a mycotoxin-degrading enzyme, increased the degradation rate of AFB_1 and zearalenone compared to the individual *Bacillus subtilis* or the combination of the three probiotics. Furthermore, the gut microenvironment, including pH, bile concentration, and the existing gut microbiota, enhances the probiotic functionality and increases the adsorption capability to mitigate mycotoxicity (Fuchs et al. 2008). These combined compound probiotic effects improve detoxification; however, there is a need to consider strain specificity, mycotoxin profiling, and host-related factors to optimize clinical efficacy.

Overall, probiotics are reported to be an effective way to remove mycotoxins using three major mechanisms: binding, biodegradation, and gut protection (Fig. 17.1). Together, these mechanisms offer a viable solution for managing fungal toxicity and underscore the close relationship between food safety and gut health. Although substantial evidence has been generated from in vitro experiments and animal models, further clinical investigations are needed to validate their effectiveness in everyday life.

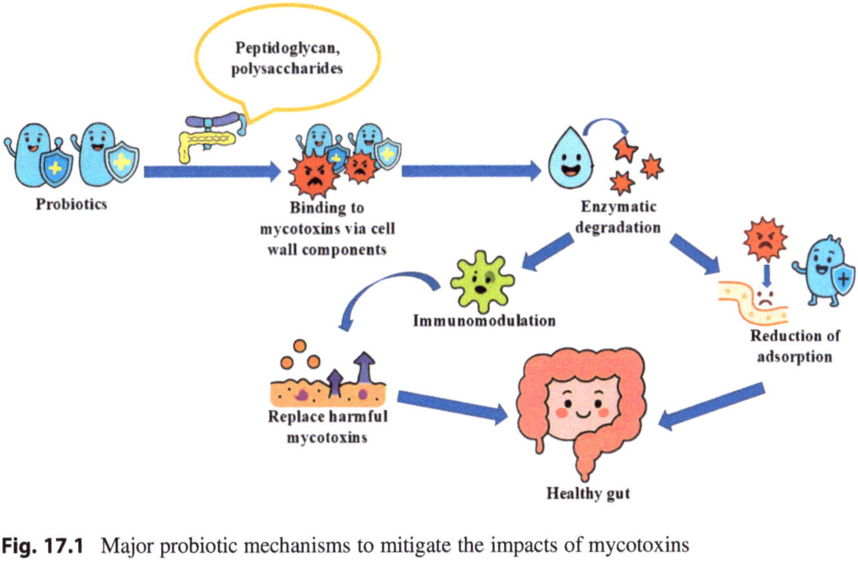

Fig. 17.1 Major probiotic mechanisms to mitigate the impacts of mycotoxins

17.4 Innovative Approach to the Use of Probiotics

The chemical adsorbents, physical removal, and enzymatic degradation are some conventional detoxification methods that encounter limitations in specificity, scalability, and in vivo applicability. However, innovative probiotic-based approaches that improve mycotoxin detoxification are possible through advancements in the fields of microbiology, biotechnology, materials science, and personalized medicine. Several new approaches have successfully and precisely addressed fungal toxicity (Table 17.3).

17.4.1 Specific Probiotic Strains

The notable diversity of microbial abilities is the basis for probiotic-based detoxification. Among the probiotics, *Lactobacillus, Bifidobacterium*, and *S. cerevisiae* (non-pathogenic yeast) can break down the mycotoxins. Over time, these capabilities have evolved in environments rich in toxins. El-Nezami et al. (1998) reported that *L. rhamnosus* GG could reduce the bioavailability of AFB_1 by up to 74% through its peptidoglycan- and polysaccharide-rich cell wall, which bound to AFB_1 in simulated intestinal conditions. Basically, the hydrophobic and electrostatic interaction with the lactone ring of AFB_1 influenced the adsorption process and the bacterial surface mechanism, which was confirmed by atomic force microscopy (Haskard et al. 2000). Likewise, *L. plantarum* EECT 749 reduced ochratoxin A (OTA) and converted it into ochratoxin α via carboxypeptidase activity (Fuchs et al. 2008). On the other hand, *S. cerevisiae* could also reduce the ZEA toxins, binding through the glucans that are present in its cell walls, and the efficacy showed best up to 80% at the optimal pH conditions (Yiannikouris et al. 2004).

The large-scale testing and metagenomic analysis of the microbes have significantly accelerated the selection of useful strains. This is due to the possibility of microbial libraries, which encompass thousands of isolates. These isolates help to identify the best strains that have a strong binding capability with the harmful mycotoxins. However, the researcher screened approximately 500 LAB, among them *L. brevis* X12 strain showed a strong binding affinity with the DON toxin. After that, this strain was used to eliminate or reduce the DON adsorption in pigs, which showed a 65% elimination rate in the intestine (Piotrowska and Żakowska 2005). Therefore, these strains could be incorporated into functional foods, dietary supplements, or animal feed, offering a scalable, cost-effective defense against mycotoxin exposure. However, the researchers may use a microbial cocktail in the future to remove several mycotoxin contaminations at a time.

17.4.2 Mycotoxins Detoxification by Next-Generation Biotics

Nowadays, researchers are developing genetically engineered probiotic strains as an innovative solution for the next generation, because of the limitations associated with natural probiotic strains. Therefore, the scientists utilized synthetic biology and

Table 17.3 Some intervention strategies for mycotoxin detoxification

Strategy	Mechanisms	Target mycotoxin (s)	Efficacy snapshot	Scalability potential	Challenges	References
Targeted probiotic strains	Binding (e.g., *L. rhamnosus* GG adsorbs AFB_1); degradation (e.g., *L. plantarum* breaks down OTA)	AFB_1, OTA, PAT	50–80% toxin reduction in vitro	High (natural strains)	Strain specificity	El-Nezami et al. (1998), Fuchs et al. (2008)
Genetic engineering	Engineered enzymes (e.g., *L. reuteri* with AFB_1 reductase; *E. coli* Nissle with laccase)	AFB_1, OTA, Fumonisins	60–70% toxin reduction in models	Moderate (regulatory)	Safety, gene stability	Yu and Pedroso (2023)
Synbiotic combinations	Prebiotics boost activity (e.g., *B. longum* + GOS); antioxidants reduce ROS (e.g., *L. casei* + quercetin)	DON, ZEA, T-2 Toxin	30–45% gut protection enhancement	High (easy formulation)	Optimal ratios unclear	Keller et al. (2015)
Advanced delivery systems	Encapsulation (e.g., *L. acidophilus* in alginate; *S. boulardii* in nanoparticles)	ZEA, OTA, Citrinin	55–90% survival/toxin reduction	Moderate (cost)	Scale-up expense	Lambert et al. (2007)
Probiotic-derived metabolites	Postbiotics bind/degrade (e.g., *L. reuteri* EPS traps DON; *B. bifidum* peptides degrade ZEA)	DON, ZEA, AFB_2	50–68% toxin reduction in vitro	High (stable, no live cells)	Production optimization	Li et al. (2011)
Personalized probiotic therapies	Tailored blends (e.g., *L. plantarum* + *B. breve* for AFB_1)	AFB_1, DON, Enniatins	40% uptake reduction in pilot	Low (diagnostics cost)	Accessibility, scalability	Marin et al. (2013)
Integration with technologies	Synergy with tools (e.g., *S. cerevisiae* + charcoal; biosensor + *L. rhamnosus*)	OTA, ZEA, AFB_1	60–70% toxin reduction in vivo	Moderate (tech complexity)	Cost, integration	Massoud and Zoghi (2022)

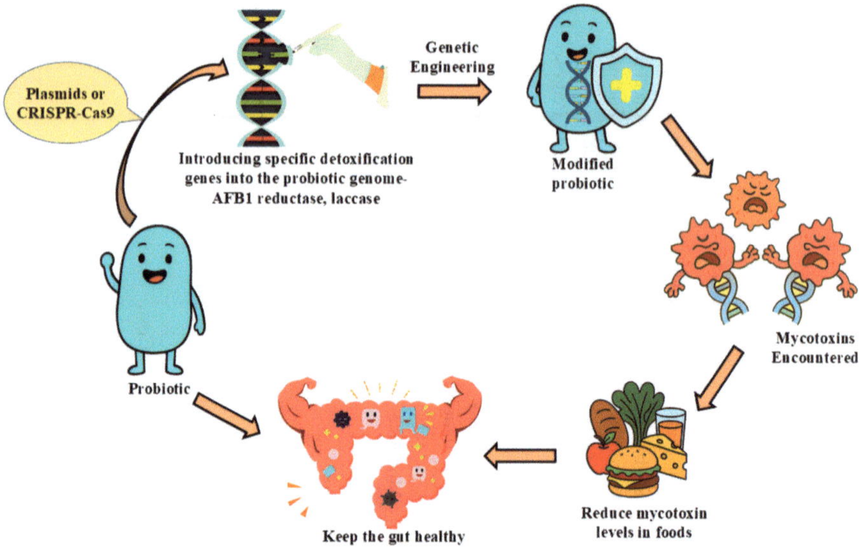

Fig. 17.2 Genetic engineering to modify probiotics and reduce mycotoxin levels

CRISPR tools to modify the probiotics, which facilitate the breakdown of mycotoxins more readily (Fig. 17.2). Numerous studies have reported that genetically engineered probiotics (*L. reuteri* and *Escherichia coli* Nissle 1917) reduced mycotoxin contamination (aflatoxins and OTA) three times faster than normal strains (Piotrowska and Żakowska 2005; Ragoubi et al. 2021). On the other hand, synbiotics are also used to remove the mycotoxins and for gut protection. A study reported that a symbiotic formulation, in combination with *B. longum* and galactooligosaccharides (GOS), significantly increased (up to 30%) SCFAs production, resulting in improved gut barrier integrity and function disrupted by DON (Bindels et al. 2015). Moreover, butyrate reduced mycotoxin levels through systemic circulation, which enhances the production of mucin and tight junction proteins (e.g., occludin) (Peng et al. 2018). Furthermore, advanced techniques, such as microencapsulation and nanotechnology using nanoparticles, have enhanced the probiotic potential of *L. acidophilus* and *S. boulardii* in the harsh gastrointestinal environment, which significantly increased the detoxification rate of mycotoxins in the gastrointestinal system (Ansari et al. 2017; Lee and Heo 2019). Additionally, the probiotic-derived postbiotic, such as exopolysaccharides (EPS) from *L. paracasei*, could prevent mycotoxin-related liver damage (Liu et al. 2018). Therefore, genetically engineered probiotic strains could be used to remove mycotoxin contamination by employing advanced delivery techniques in foods and feed supplements.

17.4.3 Incorporation with Corresponding Technologies

Probiotics could better detoxify or neutralize the harmful mycotoxins in combination with advanced technologies because probiotics do not work alone. Therefore, researchers have been using biosensors that can detect the presence of mycotoxins within the gut. This could be beneficial if put into the capsule and releases probiotics exactly at the right time when the toxins are detected. The scientists have already developed an innovative technique that can detect chemical signals within the gut and release medication (Riglar and Silver 2018). However, this is very promising because it has not yet been used in real-world treatment. It could be used in the future, specifically to target and fight against mycotoxins.

Besides its use as a biosensor, probiotics can also prevent mycotoxin contamination more effectively when combined with bentonite clay or activated charcoal or used alone. *S. cerevisiae* is a probiotic yeast that contains β-glucans as a cell wall component, among others, that effectively binds with the OTA and reduces its bioavailability in vitro (Piotrowska and Masek 2015). Likewise, if the probiotics bind with activated charcoal, it becomes stronger because probiotics can degrade the leftover mycotoxins that charcoal does not. Essentially, it functioned as a two-step detoxification process. Hamad et al. (2022) reported that the highest binding (96.80%) was for tri-mix (activated charcoal, *S. cerevisiae,* and *L. rhamnosus*) at neutral pH 6.8 for 4 h. Moreover, some minerals, such as aluminosilicate, can also serve as an effective multi-mycotoxin removal agent because they have the ability to bind with different mycotoxins simultaneously in the gastrointestinal tract (Ahn et al. 2022). The utilization of probiotics for mycotoxin detoxification in the gut is a novel approach that differs significantly from traditional methods. Moreover, these are safe and sustainable alternatives for removing mycotoxins from grains, as well as animal feeds contaminated by fungal toxins. Probiotics not only reduce mycotoxin contamination but also provide numerous positive health benefits, including improved intestinal barrier function, immunomodulation, gut microbiota homeostasis, and stress relief (Sun et al. 2023; Mafe et al. 2025). Moreover, probiotics also increased food productivity on animal farms by enhancing feed efficacy, growth, and disease resistance in a mycotoxin-contaminated environment. Therefore, probiotics have become an essential element in resolving mycotoxin-related problems, thereby enhancing food security, public health, and the agricultural field (Zoghi et al. 2022). However, the use of multiple approaches in combination is much stronger and more effective for mycotoxin detoxification in both humans and animals.

17.5 Current Obstacles and Future Prospective Research

The gastrointestinal mycotoxins detoxification by probiotics has been represented as a remarkable approach for addressing fungal toxicity. This innovative approach faces numerous challenges, including biological, technical, and socio-economic factors. Therefore, it is difficult to establish the translational laboratory research for real-world utilization. However, innovative thinking, interdisciplinary collaboration, and a future plan are necessary to effectively overcome the mentioned barriers, enabling the optimal use of probiotics in global food safety and gastrointestinal health.

17.5.1 Biological Variability and Specificity

The varying effectiveness of probiotics is a major problem. Therefore, they may not work effectively against all toxins, which depend on the specific toxins and their site of action in the body. Additionally, there are different types of mycotoxins, such as AFB_1, zearalenone, ochratoxin A, and deoxynivalenol, whose chemical structures and toxicities are clearly distinct. Therefore, it is necessary to use the multiple-strain probiotic strategy for different mycotoxins (McFarland et al. 2018). A study reported that *L. plantarum* can effectively bind to AFB_1, but does not bind to deoxynivalenol because of its size and polarity (Rämö et al. 2022). Therefore, it can be said that the activities of probiotics are strain-specific and selective. However, it is difficult to say that a single probiotic strain can remove or be effective against all mycotoxins. Moreover, the efficacy of probiotics depends on gastrointestinal conditions, such as pH, bile concentrations, and the presence of other gut microbiota, which can interfere with the detoxification of mycotoxins (Guerre 2020).

17.5.2 Scalability and Technical Barrier

Numerous studies have shown that probiotics can degrade mycotoxins in contaminated foods through various biological activities in the laboratory. However, it is very difficult to prove this in a practical setting because the food industry may not operate in the same manner as laboratory work. The mycotoxin contamination or its breakdown depends on pH, ionic strength, and temperature (Liu et al. 2022). Therefore, it is necessary to identify new, safe, and alternative probiotic strains that exhibit strong mycotoxin detoxification abilities under various environmental conditions, which can be applied practically in the food and feed industry. Moreover, it is also important to investigate whether these microbes can survive in the harsh conditions of the gastrointestinal environment and function properly within the gut before using them for detoxification in the human body. A study reported that lactic acid bacteria (LAB) can effectively degrade mycotoxins in fermented foods, thereby maintaining food quality and ensuring a healthy test result (Adebiyi et al. 2019). Additionally, some LAB strains exhibit antifungal activity, which may help prevent mycotoxin contamination. Another study also reported that *L. reuteri* R29 could reduce the DON contamination in barley by 83%. Therefore, it can be said that the use of probiotics in the real-life food industry can work as a cost-effective and useful way to control mycotoxin contamination (Oliveira et al. 2015).

17.5.3 Regulation and Safety Issues

The use of probiotics in controlling mycotoxins is also uneven, with each area perceiving it as a food additive in some cases, and in others, rigorous clinical proof is needed. Additionally, the end products of enzymatic degradation are not well studied for safety validation. A study revealed that the bacterial strain is the least preferred for examining mycotoxin-degrading products, which are used mostly due to their accessibility (Zhu et al. 2017). Eventually, human efficacy is expensive to demonstrate and raises ethical concerns, particularly with vulnerable populations (Gratz et al. 2010). The issue of safety and the lack of international standards make the commercialization even more complicated, as well as leave the researchers and manufacturers in uncertainty.

17.5.4 Accessibilities and Socioeconomic Inequality

Socioeconomic barriers prevent the use of probiotics in the fight against mycotoxins in developing regions. African, Asian, and Latin American farmers are struggling with the problem of costs and awareness as they are using the old detox methods that have disadvantages (Ezekiel et al. 2019; Njobeh and Stepman 2019). The solutions should be cost-effective and involve education to transform the cultures.

As a prospective approach, microbiome-directed probiotic regimens could provide more focused and effective detoxification, as well as artificial intelligence-based personalized probiotics. The combination of supplementary devices, e.g., biosensors, prebiotics, and dietary adsorbents, can additionally improve the results. Further efforts in the fields of research, biotechnology, and regulatory science will be needed in order to transform the probiotic-based mycotoxin detoxification into a viable therapeutic and preventative tool. Synthetic biology is capable of designing designer probiotics that are specific to particular mycotoxins by designing bacteria such as *Lactobacillus, Bifidobacterium, and E. coli Nissle 1917* (Nam et al. 2025). Mafe et al. (2025) reported that with the application of CRISPR, they increased the bacterial detox abilities and, therefore, personalized gastrointestinal security in line with personal exposure. Novel delivery systems are necessary to enhance the survival of probiotics. Probiotics are covered with microencapsulation to ensure their protection against adverse conditions and are released in the small intestine with the help of the coating, such as alginate or chitosan (Călinoiu et al. 2019). A combination of probiotics together with prebiotics, such as inulin in synbiotic formulations, increases colonization and metabolic activity. In the farm, mycotoxins can be transferred to human beings through biofertilizers or feed additives that contain probiotics. These solutions ought to be efficient in terms of cost and size to reach a broader market. Further studies on combinations of microbial consortia comprising both bacteria, yeasts, and fungi that can replicate natural gut ecosystems should be done in the future. An example of this would be the synergistic effect between a consortium of *S. cerevisiae*, *L. reuteri*, and *Bacillus subtilis* to remove several mycotoxins and improve the resilience of the gut (Lázaro et al. 2024).

Future directions of bridging lab research with real-life applications should attempt to translate translational research on high-risk populations. In areas with high exposure to mycotoxins, e.g., sub-Saharan Africa and Southeast Asia, longitudinal studies can be used to determine the impact of probiotics on such health diseases. Collaboration with local governments and non-governmental organizations may aid in making probiotics part of the national health effort, such as fortifying school feeding with fermented food. These practical data will enhance the evidence base, sponsor regulatory endorsement, and boost consumer trust. Sustainable models of production are needed to have fair access. Strains of probiotics obtained locally, produced based on local fermentation cultures (e.g., kefir, kimchi), may be more cost-effective and culturally acceptable. Renewable energy-powered bioreactor technologies could democratize production on an industrial scale, and the open-source strain libraries could enable the communities to create their own. Education communication through the use of mobile technology or community health workers will be critical to create awareness about the dangers of mycotoxins and the benefits of probiotics, especially in rural regions.

The future of probiotics in the detoxification of micro-toxins is dependent on collaboration between microbiology, toxicology, and policy. The trials can be improved with international partnerships, and the approvals can be simplified by combining the regulations. Safe development could be encouraged by incentives to private investment. Innovative solutions to the problems can provide probiotics with a global approach to combating fungal toxicity, enabling millions of people to achieve good health.

17.6 Conclusion

Mycotoxins exacerbate damage to the gut lining and increase gastrointestinal inflammation. Research has indicated that probiotics can reduce or detoxify mycotoxins or their contamination in the body, but there is a difficulty in applying these probiotics in a practical solution, which creates a gap. However, the mycotoxin detoxification by probiotics remains a promising approach. The present research is continually striving to develop new strains through a bioengineered process that can be delivered safely into the body in a more effective manner. However, the potential probiotics, their test in the human body, and their site of action where toxins are present should be focused on in future studies, which must be turned into practical solutions. Multidisciplinary research is needed to overcome the current obstacles to incorporating probiotic interventions into practical solutions that can provide positive health benefits. Additionally, the detoxification properties of these potential probiotics not only alleviate gastrointestinal problems caused by mycotoxins but also resolve other health issues in the body. Therefore, probiotics research is now emerging as a solution worldwide for its protective approach, not only for its detoxification nature. This chapter provides a clear framework for the future, along with a revolutionary approach.

17.7 Summary

Mycotoxins are generally produced by fungi as secondary metabolites, which can severely affect the gastrointestinal system through contaminated foods. Mycotoxins have a strong chemical structure that does not break down even after food processing. Therefore, they are more stable and cause severe GI disruption and accelerate the oxidative stress-related inflammation. However, probiotics are a remarkable alternative for toxin removal, rather than traditional methods such as physical removal or chemical adsorption. Additionally, probiotics offer an innovative and safe biological approach for mycotoxin detoxification through their binding ability with mycotoxins, as well as their enzymatic degradation capability, which reduces mycotoxin adsorption in the gut and restores microbial balance. The current research has developed innovative intervention strategies for detoxification, including modified probiotics, multi-strain formulations, and synbiotics, along with advanced delivery systems that may improve overall health benefits, not only in mycotoxin detoxification. However, a clinical trial is still needed to confirm their effectiveness in humans when consuming mycotoxin-contaminated foods.

Acknowledgments The authors are thankful to Dr. Pradip Ghosh, the Director of Midnapore City College, for providing the necessary facilities.

Disclosure Statement No potential conflict of interest was reported by the author(s).

Author Contributions *SM*: Writing—original draft, resources, formal analysis, conceptualization; *KP*: Writing—original draft, formal analysis, conceptualization; *TKD*: Writing—original draft, review, and editing; *SP:* Formal analysis, review, and editing; and *KG*: Writing—original draft, review and editing, formal analysis, supervision. All authors have reviewed, edited, and approved the final manuscript.

References

Adebiyi JA, Kayitesi E, Adebo OA (2019) Food fermentation and mycotoxin detoxification: an African perspective. Food Control 106:106731. https://doi.org/10.1016/j.foodcont.2019.106731

Adithi G, Rakesh S, Shruthi B, Divyashree S, Deepa N, Vanitha PR, Vasundaradevi R, Deepthi BV, Sreenivasa MY (2025) Biological decontamination of myco-toxigenic *Fusarium verticillioides* using *Lactiplantibacillus plantarum* MYSN128 isolated from traditional vellappam: implications for food safety and global mycotoxin management. Curr Microbiol 82:404. https://doi.org/10.1007/s00284-025-04401-z

Ahn JY, Kim J, Cheong DH et al (2022) An *in vitro* study on the efficacy of mycotoxin sequestering agents for aflatoxin B1, deoxynivalenol, and zearalenone. Animals 12(3):333. https://doi.org/10.3390/ani12030333

Akbari M, Tamtaji OR, Lankarani KB et al (2017) The effects of probiotics on gut permeability in health and disease. J Dig Dis 18(11):644–653. https://doi.org/10.1111/1751-2980.12539

Ansari SA, Anurag K, Chandrasekaran M (2017) Alginate–chitosan and Eudragit S100 encapsulation for enhancing probiotic viability and controlled delivery. AMB Express 7(1):78. https://doi.org/10.1186/s13568-017-0442-x

Awuchi CG, Ondari EN, Ogbonna CU et al (2021) Mycotoxins affecting animals, foods, humans, and plants: types, occurrence, toxicities, action mechanisms, prevention, and detoxification strategies—a revisit. Foods 10(6):1279. https://doi.org/10.3390/foods10061279

Baralić K, Živančević K, Bozic D et al (2023) Probiotic cultures as a potential protective strategy against the toxicity of environmentally relevant chemicals: state-of-the-art knowledge. Food Chem Toxicol 172:113582. https://doi.org/10.1016/j.fct.2022.113582

Bindels LB, Delzenne NM, Cani PD et al (2015) Towards a more comprehensive concept for prebiotics. Nat Rev Gastroenterol Hepatol 12(5):303–310. https://doi.org/10.1038/nrgastro.2015.47

Călinoiu LF, Ştefănescu BE, Pop ID et al (2019) Chitosan coating applications in probiotic microencapsulation. Coatings 9(3):194. https://doi.org/10.3390/coatings9030194

Das TK, Pradhan S, Chakrabarti S et al (2022) Current status of probiotic and related health benefits. Appl Food Res 2(2):100185. https://doi.org/10.1016/j.afres.2022.100185

Das TK, Kar P, Panchali T et al (2024) Anti-obesity potentiality of *Lactiplantibacillus plantarum* E2_MCCKT isolated from a fermented beverage, haria: a high fat diet-induced obese mice model study. World J Microbiol Biotechnol 40(6):168. https://doi.org/10.1007/s11274-024-03983-3

Das TK, Mal S, Panchali T et al (2025) The gastroprotective effect of *Lactiplantibacillus plantarum* E2_MCCKT on cold restraint stress-induced gastric ulcer mice. Food Biotechnol 39(2): 144–160. https://doi.org/10.1080/08905436.2025.2494985

Eiri A, Kaboosi H, Niknejad F et al (2024) In vitro detoxification of aflatoxin B1 by *Lactiplantibacillus plantarum* isolated from the north of Iran: a pioneering insights into the origin of fermented beverages. Folia Microbiol:1–12. https://doi.org/10.1007/s12223-024-01234-4

El-Nezami H, Kankaanpää P, Salminen S et al (1998) Ability of dairy strains of lactic acid bacteria to bind aflatoxin M1 in a food model. J Food Prot 61(2):214–216. https://doi.org/10.4315/0362-028X-61.4.466

Ezekiel C, Ortega-Beltran A, Bandyopadhyay R (2019) The need for integrated approaches to address food safety risk: the case of mycotoxins in Africa. In: Proc First FAO/WHO/AU Int Food Safety Conf, Addis Ababa, Ethiopia, pp 12–13

Fuchs S, Sontag G, Stidl R (2008) Influence of pH and bile salts on the binding of mycotoxins by probiotics. Food Chem 110(4):1227–1232. https://doi.org/10.1016/j.foodchem.2008.02.015

Gao Y, Meng L, Liu H et al (2020) The compromised intestinal barrier induced by mycotoxins. Toxins 12(10):619. https://doi.org/10.3390/toxins12100619

Gasperini AM, Faccenda D, Garcia-Cela E et al (2025) Unravelling the impact of mycotoxins on gut health: implications for inflammatory bowel disease. Curr Opin Food Sci 64:101316. https://doi.org/10.1016/j.cofs.2025.101316

Gratz S, Mykkänen H, El-Nezami H (2005) Aflatoxin B1 binding by a mixture of *Lactobacillus* and *Propionibacterium*: in vitro versus ex vivo. J Food Prot 68(11):2470–2474. https://doi.org/10.4315/0362-028X-68.11.2470

Gratz S, Wu QK, El-Nezami H et al (2007) *Lactobacillus rhamnosus* strain GG reduces aflatoxin B1 transport, metabolism, and toxicity in Caco-2 cells. Appl Environ Microbiol 73(12): 3958–3964

Gratz SW, Mykkänen H, El-Nezami HS (2010) Probiotics and gut health: a special focus on liver diseases. World J Gastroenterol 16(4):403

Greeff-Laubscher MR, Beukes I, Marais GJ et al (2020) Mycotoxin production by three different toxigenic fungi genera on formulated abalone feed and the effect of an aquatic environment on fumonisins. Mycology 11(2):105–117. https://doi.org/10.1080/21501203.2019.1604575

Grenier B, Applegate TJ (2013) Modulation of intestinal functions following mycotoxin ingestion: meta-analysis of published experiments in animals. Toxins 5(2):396–430

Guerre P (2020) Mycotoxin and gut microbiota interactions. Toxins 12(12):769. https://doi.org/10.3390/toxins12120769

Hamad GM, Amer A, El-Nogoumy B et al (2022) Evaluation of the effectiveness of charcoal, *Lactobacillus rhamnosus*, and *Saccharomyces cerevisiae* as aflatoxin adsorbents in chocolate. Toxins 15(1):21. https://doi.org/10.3390/toxins15010021

Haskard CA, Binnion C, Ahokas JT (2000) Factors affecting the sequestration of aflatoxin by *Lactobacillus rhamnosus* strain GG. Chem Biol Interact 128(1):39–49. https://doi.org/10.1016/S0009-2797(00)00186-1

Huang W, Chang J, Wang P (2018) Effect of the combined compound probiotics with mycotoxin-degradation enzyme on detoxifying aflatoxin B1 and zearalenone. J Toxicol Sci 43(6):377–385. https://doi.org/10.2131/jts.43.377

Javanshir N, Hosseini GNG, Sadeghi M et al (2021) Evaluation of the function of probiotics, emphasizing the role of their binding to the intestinal epithelium in stability and immune effects. Biol Proced Online 23:1–17. https://doi.org/10.1186/s12575-021-00160-w

Keller L, Abrunhosa L, Keller K et al (2015) Zearalenone and its derivatives α-zearalenol and β-zearalenol decontamination by *Saccharomyces cerevisiae* strains isolated from bovine forage. Toxins 7(8):3297–3308. https://doi.org/10.3390/toxins7083297

Khan R, Anwar F, Ghazali FM (2024) A comprehensive review of mycotoxins: toxicology, detection, and effective mitigation approaches. Heliyon 10:e28361. https://doi.org/10.1016/j.heliyon.2024.e28361

Kozieł MJ, Ziaja M, Piastowska-Ciesielska AW (2021) Intestinal barrier, claudins and mycotoxins. Toxins 13(11):758. https://doi.org/10.3390/toxins13110758

Lambert D, Padfield PJ, McLaughlin J et al (2007) Ochratoxin A displaces claudins from detergent-resistant membrane microdomains. Biochem Biophys Res Commun 358(2):632–636. https://doi.org/10.1016/j.bbrc.2007.04.180

Lázaro Á, Vila-Donat P, Manyes L (2024) Emerging mycotoxins and preventive strategies related to gut microbiota changes: probiotics, prebiotics, and postbiotics—a systematic review. Food Funct 15:8998–9023. https://doi.org/10.1039/D4FO01705F

Lee KY, Heo TR (2019) Survival of *Lactobacillus acidophilus* KBL409 encapsulated in alginate–chitosan beads in simulated gastric and intestinal fluids. Food Chem 290:125–131. https://doi.org/10.1016/j.foodchem.2019.04.038

Li Y, Wang Z, Beier RC (2011) T-2 toxin, a trichothecene mycotoxin: review of toxicity, metabolism, and analytical methods. J Agric Food Chem 59(8):3441–3453. https://doi.org/10.1021/jf200767

Liew WPP, Mohd-Redzwan S (2018) Mycotoxin: its impact on gut health and microbiota. Front Cell Infect Microbiol 8:60. https://doi.org/10.3389/fcimb.2018.00060

Liu Y, Ma S, Wang Q et al (2018) Protective effects of probiotic-derived metabolites against ochratoxin A-induced liver injury in mice. Front Microbiol 9:1018. https://doi.org/10.3389/fmicb.2018.01018

Liu L, Xie M, Wei D (2022) Biological detoxification of mycotoxins: current status and future advances. Int J Mol Sci 23(3):1064

Mafe AN, Nkene IH, Ali A (2025) Smart probiotic solutions for mycotoxin mitigation: innovations in food safety and sustainable agriculture. Probiotics Antimicrob Proteins:1–23. https://doi.org/10.1007/s12602-025-10569-4

Mal S, Das TK, Pradhan S et al (2025a) Probiotics as a therapeutic approach for non-infectious gastric ulcer management: a comprehensive review. Probiotics Antimicrob Proteins 17(1): 393–418. https://doi.org/10.1007/s12602-024-10348-7

Mal S, Das TK, Pradhan S et al (2025b) Fermented food-derived *Pichia kudriavzevii* A3_KGSM exhibits gastroprotective activity: a cold-restraint stress-induced mice model study. Food Biosci:108064. https://doi.org/10.1016/j.fbio.2025.108064

Marin S, Ramos AJ, Cano-Sancho G et al (2013) Mycotoxins: occurrence, toxicology, and exposure assessment. Food Chem Toxicol 60:218–237. https://doi.org/10.1016/j.fct.2013.07.047

Markowiak P, Śliżewska K (2017) Effects of probiotics, prebiotics, and synbiotics on human health. Nutrients 9(9):1021. https://doi.org/10.3390/nu9091021

Massoud R, Zoghi A (2022) Potential probiotic strains with heavy metals and mycotoxins bioremoval capacity for application in foodstuffs. J Appl Microbiol 133(3):1288–1307. https://doi.org/10.1111/jam.15685

McFarland LV, Evans CT, Goldstein EJ (2018) Strain-specificity and disease-specificity of probiotic efficacy: a systematic review and meta-analysis. Front Med 5:124. https://doi.org/10.3389/fmed.2018.00124

Nam J, Lee Y, Lee S et al (2025) Synthetic biology strategies for engineering probiotics and commensal bacteria for diagnostics and therapeutics. Biotechnol Adv:108782

Njobeh PB, Stepman F (eds) (2019) Mycotoxins: impact and management strategies, Rijeka, pp 3–22

Oatley JT, Rarick MD, Ji GE et al (2000) Binding of aflatoxin B1 to *Bifidobacteria* in vitro. J Food Prot 63(8):1133–1136. https://doi.org/10.4315/0362-028X-63.8.1133

Oliveira P, Brosnan B, Jacob F (2015) Lactic acid bacteria bioprotection applied to the malting process. Part II: substrate impact and mycotoxin reduction. Food Control 51:444–452

Peng L, Li ZR, Green RS et al (2018) Butyrate enhances the intestinal barrier by facilitating tight junction assembly via activation of AMP-activated protein kinase. Clin Sci 120(6):251–260. https://doi.org/10.1042/CS20080278

Pinton P, Nougayrède JP, Del Rio JC et al (2012) The enteric toxicity of deoxynivalenol is associated with altered intestinal cell membrane integrity. Environ Health Perspect 120(5): 577–583. https://doi.org/10.1289/ehp.1104614

Piotrowska M, Masek A (2015) *Saccharomyces cerevisiae* cell wall components as tools for ochratoxin A decontamination. Toxins 7(4):1151–1162. https://doi.org/10.3390/toxins7041151

Piotrowska M, Żakowska Z (2005) The elimination of ochratoxin A by lactic acid bacteria strains. Pol J Microbiol 54(4):279–286

Puel O, Galtier P, Oswald IP (2010) Biosynthesis and toxicological effects of patulin. Toxins 2(4): 613–631. https://doi.org/10.3390/toxins2040613

Rafai S, Moreno A, Cimbalo A et al (2025) *In vitro* evaluation of aflatoxin B1 detoxification by *Lactobacillus*, *Pediococcus*, and *Bacillus* strains. Toxins 17(8):403. https://doi.org/10.3390/toxins17080403

Ragoubi C, Quintieri L, Greco D et al (2021) Mycotoxin removal by *Lactobacillus* spp. and their application in animal liquid feed. Toxins 13(3):185. https://doi.org/10.3390/toxins13030185

Riglar DT, Silver PA (2018) Engineering bacteria for diagnostic and therapeutic applications. NPJ Syst Biol Appl 4:16. https://doi.org/10.1038/s41540-018-0063-2

Rämö S, Kahala M, Joutsjoki V (2022) Aflatoxin B1 binding by lactic acid bacteria in protein-rich plant material fermentation. Appl Sci 12(24):12769. https://doi.org/10.3390/app122412769

Shekhar R, Raghavendra VB, Rachitha P (2025) A comprehensive review of mycotoxins, their toxicity, and innovative detoxification methods. Toxicol Rep 14:101952. https://doi.org/10.1016/j.toxrep.2025.101952

Shruthi B, Deepthi BV, Adithi G, Vanitha PR, Sreenivasa MY (2025) *Saccharomyces cerevisiae* MYSY1 as a promising probiotic isolate with antifungal properties against foodborne *Aspergillus* species. Biocatal Agric Biotechnol G7:10374U. https://doi.org/10.1016/j.bcab.2025.103649

Stoev SD (2024) Food security, underestimated hazard of joint mycotoxin exposure and management of the risk of mycotoxin contamination. Food Control 159:110235. https://doi.org/10.1016/j.foodcont.2023.110235

Sun H, He Z, Xiong D et al (2023) Mechanisms by which microbial enzymes degrade four mycotoxins and application in animal production: a review. Anim Nutr 15:256–274. https://doi.org/10.1016/j.aninu.2023.09.003

Vinderola CG, Ritieni A (2014) Role of probiotics against mycotoxins and their deleterious effects. J Food Res. https://doi.org/10.5539/jfr.v4n1p10

Wang W, Huang H, Zhang X (2016) Impact of aflatoxin B1 on gut microbiota. Front Microbiol 7: 789. https://doi.org/10.3389/fmicb.2016.00789

Yiannikouris A, François J, Poughon L et al (2004) Adsorption of zearalenone by β-D-glucans in the *Saccharomyces cerevisiae* cell wall. J Food Prot 67(6):1195–1200. https://doi.org/10.4315/0362-028X-67.6.1195

Yu J, Pedroso IR (2023) Mycotoxins in cereal-based products and their impacts on the health of humans, livestock animals and pets. Toxins 15(8):480. https://doi.org/10.3390/toxins15080480

Zain ME (2011) Impact of mycotoxins on humans and animals. J Saudi Chem Soc 15(2):129–144

Zhu Y, Hassan YI, Lepp D et al (2017) Strategies and methodologies for developing microbial detoxification systems to mitigate mycotoxins. Toxins 9(4):130

Zinedine A, Soriano JM, Moltó JC et al (2007) Review on the toxicity, occurrence, metabolism, detoxification, regulations, and intake of zearalenone: an estrogenic mycotoxin. Food Chem Toxicol 45(1):1–18. https://doi.org/10.1016/j.fct.2006.07.030

Zoghi A, Todorov SD, Khosravi-Darani K (2022) Potential application of probiotics in mycotoxicosis reduction in mammals and poultry. Crit Rev Toxicol 52(9):731–741. https://doi.org/10.1080/10408444

Regulatory Perspectives: Integrating Probiotics into Clinical Practice Against Mycoses and Mycotoxin Risk Management

18

Vignesh Dhandapani, Madanraj Appiah Santharam, Ashish Chettimada, and Charith Raj Adkar Purushothama

Abstract

Probiotics are increasingly recognized as promising agents for the prevention and mitigation of mycoses and mycotoxin-related health risks due to their antifungal properties, immunomodulatory effects, and ability to bind or detoxify mycotoxins in the gastrointestinal tract. However, their clinical integration is constrained by specific regulatory gaps, including inconsistent product classification between foods, dietary supplements, and drugs; limited acceptance of strain-specific clinical evidence for disease-related claims; and the absence of harmonized pathways for probiotics targeting antifungal or antimycotoxin applications. This chapter reviews regulatory frameworks in the United States, Canada, the European Union, Japan, and India, highlighting jurisdiction-specific challenges: in the United States, probiotics may be marketed as dietary supplements but cannot carry antifungal or detoxification claims without triggering drug

V. Dhandapani
Department of Chemical and Biotechnological Engineering, Université de Sherbrooke, Sherbrooke, Canada

M. A. Santharam
Department of Immunology and Cell Biology, Faculty of Medicine and Health Sciences, Université de Sherbrooke, Sherbrooke, Canada

A. Chettimada
RNA Group, Department of Biochemistry and Functional Genomics, Faculty of Medicine and Health Sciences, Université de Sherbrooke, Sherbrooke, Canada

C. R. Adkar Purushothama (✉)
RNA Group, Department of Biochemistry and Functional Genomics, Faculty of Medicine and Health Sciences, Université de Sherbrooke, Sherbrooke, Canada

Manipal School of Life Sciences, Manipal Academy of Higher Education, Manipal, Karnataka, India
e-mail: charith.adkar@usherbrooke.ca; charith.adkar@manipal.edu

M. Y. Sreenivasa et al. (eds.), *Probiotics as Antifungals*,
https://doi.org/10.1007/978-981-92-0002-3_18

regulation; in the EU, the term "probiotic" is largely prohibited in health claims under Regulation (EC) No 1924/2006, limiting clinical translation; in Japan, Foods for Specified Health Uses (FOSHU) and Foods with Function Claims (FFC) provide a tiered but functionally restricted system; and in Canada and India, evolving guidelines face challenges in evidence thresholds, claim substantiation, and post-market oversight.

These regulatory discrepancies are particularly problematic for probiotics intended as adjunctive interventions in fungal infections or mycotoxin mitigation, which often straddle the boundary between food and pharmaceutical categories. This chapter underscores the need for risk-proportionate, indication-sensitive regulatory models, supported by standardized clinical trial guidance and harmonized definitions, to facilitate responsible innovation. Strengthened collaboration among regulators, researchers, and industry stakeholders will be critical to unlock the clinical potential of probiotics in managing mycoses and mitigating mycotoxin exposure globally.

Graphical Abstract

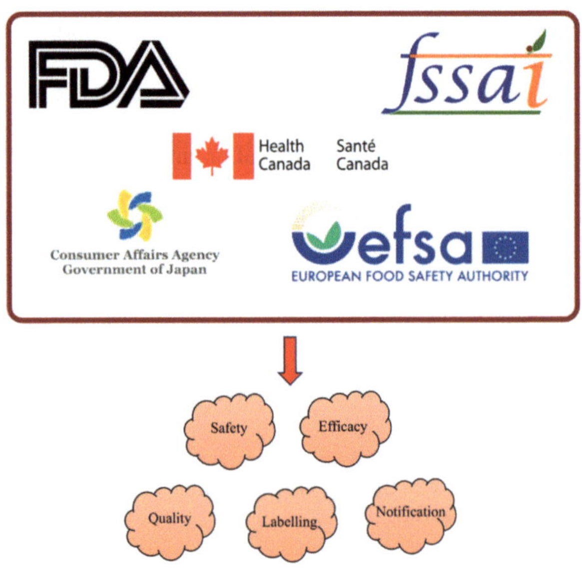

Keywords

Probiotics · Regulatory framework · Mycotoxins · Risk management · Safety and efficacy

Learning Objectives

- Describe the antifungal, immunomodulatory, and mycotoxin-mitigating mechanisms of probiotics.
- Summarize key probiotic regulatory frameworks in the United States, EU, Canada, Japan, and India.
- Identify regulatory barriers to clinical use of probiotics for mycoses and mycotoxin mitigation.
- Assess challenges in claim substantiation and evidence requirements across jurisdictions.
- Outline regulatory approaches to support safe and effective probiotic innovation.

18.1 Introduction

Fungal infections are classified as deep fungal and superficial fungal infections based on the location of the disease. Superficial infections are often highly contagious but not life-threatening. Deep fungal infections can be life-threatening and have resulted in an average of 1.5 million deaths every year globally (Dellière et al. 2020; Vitális et al. 2020). Statistics report a 50% mortality rate resulting from fungal infections even following treatment with anti-fungal agents (Brown et al. 2012). Currently, four types of drugs, including polyenes, pyrimidines, echinocandins, and azoles, are used as antifungals (Oltu et al. 2020). However, certain side effects resulting from the use of the drug and a rise in the pathogenic fungal resistance have pushed scientists to research into alternative treatments (Hu et al. 2021; Perlin 2020; Sant et al. 2016). Probiotics are defined as "living microorganisms that when administered in adequate amounts provide a health benefit to the host," according to the World Health Organization (WHO) and Food and Agriculture Organization (FAO). The most common probiotics are strains of *Bifidobacterium* (i.e., *Bifidobacterium infantis, Bifidobacterium animalis* subsp. *lactis, Bifidobacterium longum*), *Lactobacillus* (i.e., *Lactiplantibacillus plantarum, Lactobacillus casei, Lactobacillus acidophilus, Lactobacillus rhamnosus, Lactobacillus delbrueckii* subsp. *bulgaricus*), *Pediococcus acidilactici, Lactococcus lactis* subsp. *lactis, Leuconostoc mesenteroides, Bacillus subtilis, Enterococcus faecium, Escherichia coli*, and *Streptococcus thermophilus*, etc. (Divyashree et al. 2021). The yeast *Saccharomyces boulardii* also exhibits probiotic properties (Fijan 2016; Shruthi et al. 2022). Probiotics are subcategorized into probiotic foods (ingredients and supplements), probiotic drugs, direct-fed microbials (for animal use), and genetically modified probiotics. Regulations play a significant role in determining the commercialization success of a probiotic. Pharmaceutical probiotics are widely used to treat clostridia-related acute diarrhea (Goldenberg et al. 2018), irritable bowel syndrome (IBS) (Camilleri and Boeckxstaens 2017), to remove *Helicobacter pylori* (Espinoza et al. 2018), and to inhibit pathogenic fungi such as *Aspergillus fumigatus*, *Candida albicans* and *Candida glabrata* (Atanassova et al. 2003). Different regulatory agencies are responsible for regulating probiotics, such as the Food and Drug Administration (FDA) in the United States, Health Canada in Canada, the Food Safety and Standards Authority of India (FSSAI), and the Ministry of Health, Labour and Welfare (MHLW) in Japan.

In this chapter, we will outline the regulatory guidelines from a selection of key countries, including Canada, the United States, and the European Union, to provide readers with a comprehensive understanding of how probiotics are governed in clinical and food-based applications. By comparing regulatory frameworks across jurisdictions, we aim to highlight the major regulatory features that are consistently followed, including product classification, safety evaluation, strain specificity, health claim substantiation, and labeling requirements. These common regulatory elements form the foundation upon which most national frameworks are built and are especially relevant when considering the integration of probiotics into clinical practice for the prevention or mitigation of mycoses and mycotoxin-related health risks. Understanding these shared regulatory principles is essential for researchers, clinicians, and manufacturers aiming to develop, approve, or use probiotic products in global markets.

18.2 Food and Drug Administration (FDA) Regulations for Probiotics

In the United States, probiotics intended for the general healthy population are primarily regulated as dietary supplements or food products under the Dietary Supplement Health and Education Act (DSHEA) of 1994. These products do not require pre-market approval by the US Food and Drug Administration (FDA). However, manufacturers must comply with Good Manufacturing Practices (GMPs) and adhere to labeling requirements as outlined in federal regulations. The responsibility for ensuring the safety of the product and the substantiation of any health claims lies with the manufacturer. Any claims made must be supported by credible scientific evidence. Furthermore, manufacturers are prohibited from marketing dietary supplements that are adulterated (unsafe or contaminated) or misbranded (bearing false or misleading labels or claims). The FDA oversees enforcement through post-market surveillance and has the authority to act against non-compliant products.

Under the Federal Food, Drug, and Cosmetic Act, any substance intentionally added to food is considered a food additive and is subject to premarket review and approval by the FDA, unless the substance is generally recognized as safe (GRAS), among qualified experts, as having been adequately shown to be safe under the conditions of its intended use (FDA 2024c). Manufacturers can self-affirm the GRAS status of a probiotic strain or submit a GRAS notification to the FDA.

The FDA evaluates these notifications and may respond with no questions or a request for additional information or a determination that the substance does not meet GRAS criteria. For example, *Lactobacillus rhamnosus* CBT LR5 received a "no questions" letter from the FDA, affirming its GRAS status for inclusion in specific food types (Rejimus Inc. 2022). To support a GRAS determination for probiotics, manufacturers should provide detailed strain-level identification, toxicological studies, antibiotic resistance profiles, clinical data, and certificates of Analysis (COAs) confirming consistent production methods (RNI Consulting 2020).

According to the FDA, any article intended for use in the cure, mitigation, diagnosis, treatment, or prevention of disease is classified as a drug (FDA 2021). Any probiotic administered either alone to treat, cure, mitigate, or diagnose a disease or in combination with a drug to enhance its effect or alleviate the side effects should be classified as a drug. A probiotic classified as a drug would need to go through the Investigational New Drug (IND) process and subsequent approval procedures requiring extensive clinical data to support the safety and efficacy claim. The IND process is initiated by the manufacturer by submitting sufficient data and information related to the animal and pharmacological studies, manufacturing location, and clinical trial protocols. FDA reviews the IND application for safety to be tested in humans and provides a decision within 30 calendar days following the submission (Centre for Drug Evaluation, FDA 2024). Following the clinical trials, the manufacturer should submit a New Drug Application (NDA) for drugs or a Biological License Application (BLA) if it is a biologic. The drug applications are evaluated by the Centre for Drug Evaluation and Research (CDER) wing of the FDA (2024b), and the biologic applications are evaluated by the Centre for Biologics Evaluation and Research (CBER) (FDA 2025). For a probiotic to be approved as a drug by the FDA, the following stringent criteria must be met (Synapse 2025).

1. Safety and Tolerability: The probiotic must provide sufficient evidence from the pre-clinical and clinical studies that it is safe for human consumption and does not induce adverse effects.
2. Efficacy: Clinical trials with rigorous endpoints should indicate a statistically significant benefit over the standard treatments or placebo. The efficacy data should be based on well-defined endpoints and be reproducible.
3. Quality and Manufacturing control: The manufacturing process must show strict adherence to Good Manufacturing Practices (GMP). The levels of microorganisms must stay within the level indicated on the label during the entire shelf life of the product.

These criteria ensure that only the product with a greater benefit-to-risk ratio is approved as a drug. However, most probiotic products are approved as supplements and hence, do not need to go through the rigorous process. A brief outline of the process by which the probiotics are regulated in the United States is illustrated in Fig. 18.1. According to the current references, the FDA has not approved any probiotic product as a drug for the treatment of diseases. However, the FDA has approved microbiota-based therapies with similar characteristics to probiotics as drugs. For example, REBYOTA™ from Ferring's Pharmaceuticals and VOWST™ from Nestle Health Sciences have been approved as live biotherapeutic

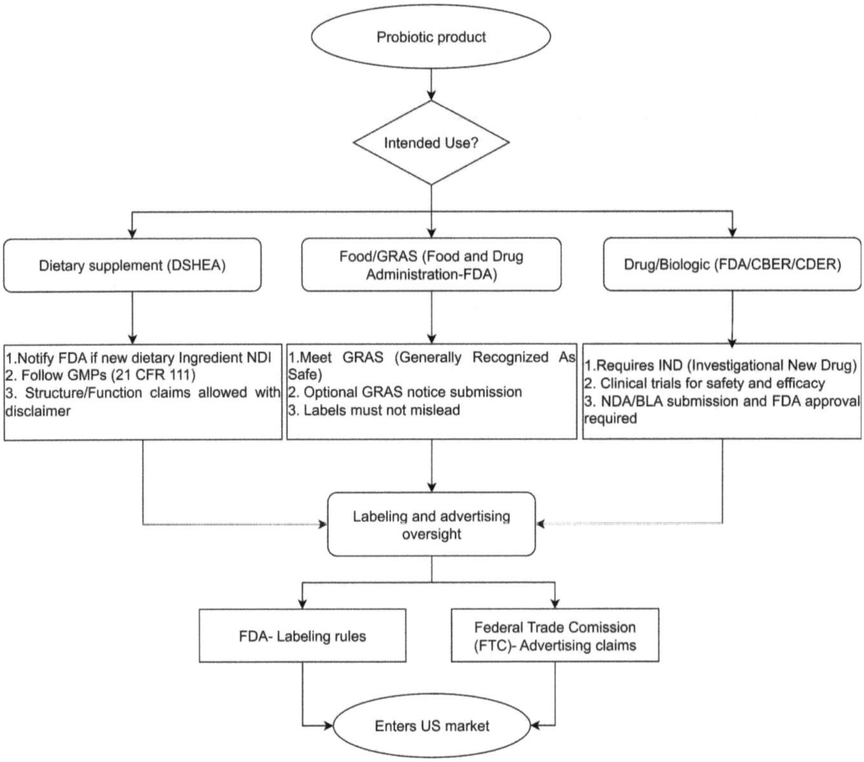

Fig. 18.1 Regulation process for probiotics in the United States

products for treating recurrent *Clostridoides difficile* infections (CDI) (Synapse 2025). These do not come under the category of conventional probiotics, and hence, all commercially available probiotics sold in the United States are approved as dietary supplements. A study conducted by Hickinson and colleagues in England, where the food of hospitalized patients over 50 years of age undergoing antibiotic treatment was supplemented with fermented milk, resulted in a significant reduction in the side effects associated with the antibiotic and diarrhea associated with *Clostridium difficile* (Hickson et al. 2007). Since the fermented milk affected the drug and condition, this would require the probiotic product to be considered as a drug and go through the pre-market approval process with the FDA.

18.3 Health Canada Regulations for Probiotics

According to Health Canada, Probiotics are live microorganisms that, when administered in adequate amounts, confer a health benefit. The usual types are bacteria (e.g., *Lactobacillus, Bifidobacterium*) or yeast. Depending on the product format and the claims made, a probiotic might be regulated as a food, a natural health product (NHP), or, in rare cases, as a drug. The acts and regulations concerning probiotics are included in the Food and Drugs Act, Food and Drug Regulations (FDR), and Natural Health Products Regulations (NHPR).

If the probiotic is added to a food product (e.g., yogurt, drink), and claims are made on the packaging or in advertising regarding its health benefits, using the term "probiotic" (or similar representations like "with beneficial probiotic cultures," "live probiotic microorganism") is considered an implied health claim. It cannot be used loosely but must be accompanied by a permitted or acceptably substantiated health claim (Health Canada 2009). General claims about probiotics contributing to healthy gut flora, etc., are called non-strain-specific claims. There is a closed list of eligible microorganisms (species) and certain conditions to meet (e.g., minimum CFU). Claims about what a specific strain does require more stringent evidence and are called strain-specific claims. As of the current guidance, Health Canada has accepted very few, or in many cases, no strain-specific claims under the food labeling route. For non-strain-specific claims, there is a required minimum number of CFU per serving size (of stated size), generally 1.0×10^9 CFU, to be present through the end of shelf-life (Health Canada 2009). The microorganism(s) must be clearly identified: genus, species, and strain identity (if strain-specific). The quantity in CFU per serving size (at the end of shelf life) must be declared (Health Canada 2009). Proper placement of identity/quantity statements close to the claim or with the nutrition facts, etc., is essential (Canadian Food Inspection Agency 2024). The probiotic strain must survive through the shelf life in the food product. Stability factors such as pH, packaging, and storage are considered relevant. The microorganism(s) used should not be "novel foods" unless they have a history of safe use or have been assessed as novel foods. If novel, they must comply with novel food regulations. Advertising claims must follow similar requirements for identity, quantity, strain information, and must not mislead. The CFIA (Canadian Food Inspection Agency) enforces compliance.

When the product is an NHP (often for supplement forms, therapeutic claims), it requires, a product licence (NPN or DIN-HM) before sale, a licensed site that manufactures, packages, labels, or imports (Health Canada 2014), the label to meet NHPR requirements, including product facts table, clear ingredient listing, and allergen/gluten info. Recent amendments have strengthened these requirements (Wray and Lewis 2002). The safety is determined by the strain identity, microbial contaminants, virulence factors, antibiotic resistance, etc. The manufacturer must ensure quality and safety. If claims are made, evidence must support them (Health Canada 2023). Probiotic NHPs are contraindicated in certain populations (e.g., immunocompromised, preterm infants) classified as vulnerable populations unless safety has been demonstrated for those groups. Health Canada has issued warnings in some cases. If a microorganism/strain has no history of safe use in food or is dealt with by novel technology, regulation as a novel food applies, and it comes under the category of Novel foods that involves pre-market safety assessment.

18.4 Regulations Governing Probiotics in India

The food Safety and Standards Authority of India (FSSAI) regulates probiotics as Foods with probiotic ingredients. Microorganism strains used in these foods are deemed to possess the probiotic property of surviving the passage through the digestive tract, be able to adhere, proliferate in the gut, and able to confer physiological benefits. Labeling of probiotics in India should include the descriptor ("Probiotic food"), genus and species including the strain designation, viable numbers at the end of the probiotic strain corresponding to the level at which the efficacy is claimed (e.g., $\geq 10^8$), recommended delivering dosage with the "best before" date, an advisory warning indicating "NOT FOR MEDICAL USE" and warning or precautions to be taken while consuming such as the side effects, contraindications. No food additives and probiotic ingredients except those listed in the Schedule VA to Schedule VF can be used in formulating the probiotic product. Probiotic microorganisms listed in Schedule VII or those approved by the FSSAI can only be used in probiotic foods. A 2016 FSSAI amendment has indicated that the viable count of probiotic organisms in the recommended serving size per day shall be $\geq 10^8$ CFU. The Food Authority may, from time to time, approve certain microorganisms after proper scientific evaluation (FSSAI 2021). Current approved strains of microorganisms that could be used in probiotic foods, as listed in Schedule VII, are *Lactobacillus acidophilus*, *Lactiplantibacillus plantarum*, *Limosilactobacillus reuteri*, *Lacticaseibacillus rhamnosus*, *Ligilactobacillus salivarius*, *Lacticaseibacillus casei*, *Levilactobacillus brevis*, *Lactobacillus johnsonii*, *Lactobacillus delbrueckii* subsp. *Bulgaricus*, *Bacillus coagulans*, *Limosilactobacillus fermentum*, *Lactobacillus caucasicus*, *Lactobacillus helveticus*, *Lactobacillus lactis*, *Lactobacillus amylovorus*, *Lactobacillus gallinarum*, *Lactobacillus delbrueckii* (*Lactobacillus delbrueckii* subsp. *delbrueckii*), *Bifidobacterium bifidum*, *Bifidobacterium lactis* (*Bifidobacterium animalis* subsp. *lactis*), *Bifidobacterium breve*, *Bifidobacterium longum* (*Bifidobacterium longum* subsp. *longum*), *Bifidobacterium animalis* (*Bifidobacterium animalis* subsp. *animalis*), *Bifidobacterium infantis* (*longum* subsp. *infantis*), *Streptococcus thermophilus* (*Streptococcus salivarius* subsp. *thermophilus*), *Saccharomyces boulardii* (*Saccharomyces cerevisiae* subsp. *boulardii*), *Saccharomyces cerevisiae* (*Saccharomyces cerevisiae* subsp. *cerevisiae*), *Lacticaseibacillus paracasei*, *Lactobacillus gasseri*, *Bacillus clausii*, established probiotic strains of *Bacillus subtilis* (FSSAI 2021).

18.5 Regulatory Framework for Probiotics in Japan

In Japan, the Consumer Affairs Agency (CAA) is the principal regulatory body now responsible for food safety standards, labeling, and regulation of functional foods, including probiotic products. The Ministry of Health, Labor and Welfare (MHLW) is also involved, especially historically and for safety/sanitary standards. Probiotics-containing foods/supplements are classified under the regulatory wings of Food for specified health uses (FOSHU) or foods with functional claims (FFC). When probiotic strains are used, especially in dairy or milk-based products (or products claiming specific health benefits), Japan now requires a filled form with a detailed description of the manufacturing facility, product, ingredients, and safety documentation regarding any non-dairy ingredients (Neo 2025). Safety documentation, such as the assessment of antibiotic resistance (probiotic strain's resistance level comparable or less than that for strains of the same species), measurement/assurance of acceptable levels of D-lactic acid production, absence of harmful secondary bile acids, consumption/safety tests in advance with no safety issues; also, post-marketing monitoring of health hazards is required (Neo 2025). Products under FOSHU must clearly carry approved health claims, recommended daily intake, usage instructions, etc. The key steps adopted by MHLW in the approval of FOSHU are shown in Fig. 18.2.

The health benefit claimed must match evidence (Przyrembel 2001). Under FFC, labeling must show that the product is a "Food with Function Claims" (機能性表示食品 "Kinōsei Hyōji Shokuhin") label prominently on the front panel, and include a notification number (Chemlinked 2024). If scientific knowledge evolves and claims are no longer appropriate, the Commissioner of the CAA can require products to cease using such claims. FOSHU requires application and government evaluation of safety and efficacy before the product is approved for the health claim (Mukherjee et al. 2022). FFC functions using a notification system. Manufacturers submit supporting scientific evidence (clinical trials or systematic reviews) to CAA. Upon notification, they may market the function claim; the government does not vet each product in advance to the same exhaustive degree as FOSHU.

Recent changes implemented by the FFC require manufacturers to notify CAA of any novel raw materials by 120 days before planned sale for supplements/tablets/powder formats. Materials with prior use in Japan require only 60 days' notice and mandatory reporting of suspected health hazards/adverse events, including cases where the causal relationship is unclear (Koe 2025). As of recent regulatory updates, Good Manufacturing Practices (GMP) are becoming mandatory for FFC products in supplement format (capsules, tablets, powders). There is a transition period and a full enforcement from September 1, 2026. Factories must ensure quality, safety, tracking of adverse events, etc. The manufacturer or seller must perform self-inspections and annual compliance reporting (Koe 2024). For dairy products (condensed milk, powdered milk, etc.) that include non-milk or non-dairy ingredients like probiotics, Japan has recently tightened rules. Probiotic strains used must meet strict safety/genetic criteria; documentation on non-dairy ingredients/mixing ratios; safety data, etc. The CAA has updated the Food Labelling Standards to include the current new requirements for FFC businesses: e.g., mandatory reporting of health hazard cases, the requirement for GMP, and stronger labeling (placement of "機能性表示食品" (Kinōsei hyōji shokuhin) label, notification numbers) (McKenzie 2025). Beginning April 1, 2025, manufacturers using novel ingredients in FFC-supplement formats (tablets, powders, capsules) must submit their notification documents 120 days before sale (Koe 2025). Clear categorization (FOSHU vs FFC vs FNFC) allows for graded regulatory burden depending on the nature of claims, and mechanisms exist for safety testing, genetic strain identification, and post-market surveillance.

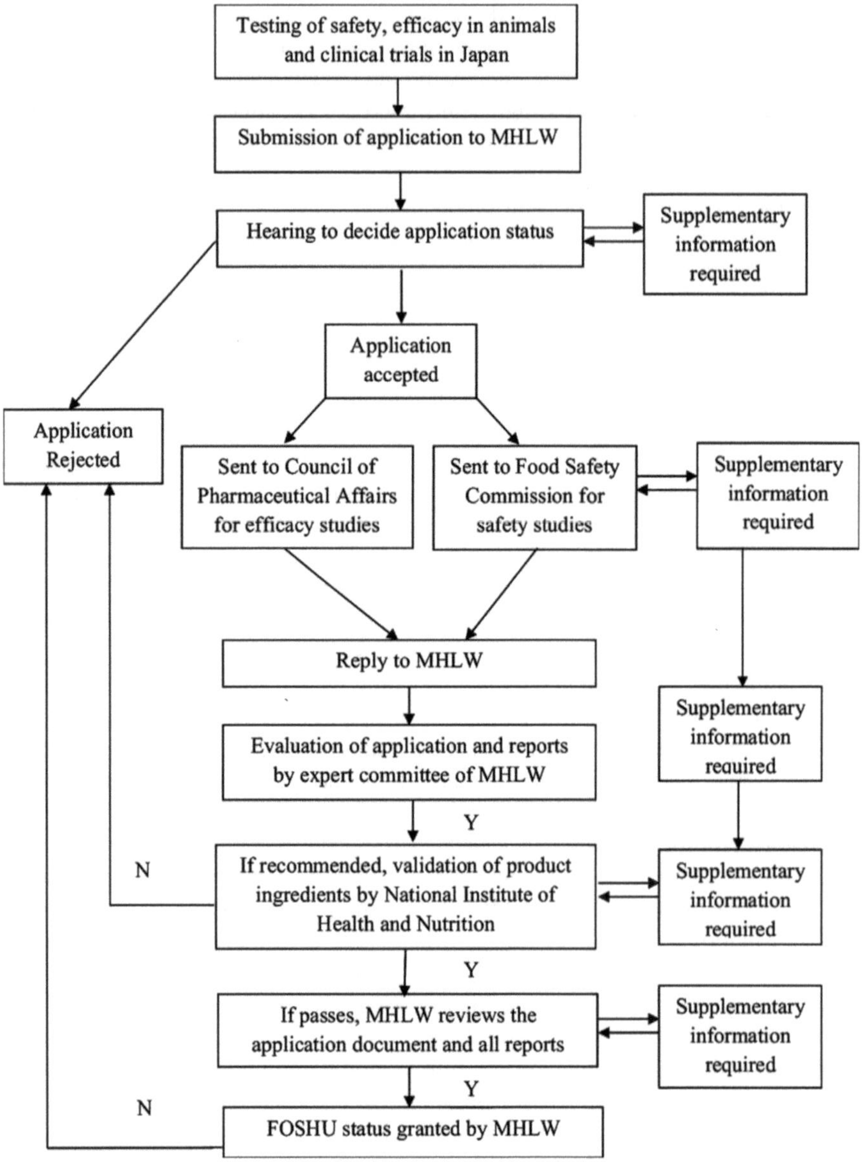

Fig. 18.2 Approval of probiotics under the FOSHU process. (Adopted from Arora et al. (2019) as per CC BY 4.0)

FFC notifications are listed, so consumers and other stakeholders can see claims and evidence submitted (Nippon 2024). Although there are advantages surrounding the Japanese regulations to regulate probiotics, certain disadvantages exist. Under FFC, because the government does *not* always fully evaluate every product's efficacy or safety before marketing, there is variable quality of the evidence submitted (Mukherjee et al. 2022). Some safety incidents (e.g., the Kobayashi red yeast rice supplement case) have highlighted risk from insufficient oversight or delayed reporting of adverse events (Koe 2024). The transition to stricter GMP requirements and novel ingredient notifications may pose compliance challenges for smaller manufacturers. Claims relating to more ambitious health effects (e.g., disease treatment or risk mitigation, detoxification, mycoses/fungal infection) are more difficult to approve, particularly under FOSHU. Under FFC, such claims would likely be disallowed or require much more evidence.

18.6 European Union Regulations for Probiotics

Probiotics in the EU are regulated largely under the general food law, health claims regulation, and the novel foods regulation, rather than by a dedicated "probiotics" regime. Key regulatory actors include the European Food Safety Authority (EFSA), the European Commission, and member-state competent authorities. The regulatory structure emphasizes safety, clear scientific substantiation of claims, and prevention of misleading labeling or advertising. The principal regulation governing the health and nutrition claims made on food products (including those with probiotic organisms) is Regulation (EC) No 1924/2006 Nutrition and Health Claims Regulation (NHCR). Under NHCR, any health claim must be authorized (via EFSA scientific assessment) before use in labeling or advertising (European Commission 2007). If a microorganism (probiotic strain) or formulation has not been used in food in the EU to a significant degree before 15 May 1997, it is considered a novel food. Such a microorganism must undergo the novel food assessment (safety, composition, use) before being placed on the market according to the Novel Food Regulation (EU) 2015/2283 (European Commission 2015). The European Food Safety Authority (EFSA) plays a central role in assessing claims and in evaluating the safety of novel microorganisms. The Qualified Presumption of Safety (QPS) framework is used for microorganisms with established safety records. If a strain is QPS-approved (at species level or similar), that helps streamline the safety assessment (European Food Safety Authority 2025). EU Regulation 1169/2011 (Food Information to Consumers) governs labeling requirements, including ingredient listing, allergen information, and ensuring that labeling is not misleading. Also, "probiotic" is interpreted under NHCR as a health claim (or implying one), so its use is heavily regulated (European IPA 2011). Very few health claims involving live microorganisms have been authorized under NHCR. The one well-known claim is that live cultures of *Lactobacillus delbrueckii* subsp. *bulgaricus* and *Streptococcus thermophilus* in yoghurt help improve lactose digestion (European IPA 2011). The term "probiotic" is generally considered a health claim in EU regulation. That is, simply stating "contains probiotics" is taken to imply a health benefit, which must be authorized. Since no broad, generic probiotics health claims have been authorized (beyond the specific yoghurt and lactose digestion claim), use of "probiotic" is risky/ unlawful unless backed by an approved claim (Food Safety Authority of Ireland

2025). Probiotic strains that are novel (not in regular food use before the cutoff date) must be evaluated and authorized under the Novel Foods Regulation. EFSA gives safety opinions, and the European Commission grants authorization, which then enters a Union-list for novel foods. Examples include *Clostridium butyricum* being assessed as a novel food (EFSA 2011). Probiotic strains must be well characterized (genus, species, strain), with evidence of safety (absence of harmful traits like antibiotic resistance, toxin production). The strain must be stable, viable, and production processes consistent For safety, the QPS list helps but does not cover all organisms (European Food Safety Authority 2011).

The EU system ensures rigorous scientific evaluation before health claims are used, which helps protect consumers from misleading statements. The Novel Foods Regulation provides a clear pathway for the assessment of new probiotic strains. QPS is useful for safety streamlining when strains have established usage and good safety history. Since very few (nearly none) generic "probiotic" health claims have been authorized, the industry is constrained in how it can label or market probiotic products. This leads to a lack of clarity for consumers and potentially limits innovation (Biosafe 2025). Lack of harmonization in how member states enforce national guidelines creates fragmentation. Consumers or companies may face different interpretations across the EU (Hancocks 2020). Table 18.1 highlights the key differences in how probiotics are regulated by various regulatory agencies.

18.7 International Harmonization Initiatives

Beyond Codex Alimentarius and industry-led efforts such as those of the International Probiotics Association, broader international initiatives led by the World Health Organization (WHO) and the Organization for Economic Co-operation and Development (OECD) offer underutilized but highly relevant frameworks for regulatory harmonization. The WHO, through its long-standing collaboration with FAO, has provided foundational guidance on probiotic evaluation, safety assessment, and functional claims, particularly in the context of food-based interventions and public health nutrition (FAO/WHO 2002; WHO 2022). These frameworks emphasize strain-level identification, documented safety histories, and evidence-based health outcomes, principles that remain directly applicable to antifungal and antimycotoxin probiotic claims.

The OECD has contributed complementary guidance through its Working Party on Biotechnology, Nanotechnology and Converging Technologies, particularly in relation to biological products, microbial risk assessment, and data transportability across jurisdictions (OECD 2022). OECD consensus documents on the safety assessment of microorganisms used in food and feed provide a potential template for internationally accepted probiotic dossiers, facilitating mutual recognition of safety data and reducing duplicative regulatory burdens. Such approaches are especially relevant for probiotics with antifungal or detoxification functions, where safety profiles are often well established, but efficacy endpoints remain heterogeneous.

Table 18.1 Key differences in the regulation of probiotics

Aspect	Canada (Health Canada)	USA (FDA)	EU (EFSA)	Japan (CAA/MHLW/PMDA)	Australia/NZ (TGA/FSANZ)	India (FSSAI/CDSCO)
Product categories	Foods; NHPs	Foods; supplements; drugs	Foods; supplements; medicines	FFC; FOSHU; pharmaceuticals	Foods; complementary medicines	Foods; health supplements; drugs
Premarket approval (food)	No (except novel foods)	No	No (novel foods only)	FFC: No; FOSHU: Yes	No	Yes, for novel probiotics
Premarket approval (supplement/NHP)	Yes (NPN)	No	Varies	Varies	Yes	Yes (FSSAI compliance)
Use of the term "probiotic"	Allowed with accepted claim	Allowed	Not allowed (no claims approved)	Allowed	Allowed	Allowed under FSSAI guidelines
Health claims	General claims allowed	Structure/function claims	Very limited	Allowed (tier-based)	Allowed	Limited, function-based claims
Clinical evidence burden	Moderate–high	Low–moderate	Very high	Tier-dependent	Moderate–high	Moderate
Novel strain assessment	Required	GRAS/NDIN	Novel food authorization	Required	Required	Mandatory safety and efficacy data
Post market surveillance	Yes	Limited	Yes	Yes	Yes	Limited

A future roadmap for harmonization could involve alignment across three regulatory layers:

(i) Codex-led global definitions and baseline safety standards for probiotics and related live microbial products.
(ii) WHO-driven public health guidance distinguishing low-risk, preventive, and adjunctive probiotic applications from high-risk therapeutic LBPs.
(iii) OECD-facilitated data harmonization, enabling cross-border acceptance of strain characterization, toxicology, and manufacturing quality data. Together, these mechanisms could support proportionate regulation while accommodating regional differences in healthcare infrastructure and disease burden.

Importantly, integrating antifungal and antimycotoxin probiotics into these international frameworks would also align with WHO priorities on food safety, antimicrobial resistance mitigation, and the prevention of fungal diseases, which are increasingly recognized as neglected global health challenges (WHO 2022). Leveraging existing WHO and OECD guidance rather than creating entirely new regulatory silos may therefore represent the most pragmatic and globally inclusive path toward harmonized probiotic regulation.

18.8 Mechanistic and Clinical Justification for Regulatory Adaptation

Regulatory adaptation for probiotics in fungal and mycotoxin contexts must be grounded in a clear understanding of mechanistic plausibility and demonstrated clinical relevance. At the molecular level, multiple probiotic strains exert antifungal activity through well-characterized mechanisms, including the production of organic acids (e.g., lactic and acetic acids), secretion of antifungal metabolites such as cyclic dipeptides and phenolic compounds, competition for adhesion sites, and disruption of fungal biofilm formation (Parolin et al. 2015; Rima et al. 2012). Certain *Lactobacillus* and *Saccharomyces* strains further modulate host immune responses by enhancing epithelial barrier integrity, upregulating antimicrobial peptides, and promoting Th1/Th17-mediated antifungal immunity, thereby reducing susceptibility to opportunistic fungal infections (Damián et al. 2022; Köhler et al. 2017; Llewellyn et al. 2017; Mazziotta et al. 2023).

In the context of mycotoxin exposure, probiotics demonstrate additional detoxification mechanisms that are distinct from conventional antifungal drugs. These include physical binding of mycotoxins to bacterial cell wall components (e.g., peptidoglycan, polysaccharides), enzymatic biotransformation into less toxic metabolites, and indirect mitigation through restoration of gut barrier function and modulation of xenobiotic-metabolizing pathways (El-Nezami et al. 2002; Gratz et al. 2005; Haskard et al. 2001). Such mechanisms are particularly relevant for chronic, low-dose dietary exposure to mycotoxins, where pharmaceutical interventions are often impractical or unavailable.

Importantly, these mechanistic findings are supported by emerging clinical and translational evidence. Randomized controlled trials have demonstrated that probiotic supplementation can reduce vaginal and gastrointestinal *Candida* colonization, decrease recurrence rates of vulvovaginal candidiasis, and improve antifungal treatment outcomes when used as adjunctive therapy (Reid et al. 2001; Russo et al. 2019). In mycotoxin-exposed populations, clinical and field studies have shown that probiotic interventions can significantly reduce urinary or faecal biomarkers of aflatoxin exposure and improve surrogate markers of gut and immune health, particularly in high-risk regions with dietary contamination (El-Nezami et al. 2006; Mohd Redzwan et al. 2016).

Beyond conventional dietary supplements, emerging therapeutic applications further underscore the need for regulatory flexibility. These include next-generation probiotics and defined microbial consortia targeting fungal dysbiosis, genetically characterized strains optimized for toxin degradation, and symbiotic formulations designed to enhance antifungal efficacy or toxin sequestration. Such innovations blur traditional food–drug boundaries while remaining mechanistically distinct from small-molecule antifungals, reinforcing the argument for proportionate, indication-sensitive regulatory pathways rather than default pharmaceutical classification.

Collectively, the convergence of mechanistic clarity, growing clinical evidence, and novel therapeutic formats provides a compelling rationale for regulatory frameworks that recognize probiotics as scientifically validated adjuncts in fungal disease management and mycotoxin risk reduction. Incorporating these data into regulatory decision-making would support evidence-based differentiation of claims, encourage standardized clinical trial designs, and facilitate responsible translation of probiotic innovations into public health and clinical practice.

18.9 Challenges

The regulatory approval of probiotics is limited primarily by stringent scientific and regulatory evidence requirements, coupled with uncertainty surrounding their mechanisms of action. Probiotic effects are highly strain-specific, meaning that efficacy demonstrated for one strain cannot be extrapolated to others, even within the same species. Regulatory agencies, particularly the European Food Safety Authority (EFSA), consider the term *"probiotic"* a health claim, requiring convincing evidence from well-designed human intervention studies that demonstrate a clear cause-and-effect relationship between a defined strain, dose, and health outcome. As a result, the majority of probiotic health claims submitted to EFSA have been rejected due to inadequate clinical endpoints, insufficient statistical power, poor strain characterization, or lack of mechanistic substantiation (Binnendijk and Rijkers 2013; Hill et al. 2014). In contrast, countries such as the United States regulate probiotics largely as dietary supplements, where pre-market efficacy approval is not mandatory; however, disease-related claims remain prohibited, creating a regulatory grey zone that limits meaningful clinical claims while allowing widespread commercialization (FDA 2024a). This global lack of harmonization creates a critical hurdle, increasing development costs and discouraging investment in large-scale, regulatory-grade clinical trials.

Additional constraints arise from manufacturing, quality control, and safety assessment challenges inherent to live microbial products. Probiotics are biologically dynamic, and issues such as strain misidentification, genetic instability, variability in colony-forming unit (CFU) counts, and loss of viability during storage undermine reproducibility and regulatory confidence (Cordaillat-Simmons et al. 2020; Douillard and de Vos 2014). Furthermore, there is no universal consensus on appropriate biomarkers or clinical endpoints for probiotic efficacy, making regulatory assessment subjective and inconsistent across agencies (Marco et al. 2021). Safety evaluation also presents challenges, particularly for novel strains or applications in immunocompromised populations, as long-term surveillance data are limited and risk–benefit profiles remain incompletely defined (Suez et al. 2019). Critically, the absence of standardized international guidelines for probiotic characterization, clinical validation, and labeling continues to restrict approval pathways, highlighting the need for coordinated regulatory frameworks that balance innovation with consumer protection.

18.10 Perspectives and Regulatory Outlook

As the global burden of mycoses and mycotoxin-related illnesses continues to rise, particularly in regions with compromised food safety and healthcare infrastructure, there is increasing interest in microbiome-centered interventions such as probiotics. Accumulating experimental and early clinical evidence suggests that selected probiotic strains can inhibit fungal pathogens, modulate host immune responses, and bind or biotransform mycotoxins in the gastrointestinal tract (Haskard et al. 2001; Rima et al. 2012). However, despite this promise, translation into clinically accepted interventions remains hindered by substantial regulatory bottlenecks, including ambiguous product classification, fragmented international frameworks, and the absence of indication-specific clinical evaluation standards (Sanders et al. 2019).

A central regulatory challenge lies in the dual identity of probiotics as foods and therapeutic agents. In many jurisdictions, probiotics are regulated either as foods, dietary supplements, or drugs, depending on their intended use and associated claims. Regulatory systems originally designed for conventional foods or pharmaceuticals are often ill-equipped to address this hybrid nature (Hill et al. 2014). In the United States, probiotics marketed as dietary supplements face relatively low premarket barriers but are prohibited from making disease-related claims without triggering drug classification under FDA regulations (FDA 2016). In contrast, within the European Union, probiotics are primarily regulated as foods, yet the European Food Safety Authority (EFSA) has consistently rejected health claims under Regulation (EC) No 1924/2006, effectively restricting even general probiotic-related claims (EFSA Panel on Dietetic Products, Nutrition and Allergies (NDA) 2011). This divergence has created a regulatory paradox that discourages the development of probiotics intended for targeted applications such as antifungal support or mycotoxin mitigation.

These challenges are becoming more pronounced with the emergence of live biotherapeutic products (LBPs), which occupy the regulatory space between probiotics and pharmaceutical biologics. Regulatory agencies such as the FDA and EMA have formally recognized LBPs as biological medicinal products intended for the prevention or treatment of disease (EMA 2019; FDA 2016). While this pathway provides regulatory clarity for high-risk or therapeutic applications, it also imposes pharmaceutical-level manufacturing controls, clinical trial requirements, and post-market surveillance that may be disproportionate for probiotic strains with long histories of safe use (Cordaillat-Simmons et al. 2020). For antifungal or antimycotoxin indications—often envisioned as adjunctive, preventive, or population-level interventions—the LBP framework may therefore act as a barrier to innovation, particularly for small and medium-sized enterprises and for deployment in low- and middle-income countries.

Another critical limitation is the lack of harmonized clinical trial guidelines specific to antifungal and antimycotoxin probiotic claims. Current studies vary widely in strain selection, dosing strategies, endpoints, and study populations, limiting reproducibility and regulatory acceptance (Sanders et al. 2019). Unlike antibacterial drug development, there is no consensus on appropriate primary endpoints for antifungal probiotic interventions, such as reductions in fungal colonization, modulation of host immune responses, prevention of recurrence, or decreased biomarkers of mycotoxin exposure. Uniform clinical guidance should incorporate strain-level genomic characterization, validated biomarkers of fungal burden or toxin exposure, standardized outcome measures for adjunctive efficacy, and safety monitoring tailored to vulnerable or immunocompromised populations.

Looking forward, adaptive regulatory models may offer a more proportionate and innovation-friendly approach. Risk-tiered claim systems, conditional or staged approvals, and expedited pathways for well-characterized probiotic strains with robust safety data could facilitate responsible product development without compromising consumer protection (Sanders et al. 2019). International harmonization initiatives led by bodies such as Codex Alimentarius, the International Probiotics Association, and cross-regional scientific working groups are essential to reduce discrepancies in definitions, safety assessments, and permissible claims (FAO/WHO 2002). Without such alignment, global development and equitable access to probiotic-based interventions for fungal and mycotoxin-related diseases will remain constrained.

Finally, ethical and public health considerations must be integral to regulatory evolution. Probiotics represent a potentially affordable and scalable intervention for populations disproportionately affected by mycotoxin exposure and opportunistic fungal infections. Regulatory systems should therefore avoid imposing unnecessarily burdensome pharmaceutical requirements on low-risk probiotic applications, particularly when intended as adjunctive or preventive measures. A proportionate, science-based regulatory framework supported by standardized clinical evidence and international cooperation will be critical to realizing the full potential of probiotics in managing mycoses and mycotoxin-related illnesses.

18.11 Conclusion

The global regulatory landscape governing probiotics is both diverse and evolving, reflecting growing scientific interest and commercial activity in this field. While regulatory frameworks vary across jurisdictions, including the United States, Canada, the European Union, India, and Japan, certain foundational principles remain consistent: product safety, strain-specific characterization, substantiated health claims, appropriate labeling, and adherence to manufacturing quality standards.

As research continues to explore the role of probiotics in mitigating fungal infections (mycoses) and managing the health risks associated with mycotoxins, it becomes increasingly clear that integrating these products into clinical practice requires more than just scientific validation. Regulatory alignment, compliance, and transparency are critical to ensuring both consumer safety and therapeutic credibility.

Probiotics intended for clinical use in combating mycoses or reducing mycotoxin exposure face heightened regulatory scrutiny, particularly when positioned as interventions with disease-preventive or therapeutic intent. These claims often cross into medical or drug territory, triggering stricter pathways for approval. As such, developers must be prepared to conduct rigorous clinical trials, establish clear mechanisms of action, and provide robust safety data to satisfy regulatory authorities.

Moreover, regional differences in how the term "probiotic" is treated, especially in the EU, where it is considered a health claim, highlight the challenges of harmonizing global product strategies. In contrast, jurisdictions like the United States offer more flexibility for marketing probiotics as dietary supplements, though without pre-market approval. Meanwhile, countries like Japan and India offer hybrid models, allowing health claims under specified regulatory systems (e.g., FOSHU, FFC, or FSSAI guidelines) that require evidence but may not demand full-scale pharmaceutical-level trials.

For clinicians and researchers, understanding these regulatory nuances is essential not only for product development but also for informed decision-making in clinical settings. As the science of probiotics matures, especially in specialized applications like fungal control and toxin neutralization, regulatory systems must also adapt to allow for innovation while maintaining rigorous standards of safety and efficacy.

Ultimately, bridging the gap between scientific potential and regulatory acceptance will be key to realizing the full clinical promise of probiotics in managing mycoses and mycotoxin-related health risks. Collaborative efforts among scientists, regulators, and industry stakeholders will be necessary to streamline approval pathways, standardize definitions, and build international consensus around the responsible use of probiotics in medicine and public health.

Acknowledgments BioRender was used to create the graphical abstract. OpenAI tools were used to assist with literature searching, language editing, and formatting standardization. All referenced literature was manually verified to ensure accuracy and validity.

Study Questions

1. What are the regulations governing approval of probiotics in the United States?
2. What are the regulations governing approval of probiotics in Canada?
3. What are the regulations governing approval of probiotics in India?
4. What are the regulations governing approval of probiotics in Japan?
5. What are the regulations governing approval of probiotics in the European Union?
6. What are the key differences in the regulations of probiotics between different countries?
7. Describe the international initiatives to harmonize the regulations for probiotics.
8. List the challenges associated with the regulation of probiotics.

References

Arora M, Arora M, Bansal P, Baldi A (2019) Need and recommendations for universal guidelines on regulatory status and quality control/safety of probiotic products. Appl Clin Res, Clin Trials Regul Aff. https://www.eurekaselect.com/article/96378 6:231

Atanassova M, Choiset Y, Dalgalarrondo M, Chobert J-M, Dousset X, Ivanova I, Haertlé T (2003) Isolation and partial biochemical characterization of a proteinaceous anti-bacteria and anti-yeast compound produced by Lactobacillus paracasei subsp. paracasei strain M3. Int J Food Microbiol 87(1–2):63–73. https://doi.org/10.1016/S0168-1605(03)00054-0

Binnendijk KH, Rijkers GT (2013) What is a health benefit? An evaluation of EFSA opinions on health benefits with reference to probiotics. Benef Microbes 4(3):223–230. https://doi.org/10.3920/BM2013.0019

Biosafe (2025) Probiotics and beyond in the EU: a regulatory maze for food and feed. Biosaf Insights. https://www.biosafe.fi/insight/probiotics-and-beyond-in-the-eu-a-regulatory-maze-for-food-and-feed/

Brown GD, Denning DW, Gow NAR, Levitz SM, Netea MG, White TC (2012) Hidden killers: human fungal infections. Sci Transl Med 4(165):165rv13. https://doi.org/10.1126/scitranslmed.3004404

Camilleri M, Boeckxstaens G (2017) Dietary and pharmacological treatment of abdominal pain in IBS. Gut 66(5):966–974. https://doi.org/10.1136/gutjnl-2016-313425

Canadian Food Inspection Agency (2024) Health claims on food labels. Canadian Food Inspection Agency. https://inspection.canada.ca/en/food-labels/labelling/industry/health-claims

Center for Drug Evaluation and Research (CDER), FDA (2024) Investigational New Drug (IND) application. U.S. Food Drug Adm. https://www.fda.gov/drugs/types-applications/investigational-new-drug-ind-application

ChemLinked (2024) [Updated] Japan Finalizes the Health Supplement Rules After Kobayashi Incident. ChemLinked. http://food.chemlinked.com/news/food-news/japan-proposes-to-amend-health-supplement-rules-after-kobayashi-incident

Cordaillat-Simmons M, Rouanet A, Pot B (2020) Live biotherapeutic products: the importance of a defined regulatory framework. Exp Mol Med 52(9):1397–1406. https://doi.org/10.1038/s12276-020-0437-6

Damián MR, Cortes-Perez NG, Quintana ET, Ortiz-Moreno A, Noguez CG, Cruceño-Casarrubias CE, Pardo MES, Bermúdez-Humarán LG (2022) Functional foods, nutraceuticals and probiotics: a focus on human health. Microorganisms 10(5):1065. https://doi.org/10.3390/microorganisms10051065

Dellière S, Rivero-Menendez O, Gautier C, Garcia-Hermoso D, Alastruey-Izquierdo A, Alanio A (2020) Emerging mould infections: get prepared to meet unexpected fungi in your patient. Med Mycol 58(2):156–162. https://doi.org/10.1093/mmy/myz039

Divyashree S, Anjali PG, Somashekaraiah R, Sreenivasa MY (2021) Probiotic properties of Lactobacillus casei and Lactobacillus plantarum. Biotechnol Rep 32:e00672. https://doi.org/10.1016/j.btre.2021.e00672

Douillard FP, de Vos WM (2014) Functional genomics of lactic acid bacteria: from food to health. Microb Cell Factories 13(Suppl 1):S8. https://doi.org/10.1186/1475-2859-13-S1-S8

EFSA Panel on Dietetic Products, Nutrition and Allergies (NDA) (2011) Guidance on the scientific requirements for health claims related to gut and immune function. EFSA J 9(4):1984. https://doi.org/10.2903/j.efsa.2011.1984

El-Nezami HS, Polychronaki NN, Ma J et al (2006) Probiotic supplementation reduces a biomarker for increased risk of liver cancer in young men from Southern China. Am J Clin Nutr 83(5): 1199–1203. https://doi.org/10.1093/ajcn/83.5.1199

El-Nezami H, Polychronaki N, Salminen S, Mykkänen H (2002) Binding rather than metabolism may explain the interaction of two food-grade lactobacillus strains with zearalenone and its derivative α-zearalenol. Appl Environ Microbiol 68(7):3545–3549. https://doi.org/10.1128/AEM.68.7.3545-3549.2002

EMA (2019) Guideline on quality, non-clinical and clinical requirements for investigational advanced therapy medicinal products in clinical trials—Scientific guideline. https://www.ema.europa.eu/en/guideline-quality-non-clinical-clinical-requirements-investigational-advanced-therapy-medicinal-productsclinical-trials-scientific-guideline

European Commission (2007) Regulation (EC) no 1924/2006 on nutrition and health claims made on foods. Nutr Health Claims. https://food.ec.europa.eu/safety/labelling-and-nutrition/nutrition-and-health-claims_en

European Commission (2015) Regulation (EU) 2015/2283 on novel foods. Nov Food. https://food.ec.europa.eu/safety/novel-food_en

Espinoza JL, Matsumoto A, Tanaka H et al (2018) Gastric microbiota: an emerging player in Helicobacter pylori-induced gastric malignancies. Cancer Lett 414:147–152. https://doi.org/10.1016/j.canlet.2017.11.009

FAO/WHO (2002) Guidelines for the evaluation of probiotics in food. FAO/WHO. https://www.fao.org/3/a0512e/a0512e.pdf

FDA (2016) Early clinical trials with live biotherapeutic products. U.S. Food Drug Adm. https://www.fda.gov/regulatory-information/search-fda-guidance-documents/early-clinical-trials-live-biotherapeutic-products

FDA (2021) Classification of products as drugs and devices and additional product classification issues. U.S. Food Drug Adm. https://www.fda.gov/regulatory-information/search-fda-guidance-documents/classification-products-drugs-and-devices-and-additional-product-classification-issues

FDA (2024a) Structure/function claims. U.S. Food Drug Adm. https://www.fda.gov/food/nutrition-food-labeling/structurefunction-claims

FDA (2024b) Center for Drug Evaluation and Research (CDER). U.S. Food Drug Adm. https://www.fda.gov/about-fda/fda-organization/center-drug-evaluation-and-research-cder

FDA (2024c) Generally Recognized as Safe (GRAS). U.S. Food Drug Adm. https://www.fda.gov/food/food-ingredients-packaging/generally-recognized-safe-gras

FDA (2025) Center for Biologics Evaluation and Research (CBER) | FDA. https://www.fda.gov/about-fda/fda-organization/center-biologics-evaluationand-research-cber

Fijan S (2016) Antimicrobial effect of probiotics against common pathogens. In: Probiotics and prebiotics in human nutrition and health. IntechOpen. https://doi.org/10.5772/63141

FSSAI (2021) Food safety and standards (health supplements, nutraceuticals, food for special dietary use, food for special medical purpose, functional food and novel food) regulations, 2016 (as amended up to 2021). Food Saf. Stand. Auth, India. https://www.fssai.gov.in/upload/uploadfiles/files/Compendium_Nutra_29_09_2021.pdf
Goldenberg JZ, Mertz D, Johnston BC (2018) Probiotics to prevent Clostridioides difficile infection. JAMA 320(5):499–500. https://doi.org/10.1001/jama.2018.9064
Gratz S, Mykkänen H, El-Nezami H (2005) Aflatoxin B1 binding by probiotic bacteria. J Food Prot 68(11):2470–2474. https://doi.org/10.4315/0362-028X-68.11.2470
Hancocks N (2020) Spanish agency approves use of term “probiotic” on food and supplements. FoodNavigator.com. https://www.foodnavigator.com/Article/2020/11/02/Spanish-agency-approves-use-of-term-probiotic-on-food-and-supplements/
Haskard CA, El-Nezami HS, Kankaanpää PE, Salminen S, Ahokas JT (2001) Surface binding of aflatoxin B1 by lactic acid bacteria. Appl Environ Microbiol 67(7):3086–3091. https://doi.org/10.1128/AEM.67.7.3086-3091.2001
Health Canada (2009) Health claims about microorganisms and use of the term “probiotic”. Health Can. https://www.canada.ca/en/health-canada/services/food-nutrition/food-labelling/health-claims/microorganisms-term-probiotic.html
Health Canada (2014) Natural health products regulations. Health Can. https://laws-lois.justice.gc.ca/eng/regulations/SOR-2003-196/
Health Canada (2023) Health product InfoWatch: October 2023. Health Can. https://www.canada.ca/en/health-canada/services/drugs-health-products/medeffect-canada/health-product-infowatch/october-2023.html
Hickson M, D’Souza AL, Muthu N, Rogers TR, Want S, Rajkumar C, Bulpitt CJ (2007) Use of probiotic Lactobacillus preparation to prevent diarrhoea associated with antibiotics. BMJ 335(7610):80. https://doi.org/10.1136/bmj.39231.599815.55
Hill C, Guarner F, Reid G et al (2014) Expert consensus on the scope and appropriate use of the term probiotic. Nat Rev Gastroenterol Hepatol 11(8):506–514. https://doi.org/10.1038/nrgastro.2014.66
Hu S, Zhu F, Jiang W et al (2021) Retrospective analysis of the clinical characteristics of Candida auris infection worldwide. Front Microbiol 12:658329. https://doi.org/10.3389/fmicb.2021.658329
Koe T (2024) FFC supplement firms mandatory to follow GMP and report adverse cases–Japan’s CAA. NutraIngredients-Asia.com. https://www.nutraingredients-asia.com/Article/2024/06/03/japan-requires-ffc-supplement-firms-to-follow-gmp-and-report-adverse-cases/
Koe T (2025) New nutra regulations across APAC in 2025. NutraIngredients-Asia.com. https://www.nutraingredients-asia.com/Article/2025/01/07/new-nutra-regulations-across-apac-in-2025
Köhler JR, Hube B, Puccia R et al (2017) Fungi that infect humans. Microbiol Spectr 5(3). https://doi.org/10.1128/microbiolspec.FUNK-0014-2016
Llewellyn A, Foey A (2017) Probiotic modulation of innate cell pathogen sensing and signaling events. Nutrients 9(10):1156. https://doi.org/10.3390/nu9101156
Marco ML, Sanders ME, Gänzle M, Arrieta MC, Cotter PD, De Vuyst L, Hill C, Holzapfel W, Lebeer S, Merenstein D, Reid G, Wolfe BE, Hutkins R (2021) The International Scientific Association for Probiotics and Prebiotics (ISAPP) consensus statement on fermented foods. Nat Rev Gastroenterol Hepatol 18(3):196–208. https://doi.org/10.1038/s41575-020-00390-5
Mazziotta C, Tognon M, Martini F et al (2023) Probiotics mechanism of action on immune cells and beneficial effects on human health. Cells 12(1):184. https://doi.org/10.3390/cells12010184
Mckenzie B (2025) Japan: amended food labeling standards revising the foods with functional claims system. https://insightplus.bakermckenzie.com/bm/consumer-goods-retail_1/japan-amended-food-labeling-standards-revising-the-foods-with-functional-claims-system
Mohd Redzwan S, Abd Mutalib MS, Wang J-S, Ahmad Z, Kang M-S, Abdul Rahman N‘A, Nikbakht Nasrabadi E, Jamaluddin R (2016) Effect of supplementation of fermented milk drink containing probiotic Lactobacillus casei Shirota on the concentrations of aflatoxin

biomarkers among employees of Universiti Putra Malaysia: a randomised, double-blind, cross-over, placebo-controlled study. Br J Nutr 115(1):39–54. https://doi.org/10.1017/S000711451500423X

Mukherjee A, Gómez-Sala B, O'Connor EM, Kenny JG, Cotter PD (2022) Global regulatory frameworks for fermented foods: a review. Front Nutr 9:902642. https://doi.org/10.3389/fnut.2022.902642

Neo P (2025) Japan tightens dairy rules to improve standardisation, probiotic fortification safety. FoodNavigator-Asia.com. https://www.foodnavigator-asia.com/Article/2025/04/22/japan-tightens-dairy-governance-process-to-improve-standardisation-probiotic-fortification-safety/

Nippon (2024) "Foods with function claims": growing Japanese market faces safety concerns. Nippon.com. https://www.nippon.com/en/japan-data/h01960/

OECD (2022) Harmonisation of regulatory oversight in biotechnology. OECD. https://www.oecd.org/biotechnology/harmonisation.htm

Oltu I, Cepoi L, Rudic V et al (2020) Current research and new perspectives in antifungal drug development. Adv Exp Med Biol 1282:71–83. https://doi.org/10.1007/5584_2019_453

Parolin C, Marangoni A, Laghi L, Foschi C, Ñahui Palomino RA, Calonghi N, Cevenini R, Vitali B (2015) Isolation of vaginal lactobacilli and characterization of anti-Candida activity. PLoS One 10(6):e0131220. https://doi.org/10.1371/journal.pone.0131220

Perlin DS (2020) Cell wall-modifying antifungal drugs. Curr Top Microbiol Immunol 425:255–275. https://doi.org/10.1007/82_2019_188

Przyrembel H (2001) Consideration of possible legislation within existing regulatory frameworks. Am J Clin Nutr 73:471s–475s. https://doi.org/10.1093/ajcn/73.2.471s

Reid G, Bruce AW, Fraser N, Heinemann C, Owen J, Henning B (2001) Oral probiotics can resolve urogenital infections. FEMS Immunol Med Microbiol 30(1):49–52. https://doi.org/10.1111/j.1574-695X.2001.tb01549.x

Rejimus Inc (2022) GRAS Notification of Lactobacillus rhamnosus CBT LR5. https://www.fda.gov/media/170163/download

Rima H, Steve L, Ismail F (2012) Antimicrobial and probiotic properties of yeasts: from fundamental to novel applications. Front Microbiol 3. https://doi.org/10.3389/fmicb.2012.00421

RNI Consulting (2020) THE PROBIOTIC SERIES - What Information is Required for a Probiotic GRAS or NDIN Dossier? RNI Consulting. https://www.rniconsulting.com/en/the-probiotic-series-what-information-is-required-for-a-probiotic-gras-or-ndin-dossier/

Russo R, Superti F, Karadja E, De Seta F (2019) Randomised clinical trial in women with recurrent vulvovaginal candidiasis: efficacy of probiotics and lactoferrin as maintenance treatment. Mycoses 62(4):328–335. https://doi.org/10.1111/myc.12883

Sanders ME, Merenstein DJ, Reid G et al (2019) Probiotics and prebiotics in intestinal health and disease: from biology to the clinic. Nat Rev Gastroenterol Hepatol 16(10):605–616. https://doi.org/10.1038/s41575-019-0173-3

Sant DG, Tupe SG, Ramana CV, Deshpande MV (2016) Fungal cell membrane—promising drug target for antifungal therapy. J Appl Microbiol 121(6):1498–1510. https://doi.org/10.1111/jam.13301

Shruthi B, Deepa N, Somashekaraiah R et al (2022) Exploring biotechnological and functional characteristics of probiotic yeasts: a review. Biotechnol Rep 34:e00716. https://doi.org/10.1016/j.btre.2022.e00716

Suez J, Zmora N, Segal E, Elinav E (2019) The pros, cons, and many unknowns of probiotics. Nat Med 25(5):716–729. https://doi.org/10.1038/s41591-019-0439-x

Synapse (2025) How many FDA approved Probiotics are there? https://synapse.patsnap.com/article/how-many-fda-approved-probiotics-are-there

Vitális E, Nagy F, Tóth Z et al (2020) Candida biofilm production is associated with higher mortality in patients with candidaemia. Mycoses 63(4):352–360. https://doi.org/10.1111/myc.13049

WHO (2022) WHO fungal priority pathogens list. World Health Organization. https://www.who.int/publications/i/item/9789240060241

Wray C, Lewis MA (2002) A review of the occurrence and control of mycotoxins in food. Food Control 13(6–7):387–399

Synergistic Approaches to Antifungal Management: Integrating Probiotics with Other Control Strategies

19

Shylaja R, D. Swathi, and M. Vinoth

Abstract

The increasing prevalence of fungal infections, particularly those caused by opportunistic and multidrug-resistant (MDR) pathogens such as *Candida*, *Aspergillus*, and *Cryptococcus* species, poses a serious global health concern, especially among immunocompromised individuals. The growing resistance to conventional antifungal drugs has intensified the search for alternative or adjunctive therapies. Promising strategies under investigation include herbal medicines, monoclonal antibodies, vaccines, cytokines, antimicrobial peptides, probiotics, and postbiotics. Among these, probiotics have gained significant attention for their ability to enhance immune responses and directly inhibit pathogenic fungi through the production of bioactive metabolites like organic acids, enzymes, and peptides. Recent studies also highlight the synergistic potential of combining probiotics with chemical antifungal agents, nanoparticles, antimicrobial peptides, and postbiotics to boost antifungal efficacy and reduce adverse effects. This chapter explores and summarizes these synergistic therapeutic strategies as innovative solutions to overcome the challenges of fungal infections and multidrug resistance.

Shylaja R (✉) · D. Swathi · M. Vinoth
Defence Institute of Biodefence Technologies, DRDO, Mysuru, Karnataka, India
e-mail: shylaja.dfrl@gov.in

M. Y. Sreenivasa et al. (eds.), *Probiotics as Antifungals*,
https://doi.org/10.1007/978-981-92-0002-3_19

Graphical Abstract

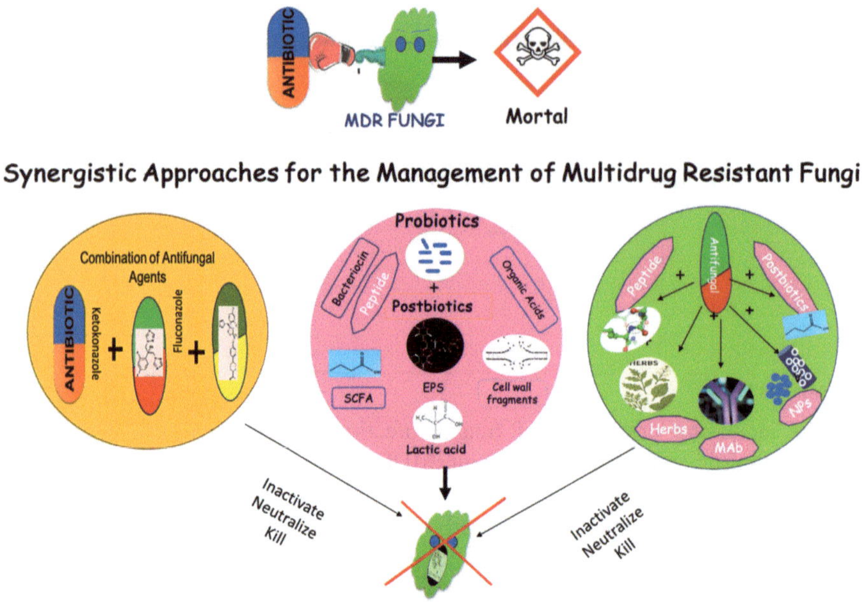

Keywords

Synergistic therapy · Probiotics · Postbiotics · Antimicrobial peptides · Antifungal resistance

Learning Objectives

- This chapter emphasizes the significance of fungal pathogens, the rise of drug-resistant fungal strains, and the resulting impact of infections caused by resistant strains.
- Existing approaches used to manage fungal infections.
- This chapter explores the application of a combination of different antifungal drugs, peptidomimetics, probiotics, postbiotics, and nanomaterials as an alternative treatment for drug-resistant fungal infections and synergistic strategies combining probiotics with other antifungal agents to mitigate fungal infections.

19.1 Introduction

Antifungal resistance is a surging global threat. Unlike bacterial infections, which have a broader arsenal of antibiotics, treatment options for invasive fungal diseases are limited to a few classes of antifungal agents. The human microbiota, particularly the bacterial flora, plays a crucial function in maintaining immune homeostasis and providing colonization resistance against opportunistic pathogens, including fungi. Broad-spectrum antibiotics, while effective against bacterial infections, often disrupt this delicate balance by significantly reducing the population of commensal bacteria. This disruption creates an ecological niche that can be exploited by fungal species, leading to the overgrowth of opportunistic fungi and a subsequent risk of fungal infections, particularly in geriatric and immunocompromised populations. This phenomenon is supported by several studies indicating a higher incidence of fungal infections following antibiotic therapy (Lackner et al. 2015; Spettel 2024). Antifungal resistance also arises due to inappropriate use of antifungals, environmental exposure to agricultural fungicides, and prolonged therapy in immunocompromised patients. Traditional antifungal agents, limited to a few drug classes such as azoles, echinocandins, and polyenes, are becoming increasingly ineffective due to resistance mechanisms including biofilm formation, efflux pump activation, and target site mutations. The efficacy of existing antifungal agents is increasingly challenged by rising antifungal resistance, treatment-associated toxicity, and the limited spectrum of activity.

These challenges have prompted exploration of adjunct or alternative strategies that enhance therapeutic outcomes while minimizing adverse effects (Mookherjee et al. 2020; Fisher et al. 2022). Synergistic therapeutic strategies that combine conventional antifungals with novel agents such as probiotics, postbiotics, peptides, peptidomimetics, nanoparticles, and monoclonal antibodies are being explored as a promising response to treat multidrug-resistant fungal strains. The integration of these agents in combination therapies is showing significant potential in preclinical studies, where they enhance efficacy, reduce resistance development, and mitigate host toxicity. These synergistic strategies represent a crucial direction for future antifungal therapeutics.

19.2 Prevalence of Fungal Infections

Fungal infections are spread worldwide, and approximately one in ten people suffer from one or the other fungal diseases at any time. Few fungal pathogens are present across the globe, whereas a few are endemic to selected regions. Fungal infections range in severity from mild superficial cases to severe, life-threatening systemic diseases (Brown et al. 2012). According to the World Health Organization, fungi cause a wide range of diseases that affect almost every organ, with high morbidity and mortality burdens accounting for a substantial percentage of the world's disease burden, with an estimated 1.7 million deaths each year, particularly in susceptible populations (WHO 2022). The most frequent fungal infections involve the skin, nails, and mucous membranes. More intense forms can penetrate the lungs, blood, and organs, causing life-threatening implications in immunocompromised individuals and have highlighted the issue of fungal infections at the top of medical research and public health agendas (Fisher et al. 2018).

Fungal infections are prevalent in subtropical and tropical areas where it is warm and humid, enabling fungi to multiply easily. Climate change, urbanization, globalization, and increased travel have contributed to the growing emergence of fungal diseases as a global public health concern. Prevalence of fungal diseases like *Aspergillosis, Candidiasis,* and *Cryptococcosis* has grown in developed nations as a result of increased numbers of immunocompromised patients. Especially in individuals with AIDS, cancer patients, or patients who have undergone organ transplants. The growth of fungal diseases is also correlated with health care procedures such as invasive surgery and the application of long-term broad-spectrum antibiotics.

Antibiotic usage, while important in the battle against bacterial infection, has the unwelcome side effect of potentially disturbing the normal microbial flora, rendering individuals more vulnerable to opportunistic fungal infections (Samonis and Kontoyiannis 2005). Fungal infections are increasingly also found among groups of people who live close to one another, i.e., prisoners, residents of nursing homes, and patients in hospitals and dormitories, where fungal spores circulate in atmospheres with impaired air exchange or dampness (Bennett et al. 2015).

19.3 Types of Fungal Infections

Fungal pathogens are diverse and include yeasts, molds, and dimorphic fungi. Fungal infections are categorized as fatal systemic infections, such as coccidioidomycosis, blastomycosis, and histoplasmosis, and opportunistic infections, namely, candidiasis and aspergillosis. A few of the common skin and nail fungal infections are ringworm and onychomycosis, as well as yeast.

Some of the most prevalent fungal infections are as follows:

(a) *Candida species* (*C. albicans, C. glabrata, C. parapsilosis*): These yeasts are ubiquitous in the human body and produce conditions like oral thrush, vaginal yeast infection, and invasive systemic infection (Nett and Andes 2016). Overgrowth is observed in immunocompromised patients following antibiotic therapy that destroys bacterial flora.
(b) *Aspergillus species* (*A. fumigatus, A. flavus*): These molds are ubiquitous in the environment and can lead to respiratory infections, particularly in immunocompromised individuals undergoing chemotherapy or organ transplantation (Denning 2003).
(c) *Cryptococcus species* (*C. neoformans*): This yeast is the most frequent causative agent of meningitis among AIDS patients (Maziarz and Perfect 2023).
(d) *Dermatophytes* (*Trichophyton spp*., *Microsporum spp.*): These are highly contagious fungi that spread easily in public places such as gyms and swimming

pools and infect superficial infections like athlete's foot, ringworm, and jock itch. (Gupta and Tu 2006).

(e) *Histoplasma*, blastomyces, *and* coccidioides: These dimorphic fungi cause systemic infections, mainly in areas like the Mississippi and Ohio River valleys (Kauffman 2017).

19.4 Important Reasons Account for the Increasing Trend of Fungal Infections

Globalization and increased human mobility have facilitated the spread of fungal species to regions where they were previously uncommon, leading to the emergence of new infections. Additionally, climate change—particularly rising temperatures and humidity—creates favorable conditions for fungal growth, spore production, and airborne dispersal. An increased number of HIV, cancer patients, older people, organ transplantation with comorbidities, poorly controlled diabetes, and other immunosuppressive therapies predispose individuals to fungal infections. Moreover, abuse of antibiotics destabilizes the normal microbiota, allowing opportunistic fungal infections, including those by *Candida* spp., to proliferate (Table 19.1). All these factors are increasing the burden of fungal diseases (Wang et al. 2021).

19.5 Significance of Fungal Infections

The severity of fungal infections cannot be overemphasized, as they have considerable public health, clinical, and economic importance.

Table 19.1 Prevalence of major fungal infections

Pathogen	Global prevalence	Population	Key findings	Study
Candida albicans	10–45% of bloodstream infections	ICU patients	High mortality, high prevalence of *Candida albicans* in ICU patients, with resistance to fluconazole	Schroeder et al. (2020)
Aspergillus fumigatus	5–23% of lung infections	Lung transplant recipients	Chronic lung infections and post-lung transplant infections	Janssens et al. (2024)
Candida auris	12% of hospital-acquired infections	Hospitalized patients	Emergence of multidrug-resistant *Candida auris* in hospital settings, with high mortality	Pandya et al. (2021)
Cryptococcus neoformans	10–15% of meningitis cases	HIV patients	Increased incidence of *C. neoformans* in HIV-positive patients, with poor survival rates. Leading cause of fungal meningitis	Mandal et al. (2024)

- *Morbidity and Mortality*: Invasive fungal infections have very high mortality rates, particularly among immunocompromised hosts. Invasive candidiasis has a mortality rate between 30% and 60%, depending on the host's underlying conditions (Pappas et al. 2018).
- *Healthcare Burden*: Treatment of fungal infections often requires prolonged use of antifungal drugs and hospitalization, which are expensive because intensive care, surgical procedures, and prolonged therapy are required.
- *Rising Fungal Pathogens*: New and novel fungal pathogens are posing challenges through drug resistance, restricted treatment alternatives, and escalating surveillance needs. MDR *Candida auris* has led to hospital outbreaks around the world, emphasizing the urgency for improved infection control practices (Lisa et al. 2025).
- *Impact on Public Health*: The increasing incidence of chronic fungal infections not only affects public health but also impacts the economy due to long hospital stays, expensive antifungal treatments, and long-term disability (Armstrong-James 2014). Fungal infections can also compromise productivity in the workforce. The effect of fungal infections is most severely experienced in less developed countries with low income and developing nations with middle income, where limited health services can be catastrophic, especially when not diagnosed or treated (Kullberg et al. 2015).

19.6 Current Antifungal Agents and Challenges

Antifungal therapy has changed over time, with multiple classes of antifungal agents being employed to treat fungal infections. Nevertheless, despite the evolution of these medications, there are still challenges in managing fungal infections because of the limited number of antifungal classes, the increasing rate of resistance, and the possibility of serious side effects.

Azoles, echinocandins, pyrimidines, and polyenes are the four classes of systemic antifungal medicines used in clinical practice currently. A few others are under development.

19.6.1 Classes of Antifungal Agents

- *Azoles (Fluconazole, Itraconazole, Voriconazole, Miconazole, and Clotrimazole)*: The most regularly administered antifungal drugs are azoles. They act by inhibiting a crucial enzyme, lanosterol 14α-demethylase, required for the synthesis of ergosterol, a key component of the fungal cell membrane, and inhibit the function of the cytochrome P450 enzyme (Odds et al. 2003).
- *Polyenes (Amphotericin B and Nystatin)*: Polyenes create pores in the fungal cell membrane by interacting with ergosterol, leading to cell death. Amphotericin B is a broad-spectrum antifungal agent for the treatment of severe, life-threatening infections such as *cryptococcal* meningitis and systemic candidiasis. It is

nephrotoxic, though, and cannot be used for long periods. Liposomal preparations of amphotericin B have been created to minimize toxicity but are still very costly and not generally available (Gallis et al. 1990).

- *Echinocandins (Caspofungin, Micafungin, Anidulafungin, and Rezafungin)*: Echinocandins block the formation of β-glucan, a critical component of the fungal cell wall. This antifungal class is most active against *Candida* species and has emerged as the treatment of choice for invasive *Candida* infections. Echinocandins are not active against molds such as *Aspergillus*, where there is a need for agents with broader activity (Denning 2003).
- *Allylamines (Terbinafine, Naftifine, Butenafine)*: This group of antifungal agents inhibits squalene epoxidase, an essential enzyme in ergosterol biosynthesis. It is primarily employed as a drug for dermatophyte infections like athlete's foot and ringworm (Ryder and Mieth 1992).
- *Pyrimidines*: A synthetic antifungal 5-fluorouracil form of flucytosine disrupts the synthesis of fungal DNA and RNA inside the fungal cells. Amphotericin B and flucytosine are used in combination to treat severe *Cryptococcal meningitis* and *Candida* infections. Resistance develops rapidly if flucytosine is used alone (Vermes et al. 2000).
- *Other Antifungals*: Other antifungal agents, such as naftifine and butenafine, inhibit squalene epoxidase, a key enzyme in fungal cell membrane synthesis (CDC 2024).

19.6.2 Challenges in Current Antifungal Therapy

- *Narrow Spectrum*: Most of the abovementioned antifungal agents possess a narrow spectrum of activity, necessitating accurate identification of the pathogen (Patterson et al. 2000).
- *Toxicity*: AMB and other drugs possess life-threatening side effects, such as nephrotoxicity and hepatotoxicity, particularly on prolonged administration (Wingard et al. 1999).
- *Resistance*: The emergence of MDR fungi, e.g., *Candida auris* and *A. fumigatus*, has diminished the efficacy of existing antifungal treatments (Chowdhary et al. 2017).
- *Cost*: Antifungal medicines, particularly liposomal amphotericin B and echinocandins, are costly and not readily available in resource-constrained environments (Ostrosky-Zeichner et al. 2007).

19.7 Development of Multidrug Resistance in Fungal Pathogens

The treatment of fungal infections has become increasingly complex and challenging due to the emergence of resistance to multiple commonly used antifungal drugs. Multidrug-resistant (MDR) fungal strains further complicate therapy, leading to increased morbidity and mortality among immunocompromised individuals, including those with HIV/AIDS, cancer, organ transplants, and diabetes mellitus (Perlin et al. 2017). The growing problem of MDR fungi has intensified efforts to develop novel antifungal agents and explore alternative therapeutic approaches. Antifungal resistance arises through several mechanisms, such as mutations in the genes of fungi that make them resistant to the drugs, excessive or improper use of antifungals in both medical and agricultural applications, and the capacity of fungi to develop biofilms, which make them resistant to treatment (Cowen et al. 2015). As resistance grows, it becomes more challenging to treat fungal infections, and infections that were previously treatable become deadly (Nwobodo et al. 2022).

19.8 Mechanisms of Multidrug Resistance in Fungi

Fungal species manifest MDR by several molecular and cellular mechanisms by which they can survive in the presence of antifungal therapy. The primary mechanisms are as follows:

- *Efflux Pumps*: Fungal pathogens tend to employ efflux pumps to eject antifungal drugs, lowering their intracellular levels. *Candida* species have evolved efflux pumps that eject azoles, making these antifungals less effective (Nwobodo et al. 2022).
- *Target Mutations*: Azole resistance is caused by point mutations in the *erg*11 gene, which codes for lanosterol demethylase in *A. fumigatus* (Xiang et al. 2013). Mutation of the gene *fks*1 in *Candida* spp. results in resistance to echinocandins, which act on the fungal cell wall.
- *Biofilm Formation*: Biofilms are tight aggregates of fungal cells embedded within a defensive extracellular matrix, play an important role in antifungal resistance. Biofilms inhibit the penetration of antifungals and shield fungi against host immune defense. This mechanism is significant in *Candida* infections, in which biofilm formation on medical devices (catheters, prosthetic implants) is frequent (Borman et al. 2020).
- *Cell Wall Alterations*: Fungi can alter their cell wall structure to avoid the action of antifungal drugs. Changes in the glucan and chitin levels of the cell wall have been seen in drug-resistant fungi, which render them less sensitive to antifungal therapy (Odds 2003).
- *Disrupted Drug Metabolism*: Certain fungi are capable of changing their internal biochemistry so that they can detoxify antifungal compounds. This is usually done through the overexpression of drug-metabolizing enzymes, which counteract the drugs' toxic effects (Parente-Rocha et al. 2017).
- *Reduced Permeability*: Modification in the cell wall or membrane permeability to antifungal drugs results in resistance. For instance, variations in the sterol content of the fungal cell membrane can decrease azole binding and activity (Ghannoum and Rice 1999).

19.9 Ranking of Fungal Pathogens According to WHO and Newly Emerging Resistant Fungal Pathogens

WHO identified, ranked, and categorized 19 fungal pathogens into three priority groups. They are critical, high, and medium risk groups. This categorization is based on factors such as the number of deaths caused by pathogen, annual incidence, global geographic distribution, changes in incidence or prevalence patterns over the past decade, average duration of hospital stay after primary diagnosis, associated complications and sequelae, rates of acquired or inherent resistance to antifungal treatments, transmission and acquisition dynamics, availability of evidence-based preventive measures, and accessibility of diagnostic tools (Fig. 19.1).

Among the major fungal pathogens, *Candida auris*, a multidrug-resistant (MDR) yeast, and members of the order *Mucorales* (commonly referred to as mucormycetes), which cause mucormycosis, gained significant importance during the COVID-19 pandemic. Mucormycosis is closely linked to immunocompromised patients with uncontrolled diabetes (Lass-Flörl and Steixner 2023).

Although the currently used antifungal azoles, polyenes, and pyrimidines are effective against a variety of fungal pathogens, there has been growing resistance against these drugs (Osset-Trénor et al. 2023). The challenges of drug resistance have prompted exploration of adjunct or alternative strategies that enhance therapeutic outcomes while minimizing adverse effects. Synergistic therapeutic strategies that combine conventional antifungals with novel agents such as probiotics, postbiotics, peptides, peptidomimetics, nanoparticles, and monoclonal antibodies are being explored as a response to the multidrug resistance challenge.

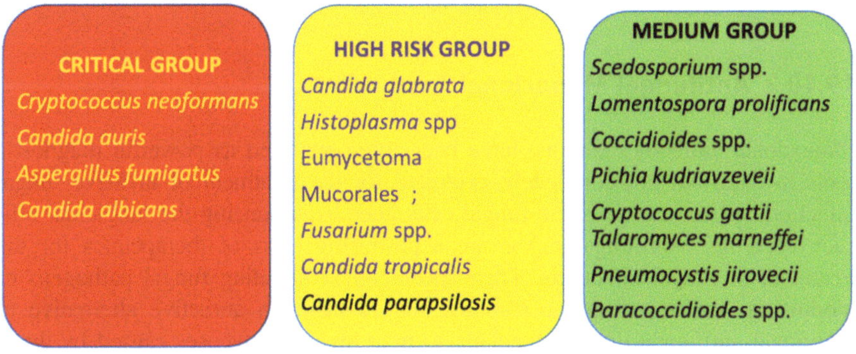

Fig. 19.1 Ranking of fungal pathogens according to WHO and newly emerging resistant fungal pathogens

19.10 Alternative Antifungal Agents and Strategies to Combat Drug Resistance

Employing combinations of two or more antifungal agents together to augment the effectiveness of the drug and to reduce resistance development is the new avenue opted by clinicians to treat drug-resistant strains. A combination of drugs has been proven to be a promising, reliable strategy to treat the challenging MDR-resistant strains with different effective modes of action. The combination of medications can slow down the development of resistance by a multi-target approach strategy.

19.10.1 Combination of Antifungal Drugs

A combination of Amphotericin B and flucytosine was recommended by the FDA and Infectious Diseases Society of America (IDSA) for the treatment of serious candidal infections involving the central nervous system (CNS), endovascular, and intra-abdominal candidiasis. Amphotericin B in combination with flucytosine or azoles to treat invasive aspergillosis, candidiasis, or cryptococcosis was reported (Pappas et al. 2016). Extensive research is being conducted to treat MDR *Candida albicans* by using a combination of drugs such as Ketoconazole and Spinosad, Nitroimidazoles and Amphotericin B, Histatin 5 and its analogs Amphotericin B, Tetrandrine and Ketoconazole, Fluconazole in combination with Thymol, and Carvacrol, Fluconazole and alverine citrate, Caspofungin and Latrunculin-A, Posaconazole + Caspofungin, or FK506 are reported by various researchers. A combination of an azole and echinocandin, ibrexafungerp plus amphotericin, proved beneficial for the treatment of azole-resistant *Aspergillus* species. Echinocandin and liposomal amphotericin B combination improves survival rates when used as first-line therapy for invasive aspergillosis, especially in patients with hematological malignancies (Pappas et al. 2016).

19.10.2 Antifungal Monoclonal Antibodies

Monoclonal antibodies (mAbs) have been more explored as powerful diagnostic tools to date, but not completely exploited for the treatment of invasive fungal infections. mAbs can sensitize fungal cells, thereby enhancing the fungicidal efficacy of drugs. Antifungal mAbs are a promising class of therapeutics for the treatment of fungal infections. Their specificity in targeting fungal pathogens or modulating the host immune response makes them an attractive alternative to traditional antifungal drugs, especially to treat invasive mycoses caused by drug-resistant strains that are difficult or impossible to treat. mAbs are being investigated for their potential to enhance the specificity and efficacy of antifungal therapy (Casadevall et al. 2004). mAbs in combination with antifungal drugs emerge as a potential treatment for systemic mycoses. A combination of mAbs with antifungals and nanoparticles is surmounting the challenges of drug resistance (Souza et al. 2024).

Mechanism of Action Monoclonal antibodies work by several mechanisms associated with fungal elimination, such as direct neutralization, immune modulation, activating macrophages and neutrophils, and also by inhibiting important enzymes required for fungal growth (Romani 2011).

Monoclonal Antibodies with Diagnostic and Therapeutic Applications

- *Anti-β-glucan mAbs*: mAbs targeting β-glucan, a major component of the cell wall of pathogenic fungi, have the ability to enhance immune clearance and provide direct antifungal activity (Capodicasa et al. 2011).
- *Anti-Hsp90 mAbs*: mAbs targeting Hsp90 inhibit the growth of various fungal species, including *Candida* and *Aspergillus* (Cowen et al. 2009).
- *mAb 7D7*: Targets the transcription factor Efg1 in *Candida albicans,* significantly inhibiting biofilm formation (a key virulence factor) and reducing overall virulence in vitro and in a murine model of systemic candidiasis (Mishra et al. 2015).

Monoclonal Antibodies in Combination with Antifungal Agents Exhibiting Synergistic Effect: H5K1

- A humanized mAb specific for β-1,3 glucans, a key component of fungal cell walls, has shown efficacy both alone and in combination with caspofungin and amphotericin B against *C. auris* (Vanzolini et al. 2021).
- *Mycograb*®: A human r-mAb generated against Hsp90 tested in vivo in combination therapy with AmB, caspofungin (CAS) or FLZ, and with r-ScFV antibody to treat *C. neoformans* more effectively (Nooney et al. 2005).
- *mAb B6.1*: In combination with amphotericin B (AmB) and fluconazole (FLZ), MAb B6.1 demonstrated a synergistic effect against candidiasis (Han et al. 2010).
- *mAb 2G8*: Inhibited the growth of *C. neoformans* by binding to the cell wall. MK-3118 in combination with amphotericin B exhibited excellent antifungal activity against CYP51 mutant, suggesting synergistic effect of monoclonal antibody with antifungal drugs could be a potent treatment for azole-resistant fungal strains (Di Mambro et al. 2019).
- *Polymeric nanoparticles (NPs)*: possess the ability to solubilize and encapsulate drugs, protect drugs from chemical and enzymatic hydrolysis in the gastrointestinal tract, offer compatibility, stability, enable controlled release, allow for surface functionalization, and hold promise for scalable, staggered industrial production. Combination of AmB with anti-chito oligomer nanoparticles with mAbs distinctly increased antifungal property, resulting in a synergistic effect multiple folds greater than that of AmB-NPs alone in the treatment of multidrug-resistant *C. neoformans*, and *C. albicans* (Souza et al. 2024).

19.10.3 Herbal Antifungal Agents

Herbal antifungal agents are plant-derived compounds and have been utilized for ages in traditional medicine. Herbal antifungal agents are popular as they are of natural origin, less toxic compared to synthetic antifungal drugs (Cowan 1999). Numerous herbal constituents exhibit strong activity against resistant fungal and yeast strains, synergistically with the antifungal drug or alone (Table 19.2). Besides their antifungal activity, most of the herbal agents possess immunomodulatory, antioxidant, and anti-inflammatory activities (Ríos and Recio 2005). Combinatorial therapy expands the spectrum of efficiency with minimum toxicity.

Mechanisms of Action Antifungal agents from plants work by different mechanisms:

- *Disruption of cell membrane*: Terpenoids, phenols, and flavonoids disrupt cell membrane integrity, causing cellular content leakage and ultimately cell death (Rao and Ravishankar 2002).
- *Enzyme Inhibition*: Inhibit essential enzymes such as squalene epoxidase, chitin synthase, and proteases required for the development of the fungal cell wall.
- *Immune Modulation*: Plant constituents activate host macrophages and T-cells, which play a vital role in fighting fungal infections (Wagner and Proksch 1985).

19.10.4 Probiotics as Antifungal Agents

Probiotics are live microorganisms that provide health benefits to the host when provided in sufficient quantities. They have been found to be effective in the management of fungal infections. Different probiotic strains have been found to possess antifungal activity against pathogens such as *Candida, Aspergillus, Fusarium,* and *Trichophyton*. Probiotics exhibit antifungal properties primarily through the secretion of various metabolites, such as organic acids (including lactic and acetic acids), hydrogen peroxide, and bacteriocins. These compounds lower the environmental pH, creating conditions that are inhospitable for many pathogenic fungi (O'Hanlon et al. 2013). Strains like *Lactobacillus* spp., *Bifidobacterium* spp., and *Saccharomyces boulardii* produce organic acids both in vitro and in vivo (Gibson et al. 1995). Lactic acid, in particular, inhibits the growth of *Candida albicans* (Lourenco et al. 2019). Additionally, *Lactobacillus* spp. forms hydrogen peroxide, which disrupts fungal cell wall synthesis and is toxic to various fungal species (Tachedjian et al. 2018). Bacteriocins, antimicrobial peptides derived from probiotics, target fungal pathogens by damaging their cell membranes (Tiwari et al. 2022).

Probiotics also regulate cytokine production, including interleukins and interferons, thereby supporting the immune response to fungal infections (Latif et al. 2023). Furthermore, probiotics strengthen the mucosal barrier, reducing the adhesion of pathogens to epithelial cells (Hardy et al. 2013). Probiotics combat fungal infections through competitive exclusion by outcompeting pathogenic fungi for nutrients and ecological niches, thereby reducing the pathogens' ability to colonize and invade the host environment (Wu et al. 2022).

Table 19.2 Herbs with conventional antifungal agents exhibiting synergistic effects

No.	Plant source and herbal compound	Conventional antifungal	Fungal pathogen	Mechanism	Reference
1	*Icariin* *Epimedium* spp. (Horny Goat Weed)	Fluconazole	*C. albicans*	Ergosterol biosynthesis inhibition and membrane disruption	Aliaghazadeh et al. (2025)
2	*Piperine* (Black Pepper)	Fluconazole	*C. albicans*	Efflux pumps, biofilm formation	Thakre et al. (2021)
3	*Curcumin* *Curcuma longa* (Turmeric)	Fluconazole	*C. albicans*	Induction of reactive oxygen species (ROS) and apoptosis	Garcia-Gomes et al. (2012)
4	*Thymol* (Thyme) *Thymus vulgaris*	Fluconazole Itraconazole	*C. albicans* *Pythium insidiosum*	Disruption of cell membrane and ergosterol biosynthesis	Kauser et al. (2024)
5	*Carvacrol* (Oregano) *Origanum vulgare*	Itraconazole Fluconazole	*C. albicans* *P. insidiosum*	Inhibition of cell wall synthesis and membrane	Mączka et al. (2023)
6	*Geraniol* *Pelargonium graveolens*	Fluconazole Itraconazole	*C. albicans* *P. insidiosum*	Disruption of cell membrane and biofilm formation	Silva-Rodrigues et al. (2024)
7	*Eugenol Syzygium aromaticum* (Clove)	Amphotericin B, Itraconazole	*C. albicans* *P. insidiosum*	Membrane disruption and inhibition of growth	Khan et al. (2019)
8	*Cinnamaldehyde* *Cinnamomum verum* (Cinnamon)	Amphotericin B Itraconazole	*C. albicans* *Fusarium* and *P. insidiosum*	Disruption of cell membrane and biofilm	Rachel et al. (2024)

The combined use of probiotics and conventional antifungal agents has demonstrated a synergistic effect in combating fungal infections beyond what either treatment could achieve alone (Table 19.3). Probiotics contribute to this enhanced effect by producing antifungal metabolites, regulating the host immune response, and strengthening the mucosal barrier, which complements the mechanisms of action of conventional antifungals. Additionally, probiotics restore the natural microbiota destroyed by antifungal therapy, thereby suppressing the risk of recurrence and secondary infections. Studies have shown that co-administration of probiotics with antifungals such as fluconazole or amphotericin B can lead to increased inhibition of fungal growth, particularly in cases involving *Candida albicans* and other opportunistic fungi. This approach not only improves treatment outcomes but may also allow for lower doses of antifungal drugs, potentially reducing their associated side effects and the risk of developing drug resistance. The combined use of various natural products, such as plant extracts, propolis, prebiotics, synbiotics, and postbiotics with probiotics, along with conventional drug therapy, serves as an effective adjuvant to standard treatments. This approach has the potential to restore drug sensitivity in multidrug-resistant (MDR) pathogens, boost host immune responses, and enhance overall clinical outcomes. The following table depicts a few of the combinations that exhibit synergistic activity against drug-resistant fungal strains.

19.10.5 Postbiotics as Antifungals

Postbiotics are a group of bioactive compounds produced and secreted by probiotic or beneficial microorganisms. These include short-chain fatty acids (SCFAs), enzymes, bacteriocins, vitamins, amino acids, exopolysaccharides (EPS), and cell-free supernatants (CFS). Postbiotics have gained attention over the past few years as potential players in the battle against fungal pathogens. In contrast to their live counterparts, postbiotics have the advantages of increased stability, a defined composition, and a possible enhanced safety profile because they lack live microorganisms. Postbiotics' exact composition relies greatly on the bacterial strain, growth medium, and fermentation conditions, resulting in a broad range of biological activities (Salminen et al. 2021).

Mechanisms of Antifungal Action of Postbiotics The antifungal activity of postbiotics involves multiple mechanisms, including direct inhibition of fungal growth, disruption of fungal cell integrity, and modulation of the immune response to enhance its capability to combat infection (Table 19.4). Postbiotics, such as specific short-chain fatty acids (e.g., butyrate and propionate) and bacteriocins, have been shown to disrupt fungal cell membranes by increasing their permeability, leading to the leakage of intracellular components and ultimately causing cell death. SCFAs affect ergosterol, a key component of the fungal cell membrane, resulting in its destabilization (Parolin et al. 2021). SCFAs can interfere with essential metabolic pathways in fungi. SCFAs can acidify the fungal cytoplasm, inhibiting enzyme

Table 19.3 Synergistic interactions of probiotics with conventional and emerging alternative antifungal compounds

Probiotic strain	Antifungal agent	Target fungi	Synergistic effect observed	Reference
L. rhamnosus GG *L. reuteri RC-14*	Fluconazole	*C. albicans*	Inhibition of biofilm formation, morphogenesis, effective against vulvovaginal candidiasis, reduces recurrence	Allonsius et al. (2017), Martinez et al. (2009)
L. plantarum	Amp B	*C. albicans,* *A. niger*	Produces antifungal cyclic dipeptides	Ström et al. (2002)
L. Crispatus *S. boulardii*	Nystatin	*C. albicans*	Modulates epithelial defense and enhances antifungal activity	Rizzo et al. (2013) Murzyn et al. (2010)
B. longum	Voriconazole	*A. fumigatus*	Enhances host immunity, potential synergy with azoles	Xiao et al. (2025)
L. rhamnosus	Fluconazole	*C. albicans*	Enhanced inhibition of *C. albicans* by the transformation of baicalin into baicalein	Zhai et al. (2025)
L.plantarum *L. rhamnosus*	Cranberry Proanthocyanidins PACs	*C. albicans*	Potential alternative therapies in treating vulvovaginal candidiasis	Wu et al. (2025)
Probiotic strain	Nanoparticle/herb/postbiotics			
L. plantarum and *L. johnsonii*	Reduced Selenium dioxide	*C. albicans*	Selenium dioxide induces the production of anti-Candida metabolites from probiotic strains	Kheradmand et al. (2014)
Lactobacillus and *Bifidobacterium* *L. salivarius*	Probiotic reduced AgNPs	*P. aphanidermam* *F. oxysporum,* *P. parasitica*	NPs with different functional groups adhere to the cell wall and facilitate easy penetration, preventing biofilm formation	Sharma et al. (2022), Abdelgadir et al. (2024)
Lactobacillus, *Bifidobacterium* *Saccharomyces*	Chitosan nanoparticles and probiotics	*C. albicans*	Boosts anti-fungal activity, increases stability of probiotics. Nps deliver probiotics to the target site, improve bioavailability and controlled release	Pillai et al. (2024)
Lactococcus lactis	Postbiotic: Indole-3-carboxaldehyde	*A. fumigatus*	3-IAld reduces aspergillosis by acting on the infection and inflammation sites	Machado et al. (2023)
S. cerevisiae	Herb: *Moringa*	*C. albicans*	In vitro inhibitory activities against *C. albicans*	
Pediococcus acidilactici	Phthalyl inulin Nps (PINs)			

Table 19.4 Antifungal postbiotics and their mode of action

Postbiotics	Target	Mechanism of action	References
Lactobacillus CFS (high lactic acid)	*C. albicans*	Cell membrane disruption Prevents growth; biofilm formation	Takano et al. (2023)
L. plantarum	*A. fumigatus, C. glabrata*	Inhibition of biofilm formation by *T. rubrum*	Srivastava et al. (2020)
Peptides from *B. subtilis*	*T. rubrum*	Interferes with cell wall synthesis	Li et al. (2021)
Bacillus metabolites	*C. albicans*	Interferes with quorum-sensing signaling and reduced hyphal formation	Wang et al. (2022)
CFS of *B. breve*	*C. albicans*	Modulation of host immune response	Mani-López et al. (2022)

activity and nutrient uptake. Bacteriocins, peptides with antimicrobial activity, can target specific fungal proteins involved in cell wall synthesis or other vital processes (Plaza-Díaz et al. 2019)

Postbiotics such as EPS and CFS inhibit the formation of biofilms and erase pre-existing biofilms, rendering the fungal cells more vulnerable to antifungal compounds and to the host immune response (Gibson et al. 2020). Postbiotics have shown significant potential in combating fungal infections through both immunomodulatory and antifungal mechanisms. They can intensify the host immune response by stimulating the synthesis of antimicrobial peptides and boosting the activity of immune cells such as macrophages and neutrophils, with short-chain fatty acids (SCFAs) like butyrate playing a key role in this process (O'Hara and Shanahan 2007). Additionally, postbiotics can disrupt fungal quorum sensing (QS), a cell-to-cell communication system that controls virulence factors like biofilm formation and hyphal growth by acting as QS inhibitors, thereby reducing fungal pathogenicity without directly killing the fungal cells (Freestone 2013). In recent years, a growing body of research has identified and characterized specific postbiotics with strong antifungal activity against pathogens, including *Candida albicans, Aspergillus fumigatus*, and dermatophytes, underscoring their potential as effective antifungal agents.

Postbiotics are safe, non-living metabolites derived from GRAS (Generally Recognized as Safe) microorganisms, reducing the risk of systemic infections in immunocompromised individuals. They are more stable during processing and storage, and when combined with standard antifungals, they enhance efficacy and reduce required dosages, minimizing side effects. Postbiotics can be easily formulated into topical creams, oral capsules, or integrated into medical devices for targeted treatment of fungal infections affecting the skin, nails, and mucous membranes (Swanson et al. 2020). Postbiotics, traditional antifungal therapies synergistically improve the efficiency of the antifungal agents, particularly against drug-resistant strains and biofilm-associated infections (Parolin et al. 2021). Incorporating postbiotics into hygiene products or dietary supplements could help prevent fungal colonization, especially in high-risk populations (Salminen et al. 2021). Additionally, postbiotics may aid in restoring gut microbiota balance and controlling fungal overgrowth, such as *Candida* spp. (Plaza-Díaz et al. 2019).

19.10.6 Antifungal Peptidomimetic Molecules

Natural antimicrobial peptides (AMPs) are produced as a first line of defense by all organisms. These peptides exhibit broad-spectrum antibacterial and antifungal activity primarily by disrupting cell membranes and targeting intracellular components. AMPs contain a combination of positively charged and hydrophobic groups that allow them to bind to the negatively charged membrane of the fungal cell, resulting in its disruption and cell death. AMPs also exhibit immunomodulatory activities. However, their therapeutic potential is limited by inherent drawbacks such as susceptibility to proteolytic degradation, high production costs, and potential host cytotoxicity. To address these challenges, peptides can be structurally modified using non-natural amino acids, with the incorporation of β-amino acids (Han et al. 2021), backbone alterations, or constrained conformations through cyclization approaches or by introducing helix motifs (Gilon et al. 1991). These modifications enhance peptide stability, bioavailability, and target specificity while retaining the key pharmacophoric features essential for antimicrobial activity. Such structurally optimized peptides that emulate the function of natural AMPs are known as peptidomimetics or synthetic peptides (Chen and Jiang 2023). Peptidomimetics are the stable, encouraging molecules for treating MDR strains of fungi individually or in synergistic association with other antifungal agents (Table 19.5). These compounds can target key fungal pathogens, leading to reduced cell wall integrity, increased membrane permeability, and disruptions in cell morphology. Furthermore, some peptidomimetics can affect fungal DNA replication, cell wall synthesis, and ergosterol synthesis, ultimately hindering fungal growth and survival. Synthetic peptides can be constructed according to the requirement. Synthetic mimics of

Table 19.5 Antifungal synthetic peptides

Synthetic peptide class	Mode of action	Fungal target	Citation
Cyclic peptidomimetics (Helix-mimics)	Disruption of cell membrane	Flucanazole resistant *C. albicans*, *C. glabrata*	Lee et al. (2019)
β-Amino acid containing peptide		*A. fumigatus*	Lai et al. (2022)
Cyclic lipopeptidomimetics		*C. albicans*	Wang et al. (2021)
Arginine-rich peptidomimetics		Broad-spectrum *C. albicans*, *A. fumigatus*	Ciociola et al. (2023)
Peptidomimetics targeting efflux pump	Efflux pump	*C. albicans* resistant to azoles, fluconazole	Jayan and Gupta (2023)

peptides constructed by Schaefer et al. 2024 with caspofungin prevented fungal infection of human epithelial cells in vitro, improved clearance of the pathogen by human macrophages, and significantly increased host survival in systemic candidiasis. Synthetic peptides Mo-CBP3-PepI and Mo-CBP3-PepIII, in combination with nystatin and itraconazole, prevent biofilm formation by drug-resistant *C. albicans* and *C. parapsilosis* strains. These findings of Lima et al. (2021) highlight the potential of Mo-CBP3-PepI and Mo-CBP3-PepIII as novel therapeutic agents and adjuvants capable of enhancing the efficacy of conventional antifungal drugs in the treatment of *C. albicans* and *C. parapsilosis* infections.

Few of the peptide studies extensively are listed below.

- *LL-37*: net positive charge of the peptide interacts with negatively charged fungal cell membrane, disrupts the fungal cell membrane, leading to membrane permeabilization, cell lysis, efflux of cellular contents, and finally death (Pinilla et al. 2022).
- *MAF-1A*: Destroys cell membrane integrity and binds to nucleic acids, disrupting gene expression (Cheng et al. 2021).
- *Thanatin*: A 21-residue peptide, highly potent in inhibiting fungal growth (Dash and Bhattacharjya 2021).
- *Protonectin*: Disrupts membrane integrity and induces ROS production in yeast cells (Wang et al. 2015).
- *Blapstin*: Disrupts the cell structure of *Candida albicans* and *Trichophyton rubrum.*
- *RQ18*: Inhibits yeast growth and kills *Candida tropicalis* cells.

19.11 Nanomaterials as Antifungal Agent

Nanomaterials, due to their nanoscale size (typically 1–100 nm), exhibit unique physicochemical and biological properties such as a high surface area-to-volume ratio, tunable properties, enhanced reactivity, strong interactions with an exceptional ability to interact with biological systems at cellular and molecular levels, making them highly attractive for antifungal applications. Their distinctive features open up new paradigms for combating fungal pathogens. Recent research has extensively examined the antifungal properties of various nanomaterials against a wide range of clinically significant fungal pathogens (Table 19.6). Driving the search for more effective alternatives to the conventional antifungal drugs, nanomaterials offer a novel and potent approach to antifungal therapy (Azeem et al. 2024).

19.12 Merits and Challenges of the Alternative Combinatorial Therapy

Synergistic therapies or combinatorial therapies are still in their infancy. Most of the combinations are in experimental or preclinical stages, with insufficient large-scale clinical trials. Clinical studies and validations are very limited. Moreover, the

Table 19.6 Major categories of nanomaterials exhibiting antifungal activity

Nanoparticle class/type	Fungal pathogens	Antifungal mechanism/ benefits	References
Metal and metal oxide NPs: AgNPs, AuNPs, CuNPs, ZnO, TiO_2, MgO	*C. albicans, A. fumigatus, C. neoformans; T. rubrum*	Disruption of fungal cells, generation of reactive oxygen species (ROS)	Chand et al. (2024)
Quantum dots (QDs): Semiconductor nanocrystals, carbon QDs, ZnO QDs	*C. albicans*	Exhibit antifungal activity by the generation of ROS and offer potential for fungal diagnostics due to unique optical properties.	Sánchez-López et al. (2020)
Liposomes and niosomes: Vesicular nanocarriers	*A. fumigatus, C. neoformans*	Improved solubility, stability, and targeted delivery of antifungal agents with reduced toxicity	de Gouveia Dias et al. (2019)
Carbon-based nanomaterials: CNTs, graphene, graphene oxide (GO), reduced graphene oxide (rGO), fullerenes	*C. albicans, A. niger*	Physical membrane disruption, oxidative stress	Thambirajoo et al. (2021)
Polymeric nanoparticles: Chitosan NPs, alginate NPs, and other biodegradable polymers	*C. albicans*	Drug encapsulation for targeted delivery; intrinsic antifungal activity, inhibits biofilm formation	Guerrero et al. (2019)

effectiveness of probiotics and postbiotics can differ depending on the strain used and the fungal pathogen targeted. Combining multiple agents may lead to unpredictable pharmacodynamic interactions or adverse effects on host cells. Obtaining approval from regulatory agencies for the standardization of concentration and combination therapies involving biological agents like probiotics can be complex. Ensuring stability and targeted delivery of live microbes or bioactive compounds remains technically challenging (Table 19.7).

19.13 Future Directions

- *Conducting large-scale multicenter clinical trials to evaluate safety*, efficacy, and optimal dosing of combination therapies.
- *Standardization of protocols* for trial design to improve reproducibility and comparability.
- To promote *public–private partnerships* and funding initiatives to support translational research to reach patients.
- Use of *advanced omics technologies* (genomics, metabolomics, proteomics) to identify and characterize probiotic strains with the highest antifungal potential.

Table 19.7 Merits and demerits of alternative antifungal agents

Alternative agents	Merits	Demerits
Monoclonal antibodies	High specificity Immune system enhancement Synergistic with conventional drugs	Expensive production Animals are required. Ethical concern Requires cold storage
Peptidomimetics	Stable and potent Broad-spectrum activity Possibility of custom designing	Possible host cytotoxicity Delivery and stability issues Limited clinical data
Probiotics	Restores microbiota Safe and well-tolerated Immune modulation	Strain-dependent efficacy Influenced by host and environmental factors
Postbiotics	Stable and easy to standardize Antifungal and anti-inflammatory Safer for immunocompromised patients	Unclear mechanisms May need high doses Limited research and clinical validation
Nanoparticles	Enhanced drug delivery and targeting Overcomes resistance Improves drug solubility and stability	Potential toxicity Expensive and complex production Regulatory and standardization challenges

- Establishing *strain libraries and databases* to catalogue functional profiles and host–pathogen interactions.
- Employing *computational modeling and AI-based predictive tools* to study interactions between combined agents before clinical testing.
- Conducting *systematic in vitro and in vivo interaction studies* to identify synergistic effects.

19.14 Conclusion

The growing challenge of antifungal resistance and the limited effectiveness of existing antifungal drugs highlight the urgent need for new therapeutic approaches. Resistance driven by microbiota disruption, improper antifungal use, and prolonged treatments has reduced the efficacy of conventional agents. Synergistic strategies that combine traditional antifungals with novel adjuncts such as probiotics, postbiotics, antimicrobial peptides, nanoparticles, and monoclonal antibodies show strong potential to enhance efficacy, limit resistance, and reduce toxicity. These approaches represent a promising direction for future antifungal therapy, warranting further clinical investigation.

19.15 Summary

Antifungal resistance has emerged as a significant global health threat, exacerbated by the limited number of available antifungal drug classes and the increasing incidence of invasive fungal infections. Disruption of the human microbiota—particularly due to broad-spectrum antibiotic use—compromises colonization resistance and immune homeostasis, creating favorable conditions for opportunistic fungal overgrowth, especially in geriatric and immunocompromised populations. Inappropriate antifungal use, environmental exposure to agricultural fungicides, and prolonged antifungal therapy further contribute to the development of resistance. Traditional antifungal agents, including azoles, echinocandins, and polyenes, are increasingly undermined by resistance mechanisms such as biofilm formation, efflux pump overexpression, and target-site mutations, along with issues of toxicity and limited spectrum of activity. These challenges have driven the exploration of alternative and adjunctive therapeutic approaches. Emerging synergistic strategies that combine conventional antifungals with novel agents—such as probiotics, postbiotics, antimicrobial peptides, peptidomimetics, nanoparticles, and monoclonal antibodies—have demonstrated promising results in preclinical studies by enhancing antifungal efficacy, reducing resistance development, and minimizing host toxicity. Collectively, these approaches represent a critical and innovative direction for the future of antifungal therapy.

Acknowledgments The authors are indebted to Dr. Joseph Kingston, Head of the FMB Division, DIBT, and Centre Head, DIBT, Mysuru, for providing support to write this chapter.

Conflict of Interest The authors do not have any conflict of interest to declare.

Author Contribution SD—Collecting literature and writing a draft; V—Collecting the literature; SR—Writing, reviewing, and editing.

Study Questions

1. How can genomics, metabolomics, and proteomics help identify probiotic strains with high antifungal potential?
2. Why are both in vitro and in vivo studies necessary to identify synergistic interactions between therapeutic agents?
3. How can computational modeling help predict interactions between agents in combination therapies before laboratory testing?
4. How can findings from preclinical interaction studies guide clinical trial design?
5. How do efflux pumps contribute to multidrug resistance in bacteria and fungi?
6. How can combination therapy be used to overcome multidrug resistance?

References

Abdelgadir A, Adnan M, Patel M et al (2024) Probiotic *Lactobacillus salivarius* mediated synthesis of silver nanoparticles (AgNPs-LS): a sustainable approach and multifaceted biomedical application. Heliyon 10(18):e03728

Aliaghazadeh B, Pascual-Ortiz M, Martínez A et al (2025) Unveiling the synergistic effect of icariin and azole drugs on *Candida albicans*. Front Oral Health 6:1517210

Allonsius CN, van den Broek MFL, De Boeck I, Kiekens S, Oerlemans EFM, Kiekens F, Lebeer S (2017) Interplay between *Lactobacillus rhamnosus* GG and *Candida* and the involvement of exopolysaccharides. Microb Biotechnol 10(6):1753–1763

Armstrong-James D (2014) A neglected epidemic: fungal infections in HIV/AIDS. Trends Microbiol 22(3):107–112. https://doi.org/10.1016/j.tim.2014.01.001

Azeem K, Irfan I, Shakir M, Rawat DS, Abid M (2024) Therapeutic potential of phytochemicals and their derivatives as antifungal candidates: recent discovery and development. In: Manzoor N (ed) Advances in antifungal drug development. Springer. https://doi.org/10.1007/978-981-97-5165-5_19

Bennett JE, Dolin R, Blaser MJ (2015) Mandell, Douglas, and Bennett's principles and practice of infectious diseases, 8th edn. Saunders

Borman AM, Hughes JM, Oliver D, Fraser M, Sunderland J, Noel AR, Johnson EM (2020) Lessons from isavuconazole therapeutic drug monitoring at a United Kingdom reference center. Med Mycol 58(7):996–999. https://doi.org/10.1093/mmy/myaa022

Brown GD, Denning DW, Gow NAR et al (2012) Hidden killers: human fungal infections. Sci Transl Med 4(165):165rv13

Capodicasa C, Chiani P, Bromuro C, Catellani M, Palma AS, Liu Y, Feizi T, Cassone A, Benvenuto E, Torosantucci A (2011) Plant production of anti-β-glucan antibodies for immunotherapy of fungal infections in humans. Plant Biotechnol J 9(1):75–86. https://doi.org/10.1111/j.1467-7652.2010.00586.x

Casadevall A, Dadachova E, Pirofski L-A (2004) Passive antibody therapy for infectious diseases. Nat Rev Microbiol 2(9):695–703. https://doi.org/10.1038/nrmicro974

Centers for Disease Control and Prevention (2024) Treating fungal diseases with antifungals. Available at: https://www.cdc.gov/fungal/treatment/index.html

Chand P, Narula K, Radhakrishnan VS et al (2024) Mechanistic insights into cellular and molecular targets of zinc oxide quantum dots (ZnO QDs) in Candida albicans. ACS Infect Dis 10(6): 1914–1934. https://doi.org/10.1021/acsinfecdis.3c00562

Chen N, Jiang C (2023) Antimicrobial peptides: structure, mechanism, and modification. Eur J Med Chem 255:115377. https://doi.org/10.1016/j.ejmech.2023.115377

Cheng R, Xu Q, Hu F et al (2021) Antifungal activity of MAF-1A peptide against *Candida albicans*. FEMS Microbiol Lett 368(2):fnab012

Chowdhary A, Sharma C, Meis JF (2017) *Candida auris*: a rapidly emerging cause of hospital-acquired multidrug-resistant fungal infections globally. PLoS Pathog 13(5):e1006290

Ciociola T, Giovati L, De Simone T et al (2023) Novel arginine- and proline-rich candidacidal peptides obtained through a bioinformatic approach. Antibiotics 12(3):472

Cowan MM (1999) Plant products as antimicrobial agents. Clin Microbiol Rev 12(4):564–582

Cowen LE, Sanglard D, Howard SJ, Rogers PD, Perlin DS (2015) Mechanisms of antifungal drug resistance. Cold Spring Harb Perspect Med 5(6):a019752

Cowen LE, Singh SD, Köhler JR Collins C, Zaas AK, Schell WA, Aziz H, Mylonakis E, Perfect JR, Whitesell L, Lindquist S (2009) Harnessing Hsp90 function as a powerful, broadly effective therapeutic strategy for fungal infectious disease. Proc Natl Acad Sci U S A 106(8):2818–2823. https://doi.org/10.1073/pnas.0813394106

Dash R, Bhattacharjya S (2021) Thanatin: an emerging host defense antimicrobial peptide with multiple modes of action. Int J Mol Sci 22(4):1522 de Gouveia Dias M (2019) Optimization of the liposomal system of DODAB: MO: development of a vaccine against systemic infections caused by *Candida albicans*. Universidade do Minho. ProQuest Dissertations and Theses Global

Di Mambro T, Guerriero I, Aurisicchio L, Magnani M (2019) The yin and yang of current antifungal therapeutic strategies: how can we harness our natural defenses? Front Pharmacol 10:80. https://doi.org/10.3389/fphar.2019.00080

Denning DW (2003) Echinocandins: a new class of antifungal. Lancet 362(9390):1142–1151

Fisher MC, Hawkins NJ, Sanglard D, Gurr SJ (2018) Worldwide emergence of resistance to antifungal drugs challenges human health and food security. Science 360(6390):739–742

Fisher MC, Alastruey-Izquierdo A et al (2022) Tackling the emerging threat of antifungal resistance to human health. Nat Rev Microbiol 20(9):557–571

Freestone PPE (2013) Communication between bacteria and their hosts. Scientifica 2013:361073. https://doi.org/10.1155/2013/361073

Gallis HA, Drew RH, Pickard WW (1990) Amphotericin B: 30 years of clinical experience. Clin Infect Dis 12(2):308–329

Garcia-Gomes AS, Curvelo JAR, Soares RMA Soares RA and Ferreira-Pereira A (2012) Curcumin acts synergistically with fluconazole to sensitize a clinical isolate of *Candida albicans* showing an MDR phenotype. Med Mycol 50(1):26–32

Ghannoum MA, Rice LB (1999) Antifungal agents: Mode of action, mechanisms of resistance, and correlation of these mechanisms with bacterial resistance. Clin Microbiol Rev 12(4):501–517. https://doi.org/10.1128/CMR.12.4.501

Gibson GR, Hutkins R et al (2020) ISAPP consensus statement on the definition and scope of postbiotics. Nat Rev Gastroenterol Hepatol 18(1):9–17

Gibson GR, Roberfroid MB (1995) Dietary modulation of the human colonic microbiota: introducing the concept of prebiotics. J Nutr 125(6):1401–1412. 412. https://doi.org/10.1093/jn/125.6.1401

Gilon C, Halle D, Chorev M, Selinger Z (1991) Backbone cyclization: a new method for conferring conformational constraint on peptides. Biopolymers 31(6):745–750. https://doi.org/10.1002/bip.360310619

Gupta AK, Tu LQ (2006) Dermatophytes: diagnosis and treatment. J Am Acad Dermatol 54(6): 1050–1055

Han Y (2010) Efficacy of combination immunotherapy of IgM MAb B6.1 and amphotericin B against disseminated candidiasis. Int Immunopharmacol 10(12):1526–1531

Han Y, Zhang M, Lai R, Zhang Z (2021) Chemical modifications to increase the therapeutic potential of antimicrobial peptides. Peptides 146:170666. https://doi.org/10.1016/j.peptides.2021.170666

Hardy H, Harris J, Lyon E Beal J, Foey AD (2013) Probiotics, prebiotics and immunomodulation of gut mucosal defences: homeostasis and immunopathology. Nutrients 5(6):1869–1912. https://doi.org/10.3390/nu5061869

Janssens I, Lambrecht BN, Van Braeckel E (2024) *Aspergillus* and the lung. Semin Respir Crit Care Med 45(1):3–20

Jayan S, Gupta P (2023) Effective combinations against efflux pump overexpressed on azole resistance *Candida* and Dermatophytes: a systematic review. Biomed Pharmacol J 16(1): 179–188

Kauffman CA (2017) Treatment of the midwestern endemic mycoses. Curr Fungal Infect Rep 11: 67–74

Kauser S, Raj N, Ahmedi S, and Manzoor N (2024) Synergistic antifungal potential of thymol and fluconazole against Candida albicans: an *in vitro* study. J Mycol Plant Pathol 54(1):1–7

Khan SN, Khan S, Misba L, Sharief M, Hashmi A and Khan AU (2019) Synergistic fungicidal activity with low doses of eugenol and amphotericin B against Candida albicans. Biochem Biophys Res Commun 518(3):459–464. https://doi.org/10.1016/j.bbrc.2019.08.053

Kheradmand E, Rafii F, Yazdi MH et al (2014) Antimicrobial effects of selenium nanoparticle-enriched probiotics. Daru 22(1):48

Kullberg BJ, Arendrup MC (2015) Invasive candidiasis. N Engl J Med 373(15):1445–1456. https://doi.org/10.1056/NEJMra1315399

Lackner M, Martin-Vicente A, Lass-Flörl C (2015) Multidrug- and cross-resistant Candida: the looming threat. Curr Fungal Infect Rep 9:23–36.

Lai Z, Yuan X, Chen H, Zhu Y, Dong N and Shan, A (2022) Strategies employed in the design of antimicrobial peptides with enhanced proteolytic stability. Biotechnol Adv 59:107962. https://doi.org/10.1016/j.biotechadv.2022.107962

Lass-Flörl C, Steixner S (2023) The changing epidemiology of fungal infections. Mol Aspects Med 94:101215. https://doi.org/10.1016/j.mam.2023.101215

Latif A, Shehzad A, Niazi S et al (2023) Probiotics: mechanism of action, health benefits and their application in food industries. Front Microbiol 14:1216674

Lee M-R, Raman N, Ortiz-Bermúdez P, Lynn DM, Palecek SP (2019) 14-Helical β-peptides elicit toxicity against *Candida albicans* by forming pores in the cell membrane and subsequently disrupting intracellular organelles. Cell Chem Biol 26(2):289–299

Li T, Li L, Du F, Sun L, Shi J, Long M, Chen Z (2021) Activity and mechanism of action of antifungal peptides from microorganisms: a review. Molecules 26(11):3438

Lima PG, Souza PFN, Freitas CDT, Bezerra LP, Neto NAS, Silva AFB, Oliveira JTA, Sousa DOB (2021) Synthetic peptides against T*richophyton mentagrophytes* and *T. rubrum*: Mechanisms of action and efficiency compared to griseofulvin and itraconazole. Life Sci 265:118803

Lisa K, Matthias E, Johannes B, Martin H, Robert K (2025) New treatment options for critically important WHO fungal priority pathogens. Clin Microbiol Infect 31(6):922–930. https://doi.org/10.1016/j.cmi.2024.03.006

Lourenco A, Pedro NA, Salazar SB, Mira NP (2019) Effect of acetic acid and lactic acid at low pH in growth and azole resistance of *Candida albicans* and *Candida glabrata*. Front Microbiol 9: 3265

Machado A, Zamora-Mendoza L, Alexis F, Álvarez-Suarez JM(2023) Use of plant extracts, bee-derived products, and probiotic-related applications to fight multidrug-resistant pathogens in the post-antibiotic era. Future Pharmacol 3(3):535–567. https://doi.org/10.3390/futurepharmacol3030034

Mączka W, Twardawska M, Grabarczyk M et al (2023) Carvacrol—A natural phenolic compound with antimicrobial properties. Antibiotics 12(5):824

Mandal M, Chatterjee R, Mukherjee S (2024) Clinicoetiological profile of meningitis cases in HIV-infected patients admitted at a tertiary care hospital in Eastern India. J Glob Infect Dis 21(4). https://doi.org/10.1177/0976001624125427

Mani-López E, Arrioja-Bretón D, López-Malo A (2022) The impacts of antimicrobial and antifungal activity of cell-free supernatants from lactic acid bacteria in vitro and in foods. Compr Rev Food Sci Food Saf 21(1):604–641. https://doi.org/10.1111/1541-4337.12872

Martinez RC, Franceschini SA, Patta MC et al (2009) Improved cure of bacterial vaginosis with single dose of tinidazole (2 g), Lactobacillus rhamnosus GR-1, and Lactobacillus reuteri RC-14: A randomized, double-blind, placebo-controlled trial. Can J Microbiol 55(2):133–138.

Maziarz EK, Perfect JR (2023) Cryptococcosis. In: Diagnosis and treatment of fungal infections, pp 245–265

Mishra NN, Ali S, Shukla PK (2015) A monoclonal antibody against 47.2 kDa cell surface antigen prevents adherence and affects biofilm formation of *Candida albicans*. World J Microbiol Biotechnol 31(1):11–21.

Mookherjee N, Anderson MA, Haagsman HP et al (2020) Antimicrobial host defence peptides: functions and clinical potential. Nat Rev Drug Discov 19(5):311–332. https://doi.org/10.1038/s41573-019-0058-8

Murzyn A, Krasowska A, Augustyniak D et al (2010) The effect of Saccharomyces boulardii on Candida albicans-infected human intestinal cell lines Caco-2 and Intestin 407. FEMS Microbiol Lett 310(1):17–23

Nett JE, Andes DR (2016) Antifungal agents: spectrum of activity, pharmacology, and clinical indications. Infect Dis Clin 30(1):51–83

Nooney L, Matthews RC, Burnie JP (2005) Evaluation of Mycograb, amphotericin B, caspofungin, and fluconazole in combination against Cryptococcus neoformans by checkerboard and time-kill methodologies. Diagn Microbiol Infect Dis 51(1):19–29. https://doi.org/10.1016/j.diagmicrobio.2004.08.013
Nwobodo DC, Ugwu MC, Anie CO, Al-Ouqaili MTS, Ikem JC, Chigozie UV and Saki M. (2022) Antibiotic resistance: the challenges and some emerging strategies for tackling a global menace. J Clin Lab Anal 36(9):e24655. https://doi.org/10.1002/jcla.24655
Odds FC, Brown AJP, Gow NAR (2003) Antifungal agents: mechanisms of action. Trends Microbiol 11(6):272–279. https://doi.org/10.1016/S0966-842X(03)00117-3
O'Hara AM, Shanahan F (2007) Mechanisms of action of probiotics in intestinal diseases. Sci World J 7:31–46
O'Hanlon DE, Le Moal V, Najjari M (2013) Vaginal pH and microbicides in relation to HIV-1 susceptibility. PLoS One 8(11):e80606
Osset-Trénor P, Pascual-Ahuir A, Proft M (2023) Fungal drug response and antimicrobial resistance. J Fungi 9(5):565
Ostrosky-Zeichner L, Marr KA, Rex JH et al (2007) Amphotericin B: time for a new "gold standard". Clin Infect Dis 44(12):1707–1715
Pandya N, Cag Y, Pandak N et al (2021) International multicentre study of *Candida auris* infections. J Fungi 7(10):878
Pappas PG, Kauffman CA, Andes DR, Clancy CJ, Marr, KA, Ostrosky-Zeichner, L et al (2016) Clinical practice guideline for the management of candidiasis: 2016 update by the Infectious Diseases Society of America. Clin Infect Dis 62(4):e1–e50. https://doi.org/10.1093/cid/civ933
Pappas PG, Lionakis MS, Arendrup MC et al (2018) Invasive candidiasis. Nat Rev Dis Primers 4: 18026. https://doi.org/10.1038/nrdp.2018.26
Parente-Rocha JA, Bailão AM, Amaral AC, Taborda CP, Paccez JD, Borges CL and Pereira M (2017) Antifungal resistance, metabolic routes as drug targets, and new antifungal agents: an overview about endemic dimorphic fungi. Mediators Inflamm 2017:9870679. https://doi.org/10.1155/2017/9870679
Parolin C, Croatti V, Laghi L, Foschi C, Ñahui Palomino RA et al (2021) *Lactobacillus* biofilms influence anti-Candida activity. Front Microbiol 12:750368. https://doi.org/10.3389/fmicb.2021.750368
Patterson TF, Kirkpatrick WR, White M et al (2000) Invasive aspergillosis: disease spectrum, treatment practices, and outcomes. Medicine (Baltimore) 79(4):250–260
Perlin DS, Rautemaa-Richardson R, Alastruey-Izquierdo A (2017) The global problem of antifungal resistance. Lancet Infect Dis 17(3):e383–e392
Pillai NS, S SA, Shanmugam R (2024) Antifungal activity of chitosan nanoparticle-incorporated probiotics. Cureus 16(9):e70093
Pinilla G, Coronado YT, Chaves G et al (2022) *In vitro* antifungal activity of LL-37 analogue peptides. J Fungi 8(11):1173
Plaza-Díaz J, Ruiz-Ojeda FJ, Vilchez-Padial LM et al (2019) Anti-inflammatory effects of probiotics and synbiotics. Nutrients 11(2):281
Rachel A, Anuradha M, Leela KV (2024) Synergistic antifungal activity of cinnamaldehyde and amphotericin B against Candida albicans biofilms. J Herb Med 45:102242
Rao SR, Ravishankar GA (2002) Plant cell cultures as chemical factories of secondary metabolites. Biotechnol Adv 20(2):101–153
Ríos JL, Recio MC (2005) Medicinal plants and antimicrobial activity. Fitoterapia 76(2):255–264
Rizzo A, Losacco A, Carratelli CR (2013) *Lactobacillus crispatus* modulates epithelial cell defense against Candida albicans. Immunol Lett 156(1–2):102–109
Romani L (2011) Immunity to fungal infections. Nat Rev Immunol 11(4):275–288
Ryder N, Mieth H (1992) Allylamine antifungal drugs. In: Current topics in medical mycology. Springer, pp 233–251. https://doi.org/10.1007/978-1-4612-2762-5_6

Salminen S, Collado MC, Endo A et al (2021) The International Scientific Association of Probiotics and Prebiotics (ISAPP) consensus statement on the definition and scope of postbiotics. Nat Rev Gastroenterol Hepatol 18(9):649–667. https://doi.org/10.1038/s41575-021-00440-6
Samonis G, Kontoyiannis DP (2005) Invasive fungal infections in the setting of cancer. Eur J Cancer 41(11):1665–1673
Sánchez-López E, Gomes D, Esteruelas G et al (2020) Metal-based nanoparticles as antimicrobial agents: an overview. Nano 10(2):292
Schaefer A, Smith J, Johnson L (2024) Synthetic mimics of peptides constructed with caspofungin prevented infection of human epithelial cells *in vitro,* improved fungal clearance by human macrophages, and significantly increased host survival in systemic candidiasis. J Antimicrob Chemother 79(2):543–557. https://doi.org/10.1016/j.jac.2024.02.001
Schroeder M, Weber T, Denker T et al (2020) Epidemiology and outcome of candidemia in critically ill patients. Ann Intensive Care 10(1):142
Sharma S, Sharma N, Kaushal N (2022) Comparative account of biogenic synthesis of silver nanoparticles using probiotics and their antimicrobial activity against challenging pathogens. BioNanoScience 12(2):833–840
Silva-Rodrigues G, Castro IM de, Borges PHG et al (2024) Geraniol potentiates the effect of fluconazole against planktonic and sessile cells of azole-resistant Candida tropicalis: *In vitro* and *in vivo* analyses. Pharmaceutics 16(8):1053
Souza CS de, Lopes VR da C, Barcellos G et al (2024) Unleashing fungicidal forces: exploring the synergistic power of amphotericin B-loaded nanoparticles and monoclonal antibodies. J Fungi 10(5):344. https://doi.org/10.3390/jof10050344
Spettel K (2024) Resistance development in fungi: an emerging public health threat. Medical University of Vienna
Srivastava N, Ellepola K, Venkiteswaran N et al (2020) *Lactobacillus plantarum* 108 inhibits *Streptococcus mutans* and *Candida albicans* mixed-species biofilm formation. Antibiotics 9(8): 478
Ström K, Sjögren J, Broberg A, Schnürer J (2002) *Lactobacillus plantarum* MiLAB 393 produces antifungal cyclic dipeptides. Appl Environ Microbiol 68(9):4322–4327
Swanson KS, Gibson GR, Hutkins R, (2020) The International Scientific Association for Probiotics and Prebiotics (ISAPP) consensus statement on the definition and scope of synbiotics. Nat Rev Gastroenterol Hepatol 17(11):687–701. https://doi.org/10.1038/s41575-020-0344-2
Tachedjian G, O'Hanlon DE, Ravel J (2018) The implausible *in vivo* role of hydrogen peroxide as an antimicrobial factor produced by vaginal microbiota. Microbiome 6(1):29. https://doi.org/10.1186/s40168-018-0418-3
Takano T et al (2023) Inhibitory effects of vaginal *Lactobacilli* on *Candida albicans*. Front Cell Infect Microbiol 13:1113401
Thakre A, Jhadav V, Kazi R, Shelar A, Patil, R et al (2021) Oxidative stress induced by piperine leads to apoptosis in *Candida albicans*. Medical Mycol 59(4):366–378. https://doi.org/10.1093/mmy/myaa058
Thambirajoo M et al (2021) Nanoparticles integrated with antibacterial properties in preventing biofilm and resistance. Antibiotics 10(11):1338
Tiwari ASK (2022) Bacteriocin-producing probiotic lactic acid bacteria in controlling dysbiosis of the gut microbiota. Front Cell Infect Microbiol 12:851140
Vanzolini T (2021) Development of new biological drugs for the treatment of fungal infections
Vermes A, Guchelaar HJ, Schoemaker RC (2000) Clinical pharmacokinetics of flucytosine. Clin Pharmacokinet 39(4):277–292
Wagner H, Proksch A (1985) Immunostimulatory drugs of fungi and higher plants. Econ Med Plant Res 1:113–153
Wang K et al (2015) Protonectin disturbs membrane integrity and induces ROS production in yeast cells. Biochim Biophys Acta Biomembr 1848(10):2365–2373
Wang W, Zhao J, Zhang Z (2022) *Bacillus* metabolites: compounds, identification and anti-*Candida albicans* mechanisms. Microbiol Res 13(4):972–984

Wang B, He X, Lu F et al (2021) Candida isolates from blood and other normally sterile foci from ICU patients: determination of epidemiology, antifungal susceptibility profile and evaluation of associated risk factors. Front Public Health 9:779590.

Wingard JR et al (1999) Clinical significance of nephrotoxicity in patients treated with amphotericin B. Clin Infect Dis 29(6):1402–1407

World Health Organization (2022) Fungal diseases. Available at: https://www.who.int/news-room/fact-sheets/detail/fungal-diseases

Wu Y, Hu S, Wu C et al (2022) Probiotics: potential novel therapeutics against fungal infections. Front Cell Infect Microbiol 11:793419. https://doi.org/10.3389/fcimb.2021.793419

Wu YR et al (2025) Synergistic inhibition of *Candida albicans* by cranberry proanthocyanidins and probiotics. Pathogens 14(4):308

Xiang MJ et al (2013) ERG11 mutations associated with azole resistance in *Candida albicans*. FEMS Yeast Res 13(4):386–393

Xiao Y et al (2025) *Bifidobacterium longum* enhances host immunity and synergizes with voriconazole. J Med Microbiol 74(1):45–52

Zhai H et al (2025) Baicalein–fluconazole synergy via probiotic-mediated bioconversion. Front Microbiol 16:15621

Development and Applications of Genetically Engineered Probiotics to Combat Pathogenic Fungi and Fungal Toxins

20

Claire L. Price, Britney Lunn, Chloe Knight, Bethan Wells, and Leanne Witts

Abstract

Fungal pathogens and mycotoxins (secondary metabolites produced by fungi) are posing a growing threat to human health, food safety, and agricultural productivity. Current antifungal treatments and decontamination strategies are limited by resistance, toxicity, and environmental concerns, prompting the need for alternative approaches. This chapter explores the emergence of genetically engineered probiotics as targeted, biologically based tools for combating pathogenic fungi. Drawing on advances in genetic engineering, including CRISPR-Cas systems, recombinant DNA technology, and synthetic biology, this chapter outlines how probiotic strains can be modified to secrete antifungal compounds, degrade fungal cell walls, and metabolize mycotoxins into non-toxic derivatives. Mechanisms such as competitive exclusion and niche occupation are also discussed. Applications are reviewed across human and animal health, crop protection, and postharvest food systems, with attention to validated cases and conceptual innovations. Regulatory, biosafety, and public perception challenges are examined as critical factors influencing real-world deployment. This chapter highlights the potential of engineered probiotics to become adaptable, sustainable interventions at the interface of microbiome science, synthetic biology, and global health.

C. L. Price (✉) · B. Lunn · C. Knight · B. Wells · L. Witts
Swansea University Medical School, Swansea University, Singleton Park Campus, Swansea, UK
e-mail: c.l.price@swansea.ac.uk

M. Y. Sreenivasa et al. (eds.), *Probiotics as Antifungals*,
https://doi.org/10.1007/978-981-92-0002-3_20

Graphical Abstract

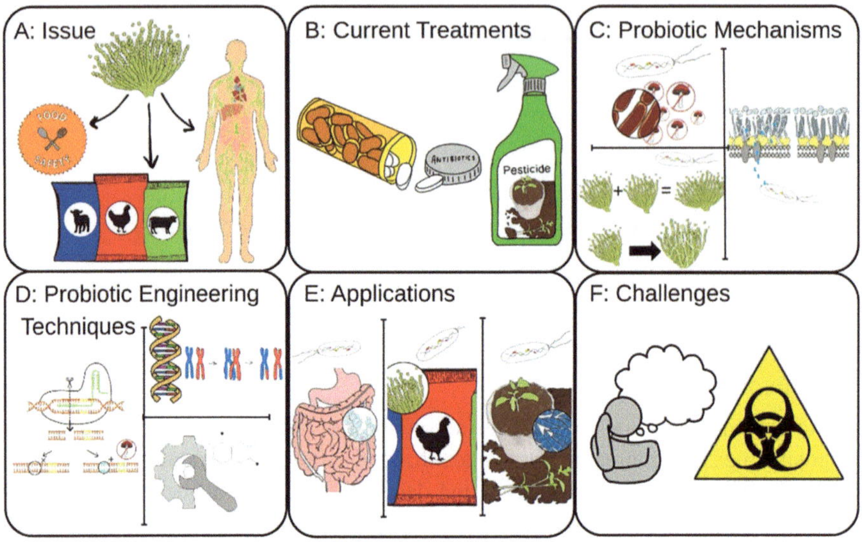

Keywords

Genetically engineered probiotics · Antifungal therapy · Mycotoxin detoxification · Synthetic biology · Microbial biocontrol

Learning Outcomes

By the end of this chapter, readers will be able to:

- Describe the threats posed by fungal pathogens and mycotoxins to human, animal, and agricultural health.
- Explain the principles and tools of genetic engineering as applied to probiotic microorganisms.
- Evaluate current and proposed mechanisms by which engineered probiotics can combat fungal infections and detoxify mycotoxins.
- Discuss the practical challenges and regulatory considerations involved in developing and applying genetically engineered probiotics.

20.1 Introduction

Fungal pathogens and the toxins they produce pose a significant threat to global health, agriculture, and food security. Opportunistic infections caused by *Candida albicans* and *Aspergillus fumigatus* are increasingly common in immunocompromised individuals, leading to severe, treatment-resistant disease (Vitiello et al. 2023). These organisms employ sophisticated mechanisms of infection, including biofilm formation, immune evasion, and tissue invasion. Alternatively, the contamination of crops and animal feed by toxigenic fungi, *Fusarium* and *Aspergillus* species, leads to the widespread presence of mycotoxins (El-Sayed et al. 2022; Krishnan et al. 2024). Aflatoxins, ochratoxins, and fumonisins are among the most prevalent, with documented effects ranging from immunosuppression and hepatotoxicity to carcinogenicity (Anumudu et al. 2024; Mafe and Büsselberg 2024). The economic ramifications of fungal contamination span agricultural productivity, food insecurity, and international trade barriers (Singh and Thakur 2023; Steinberg and Gurr 2020).

Current methods for managing fungal infections and mycotoxin contamination (ranging from antifungal drugs to chemical fungicides) are limited by resistance, toxicity, and environmental concerns (Fisher et al. 2018). Against this, probiotics have emerged as a promising alternative. Traditionally valued for their role in gut health, certain probiotic strains demonstrate natural antifungal activity and inhibit toxin-producing fungi through competitive exclusion or secretion of inhibitory metabolites (Divyashree et al. 2023; Vinayamohan et al. 2024). Advances in genetic engineering have expanded the potential of probiotics beyond their native capabilities. Tools like CRISPR (clustered regularly interspaced short palindromic repeats)-Cas (CRISPR-associated) systems, recombinant DNA technology, and synthetic biology allow researchers to design probiotics that perform specific antifungal and detoxifying functions (Liu et al. 2023a; Nazir et al. 2024). These engineered microbes have the potential to degrade fungal cell walls, produce antifungal compounds, or metabolize mycotoxins into non-toxic derivatives.

20.2 Genetic Engineering of Probiotics

Recent advances in synthetic biology and gene-editing technologies have enabled precise manipulation of probiotics to enhance their functionality. Traditionally selected for their natural health-promoting properties, microbes are now being engineered using emerging gene-editing technologies, which provide scientists with a toolkit to introduce entirely new functionalities (Fig. 20.1).

20.3 Overview of Genetic Engineering Techniques

Modern genetic engineering encompasses a variety of methods for altering microbial genomes, each offering various levels of precision, efficiency, and scalability.

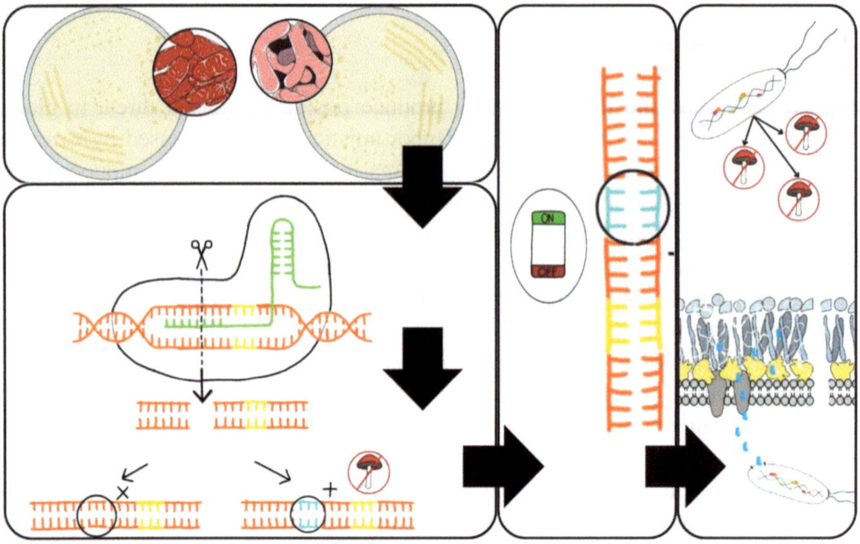

Fig. 20.1 Overview of genetically engineered design: schematic representation of the development of a genetically engineered probiotic strain. A host strain (*Lactobacillus* or *Bifidobacterium*) is selected for genetic modification. Using CRISPR-based gene editing, an antifungal peptide-encoding gene is introduced alongside biosafety elements, such as a kill switch mechanism designed to deactivate the inserted gene under specific conditions. The resulting engineered probiotic is capable of secreting antifungal compounds and degrading fungal cell walls, offering potential therapeutic applications against fungal infections

20.3.1 CRISPR-Cas Systems

One of the most powerful tools in genome editing, the CRISPR-Cas9 variant, enables precise genome modifications through a guide RNA that directs the Cas9 nuclease to a target DNA sequence (Liu et al. 2022). Engineered probiotics can be designed to knock out undesirable genes or integrate new pathways for antifungal metabolite production. Emerging versions such as base editors (making single-nucleotide changes) and CRISPRa/i (regulation of transcription via activation (a) or interference (i)) expand control over gene expression without inducing double-stranded breaks (Arroyo-Olarte et al. 2021; Zhou et al. 2020).

20.3.2 Recombinant DNA Technology

This technique involves introducing foreign DNA into probiotic cells using a plasmid vector, allowing for stable or inducible expression of antifungal peptides, detoxifying enzymes, or reporter genes. Plasmid systems can be tailored for high-copy expression, inducibility, or integration into the host genome for long-term stability (Fernández de Ullivarri et al. 2020; Liu et al. 2023b; Vachher et al. 2021).

20.3.3 Synthetic Biology Approaches

Synthetic biology involves designing novel biological circuits. Engineered probiotics can be equipped with sensor modules to detect fungal toxins or stress signals, and actuator modules that trigger the production of antifungal agents, under specific conditions (Bazhenov et al. 2023; Rottinghaus et al. 2020; Zhou et al. 2020). This programmability minimizes metabolic burden and enhances safety.

20.3.4 Promoter Engineering and Codon Optimization

Fine-tuning gene expression is essential for balancing functionality and metabolic burden. Synthetic promoters and codon optimization strategies are employed to ensure stable and efficient protein production (Bober et al. 2018; Bradley et al. 2016; Choi et al. 2025; Mauro 2018).

20.3.5 Omics-Guided Design

Advances in genomics, transcriptomics, and metabolomics allow for systems-level understanding of probiotic function. These tools aid in identifying target pathways, optimizing metabolic flux, and predicting off-target effects of engineered functions (Amer and Baidoo 2021; Kahraman Ilıkkan 2025).

20.4 Selection of Probiotic Strains

Not all microorganisms are suitable for genetic engineering or probiotic use. Selecting the right host strain is foundational to successful engineering.

20.4.1 Safety

Strains must have a well-documented history of safe use in humans or animals. Regulatory designations such as Generally Recognized as Safe (GRAS) by the FDA or Qualified Presumption of Safety (QPS) by the European Food Safety Authority are essential benchmarks to allow them to be added to food or feed (Koutsoumanis et al. 2024; Heo et al. 2022).

20.4.2 Genetic Traceability

The strain must be amenable to transformation, with tools available for stable gene insertion, inducible promoters, and selection markers, allowing for the predictable introduction and inheritance of foreign genes (Dietz-Pfeilstetter 2010).

20.4.3 Colonization and Persistence

The ability to colonize the gut, mucosal surfaces, or plant roots is critical for in situ antifungal activity (Han et al. 2021; Rau et al. 2024; Xiao et al. 2020). Adhesion factors, biofilm formation, and resistance to bile and acid are important traits in the protection of the host from pathogens and for maintaining a healthy gut microbiome. Common host organisms include *Lactobacillus plantarum*, *L. rhamnosus*, *Bifidobacterium longum*, and *Bacillus subtilis* (Kandasamy et al. 2016; Maldonado-Gómez et al. 2016; Segers and Lebeer 2014), each offering different advantages in terms of delivery, persistence, and compatibility with specific application environments.

20.5 Safety and Regulatory Considerations

The release or use of genetically modified probiotics, whether in clinical, agricultural, or food settings, raises important biosafety and ethical concerns.

20.5.1 Biosafety

Horizontal gene transfer of engineered traits to native microbiota or pathogens is a primary concern, especially those involving antibiotic resistance markers (McInnes et al. 2020; Tao et al. 2022). Strategies to mitigate this risk include the following: use of toxin-antitoxin kill switches, deletion of conjugative elements, synthetic auxotrophy (nutritional dependency), or chromosomal integration of transgenes that enhance containment (Daubin and Szöllősi 2016; Getino and de la Cruz 2018).

20.5.2 Environmental Persistence

Modified probiotics may evolve unpredictably in the environment. Biocontainment systems, such as kill switches or auxotrophy, are being developed to control probiotic lifespan and dispersal (Moe-Behrens et al. 2013; Rottinghaus et al. 2022; Wright et al. 2015).

20.5.3 Regulator Oversight

The approval of genetically modified probiotics varies globally (Spacova et al. 2023). In the European Union, they are subject to stringent GMO regulations (EFSA Panel on Dietetic Products, Nutrition and Allergies (NDA) 2021). In contrast, the United States evaluates them under the FDA's food or biologics categories, depending on intended use (Commissioner 2021).

20.5.4 Public Perception and Ethical Issues

Consumer acceptance hinges on transparency, perceived safety, and trust in regulatory oversight. Communicating the benefits and safeguards of engineered probiotics is essential for fostering acceptance and responsible adoption (Harrison et al. 2015; Trump et al. 2023).

20.6 Theoretical and Emerging Possibilities

Where direct examples are lacking, natural systems can be enhanced by genetic engineering to create strains that produce greater defenses against fungal pathogens. Areas for exploration include the following: antifungal peptides or enzymes (e.g., defensins, glucanases, proteases) (Thery et al. 2019), degrading mycotoxins through oxidoreductase or hydrolase activity (e.g., aflatoxin B1 oxidase, fumonisin esterase) (Chen et al. 2023), interrupting fungal quorum sensing by degrading signaling molecules like farnesol (Mehmood et al. 2019), and colonizing ecological niches from fungal pathogens (oral cavity, GI tract) for local delivery of antifungal effectors (Hu et al. 2017). These strategies remain unexplored in live probiotic systems but are achievable given current synthetic biology toolkits and precedents in other engineered bacteria.

20.7 Mechanisms of Engineered Probiotics Against Fungi and Toxins

Engineered probiotics may be programmed to negate fungal threats through multiple, often synergistic, mechanisms (Fig. 20.2). These include direct inhibition of fungal growth, enzymatic degradation of fungal structures, competitive exclusion of pathogenic species, and detoxification of harmful mycotoxins. While a limited number of these mechanisms have been validated, many are supported by evidence in other microbial systems and present clear opportunities for synthetic biology.

20.7.1 Secretion of Antifungal Peptides and Metabolites

One promising strategy is probiotic secretion of antifungal peptides or lipopeptides. These molecules permeate fungal membranes, interfering with cell wall synthesis or intracellular signaling (Mejía-Pitta et al. 2021). Antimicrobial peptides, such as defensins, have the potential to be credible for antifungal applications. Probiotic hosts, such as *L. plantarum* and *L. lactis,* have been engineered to produce antimicrobial peptides, showing great promise as antibacterial agents (Mejía-Pitta et al. 2021). These strategies could be extended to antifungal applications. For example, hydroxyphenylacetic acid (HPA), a monocarboxylic acid produced by *C. albicans*, was heterologously expressed in *E. coli* to produce a *C. albicans* sensor. When

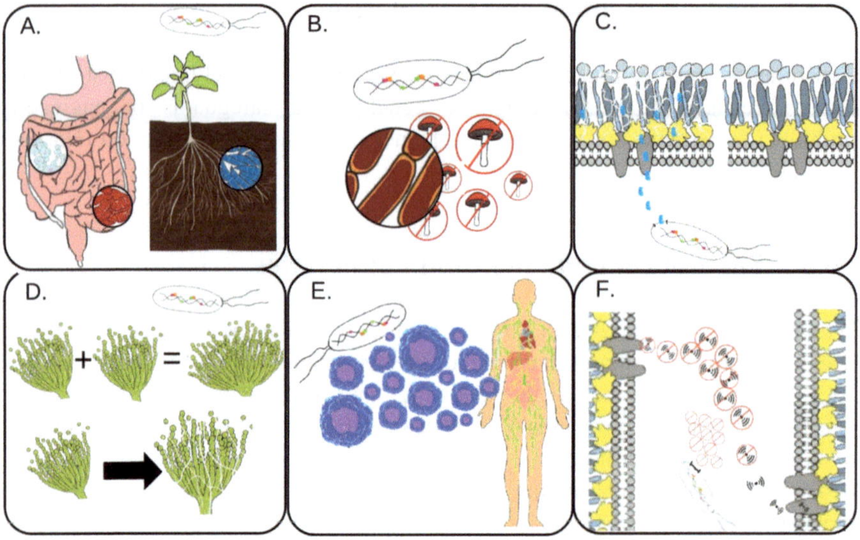

Fig. 20.2 The multifaceted antifungal mechanisms and host interactions mediated by genetically engineered probiotics. The strain inhibits fungal colonization via competitive exclusion of *Candida* in the gut and *Fusarium* in the rhizosphere (**a**) and secretes antifungal compounds such as reuterin (**b**). It produces enzymes that degrade fungal cell walls (**c**) and binds or degrades mycotoxins to reduce toxicity (**d**). The probiotic also modulates host immunity through cytokine release (**e**) and disrupts fungal quorum sensing to inhibit virulence (**f**)

coupled with an inhibitor of hypha formation, this engineered strain was able to reduce the expression of virulence factors and filamentation. Lipopeptides like surfactin, fengycin, and iturin are naturally produced by some *Bacillus* spp. (Kim et al. 2010), and have strong biosurfactant activities. Engineered production of these compounds has the potential to inhibit fungal spore adhesion, prevent biofilm formation, and facilitate immune-mediated clearance. Their amphiphilic nature allows them to integrate into lipid membranes, creating pores that lead to cell lysis (Dias and Nitschke 2023; Rodrigues and Rodrigues 2023). Engineered strains of this nature could offer a more environmentally safe alternative to current chemical treatment. Although no engineered probiotic has yet been commercialized, research has demonstrated that heterologous expression of antifungal peptides is feasible in lactic acid bacterial systems.

20.7.2 Competitive Exclusion and Niche Occupation

Competitive exclusion is a fundamental mechanism whereby probiotics occupy physical and nutritional niches that would otherwise be available to pathogens (Mitchell 2006). Engineered probiotics could be designed to augment this natural barrier function. Done through overexpressing adhesion molecules, such as mucin, to increase colonization efficiency at target sites. Another function that could be augmented using these synthetic techniques is the production of quorum-sensing inhibitors (Zawrotniak et al. 2019). Enhancing these sensors would aid their role as antifungals by blocking fungal signaling and morphogenesis (e.g., the yeast-to-hyphae transition of *C. albicans*). Alternatively, strains could be engineered to compete with natural fungal virulence factors. For instance, they could be engineered to enhance probiotic biofilm formation to outcompete fungal biofilms, particularly on in-dwelling medical devices or mucosal surfaces. These strategies have yet to be validated in engineered strains designed to prevent fungal colonization, but the underlying biological principles are well supported.

20.7.3 Enzymatic Degradation of Fungal Cell Walls

Fungal cell walls are primarily composed of chitin, β-1,3-glucans, and mannoproteins (Gow and Lenardon 2023). Engineered probiotics expressing lytic enzymes can degrade these structural components, leading to compromised fungal integrity and death (Charlet et al. 2020; Vazquez-Munoz et al. 2024). The expression of chitinase A from *Serratia marcescens* in *Bacillus subtilis* has been shown to inhibit the growth of plant-pathogenic fungi like *Fusarium oxysporum* (Okay and Alshehri 2020). Similarly, β-1,3-glucanase expression enhances activity against *Candida* biofilms (Nett et al. 2007; Tan et al. 2018). Future work in this area could concentrate on producing strains that overexpress both chitinases and glucanases. Combining these enzymes could enhance synergistic activity, providing a multipronged enzymatic attack, weakening cell walls, and enhancing host immunity or antifungal drug pretention. These enzymes may be constitutively expressed or under the control of inducible promoters that respond to fungal metabolites, allowing targeted and efficient deployment. Further research is needed to optimize expression levels, secretion efficiency, and in vivo delivery to maximize these enzymes' antifungal potential in probiotic contexts.

20.7.4 Detoxification of Mycotoxins via Metabolic Engineering

Mycotoxins are structurally diverse and highly stable, often resisting degradation by standard food processing. Engineered probiotics can be used to degrade or transform these compounds into non-toxic derivatives using specific enzymes.

20.7.4.1 Aflatoxins

While aflatoxin-degrading enzymes (e.g., oxidoreductases) have been characterized in *Pseudomonas* and *Mycobacterium* species (Maneeboon et al. 2024; Taylor et al. 2010), their expression in probiotics remains hypothetical. Engineering *Lactobacillus* strains to express such enzymes could offer a route to detoxify aflatoxin B1 in the digestive tract.

20.7.4.2 Fumonisins and Ochratoxins

Enzymes such as carboxypeptidases, hydrolases, and esterases have been studied for their ability to neutralize fumonisin B1 and ochratoxin A (Lyagin and Efremenko 2019). These enzymes have not been incorporated into probiotic systems but could be adapted using inducible or constitutive expression in *B. subtilis* or *L. plantarum.* In addition to enzymatic degradation, future strategies could involve engineering probiotics to display high-affinity binding domains that sequester toxins in the gut and prevent systemic absorption.

20.8 Synthetic Targeting and Smart Responses

Synthetic biology enables precise control and environmental responsiveness in engineered organisms and may provide a novel route to enhancing antifungal activity.

20.8.1 Chitin-Binding Domains

Chitin-binding domains could be used to target fungal cell walls. For example, *E. coli* Nissle 1917 expressing a chitin-binding domain can enhance adhesion to *Candida* hyphae (Chamas et al. 2024). While not antifungal by itself, this targeting system could localize the delivery of antifungal enzymes or compounds.

20.8.2 Sense-and-Respond Systems

Probiotic strains equipped with biosensors could detect fungal metabolites or quorum signals (e.g., farnesol) and activate antifungal responses in the presence of pathogenic fungi. Although no probiotic strain currently uses this technology against fungi, it has been applied to detect bacterial toxins (Ahmed et al. 2014; Banerjee et al. 2013), highlighting its feasibility. These advanced approaches could increase the precision, efficacy, and safety of probiotic-based antifungal therapies, reducing off-target effects and improving patient outcomes. Together, these mechanisms illustrate the potential of engineered probiotics as living therapeutics and biocontrol agents. While remaining theoretical, they offer promising directions for translational research and development.

20.9 Applications in Human and Animal Health

The application of genetically engineered probiotics to combat fungal pathogens and detoxify mycotoxins holds considerable promise across both clinical and agricultural contexts (Fig. 20.3). While practical deployment remains limited, several key areas

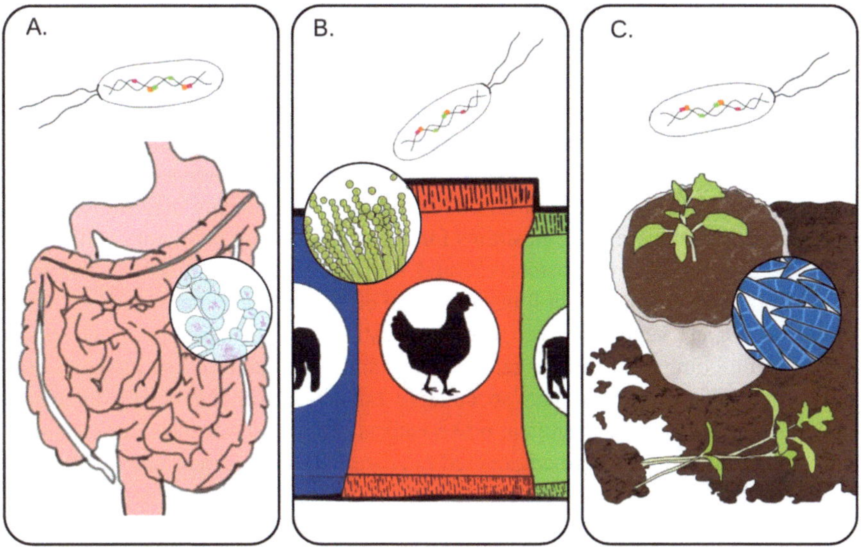

Fig. 20.3 Real-world applications of genetically engineered probiotics in human, animal, and environmental health. In the human gastrointestinal tract, engineered probiotics can target *Candida* species to help prevent or reduce fungal infections (**a**). In livestock systems, they can degrade mycotoxins present in animal feed, mitigating toxin-related health risks (**b**). In agricultural settings, these probiotics can be applied to soil to inhibit pathogenic fungi, such as *Fusarium*, supporting plant health and crop protection (**c**)

of opportunity are emerging. In human health, engineered probiotics could serve as living antifungal agents at mucosal surfaces (e.g., oral, gastrointestinal, vaginal), particularly through chitinases, glucanases, or antifungal peptides. Theoretical strategies, such as sense-and-respond circuits for detecting fungal virulence or delivering targeted responses, exemplify the future direction of microbial therapeutics. In animal health and livestock, genetically modified probiotics offer potential as feed additives to detoxify mycotoxins, reduce fungal disease burden, and enhance growth performance. Traits such as spore-forming stability, enzyme secretion, and colonization resilience are especially valuable in these environments. Although only a few strains have been tested in vivo, they represent an alternative to chemical binders and antifungal drugs. Together, these applications illustrate the dual potential of engineered probiotics, as therapeutic agents in humans and as sustainable tools in veterinary and agricultural systems. Continued progress will depend on translational research, biosafety validation, and tailored regulatory pathways.

20.10 Agricultural and Food System Applications

Engineered probiotics also offer promising solutions to fungal threats in agriculture, though their application remains theoretical. In crop protection, probiotic organisms could be engineered to secrete antifungal enzymes or peptides that inhibit pathogenic fungi on roots and leaves. While current biocontrol agents are typically non-engineered (Thambugala et al. 2020), future developments may see the introduction of genetically enhanced strains for field use, pending regulatory and ecological risk assessments. In postharvest and storage systems, engineered probiotics could be developed to prevent fungal spoilage or mycotoxin accumulation on stored grains, fruits, and vegetables. Proposed applications include surface-applied sprays, refrigeration-compatible antifungal strains, or antimicrobial biofilms integrated into packaging. These ideas remain speculative but align with increasing interest in biological alternatives to chemical preservatives. The most advanced application is in feed detoxification, where engineered *L. reuteri* strains degrade zearalenone (Liu et al. 2019), showing practical efficacy in in vitro and model digestion systems. These strains could serve as living feed additives that neutralize toxins during digestion. Despite their potential, genetically engineered probiotics in agriculture face regulatory barriers, limited field validation, and concerns about GMO deployment in open environments. Nevertheless, rising demand for sustainable food safety solutions continues to drive innovation in this space.

20.11 Challenges and Regulatory Considerations

The development and deployment of genetically engineered probiotics for antifungal and mycotoxin control raise several important challenges that must be addressed to enable safe and effective application across human, animal, and environmental contexts.

20.11.1 Biosafety and Risk Assessment

Engineered probiotics are live organisms with the capacity to interact dynamically with host microbiomes, environmental microbes, and immune systems. Key biosafety considerations include unintended gene transfer to native microbiota, particularly in the gut or soil, off-target effects on beneficial fungi or host cells, especially with broadly active enzymes like glucanases or proteases, and persistence and containment, ensuring that engineered strains do not become invasive or disrupt ecological balance. Addressing these risks will require the incorporation of biological safeguards, comprehensive in vivo testing, and clear genetic traceability of engineered constructs.

20.11.2 Regulatory Pathways

Regulation of genetically engineered microbes varies widely by jurisdiction and application. In human health, regulatory bodies such as the FDA (US) and EMA (EU) evaluate live biotherapeutics for safety, efficacy, and environmental impact. Engineered probiotics fall under investigational new drug (IND) frameworks (Spacova et al. 2023). In agriculture, genetically modified organisms intended for release in the environment face stricter requirements than feed additives used in confined settings (EFSA Panel on Additives and Products or Substances used in Animal Feed (EFSA FEEDAP Panel) et al. 2019). Currently, no engineered probiotic has received full regulatory approval for antifungal or detoxification use in humans or field agriculture, although some detoxifying strains are advancing through preclinical stages.

20.11.3 Technical and Social Barriers

Beyond formal regulation, several practical challenges remain. These include strain stability and performance must be validated across diverse conditions (e.g., gut, soil, feed matrices), manufacturing and formulation must protect engineered functions during storage and delivery, and public perception and consumer trust will be critical, especially for applications in food or probiotics labelled as "natural" or "clean label." Efforts to increase transparency, engage stakeholders, and demonstrate measurable benefits will be essential to build public confidence.

20.12 Future Directions and Challenges

The use of genetically engineered probiotics to combat fungi is a rapidly evolving field with huge potential, but with limited current real-world deployment. Future research should prioritize the translation of promising in vitro findings into in vivo models and clinical or agricultural settings. Advancements in synthetic biology, programmable biosensors, inducible expression systems, and modular enzyme libraries will enable more precise and responsive probiotic designs. Interdisciplinary efforts are needed to develop safe delivery platforms, scalable manufacturing processes, and robust biocontainment strategies. Regulatory harmonization and public engagement will be crucial in facilitating the responsible adoption of these technologies. Looking forward, engineered probiotics may become integral components of sustainable strategies for infection control, food safety, and microbiome-based health interventions across multiple domains.

20.13 Conclusion

Genetically engineered probiotics represent an emerging frontier in the fight against fungal pathogens and mycotoxins. By leveraging advances in synthetic biology, microbial ecology, and host–microbe interaction research, these living systems offer a platform for precision-targeted, sustainable interventions across health, agriculture, and food safety domains. Despite clear promise, the field remains in its early stages. Only a handful of engineered strains, such as *L. reuteri* for zearalenone detoxification (Liu et al. 2019), have demonstrated functional efficacy in controlled models. Most antifungal strategies, including enzymatic degradation of fungal cell walls or the use of sense-and-respond systems, remain under experimental development. As the synthetic biology toolbox expands, engineered probiotics could become vital components in health and food system resilience. Realizing this potential will require continued innovation, rigorous risk assessment, and proactive dialogue between scientists, regulators, and the public.

20.14 Chapter Summary

This chapter explores the emerging role of genetically engineered probiotics as innovative tools for controlling fungal pathogens and detoxifying mycotoxins. It begins by outlining the global challenges posed by fungal infections and toxin contamination, and the limitations of current treatments. This chapter then introduces key genetic engineering methods (including CRISPR, recombinant DNA, and synthetic biology) and explains how these tools can be used to enhance probiotic function. Mechanisms of action are examined, such as antifungal metabolite production, enzymatic degradation of fungal structures, competitive exclusion, and metabolic detoxification of harmful compounds. Real-world applications are discussed across human health, veterinary, and agricultural use, with an emphasis on both validated examples and future possibilities. Finally, this chapter addresses the biosafety, regulatory, and ethical considerations involved in deploying genetically modified probiotics.

Together, these insights provide a foundation for understanding the promise and complexity of engineering probiotics as targeted biological solutions for fungal and mycotoxin-related challenges.

Acknowledgments The authors would like to thank Swansea University Medical School and the members of the Biomedical Science team (both staff and students) for providing the research environment and support necessary for the development of this work. We also extend our appreciation to the peer reviewers and editors whose constructive feedback enhanced the quality and clarity of this chapter.

Conflict of Interest There is no conflict of interest.

Author Contributions CLP: Conceptualization, Writing—Original Draft, Supervision. BL, CK, LW, BW: Literature Review, Writing—Review and Editing. CK—figure creation. All authors contributed to the design and structure of the chapter, approved the final manuscript, and agreed to be accountable for the content presented.

References

Ahmed A, Rushworth JV, Hirst NA, Millner PA (2014) Biosensors for whole-cell bacterial detection. Clin Microbiol Rev 27(3):631–646

Amer B, Baidoo EEK (2021) Omics-driven biotechnology for industrial applications. Front Bioeng Biotechnol 9:613307

Anumudu CK, Ekwueme CT, Uhegwu CC, Ejileugha C, Augustine J, Okolo CA, Onyeaka H (2024) A review of the mycotoxin family of fumonisins, their biosynthesis, metabolism, methods of detection, and effects on humans and animals. Int J Mol Sci 26(1):184

Arroyo-Olarte RD, Bravo Rodríguez R, Morales-Ríos E (2021) Genome editing in bacteria: CRISPR-Cas and beyond. Microorganisms 9(4):0844

Banerjee P, Kintzios S, Prabhakarpandian B (2013) Biotoxin detection using cell-based sensors. Toxins 5(12):2366–2383

Bazhenov SV, Novoyatlova US, Scheglova ES, Prazdnova EV, Mazanko MS, Kessenikh AG, Kononchuk OV, Gnuchikh EY, Liu Y, Al Ebrahim R, Zavilgelsky GB, Chistyakov VA, Manukhov IV (2023) Bacterial lux-biosensors: constructing, applications, and prospects. Biosens Bioelectron X 13:100323

Bober JR, Beisei CL, Nair NU (2018) Synthetic biology approaches to engineer probiotics and members of the human microbiota for biomedical applications. Annu Rev Biomed Eng 20:277–300

Bradley RW, Buck M, Wang B (2016) Tools and principles for microbial gene circuit engineering. J Mol Biol 428(5):862–888

Chamas A, Svensson C-M, Maneira C, Sporniak M, Figge MT, Lackner G (2024) Engineering adhesion of the probiotic strain *Escherichia coli* Nissle to the fungal pathogen *Candida albicans*. ACS Synth Biol 13(12):4027–4039

Charlet R, Bortolus C, Sendid B, Jawhara S (2020) *Bacteroides thetaiotaomicron* and *Lactobacillus johnsonii* modulate intestinal inflammation and eliminate fungi via enzymatic hydrolysis of the fungal cell wall. Sci Rep 10(1):11510

Chen X, Abdallah MF, Chen X, Rajkovic A (2023) Current knowledge of individual and combined toxicities of aflatoxin B1 and fumonisin B1 *in vitro*. Toxins 15(11):653

Choi JY, Jang J, Park Y-C, Lee PC (2025) Systematic promoter engineering in *lactobacillus casei*: construction and screening of a synthetic aldolase promoter library for enhanced gene expression. Process Biochem 149:45–53

Commissioner, O. of the (2021) Early clinical trials with live biotherapeutic products: chemistry, manufacturing, and control information. FDA Guidance, pp 1–32

Daubin V, Szöllősi GJ (2016) Horizontal gene transfer and the history of life. Cold Spring Harb Perspect Biol 8(4):a018036

Dias MAM, Nitschke M (2023) Bacterial-derived surfactants: an update on general aspects and forthcoming applications. Braz J Microbiol 54(1):103–123

Dietz-Pfeilstetter A (2010) Stability of transgene expression as a challenge for genetic engineering. Plant Sci 179(3):164–167

Divyashree S, Shruthi B, Vanitha PR, Sreenivasa MY (2023) Probiotics and their postbiotics for the control of opportunistic fungal pathogens: a review. Biotechnol Rep 38:e00800

EFSA Panel on Additives and Products or Substances used in Animal Feed (2019) Guidance on the assessment of the safety of feed additives for the environment. EFSA J 17(4):e05648

EFSA Panel on Dietetic Products, Nutrition and Allergies (2021) Scientific and technical guidance on foods for special medical purposes in the context of Article 3 of Regulation (EU) No 609/2013. EFSA J 19(3):e06544

El-Sayed RA, Jebur AB, Kang W, El-Demerdash FM (2022) An overview of the major mycotoxins in food products: characteristics, toxicity, and analysis. J Future Foods 2(2):91–102

Fernández de Ullivarri M, Arbulu S, García-Gutiérrez E, Cotter PD (2020) Antifungal peptides as therapeutic agents. Front Cell Infect Microbiol 10:00105

Fisher MC, Hawkins NJ, Sanglard D, Gurr SJ (2018) Worldwide emergence of resistance to antifungal drugs challenges human health and food security. Science 360(6390):739–742

Getino M, de la Cruz F (2018) Natural and artificial strategies to control the conjugative transmission of plasmids. Microbiol Spectrum 6(1):MTBP-0015-2016

Gow NAR, Lenardon MD (2023) Architecture of the dynamic fungal cell wall. Nat Rev Microbiol 21(4):248–259

Han S, Lu Y, Xie J, Fei Y, Zheng G, Wang Z, Liu J, Lv L, Ling Z, Berglund B, Yao M, Li L (2021) Probiotic gastrointestinal transit and colonization after oral administration: a long journey. Front Cell Infect Microbiol 11:609722

Harrison KL, Farrell RM, Brinich MA, Highland J, Mercer M, McCormick JB, Tilburt J, Geller G, Marshall P, Sharp RR (2015) Patient perspectives on the ethical issues arising with the regulation of probiotics. Health Expect 18(2):250–261

Heo S, Kim T, Na H-E, Lee G, Park J-H, Park H-J, Jeong D-W (2022) Safety assessment systems for microbial starters derived from fermented foods. J Microbiol Biotechnol 32(10):1219–1225

Hu H-J, Zhang G-Q, Zhang Q, Shakya S, Li Z-Y (2017) Probiotics prevent *Candida* colonization and invasive fungal sepsis in preterm neonates: a systematic review and meta-analysis of randomized controlled trials. Pediatr Neonatol 58(2):103–110

Kahraman Ilıkkan Ö (2025) Lactobacillomics as a new notion in lactic acid bacteria research through omics integration. World J Microbiol Biotechnol 41(2):68

Kandasamy S, Vlasova AN, Fischer D, Kumar A, Chattha KS, Rauf A, Shao L, Langel SN, Rajashekara G, Saif LJ (2016) Differential effects of *Escherichia coli* Nissle and *Lactobacillus rhamnosus* strain GG on human rotavirus binding, infection, and B cell immunity. J Immunol 196(4):1780–1789

Kim PI, Ryu J, Kim YH, Chi Y-T (2010) Production of biosurfactant lipopeptides iturin A, fengycin, and surfactin A from *Bacillus subtilis* CMB32 for control of *Colletotrichum gloeosporioides*. J Microbiol Biotechnol 20(1):138–145

Koutsoumanis K, Allende A, Alvarez-Ordóñez A, Bolton D, Bover-Cid S, Chemaly M, De Cesare A, Hilbert F, Lindqvist R, Nauta M, Nonno R, Peixe L, Ru G, Simmons M, Skandamis P, Suffredini E, Cocconcelli PS, Fernández Escámez PS et al (2024) Update of the list of qualified presumption of safety (QPS) recommended microbiological agents intentionally added to food or feed as notified to EFSA 19: suitability of taxonomic units notified to EFSA until September 2023. EFSA J 22(1):e8517

Krishnan SV, Nampoothiri KM, Suresh A, Linh NT, Balakumaran PA, Pócsi I, Pusztahelyi T (2024) *Fusarium* biocontrol: antagonism and mycotoxin elimination by lactic acid bacteria. Front Microbiol 14:1260166

Li L, Zhu X-M, Zhang Y-R, Cai Y-Y, Wang J-Y, Liu M-Y, Wang J-Y, Bao J-D, Lin F-C (2022) Research on the molecular interaction mechanism between plants and pathogenic fungi. Int J Mol Sci 23(9):4658

Liu J, Zhang C, Lu W (2019) Biosynthesis of long-chain ω-hydroxy fatty acids by engineered *Saccharomyces cerevisiae*. J Agric Food Chem 67(16):4545–4552

Liu G, Lin Q, Jin S, Gao C (2022) The CRISPR-Cas toolbox and gene editing technologies. Mol Cell 82(2):333–347

Liu L, Helal SE, Peng N (2023a) CRISPR-Cas-based engineering of probiotics. BioDes Res 5:0017

Liu C, Ma N, Feng Y, Zhou M, Li H, Zhang X, Ma X (2023b) From probiotics to postbiotics: concepts and applications. Anim Res One Health 1(1):92–114

Lyagin I, Efremenko E (2019) Enzymes for detoxification of various mycotoxins: origins and mechanisms of catalytic action. Molecules 24(13):2362

Mafe AN, Büsselberg D (2024) Mycotoxins in food: cancer risks and strategies for control. Foods 13(21):3502

Maldonado-Gómez MX, Martínez I, Bottacini F, O'Callaghan A, Ventura M, van Sinderen D, Hillmann B, Vangay P, Knights D, Hutkins RW, Walter J (2016) Stable engraftment of *Bifidobacterium longum* AH1206 in the human gut depends on individualized features of the resident microbiome. Cell Host Microbe 20(4):515–526

Maneeboon T, Roopkham C, Mahakarnchanakul W, Chuaysrinule C (2024) Exploration of *Pseudomonas knackmussii* AD02 for the biological mitigation of post-harvest aflatoxin contamination: characterization and degradation mechanism. J Stored Prod Res 109:102470

Mauro VP (2018) Codon optimization in the production of recombinant biotherapeutics: potential risks and considerations. BioDrugs 32(1):69–81

McInnes RS, McCallum GE, Lamberte LE, van Schaik W (2020) Horizontal transfer of antibiotic resistance genes in the human gut microbiome. Curr Opin Microbiol 53:35–43

Mehmood A, Liu G, Wang X, Meng G, Wang C, Liu Y (2019) Fungal quorum-sensing molecules and inhibitors with potential antifungal activity: a review. Molecules 24(10):1950

Mejía-Pitta A, Broset E, de la Fuente-Nunez C (2021) Probiotic engineering strategies for the heterologous production of antimicrobial peptides. Adv Drug Deliv Rev 176:113863

Mitchell MA (2006) Chapter 36: Therapeutics. In: Reptile medicine and surgery, vol 2, pp 631–664

Moe-Behrens GHG, Davis R, Haynes KA (2013) Preparing synthetic biology for the world. Front Microbiol 4:5

Nazir A, Hussain FHN, Raza A (2024) Advancing microbiota therapeutics: the role of synthetic biology in engineering microbial communities for precision medicine. Front Bioeng Biotechnol 12:1511149

Nett J, Lincoln L, Marchillo K, Massey R, Holoyda K, Hoff B, VanHandel M, Andes D (2007) Putative role of β-1,3-glucans in *Candida albicans* biofilm resistance. Antimicrob Agents Chemother 51(2):510–520

Okay S, Alshehri WA (2020) Overexpression of chitinase a gene from *Serratia marcescens* in *Bacillus subtilis* and characterization of enhanced chitinolytic activity. Braz Arch Biol Technol 63:e20200061

One Health (2019) Fungal pathogens of humans, animals, and plants: report on an American academy of microbiology colloquium held on 18 Oct 2017. American Society for Microbiology, Washington (DC). https://asm.org/reports/one-health-fungal-pathogens-of-humans,-animals,-an

Rau S, Gregg A, Yaceczko S, Limketkai B (2024) Prebiotics and probiotics for gastrointestinal disorders. Nutrients 16(6):0778

Rodrigues JL, Rodrigues LR (2023) Synthetic biology approaches for biosurfactants production by lactic acid bacteria. In: Lactic acid bacteria as cell factories, pp 311–334

Rottinghaus AG, Amrofell MB, Moon TS (2020) Biosensing in smart engineered probiotics. Biotechnol J 15(10):e1900319

Rottinghaus AG, Ferreiro A, Fishbein SRS, Dantas G, Moon TS (2022) Genetically stable CRISPR-based kill switches for engineered microbes. Nat Commun 13:672

Segers ME, Lebeer S (2014) Towards a better understanding of *Lactobacillus rhamnosus* GG-host interactions. Microb Cell Factories 13(1):S7

Singh I, Thakur P (2023) Impact of fungi on the world economy and its sustainability: current status and potentials. In: Fungal resources for sustainable economy, pp 3–37. https://doi.org/10.1007/978-981-19-9103-5_1

Spacova I, Binda S, ter Haar JA, Henoud S, Legrain-Raspaud S, Dekker J, Espadaler-Mazo J, Langella P, Martín R, Pane M, Ouwehand AC (2023) Comparing technology and regulatory landscape of probiotics as food, dietary supplements and live biotherapeutics. Front Microbiol 14:1272754

Steinberg G, Gurr SJ (2020) Fungi, fungicide discovery and global food security. Fungal Genet Biol 144:103476

Tan Y, Ma S, Leonhard M, Moser D, Schneider-Stickler B (2018) β-1,3-glucanase disrupts biofilm formation and increases antifungal susceptibility of *Candida albicans* DAY185. Int J Biol Macromol 108:942–946

Tao S, Chen H, Li N, Wang T, Liang W (2022) The spread of antibiotic resistance genes in vivo model. Can J Infect Dis Med Microbiol 2022:3348695

Taylor MC, Jackson CJ, Tattersall DB, French N, Peat TS, Newman J, Briggs LJ, Lapalikar GV, Campbell PM, Scott C, Russell RJ, Oakeshott JG (2010) Identification and characterization of two families of F420H2-dependent reductases from *Mycobacteria* that catalyse aflatoxin degradation. Mol Microbiol 78(3):561–575

Thambugala KM, Daranagama DA, Phillips AJL, Kannangara SD, Promputtha I (2020) Fungi vs fungi in biocontrol: an overview of fungal antagonists applied against fungal plant pathogens. Front Cell Infect Microbiol 10:604923

Thery T, Lynch K, Arendt E (2019) Natural antifungal peptides/proteins as model for novel food preservatives. Compr Rev Food Sci Food Saf 18:12480

Trump B, Cummings C, Klasa K, Galaitsi S, Linkov I (2023) Governing biotechnology to provide safety and security and address ethical, legal, and social implications. Front Genet 13:1052371

Vachher M, Sen A, Kapila R, Nigam A (2021) Microbial therapeutic enzymes: a promising area of biopharmaceuticals. Curr Res Biotechnol 3:195–208

Vazquez-Munoz R, Thompson A, Sobue T, Dongari-Bagtzoglou A (2024) *Lactobacillus johnsonii* is a dominant *Lactobacillus* in the murine oral mucosa and has chitinase activity that compromises fungal cell wall integrity. MBio 15(10):e02416–e02424

Vinayamohan P, Joseph D, Viju LS, Baskaran SA, Venkitanarayanan K (2024) Efficacy of probiotics in reducing pathogenic potential of infectious agents. Fermentation 10(12):0599

Vitiello A, Ferrara F, Boccellino M, Ponzo A, Cimmino C, Comberiati E, Zovi A, Clemente S, Sabbatucci M (2023) Antifungal drug resistance: an emergent health threat. Biomedicine 11(4): 1063

Wang S, Fu W, Zhao X, Chang X, Liu H, Zhou L, Li J, Cheng R, Wu X, Li X, Sun C (2022) Zearalenone disturbs the reproductive-immune axis in pigs: the role of gut microbial metabolites. Microbiome 10(1):234

Wright O, Delmans M, Stan G-B, Ellis T (2015) GeneGuard: A modular plasmid system designed for biosafety. ACS Synth Biol 4(3):307–316

Xiao Y, Zhao J, Zhang H, Zhai Q, Chen W (2020) Mining *Lactobacillus* and *Bifidobacterium* for organisms with long-term gut colonization potential. Clin Nutr 39(5):1315–1323

Zawrotniak M, Wojtalik K, Rapala-Kozik M (2019) Farnesol, a quorum-sensing molecule of *Candida albicans* triggers the release of neutrophil extracellular traps. Cells 8(12):1611

Zhou Z, Chen X, Sheng H, Shen X, Sun X, Yan Y, Wang J, Yuan Q (2020) Engineering probiotics as living diagnostics and therapeutics for improving human health. Microb Cell Factories 19(1): 56

Challenges and Considerations in Probiotic Therapy 21

Maya Raman, Ruby Mishra, Sheetal Gupta, Subhaprada Nishtala, and Sanu Jacob

Abstract

Probiotic therapy is a promising strategy for modulating the human microbiome to promote health and manage disease; however, its effective application is constrained by several critical challenges. Key scientific hurdles include strain-specific efficacy, limited understanding of mechanistic pathways, and difficulties with dose optimization and maintaining viability throughout the gastrointestinal tract. Safety concerns are paramount, particularly in immunocompromised, vulnerable populations, due to the risks of opportunistic infections, the transfer of antibiotic-resistant genes, and the genetic instability of probiotic strains. Furthermore, the field is complicated by inconsistent global regulatory frameworks, inadequate quality control issues, misleading labelling, and a lack of standardized manufacturing practices. Host-specific factors, including individual microbiota composition, genetics, and lifestyle, significantly impact therapeutic outcomes,

M. Raman
Department of Food Science and Technology, Kerala University of Fisheries and Ocean Studies (KUFOS), Kochi, India

R. Mishra
Food Safety and Standards Authority of India (FSSAI), National Food Laboratory (NFL), Ghaziabad, India

S. Gupta
Food Safety and Standards Authority of India (FSSAI), National Food Laboratory (NFL), Navi Mumbai, India

S. Nishtala
Association of Food Scientists & Technologists (India), Mysuru, India

S. Jacob (✉)
Food Safety and Standards Authority of India (FSSAI), National Food Laboratory (NFL), Chennai, India
e-mail: sanu.jacob@fssai.gov.in

M. Y. Sreenivasa et al. (eds.), *Probiotics as Antifungals*,
https://doi.org/10.1007/978-981-92-0002-3_21

underscoring the need for a personalized approach. and tailored probiotic strategies. The translation of probiotic therapy as a reliable, evidence-based practice requires rigorous clinical validation, regulatory harmonization, robust manufacturing standards, and a multidisciplinary, ethically grounded approach.

Graphical Abstract

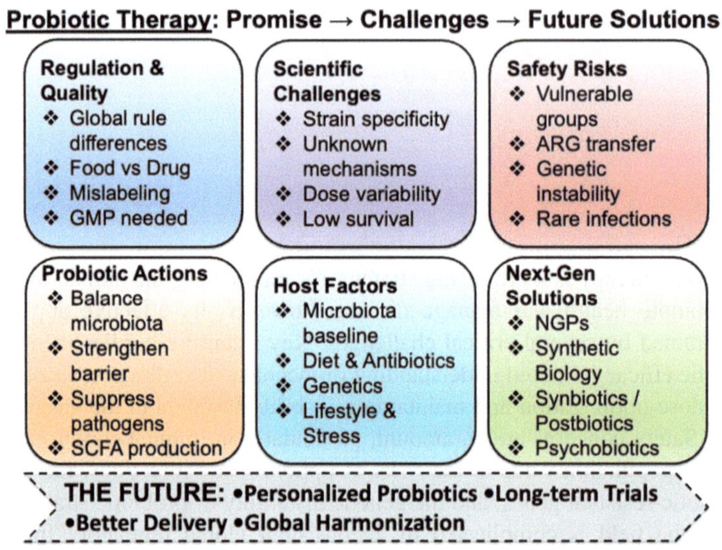

Keywords

Probiotic safety · Regulatory frameworks and challenges · Strain specificity · Microbial Interactions · Postbiotics · Safety assessment

Learning Objectives

- Assess the concept of probiotic therapy, and the scientific, safety, and regulatory challenges associated with it that limit its efficacy, consistency, and clinical translation.
- Understand the strain specificity, host-related factors, and mechanistic uncertainties on probiotic performance and therapeutic outcomes.
- Apprehend the need for standardized manufacturing, clinical validation, and personalized probiotic strategies to advance NGPs.

21.1 Introduction

Probiotic therapy has become an important part of modern approaches in maintaining health and managing diseases. As research on the human microbiome continues to advance, probiotics are increasingly valued for their health-supporting properties and strain-specific benefits.

21.1.1 Definition and Purpose of Probiotic Therapy

Probiotic therapy comprises the administration of live microorganisms that confer health benefits when consumed in adequate amounts (FAO/WHO 2002), based on the principal role of the gut microbiota in host metabolism, immune function, and physiological homeostasis. Unlike conventional drugs, probiotics modulate gut microbial community to restore balance, support disease prevention, and complement medical treatments, making them an increasingly attractive approach for managing dysbiosis-related conditions (Sanders et al. 2019).

21.1.2 Overview of Potential Health Benefits

Over the past two decades, research has demonstrated that probiotics exert diverse physiological effects, primarily by improving gut health through enhancement of epithelial barrier function, suppression of pathogen colonization, and maintenance of microbial homeostasis (Hemarajata and Versalovic 2013). Beyond the gastrointestinal tract, probiotics exert immunomodulatory effects and improve metabolic health and neurocognitive function by modulating the gut–brain axis, highlighting their broad therapeutic potential (Cryan et al. 2019).

21.1.3 Relevance in Current Clinical and Functional Food Contexts

Probiotic therapy has evolved from a niche microbiological concept to a significant component of clinical practice and the functional food industry. Probiotics are widely explored as adjuncts to antibiotics, therapeutics for inflammatory bowel diseases, and preventive interventions in neonatal and post-surgical care, and are also incorporated into functional foods and dietary supplements to meet consumer demand for natural health products (Didari et al. 2014). Regulatory bodies such as the FDA, EFSA, and FSSAI have developed frameworks to ensure probiotic safety and efficacy, despite regional differences, highlighting the growing integration of probiotics into preventive healthcare and personalized nutrition strategies (FSSAI 2021).

21.1.4 Scope and Objectives

This chapter critically reviews the scientific, clinical, and regulatory challenges of probiotic therapy, focusing on strain specificity, mechanisms of action, dosage, viability, safety risks, and host-specific influences on efficacy. It highlights regulatory gaps, quality control issues, and emerging approaches such as next-generation probiotics, synbiotics, and postbiotics, emphasizing the need for rigorous clinical validation, regulatory harmonization, and improved delivery systems.

21.2 Scientific and Clinical Challenges

Although probiotic therapy shows significant potential, its transition from laboratory research to reliable clinical practice faces various scientific and methodological challenges. These challenges arise from the biological diversity, inadequate mechanistic understanding, inconsistencies in dosage, and viability under real-world conditions.

21.2.1 Strain Specificity

21.2.1.1 Not all Probiotic Strains Are Equal

Probiotic efficacy is strain-specific, as closely associated strains may differ in genome, metabolism, adhesion, and immunomodulatory effects (Hill et al. 2014); for instance, *Lactobacillus rhamnosus* GG is effective against diarrhoeal disorders, whereas other *L. rhamnosus* strains are not.

21.2.1.2 Variability in Efficacy within the Same Species

Genomic plasticity and strain-level heterogeneity make probiotic efficacy difficult to predict, as minor genetic differences may alter functional traits, including mucin-binding, bacteriocin production, bile salt hydrolase activity, and SCFA synthesis. Even strains within the same species may yield divergent clinical outcomes

Table 21.1 Examples of strain-level variability in clinical outcomes

Species	Strain	Documented clinical outcome	Reference
Lactobacillus rhamnosus	GG	Reduces antibiotic-associated diarrhoea	Segers and Lebeer (2014)
Lactobacillus rhamnosus	GR-1	Urogenital health benefits; not effective for diarrhoea	Reid et al. (2003)
Bifidobacterium longum	BB536	Improves bowel regularity and immunity	Wong et al. (2019)
Bifidobacterium longum, *Lactobacillus* spp.	subsp. *infantis* 35,624	Modulates inflammation in IBS	O'Mahony et al. (2005)

(Table 21.1). *Lacticaseibacillus rhamnosus* GG effectively reduces antibiotic-associated diarrhoea, whereas other *L. rhamnosus* strains do not (Zheng et al. 2020).

21.2.1.3 Difficulty in Identifying Clinically Validated Strains

Several commercial probiotic products lack clinically validated strains and are supported solely by in vitro data, which may lead to frequent misidentification or incomplete labelling, hindering replication and regulatory concerns (Sanders et al. 2019). Rigorous clinical validation for strain-level identification through whole-genome sequencing, randomized controlled trials for specific indications, and verification of stability and viability throughout the shelf-life of the probiotic product are vital.

21.2.2 Mechanism of Action

Probiotics exert multifaceted effects, including modulation of the gut microbiome, SCFA production, enhancement of mucosal immunity, strengthening of gut barrier function, and interactions with epithelial or neuronal signalling. However, the mechanisms underlying specific strains or disease conditions remain poorly understood, which may contribute significantly to their specific outcomes (Ouwehand et al. 2020). Further, probiotic-host interactions occur at the microbiota level (competitive exclusion, pH modification, and bacteriocin production), the immune level (modulation of cytokine responses and induction of regulatory T cells), and the barrier level (strengthening tight junctions, mucin secretion, and epithelial turnover). This multilevel activity demonstrates that no single mechanistic model can fully explain the effects of probiotics across different conditions. Most clinical studies focus on outcomes rather than mechanisms, creating a knowledge gap between observed benefits and the underlying functional pathways. Integrative approaches utilizing metagenomics, metabolomics, and transcriptomics have been proposed to analyse probiotic–host interactions more accurately (Thu et al. 2021) (Fig. 21.1).

21.2.3 Dose and Viability

21.2.3.1 Establishing the Optimal Dose

Dose optimization is crucial, but is inconsistent. Usually, the probiotic viability in products ranges from 10^6 to 10^{11} CFU. However, the minimal effective dose varies with strain, indication, delivery matrix, and the host microbiome status (FAO/WHO 2002). The lack of standardization makes it difficult to compare clinical trials and commercial claims.

21.2.3.2 Gastrointestinal Survival of Probiotic Strains

Probiotics must survive gastrointestinal conditions, including enzymes and pH conditions. Many strains have limited survival unless encapsulated and delivered

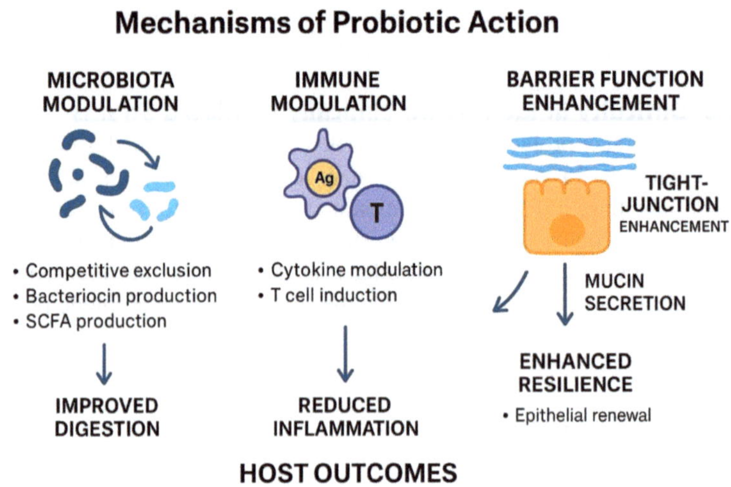

Fig. 21.1 Probable mechanisms of probiotic action

Table 21.2 Impact on probiotics under various conditions

Factor	Impact	Example
Processing	Heat/pressure may kill cells	Spray drying
Storage	Temperature/humidity reduces shelf life	Freeze-dried powders
Delivery matrix	Determines protection from GI stress	Encapsulation > yogurt > capsules
GI environment	Acid/bile reduces survival	Low survival for many *Lactobacillus* strains

(Cook et al. 2012). Failing to address this challenge diminishes the reliability of health claims.

21.2.3.3 Stability During Processing, Storage, and Shelf-Life

Probiotic viability is also affected by temperature, humidity, oxygen exposure, and time. Probiotic products affected by physical parameters may show variations in labelled and actual counts (Kim and Mills 2024). Therefore, ensuring stability during manufacturing, packaging, transport and storage is critical. However, the cost and complexity of processing techniques (freeze-drying, encapsulation, oxygen-scavenging packaging) that reduce these losses must also be addressed (Table 21.2).

21.3 Safety Considerations

21.3.1 Adverse Effects of Probiotics in Vulnerable Populations and Critically Ill Patients

Probiotics with demonstrated benefits are generally regarded as safe (GRAS) and have been used for a long time; however, accumulating evidence suggests potential risks in vulnerable populations. Clinical reports document probiotic-associated bacteraemia, fungemia, endocarditis, and increased mortality in individuals with immunosuppression, ICU admission, central venous catheters, surgery, severe illness, compromised mucosal barriers, neonates, elderly, and pregnant women (Kothari et al. 2019; Roe et al. 2022; Merenstein et al. 2023; FSAI 2024). These adverse events are chiefly attributed to bacterial translocation from the gut to sterile sites (Xu et al. 2025). Although rare, such incidents underscore the influence of host factors, gut environment, and clinical practices on probiotic safety, making results difficult to predict. Hence, the use of probiotics in high-risk or critical care settings should be guided by careful risk assessment and strict clinical supervision. Most safety data are derived from healthy adults; targeted clinical trials and post-market surveillance are, hence, essential to establish strain-specific, population-relevant safety profiles and ensure effective, safe probiotic use across diverse demographic and clinical groups (Merenstein et al. 2023).

21.3.2 Risk of Transfer of Antibiotic Resistance Genes

Probiotic strains (*Lactobacillus*, *Bifidobacterium*, *Enterococcus*) may carry intrinsic or acquired antibiotic resistance genes (ARGs) that reside on mobile genetic elements such as plasmids, transposons, or integrative conjugative elements, enabling horizontal transfer to commensals or pathogens, particularly under antibiotic selective pressure (Kothari et al. 2019; Xu et al. 2025). This raises safety concerns for vulnerable populations, highlighting the need for rigorous safety evaluation through strain-level genomic screening, phenotypic antibiotic susceptibility testing, and the exclusion of strains carrying transmissible resistance determinants (Roe et al. 2022; Xu et al. 2025).

21.3.3 Genetic Stability of Probiotic Organisms

Genetic stability is essential for sustaining the safety and functional properties of probiotic strains during commercial production and gastrointestinal transit. Long-term propagation through horizontal gene transfer, plasmid loss, mutations, or competition with the host microbiome can compromise stability. Controlled manufacturing can minimize such changes, as demonstrated by the *Lacticaseibacillus paracasei* DG strain, which retained genomic and phenotypic stability over a decade with negligible impact on performance (Sanders et al. 2010;

Korostylev et al. 2020). These considerations underscore the necessity of strain-specific safety assessment, continuous genomic monitoring and stabilization, rigorous quality control, and targeted safety data to ensure safe probiotic use in high-risk vulnerable populations (Thakur et al. 2025; Sanders et al. 2010).

21.4 Regulatory and Quality Control Issues

Regulation of probiotics is highly complex and debated, as these are classified as foods, dietary supplements, or functional ingredients despite claims that often imply therapeutic effects. The regulatory frameworks and quality-control challenges governing probiotics are highly important, and the roles of EFSA (EU), FSSAI (India), FOSHU (Japan), Codex Alimentarius (FAO-WHO), and FDA (US) in developing probiotic frameworks are noteworthy.

21.4.1 Lack of Standardized Guidelines

The rapid global expansion of probiotic products has underscored gaps and inconsistencies in regulatory oversight and quality control practices. Given that probiotics are biological organisms, their safety, efficacy, and stability require strict controls across strain identification, manufacturing, labelling, and post-market surveillance; however, several marketed products lack proper clinical validation, underscoring the need for harmonized international guidelines to ensure product quality, consumer safety, and fair trade.

21.4.1.1 Inconsistencies Between Countries

Probiotic regulation varies evidently across countries (Table 21.3), reflecting the lack of a globally harmonized framework (Spacova et al. 2023). Codex Alimentarius has to establish a formal definition or unified guidelines for probiotics, and efforts aiming to address this gap are ongoing. In the United States, probiotics are regulated as foods or dietary supplements, where compliance with GMPs is required. Pre-market evaluation of quality and efficacy is not a mandate; only structure-function claims are permitted, and disease-related claims are prohibited, even though

Table 21.3 Regulatory classification of probiotic products across regions

Region/authority	Regulatory category
FDA (US)	Dietary supplement or food; biological drug (if claims are therapeutic)
EFSA (EU)	Food or food supplement
FSSAI (India)	Functional food, health supplement
FOSHU (Japan)	Food for specified health uses
Codex Alimentarius (FAO-WHO)	Food standard

some probiotics may exhibit these properties. In the European Union, probiotics are regulated under food and food supplements, with all health claims subject to EFSA authorization; however, inconsistent acceptance of the term "probiotic" across member states remains a challenge. In India, FSSAI, in coordination with ICMR, has issued specific guidelines requiring the use of approved, non-GMO strains listed in Schedule VII, minimum viable counts, and clear labelling, though classification as foods or health supplements is still inconsistent. Canada regulates probiotics as foods or Natural Health Products under a distinct regulatory pathway. Collectively, these divergent national approaches result in variability in safety evaluation, health claim substantiation, labelling, and post-market surveillance, creating significant challenges for multinational product development, cross-border clinical research, and global harmonization of scientific evidence.

21.4.1.2 Classification: Drug, Food, or Supplement?

Probiotics, defined as "live microorganisms that when administered in adequate amounts, confer a health benefit on the host," were established by the WHO and FAO. The term "probiotic" is often misused and applied to products that do not meet the requirements. This issue is exacerbated by the "probiotic umbrella" concept, which generalizes evidence from individual strains to others, undermining strain specificity and obscuring critical factors such as product identity, dose, intake duration, and strain combinations (De Simone 2019). One of the chief regulatory challenges is unfitting classification. Depending on the intended use, formulation, and claims, a probiotic can be classified as a drug intended to treat or mitigate a disease and require clinical trials; a dietary supplement; a food or food supplement; or a natural health product (Spacova et al. 2023). For instance, *Lactobacillus rhamnosus* GG is marketed as a dietary supplement in many countries but faces drug-level regulatory challenges in several parts of Europe, while *Saccharomyces boulardii* CNCM I-745 is sold as an over-the-counter probiotic in some regions; it is regulated as a drug in Brazil and Italy due to its clinical application in treating acute diarrhoea (Rai et al. 2024). The probiotic framework may also extend to traditional fermented foods containing live microbes, as evidence from several regions indicates that certain fermented dairy products are associated with a reduced risk of specific diseases. Consequently, probiotics span multiple product categories, complicating their regulatory classification. The regulatory classification of probiotic products is primarily determined by their health claims: products making structure-function claims are generally regulated as foods or dietary supplements, whereas those claiming disease prevention or treatment are classified as drugs or biotherapeutics, such as faecal microbiota products under the US FDA. Additionally, novel engineered or next-generation probiotic strains are often subject to more stringent biotechnology or pharmaceutical regulatory frameworks that dictate the scope of safety assessments, quality controls, and regulatory approvals. A greater concern is that the same probiotic strain may be regulated differently across regions, thereby complicating commercialization.

21.4.2 Labelling and Claims

21.4.2.1 Misleading or Unverified Health Claims

Unsubstantiated or exaggerated health claims remain a major concern in the probiotic market, with many products promoting benefits for immunity, weight management, or mental health despite limited or inconsistent scientific support. Regulatory bodies such as EFSA have rejected most probiotic health claims due to insufficient mechanistic or clinical evidence, highlighting the gap between marketing and scientific validation (De Simone 2019), while FOSHU remains more stringent, approving only those products with replicable clinical validations (Marteau and Shanahan 2003). Agencies continue to warn against unapproved disease-related claims that may mislead consumers and pose risks to vulnerable populations.

21.4.2.2 Mismatch Between Label Content and Actual Strain

Multiple studies consistently demonstrate discrepancies between labelled and actual probiotic content, including misidentified strains, undeclared contaminants, and viable cell counts lower than claimed, as revealed by DNA-based analyses and culture methods (McFarland 2010; Lewis et al. 2016; Patel and Denning 2021). Probiotic efficacy and safety are strain-specific. These inaccuracies may undermine the product's credibility and may pose health risks, particularly in vulnerable populations. An example of misleading labelling is shown below:

Actual Probiotic Label Claims		Actual Laboratory Findings
Strain A: 10^9 CFU Strain B: 10^8 CFU Health claim: Immunity	→	Strain A: 3×10^7 CFU Strain B: Not detected Health claim: Contaminant strain present

21.4.3 Quality Control Challenges

21.4.3.1 Contamination Risks

Probiotics products are susceptible to contamination at multiple stages of production and distribution, including fermentation, drying, packaging, storage, and transport, with reported contaminants ranging from pathogenic microorganisms (*Enterobacter* spp., *Bacillus cereus*) to spoilage microorganisms and environmental residues. Inconsistent contamination thresholds and regulatory enforcement across countries heighten the risks, and although rare, documented cases of probiotic-associated bloodstream infections in vulnerable populations emphasize the need for standardized microbiological testing (Salminen et al. 2021; Ahire et al. 2023).

21.4.3.2 Batch-to-Batch Variability

Ensuring batch-to-batch consistency in probiotic viability and strain composition remains technically challenging, as variations in growth conditions, raw material quality, freeze-drying efficiency, storage, oxygen exposure, and moisture control can significantly affect CFU counts, metabolic activity, and stability. In the absence of harmonized quality standards, reliance on variable internal manufacturing protocols leads to inconsistencies that compromise therapeutic efficacy and shelf life (Marchesi and Nader-Macías 2023).

Fermentation Temperature	→	Freeze-drying efficiency	→	Packaging & Transport
⇓		⇓		⇓
Change in CFUs		Loss of viability		Moisture/oxygen stress

21.4.3.3 Good Manufacturing Practices (GMP)

Compliance with good manufacturing practices (GMP) is critical for ensuring the safety, purity, potency, and consistency of probiotic products, yet many supplements are not produced under pharmaceutical-grade conditions, leading to variable quality (Table 21.4). Key GMP requirements for probiotics include the use of genuine strains from recognized culture collections; control of aerobic and anaerobic conditions; validated stability and shelf-life testing; routine molecular or metagenomic verification of strain identity; comprehensive testing for endotoxins, toxins, and allergens; and adherence to recognized regulatory frameworks such as USFDA cGMP, EU hygienic food standards, or FSSAI Schedule 4 guidelines (Ahire et al. 2023).

21.4.4 Roles of Major Regulatory Bodies

Probiotic regulatory inaccuracies vary considerably worldwide, with major agencies shaping standards for safety, efficacy, and labelling through distinct frameworks (Table 21.5). In the United States, the US FDA regulates probiotics as foods, dietary supplements, or biological drugs when therapeutic claims are made, requiring compliance with GRAS assessments, cGMP, and IND approval where applicable; similarly, the European Union, through EFSA, regulates probiotics under food and novel food legislation, granting Qualified Presumption of Safety (QPS) only to well-characterized strains and authorizing health claims based on robust, strain-specific clinical evidence. In India, FSSAI guidelines mandate strain-level identification, deposition in international culture collections, minimum viable counts at the end

Table 21.4 Essential quality control tests for probiotic products

Category	QC tests	Purpose
Identity	16S rRNA sequencing, whole genome sequencing	Confirms correct strain
Purity	Microbial contamination assays	Ensures absence of pathogens
Viability	CFU counts under stress conditions	Ensures the claimed dose
Stability	Accelerated aging tests	Predicts shelf-life
Safety	Haemolysis, antibiotic resistance profiling	Confirms GRAS/QPS suitability

Table 21.5 Major regulatory bodies, with their regulatory basis, key requirements/features, and types of claims allowed in probiotic products

Regulatory body/region	Regulatory basis	Key requirements/ features	Type of claims allowed
Codex Alimentarius (UN, international)	Global food standards	Provides harmonized guidelines for identity, purity, safety evaluation, and labelling of probiotics	No specific health claims; supports standardized international trade
USFDA (United States)	GRAS, cGMP, Dietary Supplement Health and Education Act (DSHEA), IND for therapeutic claims	Classifies probiotics as foods, dietary supplements, or biologics; safety review via GRAS; strict requirements for products claiming disease treatment	Structure/function claims for supplements; therapeutic claims only with IND approval
EFSA (European Union)	Novel Food Regulation (EU 2015/2283), QPS Framework	Highly stringent assessment of clinical evidence; QPS approval for safe strains; most health claims rejected due to insufficient strain-specific data	Only claims supported by robust clinical evidence; very limited approvals
FSSAI (India)	Food Safety and Standards (Health Supplements, Nutraceuticals, Food for Special Dietary Use, Food for Special Medical Purpose, and Prebiotic and Probiotic Food) Regulations, 2022 [FSSAI (Nutra) Regulations, 2022], FSSAI Probiotic Guidelines (2021) & Food Safety And Standards (Health Supplements, Nutraceuticals, Food for Special Dietary Use, Food for Special Medical Purpose, Functional Food And Novel Food) Regulations, 2016	Requires strain-level identification, minimum CFU at the end of shelf life, deposition in an international culture repository, and scientific substantiation of health claims	General wellness claims allowed; specific health claims require strong evidence
FOSHU (Japan)	Food for Specified Health Uses (FOSHU)	One of the earliest functional food regulatory systems mandates human studies demonstrating specific physiological benefits	Allows explicit health claims only after approval based on clinical evidence

of shelf-life, and scientific substantiation of claims. In Japan, FOSHU requires controlled human studies to support specified health benefits. At the international level, Codex Alimentarius provides reference guidance on probiotic identity, safety, quality, and labelling, collectively underscoring the need for harmonized definitions, validated clinical evidence, and rigorous quality control to ensure safe and effective probiotic products across regions.

21.5 Host-Specific Factors

Host-specific factors critically influence probiotic efficacy and safety, as variations in gut microbiota composition, host genetics, and diet modulate strain colonization, persistence, and functional activity. Accounting for these individual variations is the key to designing precision probiotic interventions (Zhernakova et al. 2024).

21.5.1 Individual Microbiota Composition

The composition and diversity of the gut microbiota determine probiotic colonization and functional outcomes: high-diversity, balanced communities typically resist external strains, whereas dysbiotic or low-diversity microbiomes are more permissive (Wu et al. 2024). Colonization success is shaped by ecological compatibility between probiotic strain and resident taxa (*Bacteroides* or *Prevotella*) (Jiang et al. 2025).

21.5.2 Genetic and Lifestyle Influences

21.5.2.1 Diet, Antibiotics, Stress, and Environmental Factors

Diet and lifestyle are pivotal in shaping probiotic outcomes. Fibre-rich diets promote beneficial short-chain fatty acid (SCFA) production, creating a supportive environment for probiotic growth, whereas high-fat, low-fibre diets diminish microbial diversity and reduce probiotic persistence (Sarita et al. 2025; Manan 2025). Antibiotic use transiently suppresses native microbes, often impairing probiotic engraftment, though targeted supplementation can accelerate post-antibiotic recovery. Beyond diet, psychosocial stress, circadian disruption, and environmental exposures, these factors may also compromise gut barrier function and immune–microbe balance, thereby altering probiotic efficacy (Sarita et al. 2025). These dynamic lifestyle factors highlight the need to contextualize probiotic therapy within broader health and behavioural patterns.

21.5.2.2 Host Genetics Modulating Microbial Interactions

Host genetic disparity shapes the gut microbiome and immune tolerance through polymorphisms in genes that regulate mucosal immunity, glycan expression, and other biomarkers, including TLR, MUC, FUT2, and ABO (Zhernakova et al. 2024)

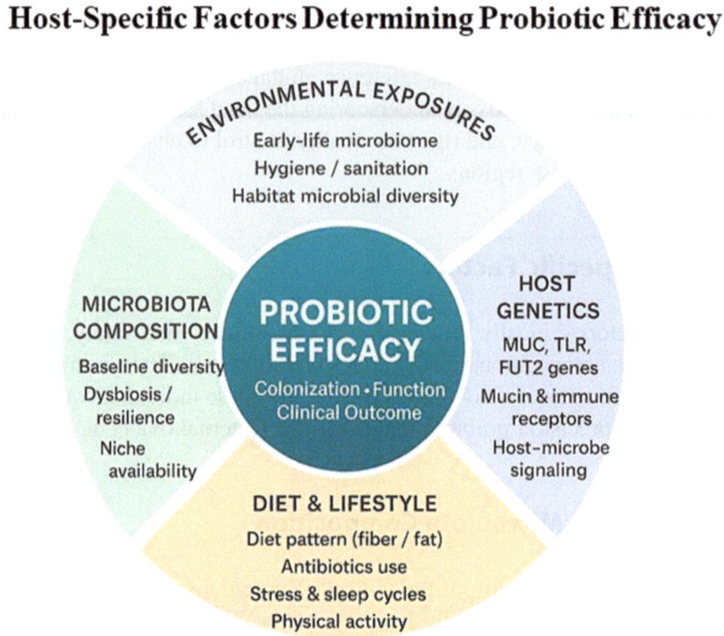

Fig. 21.2 Host-specific factors determining the probiotic efficacy

(Fig. 21.2 and Table 21.6). In particular, FUT2- and ABO-dependent secretor status regulates the availability of host glycans that aid in microbial adhesion and nutrient disposal, influencing microbiota diversity and stability; thus, highlighting the need for genomics-guided precision probiotic design (Leeflang et al. 2025).

21.5.3 Case Study: Discovery of a Probiotic Microorganism Leading to the Popular Drink "Yakult"

Birth of Yakult The discovery of *Lacticaseibacillus casei* Shirota (formerly known as *Lactobacillus casei* Shirota) is a fascinating story that highlights the combination of microbiology, health science, and innovative entrepreneurship that led to the creation of Yakult, the world's most popular probiotic drink (Table 21.7). In the early twentieth century, scientists realized the connection between gut bacteria and human health, and around this time, Élie Metchnikoff, a Russian Nobel laureate (1908), proposed that certain beneficial bacteria could promote longevity by suppressing harmful intestinal microbes. His work inspired many researchers worldwide, including a young Japanese microbiologist named Dr. Minoru Shirota. Dr. Minoru Shirota (1899–1982) studied at the Department of Microbiology, School of Medicine, Kyoto Imperial University, Japan (now Kyoto University), and discovered the Shirota strain in 1930. He successfully isolated and cultured the specific lactic acid bacteria strain, which he named *Lactobacillus casei Shirota*, now

Table 21.6 Summary of key host determinants of probiotic success and their clinical implications

Host determinant	Mechanistic role/influence	Clinical implications	Supporting evidence
Baseline microbiota composition	Determines probiotic colonization potential and metabolic integration through competitive exclusion and ecological compatibility	High microbial diversity may resist colonization; dysbiotic or low-diversity states permit engraftment. Tailored probiotic selection based on microbiome profiling enhances response	Wu et al. (2024), Jiang et al. (2025)
Inter-individual microbial heterogeneity	Drives variable responses to identical probiotic formulations due to distinct microbial network structures	Explains “responders” vs. “non-responders” in clinical trials; necessitates personalized probiotic interventions	Sarita et al. (2025)
Dietary pattern	Modulates gut nutrient environment and SCFA production, influencing probiotic survival and activity	Fibre-rich diets improve colonization and metabolic outcomes; high-fat diets reduce diversity and efficacy	Manan (2025), Sarita et al. (2025)
Antibiotic exposure	Disrupts resident microbiota and alters colonization niches	Post-antibiotic supplementation may aid microbial recovery; timing and strain selection are critical for success	Manan (2025)
Psychosocial stress and lifestyle factors	Affect intestinal permeability, immune signalling, and microbial turnover	Stress and poor sleep reduce probiotic persistence and gut barrier integrity; lifestyle modulation enhances outcomes	Ma et al. (2023)
Host genetic polymorphisms (e.g. FUT2, ABO, TLR, MUC genes)	Regulate mucosal glycan patterns, immune recognition, and bacterial adherence	Genetic screening may predict probiotic compatibility and guide precision formulations	Zhernakova et al. (2024)
Environmental exposures	Influence microbial acquisition and immune training, particularly in early life	Environmental microbiome diversity may condition probiotic responsiveness later in life	Leeflang et al. (2025)
Metabolic and immune phenotype	Host metabolic rate and immune tone shape microbial stability and resilience	Probiotic effects may vary between inflammatory and metabolically stable states; supports integrated multi-omics profiling	Sarita et al. (2025)

reclassified as *Lacticaseibacillus casei Shirota* in 2020. This strain has strong acid and bile tolerance, allowing it to survive stomach acid and reach the intestines alive. It also has a high colonization ability, antagonistic activity (suppressing pathogens), and induces health-promoting effects. The strain is the first scientifically selected and

Table 21.7 Summary timeline of Yakult's journey to commercialization

Year	Event
1899	Birth of Dr. Minoru Shirota
1920s	Research on intestinal bacteria at Kyoto Imperial University
1930	Isolation of *Lactobacillus casei Shirota* strain
1935	Launch of Yakult probiotic drink
1955	Establishment of Yakult Honsha Co., Ltd.
1960s–2000s	Global expansion and scientific validation
2020	Reclassified as *Lacticaseibacillus casei Shirota*
Today	Yakult is consumed by millions across 40+ countries

cultured probiotic bacterium designed to improve human health through the gut. Dr. Shirota envisioned a "preventive medicine" approach, promoting health through everyday diet rather than treating disease after it occurs and in 1935, he and his team developed the fermented milk drink containing live *L. casei* Shirota or the Yakult. The name "Yakult" is derived from the Esperanto word "Jahurto," meaning yogurt. Dr. Shirota also founded Yakult Honsha Co., Ltd., to commercialize this probiotic beverage. His vision was summarized as the "Shirota-ism" philosophy, aiming to provide good health at a reasonable cost through scientifically proven techniques.

Over the decades, Yakult's *L. casei* Shirota strain became one of the most extensively studied probiotics. More than 90 years of research have confirmed its beneficial effects, including improving intestinal flora balance, boosting immunity, lowering the risk of gastrointestinal infections, and supporting overall digestive health. Yakult expanded globally from the 1960s and is currently sold in over 40 countries, consumed by millions daily. The strain is still maintained and studied under strict quality control at Yakult's research centres in Japan and abroad.

21.6 Emerging Considerations

Probiotic therapy is entering an advanced era, driven by microbiome science, systems biology, and precision medicine. Probiotics are frequently understood through strain-specific functions, host–microbe interactions, and context-dependent outcomes, paving the way to personalized responses (Manan 2025). Research focusing on next-generation commensals and engineered microbes with potential applications in promoting human health and preventing metabolic, neurological, and immune-related disorders is gaining enormous attention (Jan et al. 2024) (Fig. 21.3); and hence, the importance of safety assessments, regulatory frameworks, and translational strategies is shifting towards evidence-based, precision-driven interventions.

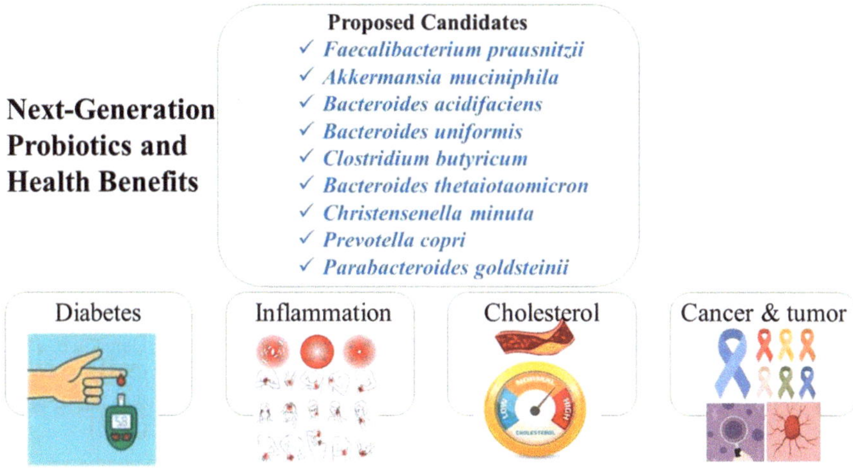

Fig. 21.3 Next-generation probiotics for the prevention of diseases

21.6.1 Next-Generation Probiotics

Advances in metabolomics, gnotobiotic models, and next-generation sequencing have significantly improved the understanding of probiotic characters such as colonization resistance, microbial susceptibility, biogeographical distribution, and inter-individual microbiome variability, thereby influencing therapeutic outcomes (Zmora et al. 2018). Next-generation probiotics (NGPs), a term introduced by O'Toole et al. (2017), encompass novel commensals, strict anaerobes, and genetically engineered microbes with well-defined mechanisms of action. Classified as live biotherapeutic products (LBPs) by the US FDA, these organisms require rigorous safety, mechanistic, and regulatory evaluation (Yahfoufi et al. 2018). NGPs, including *Akkermansia muciniphila*, *Faecalibacterium prausnitzii*, *Bacteroides* spp., *Clostridium butyricum*, and *Christensenella minuta*, show potentialities in modulating mucin dynamics, short-chain fatty acid (SCFA) production, immune responses, bile acid metabolism, and metabolic health, with applications in obesity, diabetes, inflammatory bowel disease, and cancer immunotherapy (Jan et al. 2024) (Table 21.8). Engineered or customized *E. coli* or *L. lactis* strains were reported to deliver therapeutic molecules directly in situ. The Human Genome Project has also greatly advanced the discovery and development of NGPs (Ronaghi et al. 1998).

21.6.1.1 Applications of Next-Generation Probiotics

The applications of major NGPs are as follows:

Akkermansia muciniphila is a mucin-degrading bacterium of the phylum *Verrucomicrobia*; it colonizes the mucus-epithelium interface and is enriched in metabolically healthy individuals. By degrading mucin, it enhances gut barrier integrity by upregulating tight junction proteins, thereby reducing metabolic endotoxemia and low-grade inflammation associated with diet-induced obesity

Table 21.8 Next-generation probiotic strains and their benefits

S. no	Strain	Health benefits	References
1.	*Bacteroides fragilis*	Isolated from human faeces; enhances phagocytosis and polarizes M1 macrophages	Deng et al. (2016)
		Isolated from human faeces; prevents antibiotic-associated diarrhoea	Zhang et al. (2018)
		Isolated from the human gut; investigated in oxazolone-induced experimental colitis	An et al. (2014)
		Isolated from human faeces; prevents *Vibrio parahaemolyticus* infection	Li et al. (2017)
		Reduces intestinal inflammation	Hamady et al. (2011)
		Isolated from human faeces; prevention of cancer	Ulsemer et al. (2016)
		Enhances the phagocytic functions of macrophages, polarizing them to an M1 phenotype	
2.	*Lactococcus lactis*	Isolated from GMO food; reduces inflammatory bowel diseases	Motta et al. (2012)
		Isolated from faeces; modulates oxidative stress, inhibits ferroptosis, and addresses dysbiosis	Boesmans et al. (2018)
3.	*Faecalibacterium prausnitzii*	Isolated from human faeces; reduces IBD, asthma, eczema, and type 2 diabetes	Rossi et al. (2016)
		Isolated from the human gut: Reduced hyperlipidaemia, prediabetes and type 2 diabetes, non-alcoholic fatty liver disease, and inflammatory bowel disease	Khan et al. (2023)
		Isolated from the human gut; developed LBP	Kumari et al. (2021)
		Anti-tumour response	Chaput et al. (2017)
4.	*Clostridium butyricm*	Isolated from humans: Prevention of pouchitis and modulation of gut microbiome in ulcerative colitis patients	Yasueda et al. (2016)
5.	*Bacteroides acidifaciens*	Isolated from mouse faeces; clearance of infectious agents	Yanagibashi et al. (2013)
6.	*Oscillospira* sp.	Isolated from the human gut; improved diabetes, obesity, and reduced systemic chronic inflammation	Yang et al. (2021)
7.	*Akkermansia muciniphila*	Isolated from the human gut; reduces fat mass, plasma glucose, gut permeability, and metabolic inflammation	Cani and Van Hul (2015)
		Isolated from human faeces; prevents obesity-associated disorders	Gauffin Cano et al. (2012)
		Isolated from the human gut; developed LBP	Kumari et al. (2021)
		Reduce systemic inflammation and cancer risk	Gopalakrishnan et al. (2018)

(continued)

Table 21.8 (continued)

S. no	Strain	Health benefits	References
		Enhance cancer immunotherapy	Routy et al. (2018)
		Combat cancer disease	Matson et al. (2018)
8.	*Bacteroides uniformis*	Isolated from the human gut; developed LBP	Kumari et al. (2021)
9.	*Ruminococcus bromii*		
10.	*Anaerobutyricum hallii*		
11.	*Roseburia intestinalis*		
12.	*Bifidobacterium* sp.	Promotes antitumor immunity and facilitates anti-PD-L1 efficacy	Sivan et al. (2015)
13.	*Bifidobacterium longum*	Isolated from human blood; robust $CD8^{+}$ T cell response and better prognosis in HBV-related hepatocellular carcinoma	Rong et al. (2017)
14.	*Enterococcus hirae*		
15.	*Lactobacillus rhamnosus*	Isolated from human blood; improves the immune system	Roller et al. (2007)
16.	*Bifidobacterium lactis*		
17.	*Clostridium butyricum*	Modulated gut microbiome	Shimbo et al. (2005)
		Antitumor effects by release of TRAIL through MMP-8 and novel intravesical therapy for bladder cancer	Shinnoh et al. (2013)
		Reduces the incidence of diarrhoea in digestive diseases, including inflammatory bowel disease	Tian et al. (2019)
18.	*Eubacterium limosum*	Increases mucosal integrity and shows anti-inflammatory effects through TLR4	Kanauchi et al. (2006)
19.	*Eubacterium hallii*	Improves insulin sensitivity and increases energy metabolism in severely obese and diabetic individuals	Udayappan et al. (2016)
20.	*Enterococcus hirae*	Enhance cancer immunotherapy	Daillère et al. (2016)
21.	*Barnesiella intestinihominis*		

(Derrien et al. 2017). Amuc1100 activates TLR2-mediated signalling to modulate host immunity and reinforce epithelial function, with pasteurized cells exhibiting superior metabolic benefits (Plovier et al. 2017). Clinical studies demonstrated improved insulin sensitivity, reduced insulinemia, and lowered plasma cholesterol following supplementation (Depommier et al. 2019). Additionally, *A. muciniphila* confers colonization resistance against *Candida albicans* by promoting SCFA production and inhibiting fungal hyphal transition (Wang et al. 2023), positioning it as a leading NGP for metabolic disorders. *Faecalibacterium prausnitzii* (Firmicutes) is a major butyrate-producing bacterium, constituting about 5% of the total gut microbiota. It plays a critical role in colonocyte energy metabolism and maintenance of epithelial integrity, with butyrate acting as a histone deacetylase inhibitor that exerts immunomodulatory effects and strengthens gut barrier function (Louis and Flint 2017). Reduced abundance of *F. prausnitzii* is a characteristic microbial signature of inflammatory bowel diseases, including Crohn's disease and ulcerative colitis, and is associated with postoperative recurrence risk (Pascal et al. 2017). In addition to butyrate production, *F. prausnitzii* secretes microbial anti-inflammatory molecules (MAM) that inhibit NF-κB signalling and suppress pro-inflammatory cytokine production (Quévrain et al. 2016). However, its clinical translation is challenged by its extreme oxygen sensitivity, prompting the development of encapsulation, co-culture, and defined microbial consortia strategies as potential research topics to enhance probiotic stability and therapeutic delivery (Khan et al. 2023). Non-enterotoxigenic *Bacteroides* spp. (NTBF) exert potent immunoregulatory effects through the production of polysaccharide A (PSA) via TLR2 to promote IL-10 secretion, regulatory T-cell induction, and balanced Th1/Th2 immune responses (Ochoa-Repáraz et al. 2010). In murine models, PSA has been reported to confer protection against colitis and to modulate gut–brain immune crosstalk, highlighting its systemic immunological relevance (Erny et al. 2015). *B. xylanisolvens* and *B. thetaiotaomicron* degrade xylans and complex dietary fibres, producing SCF, supporting mucosal integrity and metabolic homeostasis (Mirande et al. 2010). These species have been deemed safe under defined conditions of use, leading to the first regulatory approvals of *Bacteroides* spp. for food applications (EFSA 2015; Marco et al. 2021). The close phylogenetic relationship between beneficial and pathogenic *Bacteroides* emphasizes the need for stringent genomic screening, toxin profiling, and antimicrobial susceptibility testing to ensure safety (Erny et al. 2015). *Christensenella minuta* is a low-abundance gut commensal regularly associated with a lean phenotype and favourable metabolic profiles, and exhibits syntrophic co-occurrence with methanogenic archaea (*Methanobrevibacter* spp.), suggesting coordinated effects on energy-harvesting efficiency and intestinal gas metabolism (Waters and Ley 2019). In vivo colonization studies demonstrate that *C. minuta* reduces adiposity and weight gain while modulating bile acid metabolism and SCFA production (Morotomi et al. 2012; Goodrich et al. 2014). Human cohort analyses revealed its depletion in the population with obesity and metabolic syndrome. Further, faecal microbiota transplantation experiments demonstrated transmissible lean phenotypes from humans to mice (Goodrich et al. 2014; Beaumont et al. 2016; Waters and Ley 2019). *Eubacterium hallii, Roseburia intestinalis,* and *Bifidobacterium longum subsp. Infantis, Parabacteroides distasonis, Prevotella copri* are other potent NGPs that exhibit antifungal activity through SCFA production, mucosal barrier reinforcement, immune modulation, and indirect control of opportunistic fungi (Udayappan et al. 2016; Tamanai-Shacoori et al. 2017; Wang et al. 2023). Challenges, such as strain-level variability, oxygen sensitivity, and manufacturing complexity, limit large-scale clinical interventions, underscoring the need for synthetic biology, which provides a transformative platform to overcome these hurdles.

21.6.1.2 Synthetic Biology Approach to NGPs

Synthetic biology is a powerful tool to enhance the stability, safety, and therapeutic efficacy of NGPs through targeted genetic, metabolic, and delivery-system engineering. Engineered strains can be programmed to produce beneficial metabolites such as SCFAs, anti-inflammatory peptides, and other bioactive to modulate host immunity and inhibit pathogens (Canani et al. 2011). Probiotics, such as *Escherichia coli* Nissle 1917 and *Lactobacillus reuteri* were engineered to secrete antifungal peptides, quorum-sensing inhibitors, and cell wall-degrading enzymes, enabling selective suppression of fungal pathogens while preserving commensals (Isabella et al. 2018; Zhou et al. 2022). Synthetic gene circuits allow stimulus-responsive, controlled release of therapeutics, minimizing off-target effects (Riglar and Silver 2018). Metabolic engineering of butyrate-producing taxa (*Faecalibacterium*, *Roseburia*) and overexpression of microbial anti-inflammatory molecules further expand therapeutic scope (Louis and Flint 2017; Quévrain et al. 2016). Synthetic biology enables the development of stress-tolerant strains, disease-sensing biosensors, and defined microbial consortia, while incorporating advanced biocontainment and biosafety strategies to ensure controlled functionality and prevent unintended gene transfer (Stirling et al. 2017; Nyerges et al. 2019).

21.6.2 Synbiotics, Postbiotics, and Psychobiotics

Synbiotics (probiotics with complementary prebiotics) enhance microbial survival, colonization, and functionality, including tailored pairing of NGPs with specific substrates to overcome stability and other challenges (Markowiak and Śliżewska 2017; Salminen et al. 2021). Precision synbiotic formulations are increasingly explored for targeted modulation of the gut microbiome in obesity, inflammatory bowel disease, and metabolic disorders (Salminen et al. 2021). Postbiotics, comprising non-viable microbial cells, metabolites, and structural components (SCFAs, exopolysaccharides, and peptidoglycans), offer stable, safe, and regulatory-friendly alternatives with barrier-protective, immunomodulatory, and antifungal effects (Salminen et al. 2021). Collectively, these expand the therapeutic scope of NGPs, supporting microbiome-based interventions for gut, metabolic, neurobehavioral, and antifungal health outcomes. Psychobiotics are live microorganisms (probiotics) or their food sources (prebiotics) that, when ingested, confer mental health benefits by influencing the brain–gut axis, potentially easing anxiety, depression, and stress by producing neurotransmitters such as serotonin and GABA, or by reducing inflammation. They work through neural, immune, and metabolic pathways, affecting stress responses and mood (Cheng et al. 2019; Kamal et al. 2025).

21.6.3 Ethical and Cultural Considerations

The development and application of NGPs raise important ethical and cultural considerations. The ethical issues related to patient safety, risk assessment, equitable access, data ownership, benefit sharing, and bioprospecting rights must be addressed through regulatory frameworks (Riglar and Silver 2018). Cultural acceptance is shaped by dietary customs and local health beliefs, which influence public perception and the adoption of microbiome-based interventions (McFarland 2015; Marco et al. 2021). Integration of probiotic therapy varies across traditional and modern healthcare systems due to differences in medical philosophy, infrastructure, and cultural practices. Promoting holistic health concepts, preventive care, and culturally aligned education is essential for successful translation and acceptance of NGPs alongside conventional medicine (O'Toole et al. 2017). Transparent communication, community engagement, and culturally sensitive health strategies are critical for building trust and supporting precision microbiome therapies (Knoppers et al. 2015).

21.7 Research Gaps and Future Directions

Probiotic therapy faces critical research gaps, including a lack of large-scale, long-term human trials, particularly in vulnerable populations, and significant discrepancies across global regulatory frameworks. Furthermore, the development of personalized probiotic interventions, effective delivery systems, and standardized protocols for strain selection, dosage, duration, and outcome assessment remains limited and could be explored in the future (Fig. 21.4).

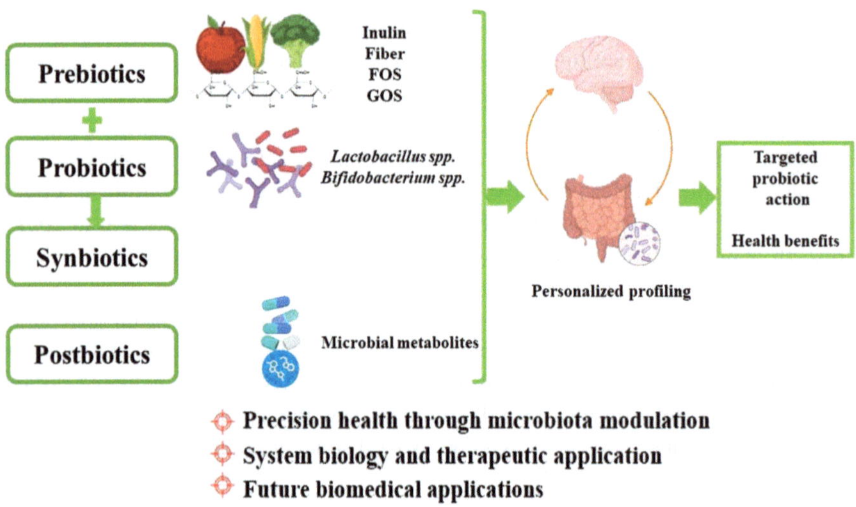

Fig. 21.4 Probiotics, prebiotics, synbiotics, and postbiotics in future applications

21.7.1 Need for Long-Term Clinical Trials

Most clinical evaluations of probiotics focus on short-term outcomes and insufficient long-term data (Merenstein et al. 2023). Extended monitoring is vital to identify delayed or cumulative effects (Al-Madhagi and Alramo 2023). NGPs may persist in the host for a longer period, raising concerns about microbiome imbalance and prolonged ecological alterations. Rare invasive infections and subtle dysbiosis-related effects could only be detected through long-term follow-up (Merenstein et al. 2023). Probiotic administration following antibiotic therapy has been associated with permanent changes in the gut microbiome, increased antibiotic-resistant genes, and altered functional gene profiles (Zavišić et al. 2023). Preterm infants, pregnant women, the elderly, and immunocompromised individuals may exhibit delayed metabolic, immune, or developmental effects (Merenstein et al. 2023). Accordingly, long-term clinical trials are warranted in high-risk groups.

21.7.2 Integration with Personalized Medicine

Advances in genetic profiling and microbiome analysis are shifting probiotic therapy towards a personalized approach, emphasizing genetic, metabolic, and microbiome profiles (Anjum and Ouwehand 2023; Manan 2025). Metagenomics, metabolomics, and transcriptomics have enabled targeted identification of probiotic strains with specific functional benefits (Manan 2025). However, partial knowledge of host-microbe interactions and inter-individual microbiome variability obstructs the standardization of personalized probiotic strategies. Further, high cost and restricted access further limit clinical translation (Manan 2025). Addressing these challenges requires coordinated efforts among researchers, clinicians, and regulatory bodies. With rigorous clinical validation, personalized probiotics have strong potential to become integral components of precision medicine (Armet et al. 2022).

21.7.3 Development of Global Regulatory Frameworks

Global probiotic regulation remains fragmented due to inconsistent definitions, classifications, and evidence across jurisdictions, with these being variably regulated as foods, supplements, or therapeutics (Arora and Baldi 2015). The absence of a universal definition and divergent categorization frameworks creates regulatory ambiguity and uneven oversight. Scientific evidence for strain-specific efficacy, appropriate dosing, manufacturing consistency, and safety is lacking. Quality control challenges, imprecise labelling, insufficient viability data, and varying manufacturing practices compromise regulatory compliance and consumer confidence (Anjum and Ouwehand 2023). Further, engineered strains, microbial consortia, and personalized therapies are beyond regulatory systems, adding to the challenges. Global regulatory harmonization (relevant to strains, manufacturing, and health claims) is urgent for consumer acceptance of the probiotic therapy (Merenstein et al. 2023).

21.7.4 Improved Delivery Systems (Microencapsulation, Nano-carriers)

Encapsulation, hydrogels, nanocoating, and core–shell structures are used to deliver probiotics safely during manufacturing, storage, and gut transit, which remain key challenges. The challenges include ensuring viability, targeted delivery, and colonization in the correct gut niche, developing scalable, cost-effective production methods, and validating performance in a realistic human system (Ramírez et al. 2025; Sun et al. 2025). Future efforts should focus on smart-responsive carriers (pH or enzyme), using biocompatible, food-grade materials, customizing systems for specific probiotic strains and host microbiome, validating delivery effectiveness in vivo, and establishing manufacturing and regulatory pathways to support transition from laboratory to market (Jan et al. 2024; Ramírez et al. 2025).

21.8 Conclusion

Probiotic therapy represents a promising complex approach to health management; yet it is constrained by scientific, regulatory, and host-specific challenges. The effectiveness of the therapy is strain-dependent, while mechanistic understanding, dose optimization, and ensuring its viability through the gastrointestinal tract limit its acceptance. Safety concerns in vulnerable populations and the risk of antibiotic resistance gene transfer are critical considerations. Inconsistent global regulations, a lack of standardized manufacturing practices, and quality control issues impede its clinical translation and consumer assurance. Further, inter-individual differences in microbiome, genetics, and lifestyle influence probiotic outcomes, highlighting the need for personalized strategies. Despite these constraints, validated probiotics offer significant benefits in modulating the gut microbiome, regulating the immune system, and improving metabolic health. Progress in probiotic therapy requires evidence-based and ethically guided developments supported by multidisciplinary research and strong clinical validation. Harmonized regulatory frameworks and transparent labelling are also essential for safe application.

References

Ahire JJ, Rohilla Alka Kumar, Vikram, Tiwari Amit (2023) Quality Management of Probiotics: ensuring safety and maximizing health benefits. Curr Microbiol 81(1):1

Al-Madhagi H, Alramo A (2023) Clinical trials of probiotics: current outlook. Food Sci Nutr 12(3): 2192

An D, Oh SF, Olszak T, Neves JF, Avci FY, Erturk-Hasdemir D et al (2014) Sphingolipids from a symbiotic microbe regulate homeostasis of host intestinal natural killer T cells. Cell 156(1): 123–133

Anjum M, Ouwehand AC (2023) Probiotic development: research gaps and opportunities. Food Sci Nutr Cases 2023:fsncs20230004

Armet AM, Deehan EC, O'Sullivan AF, Mota JF, Field CJ, Prado CM et al (2022) Rethinking healthy eating in light of the gut microbiome. Cell Host Microbe 30(6):764–785
Arora M, Baldi A (2015) Regulatory categories of probiotics across the globe: a review representing existing and recommended categorization. Indian J Med Microbiol 33:S2–S10
Beaumont M, Goodrich JK, Jackson MA, Yet I, Davenport ER, Vieira-Silva S et al (2016) Heritable components of the human fecal microbiome are associated with visceral fat. Genome Biol 17(1): 189
Boesmans L, Valles-Colomer M, Wang J, Eeckhaut V, Falony G, Ducatelle R et al (2018) Butyrate producers as potential next-generation probiotics: safety assessment of the administration of *Butyricicoccus pullicaecorum* to healthy volunteers. mSystems 3(6):10–1128
Canani RB, Di Costanzo M, Leone L, Pedata M, Meli R, Calignano A (2011) Potential beneficial effects of butyrate in intestinal and extraintestinal diseases. World J Gastroenterol: WJG 17(12): 1519
Cani PD, Van Hul M (2015) Novel opportunities for next-generation probiotics targeting metabolic syndrome. Curr Opin Biotechnol 32:21–27
Chaput N, Lepage P, Coutzac C, Soularue E, Le Roux K, Monot C et al (2017) Baseline gut microbiota predicts clinical response and colitis in metastatic melanoma patients treated with ipilimumab. Ann Oncol 28(6):1368–1379
Cheng L, Liu Y, Wu C, Wang S, Tsai Y (2019) Psychobiotics in mental health, neurodegenerative and neurodevelopmental disorders. J Food Drug Anal 27(3):632–648
Cook MT, Tzortzis G, Charalampopoulos D, Khutoryanskiy VV (2012) Microencapsulation of probiotics for gastrointestinal delivery. J Control Release 162(1):56–67
Cryan JF, O'Riordan KJ, Cowan CS, Sandhu KV, Bastiaanssen TF, Boehme M et al (2019) The microbiota-gut-brain axis. Physiol Rev. https://doi.org/10.1152/physrev.00018.2018
Daillère R, Vétizou M, Waldschmitt N, Yamazaki T, Isnard C, Poirier-Colame V et al (2016) *Enterococcus hirae* and *Barnesiella intestinihominis* facilitate cyclophosphamide-induced therapeutic immunomodulatory effects. Immunity 45(4):931–943
De Simone C (2019) The unregulated probiotic market. Clin Gastroenterol Hepatol 17:809–817
Deng H, Li Z, Tan Y, Guo Z, Liu Y, Wang Y et al (2016) A novel strain of *Bacteroides fragilis* enhances phagocytosis and polarises M1 macrophages. Sci Rep 6(1):29401
Depommier C, Everard A, Druart C, Plovier H, Van Hul M, Vieira-Silva S et al (2019) Supplementation with *Akkermansia muciniphila* in overweight and obese human volunteers: a proof-of-concept exploratory study. Nat Med 25(7):1096–1103
Derrien M, Belzer C, de Vos WM (2017) *Akkermansia muciniphila* and its role in regulating host functions. Microb Pathog 106:171–181
Didari T et al (2014) Effectiveness of probiotics in the treatment of inflammatory bowel disease: a systematic review. World J Gastroenterol 20(17):5404–5417
EFSA Panel on Dietetic Products, Nutrition and Allergies (NDA) (2015) Scientific Opinion on the safety of 'heat-treated milk products fermented with *Bacteroides xylanisolvens* DSM 23964'as a novel food. EFSA J 13(1):3956
Erny D, Hrabě de Angelis AL, Jaitin D, Wieghofer P, Staszewski O, David E et al (2015) Host microbiota constantly control maturation and function of microglia in the CNS. Nat Neurosci 18(7):965–977
FAO/WHO Codex Alimentarius Commission (2002) Guidelines for the evaluation of probiotics in food. Codex Alimentarius, Rome
FSAI (2024) Assessment of the safety of "probiotics" in food supplements. Food Safety Authority of Ireland
FSSAI (2021) Food safety and standards (probiotic foods) guidelines, 2021. Government of India, New Delhi
Gauffin Cano P, Santacruz A, Moya Á, Sanz Y (2012) *Bacteroides uniformis* CECT 7771 ameliorates metabolic and immunological dysfunction in mice with high-fat-diet induced obesity. PLoS One. https://doi.org/10.1371/journal.pone.0041079

Goodrich JK, Waters JL, Poole AC, Sutter JL, Koren O, Blekhman R et al (2014) Human genetics shape the gut microbiome. Cell 159(4):789–799

Gopalakrishnan V, Spencer CN, Nezi L, Reuben A, Andrews MC, Karpinets TV et al (2018) Gut microbiome modulates response to anti–PD-1 immunotherapy in melanoma patients. Science 359(6371):97–103

Hamady ZZ, Scott N, Farrar MD, Wadhwa M, Dilger P, Whitehead TR et al (2011) Treatment of colitis with a commensal gut bacterium engineered to secrete human tgf-β1 under the control of dietary xylan. Inflamm Bowel Dis 17(9):1925–1935

Hemarajata P, Versalovic J (2013) Effects of probiotics on the microbiome and human health. J Clin Gastroenterol 47(Suppl):S6–S11

Hill C, Guarner F, Reid G, Gibson GR, Merenstein DJ, Pot B et al (2014) The International Scientific Association for Probiotics and Prebiotics consensus statement on the scope and appropriate use of the term "probiotic". Nat Rev Gastroenterol Hepatol 11(8):506–514

Isabella VM, Ha BN, Castillo MJ, Lubkowicz DJ, Rowe SE, Millet YA et al (2018) Development of a synthetic live bacterial therapeutic for the human metabolic disease phenylketonuria. Nat Biotechnol 36(9):857–864

Jan T, Negi R, Sharma B, Kumar S, Singh S, Rai AK et al (2024) Next generation probiotics for human health: An emerging perspective. Heliyon 10(16):e35980

Jiang X, Zheng Y, Sun H, Dang Y, Yin M, Xiao M, Wu T (2025) Fecal microbiota transplantation improves cognitive function of a mouse model of Alzheimer's disease. CNS Neurosci Ther 31(2):e70259

Kamal N, Saharan BS, Duhan JS, Kumar A, Chaudhary P, Goyal C, Kumar M, Goyat N, Sindhu M, Mudgil P (2025) Exploring the promise of psychobiotics: bridging gut microbiota and mental health for a flourishing society. Med Microecol 23:100118

Kanauchi O, Fukuda M, Matsumoto Y, Ishii S, Ozawa T, Shimizu M et al (2006) *Eubacterium limosum* ameliorates experimental colitis and metabolite of microbes attenuates colonic inflammatory action with increase of mucosal integrity. World J Gastroenterol: WJG 12(7):1071

Khan MT, Dwibedi C, Sundh D, Pradhan M, Kraft JD, Caesar R et al (2023) Synergy and oxygen adaptation for development of next-generation probiotics. Nature 620(7973):381–385

Kim YT, Mills DA (2024) Exploring the gut microbiome: probiotics, prebiotics, synbiotics, and postbiotics as key players in human health and disease improvement. Food Sci Biotechnol 33(9):2065–2080

Knoppers BM, Zawati MNH, Sénécal K (2015) Return of genetic testing results in the era of whole-genome sequencing. Nat Rev Genet 16(9):553–559

Korostylev N, Prazdnova E, Mazanko M, Meskhi B, Rudoy D, Ermakov A, Olshevskaya A, Chistyakov A, Zharov V (2020) Genetic basis of the probiotic properties of lactobacillus. E3S Web Conf 203:04016. https://doi.org/10.1051/e3sconf/202020304016

Kothari D, Patel S, Kim SK (2019) Probiotic supplements might not be universally-effective and safe: a review. Biomed Pharmacother Biomed Pharmacother 111:537–547

Kumari M, Singh P, Nataraj BH, Kokkiligadda A, Naithani H, Ali SA et al (2021) Fostering next-generation probiotics in human gut by targeted dietary modulation: An emerging perspective. Food Res Int 150:110716

Leeflang J, Wright JA, Worthley DL, Din MO, Woods SL (2025) Evolutionary adaptation of probiotics in the gut: selection pressures, optimization strategies, and regulatory challenges. npj Biofilms Microbiomes 11(1):96

Lewis Z, Shani G, Masarweh C et al (2016) Validating bifidobacterial species and subspecies identity in commercial probiotic products. Paediatric Research 79:445–452

Li Z, Deng H, Zhou Y, Tan Y, Wang X, Han Y et al (2017) Bioluminescence imaging to track *Bacteroides fragilis* inhibition of *Vibrio parahaemolyticus* infection in mice. Front Cell Infect Microbiol 7:170

Louis P, Flint HJ (2017) Formation of propionate and butyrate by the human colonic microbiota. Environ Microbiol 19(1):29–41

Ma T, Shen X, Shi X, Sakandar HA, Quan K, Li Y et al (2023) Targeting gut microbiota and metabolism as the major probiotic mechanism-An evidence-based review. Trends Food Sci Technol 138:178–198

Manan MA (2025) The role of probiotics in personalized therapeutics: advances in gut microbe-driven interventions. Microbe 8:100497

Marchesi A, Nader-Macías MEF (2023) Probiotic lactic acid bacteria for vaginal application. Optimization of biomass production and freeze-drying conditions. Appl Microbiol 3(2): 519–535

Marco ML et al (2021) Health benefits of fermented foods: microbiota and beyond. Curr Opin Biotechnol 70:165–171

Markowiak P, Śliżewska K (2017) Effects of probiotics, prebiotics, and synbiotics on human health. Nutrients 9(9):1021

Marteau P, Shanahan F (2003) Basic aspects and pharmacology of probiotics: an overview of pharmacokinetics, mechanisms of action and side-effects. Best Pract Res Clin Gastroenterol 17(5):725–740

Matson V, Fessler J, Bao R, Chongsuwat T, Zha Y, Alegre ML et al (2018) The commensal microbiome is associated with anti–PD-1 efficacy in metastatic melanoma patients. Science 359(6371):104–108

McFarland LV (2010) Systematic review and meta-analysis of *Saccharomyces boulardii* in adult patients. World J Gastroenterol 16:2202–2222

McFarland LV (2015) From yaks to yogurt: the history, development, and current use of probiotics. Clin Infect Dis 60(suppl_2):S85–S90

Merenstein D, Pot B, Leyer G, Ouwehand AC, Preidis GA, Elkins CA, Hill C, Lewis ZT et al (2023) Emerging issues in probiotic safety: 2023 perspectives. Gut Microbes 15(1):2185034

Mirande C, Kadlecikova E, Matulova M, Capek P, Bernalier-Donadille A, Forano E, Béra-Maillet C (2010) Dietary fibre degradation and fermentation by two xylanolytic bacteria *Bacteroides xylanisolvens* XB1AT and *Roseburia intestinalis* XB6B4 from the human intestine. J Appl Microbiol 109(2):451–460

Morotomi M, Nagai F, Watanabe Y (2012) Description of *Christensenella minuta* gen. nov., sp. nov., isolated from human faeces, which forms a distinct branch in the order Clostridiales, and proposal of Christensenellaceae fam. nov. Int J Syst Evol Microbiol 62(1):144–149

Motta JP, Bermúdez-Humarán LG, Deraison C, Martin L, Rolland C, Rousset P et al (2012) Food-grade bacteria expressing elafin protect against inflammation and restore colon homeostasis. Sci Transl Med 4(158):158ra144

Nyerges Á, Balint B, Cseklye J, Nagy I, Pál C, Fehér T (2019) CRISPR-interference-based modulation of mobile genetic elements in bacteria. Synth Biol 4(1):ysz008

O'Mahony L, McCarthy J, Kelly P, Hurley G, Luo F, Chen K et al (2005) *Lactobacillus* and *Bifidobacterium* in irritable bowel syndrome: symptom responses and relationship to cytokine profiles. Gastroenterology 128(3):541–551

O'Toole PW, Marchesi JR, Hill C (2017) Next-generation probiotics: the spectrum from probiotics to live biotherapeutics. Nat Microbiol 2(5):1–6

Ochoa-Repáraz J, Mielcarz DW, Wang Y, Begum-Haque S, Dasgupta S, Kasper DL, Kasper LH (2010) A polysaccharide from the human commensal *Bacteroides fragilis* protects against CNS demyelinating disease. Mucosal Immunol 3(5):487–495

Ouwehand AC et al (2020) Strain-specificity and effects of probiotics: a critical review. Food Res Int 137:109491

Pascal V, Pozuelo M, Borruel N, Casellas F, Campos D, Santiago A et al (2017) A microbial signature for Crohn's disease. Gut 66(5):813–822

Patel R, Denning DW (2021) Quality issues with probiotics for human use. Clin Infect Dis 72(8): 1295–1301

Plovier H, Everard A, Druart C, Depommier C, Van Hul M, Geurts L et al (2017) A purified membrane protein from *Akkermansia muciniphila* or the pasteurized bacterium improves metabolism in obese and diabetic mice. Nat Med 23(1):107–113

Quévrain E, Maubert MA, Michon C, Chain F, Marquant R, Tailhades J et al (2016) Identification of an anti-inflammatory protein from *Faecalibacterium prausnitzii*, a commensal bacterium deficient in Crohn's disease. Gut 65(3):415–425

Rai A, Singh A, Baliyan A, Choudhary DS, Choudhary P, Chaudhary R et al (2024) *Saccharomyces boulardii* CNCM I-745: an efficient yeast probiotic with unique properties. Int J Sci Study 12(4):46

Ramírez A, Bermúdez-Luque A, Román-Camacho JJ, Martín-García JM, Moreno-García J, Güngörmüşler M, Ruiz-Castilla FJ (2025) Exploring the future of probiotics with innovations in delivery systems and market insights. Food Biosci 71:107029

Reid G, Charbonneau D, Erb J, Kochanowski B, Beuerman D, Poehner R, Bruce AW (2003) Oral use of *Lactobacillus rhamnosus* GR-1 and L. fermentum RC-14 significantly alters vaginal flora: randomized, placebo-controlled trial in 64 healthy women. FEMS Immunol Med Microbiol 35(2):131–134

Riglar DT, Silver PA (2018) Engineering bacteria for diagnostic and therapeutic applications. Nat Rev Microbiol 16(4):214–225

Roe AL, Boyte ME, Elkins CA, Goldman VS, Heimbach J, Madden E, Oketch-Rabah H, Sanders ME, Sirois J, Smith A (2022) Considerations for determining safety of probiotics: a USP perspective. Regul Toxicol Pharmacol 136:105266

Roller M, Clune Y, Collins K, Rechkemmer G, Watzl B (2007) Consumption of prebiotic inulin enriched with oligofructose in combination with the probiotics *Lactobacillus rhamnosus* and *Bifidobacterium lactis* has minor effects on selected immune parameters in polypectomised and colon cancer patients. Br J Nutr 97(4):676–684

Ronaghi M, Uhlén M, Nyrén P (1998) A sequencing method based on real-time pyrophosphate. Science 281(5375):363–365

Rong Y, Dong Z, Hong Z, Jin Y, Zhang W, Zhang B et al (2017) Reactivity toward *Bifidobacterium longum* and *Enterococcus hirae* demonstrate robust CD8+ T cell response and better prognosis in HBV-related hepatocellular carcinoma. Exp Cell Res 358(2):352–359

Rossi O, Berkel LV, Chain F, Khan MT, Taverne N et al (2016) *Faecalibacterium prausnitzii* A2-165 has a high capacity to induce IL-10 in human and murine dendritic cells and modulates T cell responses. Sci Rep 6(4):18507

Routy B, Le Chatelier E, Derosa L, Duong CP, Alou MT, Daillère R et al (2018) Gut microbiome influences efficacy of PD-1–based immunotherapy against epithelial tumors. Science 359(6371):91–97

Salminen S, Collado MC, Endo A, Hill C, Lebeer S, Quigley EM, Sanders ME et al (2021) The international scientific Association of Probiotics and Prebiotics (ISAPP) consensus statement on the definition and scope of postbiotics. Nat Rev Gastroenterol Hepatol 18(9):649–667

Sanders ME, Akkermans LM, Haller D, Hammerman C, Heimbach J, Hörmannsperger G, Huys G, Levy DD, Lutgendorff F, Mack D, Phothirath P, Solano-Aguilar G, Vaughan E (2010) Safety assessment of probiotics for human use. Gut Microbes 1(3):164–185

Sanders ME, Merenstein DJ, Reid G, Gibson GR, Rastall RA (2019) Probiotics and prebiotics in intestinal health and disease: from biology to the clinic. Nat Rev Gastroenterol Hepatol 16(10): 605–616

Sarita B, Samadhan D, Hassan MZ, Kovaleva EG (2025) A comprehensive review of probiotics and human health-current prospective and applications. Front Microbiol 15:1487641

Segers ME, Lebeer S (2014) Towards a better understanding of *Lactobacillus rhamnosus* GG-host interactions. Microb Cell Factories 13(Suppl 1):S7

Shimbo I, Yamaguchi T, Odaka T, Nakajima K, Koide A, Koyama H, Saisho H (2005) Effect of *Clostridium butyricum* on fecal flora in *Helicobacter pylori* eradication therapy. World J Gastroenterol 11(47):7520

Shinnoh M, Horinaka M, Yasuda T, Yoshikawa S, Morita M, Yamada T et al (2013) *Clostridium butyricum* MIYAIRI 588 shows antitumor effects by enhancing the release of TRAIL from neutrophils through MMP-8. Int J Oncol 42(3):903–911

Sivan A, Corrales L, Hubert N, Williams JB, Aquino-Michaels K, Earley ZM et al (2015) Commensal *Bifidobacterium* promotes antitumor immunity and facilitates anti–PD-L1 efficacy. Science 350(6264):1084–1089

Spacova I, Binda S, Ter Haar JA, Henoud S, Legrain-Raspaud S, Dekker J et al (2023) Comparing technology and regulatory landscape of probiotics as food, dietary supplements and live biotherapeutics. Front Microbiol 14:1272754

Stirling F, Bitzan L, O'Keefe S, Redfield E, Oliver JW, Way J, Silver PA (2017) Rational design of evolutionarily stable microbial kill switches. Mol Cell 68(4):686–697

Sun C, Zhang Z, Sun Y, Sun X, Jin Y, Zhu J, Yu J, Wu T (2025) Enteric delivery of probiotics: challenges, techniques, and activity assays. Foods (Basel, Switzerland) 14(13):2318

Tamanai-Shacoori Z, Smida I, Bousarghin L, Loreal O, Meuric V, Fong SB et al (2017) Roseburia spp.: a marker of health? Future Microbiol 12(2):157–170

Thakur A, Sharma P, Sharma AJ, Sharma N, Singh S et al (2025) Probiotics: microbes for human health and beyond. Food Materials Research 5:e001

Thu TV, Chung MJ, Singh S, Lee YH, Park YH (2021) Multi-omics approaches for understanding host–microbe interactions in probiotic research. J Microbiol 59(6):525–539

Tian Y, Li M, Song W, Jiang R, Li YQ (2019) Effects of probiotics on chemotherapy in patients with lung cancer. Oncol Lett 17(3):2836–2848

Udayappan S, Manneras-Holm L, Chaplin-Scott A, Belzer C, Herrema H, Dallinga-Thie GM et al (2016) Oral treatment with *Eubacterium hallii* improves insulin sensitivity in db/db mice. npj Biofilms Microbiomes 2(1):1–10

Ulsemer P, Toutounian K, Kressel G, Goletz C, Schmidt J, Karsten U et al (2016) Impact of oral consumption of heat-treated *Bacteroides xylanisolvens* DSM 23964 on the level of natural TFα-specific antibodies in human adults. Benefic Microbes 7(4):485–500

Wang F, Wang Z, Tang J (2023) The interactions of *Candida albicans* with gut bacteria: a new strategy to prevent and treat invasive intestinal candidiasis. Gut Pathogens 15(1):30

Waters JL, Ley RE (2019) The human gut bacteria *Christensenellaceae* are widespread, heritable, and associated with health. BMC Biol 17(1):83

Wong CB, Odamaki T, Xiao JZ (2019) Beneficial effects of *Bifidobacterium longum* subsp. longum BB536 on human health: modulation of gut microbiome as the principal action. J Funct Foods 54:506–519

Wu G, Xu T, Zhao N, Lam YY, Ding X, We D et al (2024) A core microbiome signature as an indicator of health. Cell 187(23):6550–6565

Xu R, Yifeng Y, Tingtao C (2025) Exploring the dark side of probiotics to pursue light: intrinsic and extrinsic risks to be opportunistic pathogens. Curr Res Food Sci 10:101044

Yahfoufi N, Alsadi N, Jambi M, Matar C (2018) The immunomodulatory and anti-inflammatory role of polyphenols. Nutrients 10(11):1618

Yanagibashi T, Hosono A, Oyama A, Tsuda M, Suzuki A, Hachimura S et al (2013) IgA production in the large intestine is modulated by a different mechanism than in the small intestine: *Bacteroides acidifaciens* promotes IgA production in the large intestine by inducing germinal center formation and increasing the number of IgA+ B cells. Immunobiology 218(4):645–651

Yang J, Li Y, Wen Z, Liu W, Meng L, Huang H (2021) *Oscillospira*-a candidate for the next-generation probiotics. Gut Microbes 13(1):1987783

Yasueda A, Mizushima T, Nezu R, Sumi R, Tanaka M, Nishimura J et al (2016) The effect of *Clostridium butyricum* MIYAIRI on the prevention of pouchitis and alteration of the microbiota profile in patients with ulcerative colitis. Surg Today 46(8):939–949

Zavišić G, Popović M, Stojkov S, Medić D, Gusman V, Jovanović Lješković N, Jovanović Galović A (2023) Antibiotic resistance and probiotics: knowledge gaps, market overview and preliminary screening. Antibiotics 12(8):1281

Zhang W, Zhu B, Xu J, Liu Y, Qiu E, Li Z et al (2018) *Bacteroides fragilis* protects against antibiotic-associated diarrhoea in rats by modulating intestinal defences. Front Immunol 9:1040

Zheng J, Wittouck S, Salvetti E, Franz CMAP, Harris HMB, Mattarelli P, O'Toole PW et al (2020) A taxonomic note on the genus *Lactobacillus*: description of 23 novel genera, emended description of the genus Lactobacillus Beijerinck 1901, and union of Lactobacillaceae and Leuconostocaceae. Int J Syst Evol Microbiol 70(4):2782–2858

Zhernakova DV, Wang D, Liu L, Andreu-Sánchez S, Zhang Y, Ruiz-Moreno AJ et al (2024) Host genetic regulation of human gut microbial structural variation. Nature 625(7996):813–821

Zhou J, Li M, Chen Q, Li X, Chen L, Dong Z et al (2022) Programmable probiotics modulate inflammation and gut microbiota for inflammatory bowel disease treatment after effective oral delivery. Nat Commun 13(1):3432

Zmora N, Zilberman-Schapira G, Suez J, Mor U, Dori-Bachash M, Bashiardes et al (2018) Personalised gut mucosal colonisation resistance to empiric probiotics is associated with unique host and microbiome features. Cell 174(6):1388–1405

Case Studies: Success Stories of Probiotics in Fungal Defence

22

D. Yogesh and D. Nirmala Devi

Abstract

Incidence of fungal infections is on the rise worldwide for various reasons over time, and is one of the serious health threats to humans. Extensive use of antifungal agents, as well as the development of resistance to these drugs, poses an additional risk to the world. Use of probiotics against these fungal infections has become a new strategy as well as a promising measure for the development of treatment for such fungal infections. In this chapter, several case studies have been discussed to elaborate and convince the scientific world about promising strategies to use probiotics for fungal defence. Several in vitro studies, animal experiments, microbiological studies, and clinical trials have been carried out for the assessment of the effectiveness of probiotics against fungal infections. Probiotic-based strategies have also proved successful for enhancing disease resistance, growth, and overall performance in animal husbandry, poultry, and aquaculture systems. Probiotics have proven to be safe and effective alternatives to treat fungal infections.

D. Yogesh (✉) · D. Nirmala Devi
Postgraduate Department of Microbiology, Maharani's Science College for Women, Mysuru, Karnataka, India

M. Y. Sreenivasa et al. (eds.), *Probiotics as Antifungals*,
https://doi.org/10.1007/978-981-92-0002-3_22

Graphical Abstract

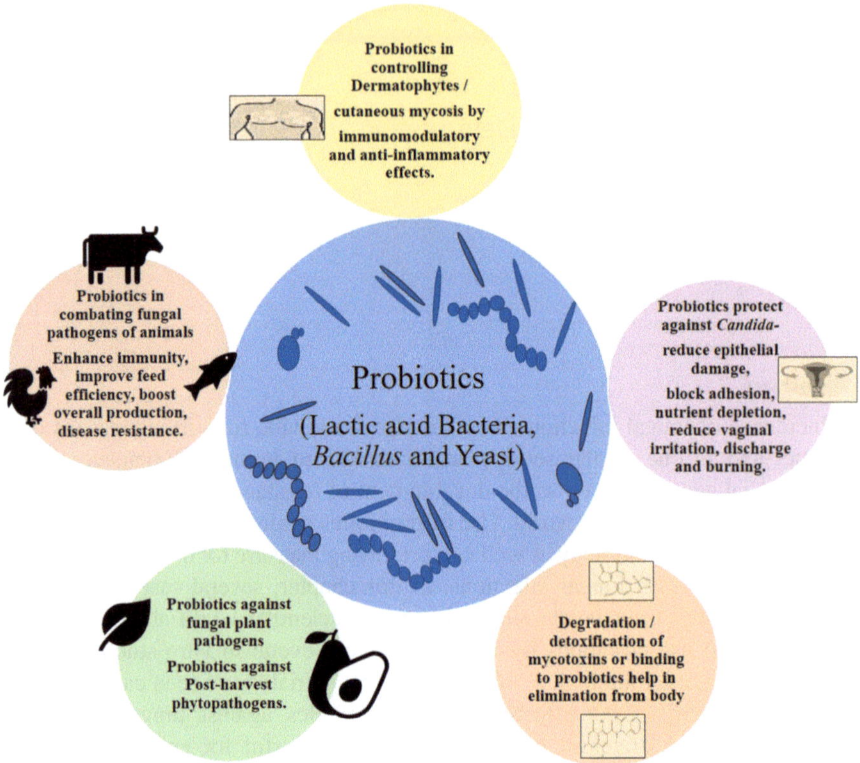

Keywords

Probiotics · Fungal infections · Candidiasis · Lactic acid bacteria · Fungal pathogens

Learning Objectives

- Importance of probiotics
- Case studies with respect to probiotics in fungal infections
- Future prospects of probiotics in fighting fungal infections and mycotoxins

22.1 Introduction

Estimates of the global burden of fungal diseases suggest approximately 6.5 million cases of invasive fungal infections and 3.8 million deaths yearly (Denning 2024). There are numerous fungal infections in the human population, with the most common and significant invasive fungal infections including opportunistic invasive mycoses such as Candidiasis, Aspergillosis, Cryptococcosis, Pneumocystis, and Mucormycosis. There are also endemic dimorphic mycoses such as Coccidioidomycosis, Blastomycosis, Histoplasmosis, Penicilliosis, and Paracoccidioidomycosis. There is no wide recognition for the impact of fungal infections on human health and deaths, as these infections are often overlooked (Brown et al. 2012). One of the important concerns is that the high rate of mortality, which is associated with invasive fungal infections, usually exceeds 50% even with the availability of numerous antifungal drugs (Brown et al. 2012).

Fungal infections of various types, based on location as well as severity, can be superficial or deep fungal infections. Those common on the external surfaces such as skin, nails, and scalp are superficial infections while some infections can be deep and life-threatening. In severe cases, fungal cells can be seen in blood, as in case of fungimia (Dellière et al. 2020; Vitalis et al. 2020). There are several antifungal drugs that are in clinical practice, which include polyenes, echinocandins, azoles, and pyrimidines (Oltu et al. 2020). These drugs are not free of side effects and have certain limitations (Sant et al. 2016; Deepa et al. 2016; Perlin 2020; Wu et al. 2022). Antimicrobial resistance in pathogenic fungi has become serious, and multidrug resistance in *Candida* has caused alarm worldwide (Hu et al. 2021). Hence, it is important to develop safe and effective alternatives to treat fungal infections.

By definition, probiotics are viable microorganisms that confer health benefits when administered to the host in adequate amounts. They provide a safe and healthy way to improve health and overall well-being of humans and animals. In vitro studies and clinical Investigations on the antifungal effects of probiotics in the oral cavity, urogenital tract, and gastrointestinal tract of humans have been performed to have a possible strategy for treatment against fungal infections. Several researchers have investigated the use of probiotics in combating *Candida* infection worldwide (Matsubara et al. 2016). Fungal colonization in the Gastro-Intestinal (GI) tract is known to delay ulcer and inflammation healing (Trojanowska et al. 2006).

Microbial commensals inhabiting the gastrointestinal tract or skin are known to play a vital role in both health and disease. Use of probiotics has seen an increase as they have asserted a good safety profile in the age of the antibiotic resistance problem. Oral and topical probiotics can be tested and studied for the treatment or prevention modality in cutaneous inflammations. Hence, it helps in bringing down the requirement for topical or systemic steroids as well as antimicrobial agents. In this age of antimicrobial resistance, tempering microbiota with suitable probiotics is one of the safe and well-tolerated approaches. However, evidence regarding the mechanism of action, safety profile, and efficacy of these probiotics requires critical validation and further studies (Dass et al. 2007; Gowda et al. 2024). There are multiple factors, such as age, diet, antibiotics, and environmental factors, and many more that influence the microbial composition in the gut. This may lead to the activation of the immune system, leading to immunomodulatory as well as shielding effects on other sites, such as the skin. Probiotics are known to have immunomodulatory and anti-inflammatory effects. Hence, the role played by gut microbiota has an important effect on influencing immunological tolerance (Gowda et al. 2024). Probiotics are known to upregulate the production of Tregs and cytokines, which are anti-inflammatory and are dispensed to the sites of inflammation. The other mechanisms by which probiotics work include intestinal epithelial stimulation for mucin production, and pathogenic bacteria are inhibited by molecular mimicry (Mack et al. 2003).

22.2 Case Studies

22.2.1 *Lactobacillus rhamnosus* GG (LGG) Against *Candida albicans*

Candida, an opportunistic pathogen, may be associated with gastric ulcers, known for delays in healing of ulcers and to aggravate the disease condition (Brzozowski et al. 2005). Therapy using probiotics is one effective method in the treatment of fungal colonization of the Gastro-Intestinal (GI) tract. The lactic acid bacteria, such as *Lactobacillus acidophilus,* used in therapy, have been reported to shorten the duration of fungal colonization of the mucosa. The enhanced *Candida* clearance is associated with IL-4, IFN-γ response.

The effect of *Lactobacillus rhamnosus* was studied, where the protective property of *Lactobacillus rhamnosus* GG (LGG) against *Candida albicans SC5314,* causing epithelial damage, was mediated by a synergy of reduced adhesion and glucose depletion. In vitro model of oral candidiasis using RHOEs and TR146 cell monolayers demonstrated that the probiotic LGG protects oral epithelial tissue from *C. albicans,* which is known to cause damage (Mailänder-Sánchez et al. 2017). The antifungal defence mechanism of LGG is mediated by blocking adhesion and nutrient depletion, which led to a shift in metabolism of *C. albicans* and resulted in impaired hyphae formation, resulting in reduced damage. This study provided insight into the mechanism by which functionally probiotic lactobacilli protect oral epithelial cells against *C. albicans* damage in vitro.

22.2.2 *Lactobacillus plantarum* Strains Against *Fusarium*—In Vitro Studies in Malting of Barley

The antifungal potential of *Lactobacillus plantarum* (strains VTT E-78076 (E76) and VTT E-79098 (E98)) was studied against *Fusarium* species. This was demonstrated by Laitila et al. (2002) using in vitro screening methods such as automated turbidometry, impedimetric (direct and indirect) assay methods. Results indicated that the cell-free extract of *Lactobacillus plantarum* was effective against four species of *Fusarium,* including *Fusarium culmorum, F. avenaceum,*

F. oxysporum, and F. graminearum. The cells and spent medium of *Lactobacillus plantarum* E76 culture had antifungal potential when examined in laboratory-scale malting with naturally contaminated barley. The growth of *Fusarium* was restricted when *Lactobacillus plantarum* E76 was added to the steeping water. Conclusions of the study were that the *Lactobacillus plantarum* strains (E76 and E98) had a fungistatic effect on different plant pathogenic, toxigenic, and gushing-active *Fusarium* fungi and could be employed as a natural, food-grade biocontrol agent in the management of problems associated with *Fusarium* during cereal germination.

22.2.3 *Bacillus subtilis* SB8, a Probiotic Bacterium for Sustainable Fungal Disease Management

Post-harvest phytopathogens pose a critical challenge in agriculture, resulting in unsafe conditions as well as loss of the produce. Fungal diseases in berries such as strawberries are one of the key economic hurdles; hence, these are highly prized. Menéndez-Cañamares et al. (2024) have demonstrated that *B. subtilis* is Generally Recognized as Safe (GRAS). This strain displayed its antagonistic effects in an in vivo assay using the SB8 strain, composed of an alginate-based edible coating, which indicated a significant reduction in *Botrytis cinerea* infection. The study highlights the potential strategies for strawberry production to enhance its safety and sustainability, to combat fungal diseases. The genes responsible for its capability to produce antimicrobial compounds and traits with respect to probiotic properties were revealed by genome sequencing of the *B. subtilis* SB8.

22.2.4 Vulvovaginal Candidiasis and Probiotics

Vulvovaginal candidiasis (VVC) is one of the most common infections in women and a greater prevalence is seen in pregnant women with elevated risks attributed to repeated incidence of infections (Aguin and Sobel 2015; Jeanmonod et al. 2025). With the hypothesis that the disruption of vaginal lactobacilli leads to recurrences of this candidiasis in pregnant women due to overgrowth of vaginal *Candida*, while probiotics may restore such a disruption. The effects of a multi-species lactobacilli probiotic product SynForU-HerCare were evaluated against *Candida* in the high and low-vaginal as well as cervicovaginal regions of pregnant women with vaginal candidiasis (VC), and the assessments for inflammatory responses in these areas. The multi-species probiotic product was composed of around six lactic acid bacteria, namely *Lactiplantibacillus* LP115, *Lactobacillus helveticus* LA25, *Lacticaseibacillus rhamnosus* LRH10, *Lacticaseibacillus paracasei* LPC12, *Limosilactobacillus fermentum* LF26, and *Lactobacillus delbrueckii* subsp. *Lactis* LDL114. The product demonstrated to have beneficial effects in improving symptoms such as vaginal irritation, discharge, and burning in pregnant women with VC upon administration for 8 weeks, along with improvement in parameters such as emotional and social quality of life. It was found to decrease the recurrence of VC in pregnant women (Ang et al. 2021). A previous study involving 78 pregnant women with VC showed that the oral administration of SynForU-HerCare for the duration of 8 weeks reduced the symptoms and recurrences of VC (Ang et al. 2021, 2022). This is a promising strategy for the maintenance of vaginal health during pregnancy. The study suggests that probiotics can be a key alternative to treat vaginal candidiasis in pregnant women. In particular, this observation can be attributed to the inhibition of *Candida*, and helps in the prevention of recurrences and a decrease in vaginal inflammation.

22.2.5 Probiotics in Combating Fungal Pathogens of Animals

Contemporary veterinary research has shifted towards the development of sustainable, health-focused practices in animal husbandry, encouraging the use of probiotics for better animal health. In parallel, there is also increasing interest in developing antimicrobial-free strategies to safely enhance growth and productivity in poultry as well as aquaculture systems. The use of probiotics as feed supplements in the livestock and poultry sector has gained considerable attention owing to their beneficial health effects, ease of administration, safety, and potential to enhance overall productivity. Probiotic feed additives comprise live microorganisms that help maintain a healthy intestinal microflora, along with supporting essential metabolic and immune functions. They serve as therapeutic agents by enhancing immunity, improving feed efficiency, and boosting overall production potential (Singh et al. 2024). The frequently used probiotics include *Lactobacillus, Bifidobacterium, Streptococcus, Pediococcus, Enterococcus, Bacillus, Saccharomyces,* and other genera as single or multi-strain preparations that benefit the health and immunity of the host. Administering a consortium of such multi-strain probiotic preparations is a safe and healthy alternative to conventional antibiotic growth promoters, improving the productivity in ruminants, pigs, fish, and poultry (Hoseinifar et al. 2018; Lambo et al. 2021).

There is an ever-increasing global demand for animal feed in the livestock sector to accelerate animal growth and prevent disease. Integrating probiotic-based feed additives is of prime importance as they provide a safe and healthy way to enhance productivity, improve health, and overall well-being of the animals. Lactic acid bacteria (LAB) are a versatile group of organisms with the potential to produce several metabolites along with a range of bioactivities. They are Gram-positive, non-spore-forming, that are acid-tolerant, and facultatively anaerobic. Several genera of LAB, including *Lactobacillus, Lactococcus, Leuconostoc, Pediococcus, Streptococcus,* and others have been extensively researched for their probiotic properties (Goldstein et al. 2015). With their ability to colonize the host gut, these LAB participate in the host's physiological metabolism, offering many benefits to the host. They aid in feed digestion, improve nutrient absorption, overall nutrient utilization, modulate and create a balanced gut microbiome and regulate host immune responses. They produce organic acids, short chain fatty acids (SCFAs), amino acids, exopolysaccharides (EPSs), antimicrobial proteins such as bacteriocins and vitamins that strengthen the gut barrier and help animals fight pathogens (Tang et al. 2023). These metabolites have been attributed to the antimicrobial, antioxidant, anti-cancer, and anti-inflammatory activities, making the application of LAB as animal feed additives a sustainable alternative to antimicrobial growth promoters. This helps to overcome the challenge of antibiotic resistance among animals and the consumers of such animal products (Werning et al. 2022; Lambo et al. 2021). Application of probiotic supplements to dairy cattle has been linked to enhanced milk yield and a decrease in veterinary expenses (Lozupone et al. 2012). The synergistic effects of low concentrations of lactic acid and hydrogen peroxide were investigated by Zhang et al. (2019), showing a broad-spectrum antimicrobial activity against several bacterial as well as fungal pathogens, including *Aspergillus oryzae* and *Penicillium citrinum*. Pathogen lysis was brought about by a combination of DNA damage, alterations in membrane permeability, and cellular content leakage. Such probiotic-based activities could be safe antimicrobials for agriculture and feed production, offering a natural pathogen control method that aligns with sustainable practices.

Bacillus velezensis strains have gained wide recognition as a probiotic for agricultural applications for improving animal growth performance, improving immunity, and their overall health benefits. Strong probiotic potential of *Bacillus velezensis* JW was evident on genomic analysis, which revealed the presence of several gene clusters such as polyketide synthase (PKS), non-ribosomal peptide synthase (NRPS), as well as fungicin biosynthetic gene clusters. These gene clusters encode an array of antibacterial metabolites, including lipopeptides and polyketide compounds, which might have inhibitory effects on fungal pathogens (Yi et al. 2018; Zhang et al. 2025). Certain strains of *Bacillus velezensis* produce potent lipopeptides, including surfactin, iturin, and fengycin, which rupture fungal cell membranes and effectively suppress a variety of fungal pathogens such as Fusarium and Alternaria. Notably, iturin and fengycin weaken the structural framework of the fungal cell wall, targeting the synthesis of β-1,3-glucan and ultimately causing fungal cell lysis. Yu et al. (2025) suggested that *B. velezensis* Y01 effectively improved growth performance, intestinal microbiota, and immune function in Chinese local breeds of chicken. Dietary supplementation of Langya chicks with *B. velezensis* Y01 resulted in increased body weight, increased serum immunoglobulin G (IgG) levels, IgM, and interleukin-2 (IL-2). It also enriched the abundance of caecal microbial communities such as Firmicutes and Bacteroidetes, resulting in the production of certain beneficial metabolites that play pivotal roles in reducing pathogen infections and ultimately enhancing the immunity of the poultry chicken.

Probiotic-based strategies have proved successful for enhancing disease resistance in aquaculture systems. Opportunistic fungal infections are a threat to natural and commercial fish farms, causing great losses of fish and their eggs (Abolude et al. 2013). Aspergillosis caused by *Aspergillus flavus* and the mycotoxins produced shows localized as well as systemic effects. Abdel-Tawwab et al. (2020) showed the effects of dietary *Saccharomyces cerevisiae*, supplementation on the improved growth performance, haematological parameters, antioxidant status, and immune responses of Nile Tilapia (*Oreochromis niloticus*), along with their resistance to infection by the pathogenic fungus *Aspergillus flavus*. This was evident with no mortality in fish fed with 6–10 g of probiotic yeast/kg after *A. flavus* challenge in comparison to the highest mortality rate seen in the control-fed fish.

22.2.6 Lactic Acid Bacteria for Degradation and Elimination of Mycotoxins

Mycotoxins are common contaminants in food and feed. Aflatoxin B_1 is an important toxin found as a contaminant known to cause Cancer (Gratz et al. 2007). Lactic acid bacteria are known to reduce the bioavailability of aflatoxins, and also *Bifidobacteria* adsorb mycotoxins. In simulated gastrointestinal conditions, *Lactobacillus plantarum* and *Lactobacillus rhamnosus* GG have shown high affinity for aflatoxin B1, resulting in decreased bioavailability by up to 50–80% (Eiri et al. 2024). The peptidoglycan and cell wall polysaccharide of *Lactobacillus rhamnosus* GG is known to bind aflatoxin B1 (AFB1) and reduce its bioavailability (by up to 74%) as studied with simulated intestinal conditions (El-Nezami et al. 1998). The electrostatic and hydrophobic interactions between the lactone ring of AFB1 and the bacterial surface are responsible for the adsorption, and this was confirmed using atomic force microscopy (Haskard et al. 2000). Similarly, *Lactobacillus plantarum* CECT 749 forms a non-toxic ochratoxin α from the degradation of ochratoxin A (OTA) with the help of carboxypeptidase activity (Fuchs et al. 2008). On the other hand, *Saccharomyces cerevisiae,* through glucan-mediated binding, sequesters another toxin zearalenone (ZEA), with effectiveness cresting at 80% with the optimal pH conditions.

Another mycotoxin, that is, ochratoxin A, was shown to adsorb to *Bifidobacterium bifidum* (Massoud and Zoghi 2022). The combination of *Lactobacillus fermentum* (LC5/a) with chlorophyll has demonstrated synergistic detoxification of Aflatoxin B1 in a mouse model. This strain LC5/a, was also known for the production and regulation of proinflammatory cytokines. This adsorption of toxins may help in the excretion of toxins through faeces instead of getting absorbed (Kumara et al. 2020). Therefore, probiotic strains could be the components of functional foods, supplements in feeds for animals or poultry, which may offer a cost-effective as well as scalable strategy against exposure to mycotoxins (Kumara et al. 2020).

22.3 Summary

Fungal infections have substantially contributed to both morbidity and mortality in humans. There is an urgent need for competent diagnostic tests, effective and safe new drugs and vaccines, and a drive towards research in the pathophysiology of human fungal infections. Antifungal activity of many probiotic bacteria has been demonstrated mainly in vitro and in some in vivo studies. However, the majority of the studies focus mainly on candidiasis rather than other fungal pathogen infections.

This may be due to the fact that severe cases of fungal infections are neglected or overlooked. Antifungal activity of probiotic bacteria against dermatophytes, as well as other fungal members such as *Penicillium*, *Aspergillus*, and *Mucor*, has been demonstrated by many research groups. Yet, further in vivo demonstration, clinical trials, and formulations are necessary for their effective use against fungal infections. Many probiotics are also known to degrade or modify mycotoxins to harmless or less toxic compounds, exerting a protective action. There is a need for more detailed in vivo studies and clinical trials in order to authenticate the effect of probiotics in fungal defence.

22.4 Conclusion

Fungi are one of the main reasons for many infections in humans, animals, and plants, as well as losses in agriculture. Probiotics are the potential answers for problems associated with fungi. There are various mechanisms, such as immune modulation, competitive exclusion, metabolite production, and toxin binding, by which probiotics fight against fungal pathogens. These are supported by many in vitro and in vivo evaluations on the utility of probiotics against fungi. More research is necessary for a deeper understanding of their effect and applications. This will help in exploring and exploiting their ability to fight the fungal infections strategically. Probiotics will be one of the better alternatives for antifungal drugs as well as weapons to fight against fungi.

Conflict of Interest The authors confirm that they have no conflict of interest.

Authors' Contribution Both the authors have contributed equally in preparing this chapter.

References

Abdel-Tawwab M, Adeshina I, Issa ZA (2020) Antioxidants and immune responses, resistance to *Aspergillus flavus* infection, and growth performance of Nile tilapia (*Oreochromis niloticus*) fed diets supplemented with yeast (*Saccharomyces cerevisiae*). Anim Feed Sci Technol 263:114484

Abolude DS, Opabunmi OO, Davies OA (2013) Fresh water fungi associated with eggs and broodstock of African catfish (*Clarias gariepinus*) in fish hatchery farms, Nigeria. J Res Environ Sci Toxicol 2(7):131–135

Aguin TJ, Sobel JD (2015) Vulvovaginal candidiasis in pregnancy. Curr Infect Dis Rep 17:462

Ang XY, Chung FYL, Lee BK, Azhar SNA, Roslan NS, Ahmad N, Yusof SM, Abdullah N, Rahman NNNA (2021) *Lactobacilli* reduce recurrences of vaginal candidiasis in pregnant women: a randomized, double-blind, placebo-controlled study. J Appl Microbiol 132(4): 3168–3180

Ang XY, Mageswaran UM, Chung YLF, Lee BK, Azhar SNA, Roslan NS et al (2022) Probiotics reduce vaginal candidiasis in pregnant women by modulating *Candida* and *Lactobacillus* abundance. Microorganisms 10:285. https://doi.org/10.3390/microorganisms10020285

Brown GD, Denning DW, Gow NAR, Levitz SM, Netea MG, White TC (2012) Hidden killers: human fungal infections. Sci Transl Med 4(165):165rv13

Brzozowski T, Zwolińska-Wcisło M, Konturek PC et al (2005) Influence of gastric colonization with *Candida albicans* on ulcer healing in rats. Scand J Gastroenterol 40:286–296

Dass RS, Sreenivasa MY, Janardhana GR (2007) High incidence of Fusarium verticillioides in animal and poultry feed mixtures produced in Karnataka, India. Plant Pathol J 6(2):174–178

Deepa N, Charith Raj AP, Sreenivasa MY (2016) Nested PCR method for the early detection of fumonisin-producing Fusarium verticillioides in pure cultures, cereal samples and plant parts. Food Biotechnol 30(1):18–29

Dellière S, Rivero-Menendez O, Gautier C, Garcia-Hermoso D, Alastruey-Izquierdo A, Alanio A (2020) Emerging mould infections. Med Mycol 58:156–162. https://doi.org/10.1093/mmy/myz074

Denning DW (2024) Global incidence and mortality of severe fungal disease. Lancet Infect Dis 24(7):e428–e438

Eiri A, Kaboosi H, Niknejad F et al (2024) *In vitro* detoxification of aflatoxin B1 by *Lactiplantibacillus plantarum*. Folia Microbiol. https://doi.org/10.1007/s12223-024-01234-4

El-Nezami H, Kankaanpää P, Salminen S, Ahokas J (1998) Ability of dairy lactic acid bacteria to bind aflatoxin M1. J Food Prot 61(2):214–216

Fuchs S, Sontag G, Stidl R (2008) Influence of pH and bile salts on mycotoxin binding. Food Chem 110(4):1227–1232

Goldstein EJC, Tyrrell KL, Citron DM (2015) *Lactobacillus* species. Clin Infect Dis 60(Suppl 2): S98–S107

Gowda V, Sarkar R, Verma D, Das A (2024) Probiotics in dermatology. Indian Dermatol Online J 15:571–583

Gratz S, Wu QK, El-Nezami H, Juvonen RO, Mykkänen H, Turner PC (2007) *Lactobacillus rhamnosus* GG reduces aflatoxin B1 toxicity. Appl Environ Microbiol 73(12):3958–3964

Haskard CA, Binnion C, Ahokas JT (2000) Sequestration of aflatoxin by *Lactobacillus rhamnosus*. Chem Biol Interact 128(1):39–49

Hoseinifar SH, Sun YZ, Wang A, Zhou Z (2018) Probiotics in aquaculture. Front Microbiol 9:2429

Hu S, Zhu F, Jiang W, Wang Y, Quan Y, Zhang G et al (2021) Retrospective analysis of the clinical characteristics of *Candida auris* infection worldwide from 2009 to 2020. Front Microbiol 12: 658329

Jeanmonod R, Chippa V, Jeanmonod D (2025) Vaginal Candidiasis. [Updated 2024 Feb 3]. In: StatPearls [Internet]. Treasure Island (FL), StatPearls Publishing. Available from: https://www.ncbi.nlm.nih.gov/books/NBK459317/

Kumara SS, Gayathri D, Hariprasad P, Venkateswaran G, Swamy CT (2020) Aflatoxin detoxification by *Lactobacillus fermentum*. Toxicon 187:214–222

Laitila H-L, Alakomi L, Raaska T, Mattila-Sandholm A, Haikara A (2002) Antifungal activity of *Lactobacillus plantarum*. J Appl Microbiol 93(4):566–576

Lambo MT, Chang X, Liu D (2021) Multistrain probiotics in livestock. Animals 11:2805

Lozupone CA, Stombaugh JI, Gordon JI, Jansson JK, Knight R (2012) Human gut microbiota diversity. Nature 489:220–230

Mack DR, Ahrne S, Hyde L, Wei S, Hollingsworth MA (2003) Mucin secretion and *Lactobacillus*. Gut 52:827–833

Mailänder-Sánchez D, Braunsdorf C, Grumaz C et al (2017) Antifungal defense of *Lactobacillus rhamnosus* GG. PLoS One 12(10):e0184438

Massoud R, Zoghi A (2022) Probiotics for mycotoxin removal. J Appl Microbiol 133(3): 1288–1307

Matsubara VH, Bandara HM, Mayer MP, Samaranayake LP (2016) Probiotics as antifungals. Clin Infect Dis 62(9):1143–1153

Menéndez-Cañamares S, Blázquez A, Albertos I, Poveda J, Díez-Méndez A (2024) *Bacillus subtilis* in strawberry disease control. Biol Control 196:105572

Oltu I, Cepoi L, Rudic V et al (2020) Antifungal drug development. Adv Exp Med Biol 1282:71–83

Perlin DS (2020) Cell wall-modifying antifungal drugs. Curr Top Microbiol Immunol 425:255–275

Sant DG, Tupe SG, Ramana CV, Deshpande MV (2016) Fungal cell membrane as a drug target. J Appl Microbiol 121:1498–1510

Singh AP, Shrivastava K, Tiwari A et al (2024) Probiotics in livestock farming. Asian J Dairy Food Res:1–9

Tang H, Huang W, Yao YF (2023) Metabolites of lactic acid bacteria. Microb Cell 10(3):49–62

Trojanowska D, Konturek P, Pajdo R, Drozdowicz D, Kwiecień S (2006) Probiotics and fungal colonization. J Physiol Pharmacol 57:35–49

Vitalis E, Nagy F, Tóth Z et al (2020) *Candida* biofilm and mortality. Mycoses 63:352–360

Werning ML, Hernandez-Alcantara AM, Ruiz MJ et al (2022) Exopolysaccharides from lactic acid bacteria. Foods 11(9):1284

Wu Y, Hu S, Wu C, Gu F, Yang Y (2022) Probiotics against fungal infections. Front Cell Infect Microbiol 11:793419

Yi Y, Zhang Z, Zhao F et al (2018) Probiotic potential of *Bacillus velezensis*. Fish Shellfish Immunol 78:322–330

Yu L, Zhang L, Zhang S et al (2025) Dietary *Bacillus velezensis* supplementation. BMC Microbiol 25:288

Zhang C, Zhang S, Liu W, Guo T, Gu R, Kong J (2019) Lactic acid–hydrogen peroxide consortium. Appl Biochem Biotechnol 189(3):822–833

Zhang M, Zhang Y, Zhao Z, Deng F, Jiang H, Liu C, Li Y, Chai J (2025) Bacterial–fungal interactions. Life 15(8):1242

Innovations and Emerging Trends in Probiotic Antifungal Research

23

Deepthi B V, K. Poornachandra Rao, Premila N Achar, and M. Y. Sreenivasa

Abstract

Probiotics have played a vital role in therapeutic science by providing beneficial health effects. The increasing resistance to antifungal drugs and chemical preservatives has accelerated interest in probiotics as alternative and adjunct antifungal strategies. Emerging trends in probiotic-based antifungal research focus on the development of advanced, mechanism-based, and precision-oriented approaches. Next-generation probiotics and live biotherapeutic products signify a shift from traditional probiotic strains to precisely characterized, functionally defined microorganisms with improved efficacy and safety. Postbiotics derived from probiotics have emerged as promising antifungal agents due to their stable and targeted antifungal activity. Omics-driven strain characterization has provided a deeper understanding of antifungal mechanisms, host-microbe interactions, and functional potential. Bioengineering of probiotic strains enhances antifungal efficacy, tolerance to environmental stresses, and targeted delivery capabilities. Furthermore, microbiome-tailored personalized probiotics are emerging as precision-based interventions that account for host-specific therapies with improved outcomes. Additionally, technologies such as artificial intelligence are increasingly utilized to optimize probiotic strain selection and

D. B V (✉) · M. Y. Sreenivasa
Applied Mycology Laboratory, Department of Studies in Microbiology, University of Mysore, Mysuru, Karnataka, India
e-mail: deepthibv@microbiology.uni-mysore.ac.in; deepthi1806@gmail.com

K. Poornachandra Rao
Section In charge – Microbiology, Department of Microbiology and Chemistry, United Integrated Laboratories, LLC, Barka, Oman

P. N Achar
Department of Molecular and Cellular Biology, Kennesaw State University, Biology, Kennesaw, GA, USA

M. Y. Sreenivasa et al. (eds.), *Probiotics as Antifungals*,
https://doi.org/10.1007/978-981-92-0002-3_23

delivery mode and to predict active antifungal compounds. These advanced probiotics are designed to enhance survivability, bioavailability, and biodistribution, to provide site-specific targeted therapy and achieve synergistic effects through multi-strain formulations. Collectively, these innovations highlight a transformative shift toward next-generation probiotic antifungal therapeutics with significant translational potential.

Graphical Abstract

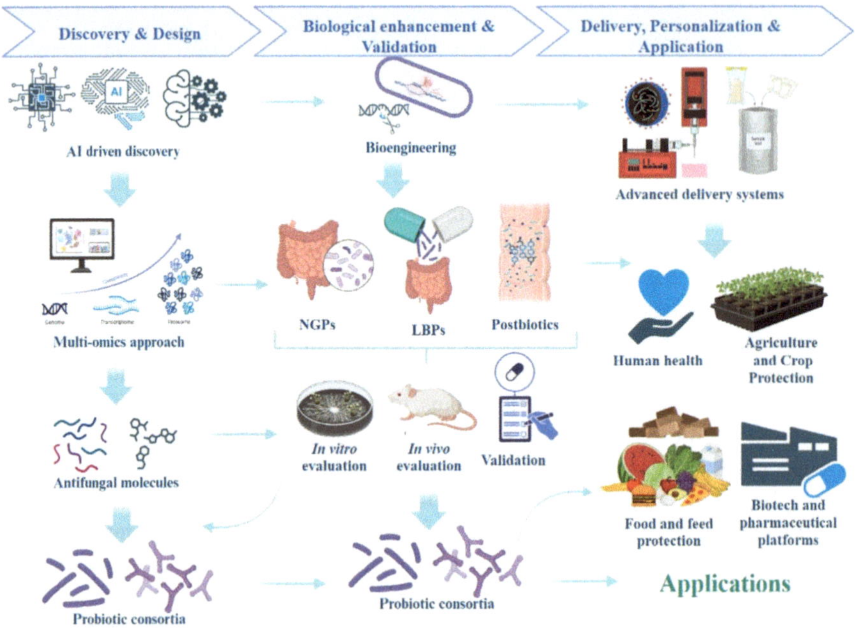

Keywords

Next-gen probiotics · Live biotherapeutic products · Omics · Postbiotics · Personalized probiotics · Artificial intelligence

Learning Objectives

By the end of this chapter, readers will be able

- To evaluate emerging probiotic-based antifungal therapies and their potential clinical and translational applications.

(continued)

- To delineate the rationale and techniques used to enhance the efficacy, stability, and targeted action of probiotics.
- To understand the challenges and limitations associated with advanced probiotic research and explore potential strategies to overcome them.

23.1 Introduction

Probiotics are live microorganisms that are non-pathogenic and confer health benefits to the host (Martins et al. 2005; FAO 2006). Probiotics can be classified into bacterial probiotics, yeast or fungal probiotics, and next-generation probiotics. Probiotic microorganisms have been identified in animals, including the human gut, fermented foods, the intestinal tracts of marine and freshwater fish, and human breast milk (Fontana et al. 2013). The studies on probiotics have been reported over the last few decades. However, in recent years, probiotics have gained immense global importance due to advancements in various areas and techniques of biological sciences. Several studies have demonstrated the role of probiotic interventions in the treatment of various conditions such as chronic gastrointestinal inflammatory disorders, diarrheal diseases, lactose intolerance, allergic responses, metabolic diseases, and cancer (Grover et al. 2012). In addition to these conditions, probiotics have also been explored for their potential against fungal infections (Arora and Baldi 2015). This interest has increased due to the rising incidence of fungal infections and systemic mycoses worldwide. Pathogenic fungi such as *Candida*, *Aspergillus, Cryptococcus*, and dermatophytes pose significant threats to human health, while plant and environmental fungi, including *Rhizoctonia*, *Botrytis*, *Phytophthora*, *Alternaria*, *Aspergillus*, *Penicillium*, and *Fusarium*, cause substantial agricultural and economic losses. The alarming rise of multidrug-resistant fungal pathogens, exemplified by *Candida auris*, further underscores the urgent need for safer and more sustainable antifungal alternatives.

In this context, probiotics have emerged as promising biotherapeutic agents having antifungal potential with minimal or no side effects. A wide range of probiotic microorganisms, including lactic acid bacteria such as *Lactobacillus rhamnosus*, *L. plantarum*, *L. reuteri*, as well as *Bacillus coagulans* (presently named *Heyndrickxia coagulans*), *Bifidobacterium* species, have demonstrated antifungal activity through multiple mechanisms. Additionally, next-generation probiotics such as *Akkermansia muciniphila* and *Faecalibacterium prausnitzii* represent a paradigm shift toward functionally defined and mechanistically characterized microbial therapeutics. The antifungal mechanisms of probiotics are multifaceted and include competitive exclusion of harmful microbiota in the host through colonization resistance, production of antifungal metabolites, modulation of host immune responses, and interference with quorum-sensing pathways. Advances in omics technologies, bioengineering approaches, and artificial intelligence-driven

Table 23.1 Advancements in probiotic research using advanced techniques

Techniques	Advancement in probiotics	Reference
Bioinformatics and genetic sequencing	Next-generation probiotics	O'Toole et al. (2017)
Genetic engineering	Live biotherapeutic products	Singh and Natraj (2021)
Metabolites derived from probiotics	Postbiotics	Thorakkattu et al. (2022)
Omics	Omics-based strain characterization	Verma et al. (2025)
Microbiome tailoring	Personalized probiotic therapy	Duhan et al. (2025)
Artificial intelligence	Discovery of probiotic metabolites	Jo et al. (2025)
Encapsulation—micro and nanotechnology	Advanced delivery system	Cheng et al. (2025)
Multi-strain synergic effect	Probiotic-consortia	Boranbayeva et al. (2025)

strain selection have further enhanced our understanding of probiotic-fungal interactions and facilitated the development of precision-oriented, next-generation probiotic formulations. These probiotics can be used as natural products, medical foods, dietary supplements, drugs, live biotherapeutic agents, and functional foods (Arora and Baldi 2015).

Beyond human therapeutics, bacterial probiotics have gained considerable importance in agriculture, veterinary medicine, aquaculture, food preservation, and industrial applications. Their versatility, safety profile, and multifunctional attributes position probiotics as promising alternatives to chemical antifungals. Collectively, these advancements highlight a transformative shift toward probiotic-based antifungal strategies with significant translational and commercial potential. Recent advancements in probiotic biology achieved using advanced and high-throughput techniques are summarized in Table 23.1.

23.2 Next-Generation Probiotics (NGPs)

Next-generation probiotics (NGPs) represent one of the most advanced developments in probiotic research. Traditionally, probiotics have been primarily used to maintain intestinal health through the administration of selected bacterial or yeast species. However, with the advent of advanced technologies such as bioinformatics and high-throughput genetic sequencing, probiotic research has evolved toward new horizons, including the development of NGPs and live biotherapeutic products (LBPs) (O'Toole et al. 2017; Singh and Natraj 2021). NGPs differ from traditional probiotics, which are mainly used to support gut health by modulating beneficial microbiota. In contrast, NGPs include microorganisms specifically designed to target particular diseases, based on comparative analyses of the microbiota in healthy and diseased individuals (O'Toole et al. 2017). When administered in appropriate quantities, NGPs exert beneficial health effects on the host (Martin and Langella 2019). Several NGPs have been extensively studied and show promising therapeutic potential, including *Faecalibacterium prausnitzii, Akkermansia muciniphila, Eubacterium hallii, Bacteroides fragilis, Oscillospira* species, and *Roseburia* species, in the management of various diseases (Yang et al. 2021; Cani et al. 2022; Al-Fakhrany and Elekhnawy 2024).

NGPs are an advanced alternative to classical probiotics and differ in several key aspects. Notably, NGPs are designed using genetic sequencing and bioinformatics approaches to target specific diseases; they encompass a wider range of microbial species with precisely defined mechanisms of action; and they are subject to stringent regulatory frameworks governing safety, toxicology, and efficacy, which require validation through preclinical and clinical studies (Arora and Baldi 2015; Singh and Natraj 2021; Zhang et al. 2022). Consequently, NGPs are increasingly recognized as promising biotherapeutic tools.

The development of NGPs for specific diseases involves several critical steps. Initially, host-microbiota interactions and disease-associated microbial signatures are identified, followed by screening and isolation of NGP strains using next-generation sequencing-based approaches. These NGPs are subsequently evaluated for safety and efficacy through in vitro and in vivo models, and their mechanisms of action are systematically elucidated. Despite their beneficial attribute, the isolation and production of NGPS present several challenges as discussed below.

(i) *Selection of appropriate microbial strain*: The strain selected should be safe, exhibit health-beneficial properties and should come under the human consumptive category (Stanton et al. 2003).
(ii) *Survivability of the NGPs*: NGPs should be able to survive harsh conditions both externally during storage and internally within the microenvironments of the stomach and small intestine (Fuller 2012).
(iii) *Production of NGPs on a large commercial scale*: Large-scale production of NGPs presents greater challenges compared to small-scale laboratory cultivation (Figueroa-Gonzalez et al. 2011; Foligne et al. 2013).
(iv) *Shelf life and stability*: NGPs must maintain their viability and efficacy during storage (Rokka and Rantamäki 2010; Ranjan et al. 2022).

To overcome the challenges associated with NGPs, emphasis should be placed on the establishment of robust safety assessment methods and the development of human-relevant models using Artificial Intelligence (AI)-based tools to confirm the efficacy of developed NGPs (Carlino et al. 2024). In addition, encapsulation technologies can protect probiotics from external stress factors and enhance their stability, thereby ensuring targeted delivery (Gullifa et al. 2023; More et al. 2023).

Despite NGPs being a relatively recent advancement and facing several challenges, numerous studies have reported the efficacy of various NGPs in the management of diverse disease conditions. These include gastrointestinal disorders such as inflammatory bowel disease (Lee et al. 2017; Gao et al. 2017; Martoni et al.

2020; Bonfrate et al. 2020; Leylabadlo et al. 2022; Torp et al. 2022; Mazziotta et al. 2023; Hizay et al. 2023), cancer (Yu and Li 2016), skin diseases (Torii et al. 2011; Kim et al. 2010; Gueniche et al. 2010; Gudadappanavar et al. 2017), cardiovascular diseases (Hijova 2023; Chang et al. 2019), neuropsychological disorders (Akkasheh et al. 2016; Yong et al. 2020), infectious diseases (Gautier et al. 2021; Jiang et al. 2021; Qi et al. 2021; Mazziotta et al. 2023), and metabolic disorders including obesity (Kim et al. 2013; Ritze et al. 2014; Palade et al. 2022; Jiang et al. 2022; Kim et al. 2022).

Further, omics-based studies have revealed that patients with non-alcoholic fatty liver disease (NAFLD) exhibit reduced levels of hyodeoxycholic acid. Administration of hyodeoxycholic acid has been shown to improve NAFLD by increasing the abundance of *Parabacteroides distasonis* and modulating the gut-liver axis in high-fat-diet fed mice (Inc et al. 2023). Similarly, individuals with depression have been reported to show decreased levels of homovanillic acid, and supplementation with *Bifidobacterium longum* increased homovanillic acid levels, thereby alleviating depression-like symptoms in mice exposed to unpredictable stress and corticosterone-induced depression (Zhao et al. 2024).

Although several studies demonstrate the therapeutic potential of NGPs across multiple disease conditions, validation of these findings in human clinical trials remains limited (Al-Fakhrany and Elekhnawy 2024). In addition, certain NGPs, including some *Lactobacillus* and *Bifidobacterium* species, are not universally categorized as Generally Regarded as Safe (GRAS), highlighting the need for comprehensive safety evaluation prior to clinical application (Franklin and Ericsson 2017; Merenstein et al. 2023).

23.3 Live Biotherapeutic Products (LBPs)

As discussed earlier in this chapter, probiotic research has witnessed the emergence of several new technologies, among which live biotherapeutic products (LBPs) represent a significant advancement (Steidler 2003). LBPs are microorganisms or biological products that function in a manner similar to next-generation probiotics (Singh and Natraj 2021). Unlike conventional therapeutic approaches, LBPs are living entities that actively grow and interact with their host systems (Vargason and Anselmo 2021). According to the US Food and Drug Administration (FDA), LBPs are defined as live microorganisms, either bacteria or yeast, excluding filterable viruses and oncolytic bacteria, that are intended for the prevention or treatment of disease in humans, are not administered by injection, and are not classified as vaccines (FDA 2016). Although NGPs are sometimes considered a subset of LBPs, regulatory authorities distinguish between the two and recommend that they be classified separately (FDA 2016; O'Toole et al. 2017).

LBPs may be formulated as either a single microbial strain or as multi-strain microbial consortia. Their therapeutic effects may arise from inherent functional properties or through genetic engineering-based modifications (O'Toole et al. 2017). The FDA formally established the LBP category and issued regulatory guidelines in

2012, while the European Pharmacopoeia defined quality requirements for LBPs as drug products in 2019. However, in both Europe and the United States, LBPs must be registered as medicinal products prior to market authorization (Cordaillat-Simmons et al. 2020).

LBPs exhibit a wide range of beneficial activities, including nutrient and toxin metabolism, regulation of the mucosal immune system, functioning as living diagnostics, and in situ production of therapeutic molecules (Riglar and Silver 2018; Charbonneau et al. 2020). Recurrent *Clostridium difficile* infection has been treated using both single-strain and consortium-based LBPs in clinical trials (Blount et al. 2019; Garber 2020). Additionally, more than 50% of LBPs evaluated in clinical studies, often in combination with immunomodulators, are targeted toward cancer therapy (Wang et al. 2021).

Despite their therapeutic potential, LBPs face multifactorial challenges during oral administration and gastrointestinal transit. Limitations in clinical trials include the lack of quantitative approaches to characterize LBP components and the absence of effective strategies for controlled delivery to the gut (Heavey et al. 2022). However, these challenges can be addressed through the identification of genetic and functional components of LBPs using omics-based approaches, as well as through advanced pharmaceutical formulations and genetically engineered delivery strategies (Heavey et al. 2022).

23.4 Probiotic-Derived Postbiotics as Antifungals

In addition to probiotics and next-generation probiotics (NGPs), which confer health benefits primarily by improving disease conditions through modulation of the gut microbiome, researchers have also reported similar beneficial roles for postbiotics. Postbiotics are metabolites or structural components derived from or produced by probiotic microorganisms. Compared to live probiotics, postbiotics are less sensitive to environmental factors such as pH, temperature, and processing conditions. These properties contribute to their enhanced stability and prolonged health-promoting effects.

The composition of postbiotics can be highly variable and complex, as they often consist of a mixture of metabolites that exert functions through multiple mechanisms and biological pathways (Liu et al. 2023). Furthermore, future research should consider factors that influence the efficacy of postbiotics in human health, as multiple host and environment-related parameters play critical roles in determining their functional outcomes. These factors include physiological status, diet, lifestyle, age, environmental exposure, sex, and host species (Zhao et al. 2024).

Well-studied postbiotics include short-chain fatty acids (SCFA), lactic acid, exopolysaccharides (EPS), pyranose derivatives, cell wall components, enzymes, volatile compounds, cell-free supernatants (CFSs), and other metabolites isolated from various probiotic microorganisms. Extensive studies have demonstrated the diverse biological efficacies of these postbiotics. For instance, EPS exhibit anti-inflammatory, antitumor, antioxidant, immunomodulatory, antihypertensive, antibacterial, and gastrointestinal health-promoting activities (Garai-ibabe et al. 2010; Makino et al. 2016; Wang et al. 2018; Rajoka et al. 2020; Thorakkattu et al. 2022). SCFAs have been shown to improve poultry productivity, regulate carbohydrate metabolism and cholesterol synthesis, exert anti-inflammatory effects, and promote immune cell differentiation (Kasahara et al. 2018; Li et al. 2021; Xie et al. 2022; Rodrigues et al. 2023). Microbial enzymes contribute to host nutrition by facilitating digestion and also exhibit antioxidant activity (Kullisaar et al. 2002; Izuddin et al. 2020). Cell-free supernatants stimulate host immune responses, exhibit antimicrobial activity, alleviate irritable bowel syndrome and type 2 diabetes, and influence immune cell production (Scillato et al. 2021; Sornsenee et al. 2021; Lee et al. 2022a, b). In addition, cell wall components/fragments exert anti-inflammatory effects and modulate the immune system (Jensen et al. 2010; Roselli et al. 2016; Matsuzaki et al. 2022), while bacterial lysates activate both T and B lymphocytes, reduce the risk of respiratory infection, and improve immune function and gut health (Navarro et al. 2011; Suarez et al. 2020; Taube 2022).

Several studies have also reported the antifungal and bioprotective potential of postbiotics. Postbiotics derived from *Lactobacillus* species act as natural preservatives by inhibiting the growth of both spoilage and pathogenic microorganisms (Deepthi et al. 2016; Poornachandra Rao et al. 2019; Kaya et al. 2024; Vanitha et al. 2024; Adithi et al. 2025). In a study by Kaya et al. (2024), the incorporation of 3% *Lactiplantibacillus plantarum* Y48 postbiotic in combination with 0.1% potassium sorbate into bread effectively inhibited the growth of *Aspergillus niger* and *Penicillium expansum*. Similarly, *L. plantarum* postbiotics have demonstrated antifungal activity against mycotoxigenic *Fusarium* sp. (Deepthi et al. 2016, 2017) and against *Candida* species, including *Candida albicans* and *Candida kefyr*, members of the human gut microbiota, by significantly reducing fungal growth (Garcia-Gamboa et al. 2024). Furthermore, antifungal postbiotics produced by bacterial probiotics also include biosurfactants and bacteriocins (Peleg et al. 2010).

23.5 Omics-Driven Strain Characterization

Omics refers to the comprehensive analysis of large-scale biological data related to the structure, function, and dynamics of biological systems at specific molecular levels (Verma et al. 2025). In recent years, omics technologies have evolved significantly and are now widely applied across various fields of biological research. These technologies have also substantially influenced probiotic research by enabling the investigation of multiple aspects of probiotic biology and functionality. The major omics approaches include genomics, transcriptomics, proteomics, and metabolomics (Dai and Shen 2022).

Genomics involves the study of the complete set of genes within a probiotic organism and aids in identifying genes associated with important traits such as environmental tolerance, survivability, and safety (Verma et al. 2025). Transcriptomics focuses on the analysis of RNA expression profiles and provides insights into gene regulation and host-probiotic interactions, particularly in relation to host immune responses (Milward et al. 2016; Verma et al. 2025). Proteomics enables the identification and characterization of proteins involved in microbial interactions and in the adaptation of probiotic microorganisms to external and host gut environment (Verma et al. 2025). Finally, metabolomics reveals bioactive compounds produced by probiotics and elucidates their roles in host health and disease modulation (Kwoji et al. 2023).

To cite a few examples of probiotic strain characterization using multi-omics technologies, whole-genome sequencing of *Limisilactobacillus reuteri* PNW1 revealed genes responsible for the production of bioactive peptides, contributing to the determination of probiotic potency (Alayande et al. 2020). Furthermore, using transcriptomic analysis, a novel probiotic strain isolated from the gut of the tribal population of Odisha was characterized, identified, and designated as *Lactiplantibacillus plantarum* strain ILSF15 (Arjun et al. 2024). Similarly, a metabolomic study on *Lactobacillus plantarum* MLK 14-2 and *L. plantarum* KCCM 11322 elucidated strain-specific changes in metabolite profiles resulting from fermentation (Park et al. 2016). A comprehensive multi-omics investigation integrating metagenomics, metabolomics, transcriptomics, and proteomics on *Lacticaseibacillus casei* Shirota in a mouse model of the gut-liver axis revealed diverse host-probiotic interactions and functional pathways involved in health modulation (Lv et al. 2024).

Although multi-omics approaches offer significant advantages in probiotic research, a few limitations must be considered. For instance, batch effects can introduce variability due to differences in data generated across laboratories, experimental conditions, instrumentation, and analytical methodologies (Verma et al. 2025).

23.6 Bioengineering of Probiotic Strains

As probiotics are gaining immense importance in the biomedical field for the treatment of various diseases, probiotic research has witnessed several advancements, including the bioengineering of probiotic strains to overcome the limitations of conventional probiotic approaches (Hua et al. 2024). Engineered probiotics often outperform traditional probiotics by offering enhanced functionality while minimizing the risk of pathogenicity associated with the microbial strains being modified (Mathipa and Thantsha 2017). Compared to conventional probiotics, engineered probiotics provide novel therapeutic prospects by regulating metabolic functions, improving gut colonization, and enabling targeted delivery of bioactive molecules (Duhan et al. 2025). The first genetically engineered live biotherapeutic product (LBP) involved *Lactococcus lactis* engineered to secrete interleukin-10 for the treatment of intestinal inflammation in animal models and subsequent evaluation in clinical subjects (Steidler et al. 2000, 2003; Deepa et al. 2016; Braat et al. 2006).

Probiotics can be biologically engineered using advanced biotechnological tools such as AI-driven microbiome analysis, synthetic biology, personalized probiotic design, and CRISPR-based genome editing (Makhkamova 2024). Several genetically engineered beneficial bacteria with reported therapeutic efficacy include *Streptococcus gordonii*, *Clostridium sporogenes*, *Salmonella* Typhimurium, *Bacillus subtilits*, *Listeria monocytogenes*, *Bacteroides ovatus*, and *Escherichia coli* BL21 (DE3) (Riglar and Silver 2018).

Bioengineered microbes or probiotics exhibit enhanced stress tolerance compared to traditional probiotics, which often display limited resistance to environmental stresses (Desmond et al. 2004). Engineered probiotics have also been explored for the detection of liver cancer metastasis (Danino et al. 2015). Further, *Lactobacillus* strains have been genetically engineered to secrete glucagon-like peptide-1 (GLP-1), which acts as an antihyperglycemic agent in diabetic rat models (Nie et al. 2022). In addition, engineered probiotics have recently been developed as potential therapeutic tools against SARS-CoV-2 infection by targeting angiotensin-converting enzyme 2 (ACE2) receptors or by secreting ACE-2, thereby preventing viral entry into host cells and reducing infection (Suresh et al. 2021).

Although engineered probiotics exhibit several beneficial properties for human health, careful consideration must be given to biosafety concerns related to both the host and the genetically modified organisms (Debnath et al. 2022). Regulatory aspects also require close attention, as any bioengineered probiotic must obtain appropriate regulatory approval prior to commercial release (Jain et al. 2024). Detailed information regarding health benefits, potential adverse effects, and the consequences of long-term use must be thoroughly evaluated and documented (Azadi and Ho 2010). In addition to biosafety and regulatory challenges, engineered cells face several technical limitations, including dependence on transformation efficiency, limited host range, non-seamless genome editing, and the potential for genetic instability or unintended mutations (Zuo et al. 2019).

23.7 Microbiome-Tailored Personalized Probiotic Therapies

With advancements in probiotic research, a novel therapeutic strategy, personalized probiotic therapy has evolved. Personalized probiotic therapy involves the design of probiotic interventions based on an individual's gut microbiome composition (Duhan et al. 2025). Administration of an adequate dosage of personalized probiotic formulations is a critical determinant of their health benefits (Kashyap et al. 2017). The increasing number of microbiome studies has facilitated the development of tailored fermented foods, such as kefir and yogurt, supplemented with specific probiotic strains selected according to an individual's gut microbiome profile (Hemmati et al. 2024). Evidence from the literature highlights the significant role of personalized probiotic therapy in the management of metabolic syndromes and inflammatory bowel diseases by correcting dysbiosis and restoring microbial balance (Manan 2025). Indeed, based on individual deficiencies in specific gut microbial populations, customized multi-strain probiotic formulations have been developed to treat conditions such as diabetes, cardiovascular diseases, and IBS (Hemmati et al. 2024; Duhan et al. 2025). Further, multi-omics technologies are increasingly employed to enable and optimize personalized-probiotic therapies (Lagoumintzis and Patrinos 2023).

Personalized probiotic therapy provides health benefits through targeted treatment strategies. Personalized probiotic formulations developed for conditions such as diarrhea and constipation have been shown to alleviate symptoms by enhancing short-chain fatty acids-producing microbial families (Min et al. 2024). Other studies on personalized probiotic therapy have reported antifungal efficacy against *Trichophyton tonsurans* using a probiotic formulation containing *Lactiplantibacillus plantarum* MYS N7 (Vanitha et al. 2023), and anti-inflammatory effects on colon cancer cells using *Lactobacillus* spp. cocktail (Aghamohammad et al. 2022), and protective effects against upper respiratory tract infections using probiotics supplemented with *Lacticaseibacillus casei* (Hor et al. 2018). Additionally, several clinical case studies supporting the efficacy of personalized probiotic therapies have been reviewed by Manan (2025).

As with any emerging therapeutic approach, microbiome-tailored personalized probiotic therapy faces several challenges that limit its widespread application (Abouelela and Helmy 2024). These challenges include safety concerns, social implications, ethical considerations, and the need for robust regulatory frameworks to ensure efficacy and patient safety (Duhan et al. 2025).

23.8 Artificial Intelligence (AI)-Driven Discovery of Probiotic Antifungal Metabolites

With the expanding knowledge of the role of probiotics in human and animal health, the demand for precision probiotics and targeted probiotic therapies is increasing rapidly. However, the identification of specific strains, comprehensive probiotic characterization, and the design of personalized probiotic medicines are time-consuming and labor-intensive processes. To meet this growing demand and to overcome the limitations of conventional, manually designed probiotics, there is an alternative approach that enables systematic screening, selection, optimization, and rational design of probiotic candidates using data-driven strategies. One such promising approach is artificial intelligence (AI) technology, which can efficiently predict probiotic strains, elucidate host–probiotic interactions, and assist in identifying functional metabolites relevant to specific disease conditions (Jo et al. 2025).

The application of AI in probiotic research includes machine learning (ML), deep learning (DL), and hybrid modeling approaches to analyze complex biological datasets, such as omics data (genomics, transcriptomics, proteomics, metabolomics) and microbiome profiles (Patil et al. 2025). Common ML and DL methods employed include Bayesian networks, artificial neural networks, support vector machines, convolutional neural networks, and random forest algorithms (Khanal et al. 2025). These AI-based tools facilitate the identification of microbial strains, prediction of host-probiotic interactions, and functional profiling of probiotic candidates. AI-driven antifungal potency prediction against *Ganoderma* species for food industry applications has been reported by Khanal et al. (2025). Moreover, the integration of AI is contributing significantly to the accelerated discovery and development of antifungal therapies/drugs (Li et al. 2025).

Despite its numerous advantages, the application of AI in probiotic research is associated with technical challenges and ethical concerns that must be addressed to fully realize its potential. A major limitation is the quality, completeness, and standardization of biological datasets used to train AI tools or models (Orkkatteri Krishnan et al. 2025). Additional challenges include variability in experimental designs, differences in sequencing platforms, poor data interoperability, and limited model generalizability, which can restrict the broader applicability of AI-based predictions.

23.9 Encapsulation and Advanced Delivery Systems

Targeted delivery remains one of the major challenges in probiotic research. This is primarily because probiotics often exhibit poor biodistribution, limited stability, reduced effectiveness, and low survivability in a harsh gastrointestinal environment (Wilczewska et al. 2012; Cheng et al. 2025). As with other limitations in probiotic research, this challenge has been addressed through probiotic modification strategies, particularly by reducing the particle size. Nanotechnology-based approaches, such as micro- and nano-capsulation, have emerged as promising solutions to overcome these constraints (Zhang and Uludag 2009). Several studies have demonstrated that nano-encapsulated probiotics exhibit enhanced stability, improved bioavailability, and increased protection against adverse gastrointestinal conditions, including gastric acidity and digestive enzymes, with minimal or no associated side effects (Mohamadzadeh et al. 2024).

For the delivery system to be effective, critical parameters extend beyond the selection of the carrier material and its interaction with the probiotic or therapeutic agent. These parameters also include the impact of the delivery system on key probiotic properties such as absorption, solubility, bioavailability, and intestinal permeability (Yerlikaya et al. 2021). The efficiency and survivability of encapsulated probiotics are influenced by multiple factors, including exposure to oxygen, low pH conditions, the nature of the encapsulating material, the microbial strain being encapsulated, digestive enzymes, heat treatment, and the encapsulation technique employed (Martin et al. 2015; Sipailiene and Petraityte 2018). Commonly used delivery material for probiotic encapsulation includes cellulose, polyvinyl alcohol-based nanofibers, polysaccharides (Sinha and Kumria 2001), chitosan, and calcium alginate (Ceylan et al. 2019; Lang et al. 2023; Wang et al. 2025).

Several examples of probiotic encapsulation with notable advantages have been reported. *Lactobacillus plantarum* CICC 6240 was encapsulated into spherical microgels using cellulose, which enabled the controlled release of the probiotic in the intestine (Li et al. 2016). Gum Arabic, a polysaccharide composed of glucuronic acid, rhamnose, arabinose, and galactose, has also been widely used as an encapsulating agent. For example, a gum arabic-based solid lipid microparticle was developed to encapsulate *L. plantarum* NRRL B-4496, which prevented cell inactivation in the host gut, thereby protecting the probiotic from both external and internal environmental stresses (Azeem et al. 2023). In addition, starch and its derivatives have been employed as encapsulation materials to protect probiotics such as *Lactobacillus brevis* MTCC 01, *Lactobacillus plantarum* MTCC 021, and *Lactobacillus casei* MTCC 297 throughout the storage period. Encapsulation using starch-based materials significantly enhanced probiotic survivability in simulated intestinal fluid (Ashwar et al. 2018). The commonly used techniques for probiotic encapsulation include extrusion (Zhou et al. 2017), co-extrusion (Mei et al. 2014), emulsion-based methods (Zhang et al. 2025), and spray drying (Zaeim et al. 2019). In addition to these established techniques, newer encapsulation strategies are being developed to achieve more precise and efficient delivery systems for probiotics (Wang et al. 2025).

23.10 Multi-probiotic Consortia

Earlier sections have focused on single-strain microorganisms or their derived products as probiotic agents capable of modulating the gut microbiome and exerting therapeutic effects. Recent research has shifted focus toward multi-strain probiotic consortia, as they are expected to confer broader therapeutic benefits due to the synergistic interactions among multiple microbial strains that target diverse pathways involved in disease conditions (Boranbayeva et al. 2025). The synergistic effects of multi-strain consortia enable enhanced stability, improved efficacy, and a broader spectrum of biological activities. These properties are particularly crucial for effectively addressing the complex interactions among the host, the environment, and pathogenic microorganisms (Boranbayeva et al. 2025). Further, multi-strain consortia may consistently exhibit health-beneficial effects due to their greater resilience to environmental stresses.

Further, lactic acid bacteria synergistically enhance efficacy by individually targeting specific immune mechanisms or virulence factors, thereby protecting the host against pathogens. For instance, the LAB consortium KMK3, consisting of *Lactobacillus brevis* (TN 9), *Ligilactobacillus salivarius* (F9/2), and *Lactobacillus salivarius* (TL4/1), isolated from the poultry gut, was shown to improve poultry health and combat infectious diseases (Boranbayeva et al. 2025). Similarly, the probiotic consortium composed of EcNPAQ, consisting of *Escherichia coli* Nissle 1917 derivatives (EcNLA and EcNPD), displays antitumor properties by suppressing the tumor progression. This consortium integrates multiple tumor microenvironment signals, including hypoxia, elevated lactate levels, and pH variations, to achieve targeted therapeutic effects (Guo et al. 2025).

In marine ecosystems, corals maintain symbiotic associations with diverse microorganisms, including archaea, dinoflagellates, bacteria, viruses, and fungi. Environmental stresses such as oil spills can disrupt these symbiotic relationships (Rohwer et al. 2002; Rosenberg et al. 2007; Peixoto et al. 2017); however, the application of multi-domain microbial consortia has been proposed as a strategy to restore coral health. Additionally, Lee et al. (2024) reported that the bacterial consortium MPRO, comprising *Lactiplantibacillus plantarum* Hy7712, *Lacticaseibacillus casei* HY2782, and *Bifidobacterium animalis* subsp. *lactis* HY8002 demonstrated significantly higher anti-inflammatory activity than the corresponding single strains in an experimental mouse model.

Overall, these studies underscore the enhanced therapeutic potential of multi-strain and multi-domain probiotic consortia due to their synergistic interactions and broader functional capabilities. Rational design and validation of such consortia may enable more effective and resilient probiotic-based interventions across diverse applications.

23.11 Conclusion

The evolving probiotic research demonstrates a clear shift toward advanced, mechanism-driven, and precision-based therapeutic strategies with applications extending beyond human health to veterinary, agricultural, and environmental sectors. Although these innovative probiotic approaches show significant potential, several challenges remain, including safety validation, regulatory standardization, scalability, and clinical translation. Addressing these challenges through interdisciplinary research and robust clinical evaluation will be critical to fully realizing the therapeutic potential of advanced probiotic technologies in the future.

23.12 Chapter Summary

This chapter provides a comprehensive overview of recent advances in probiotic research aimed at overcoming the limitations of traditional probiotic approaches. The integration of high-end technologies such as genomics, transcriptomics, proteomics, metabolomics, artificial intelligence, and bioengineering has significantly expanded the scope of probiotic science. These technological advancements have facilitated the development of next-generation probiotics, live biotherapeutic products, engineered probiotic strains, AI-driven discovery of antifungal probiotics, omics-based strain characterization, advanced encapsulation strategies, and multi-strain probiotic consortia. Collectively, these emerging approaches enhance targeted therapy, improve strain selection and functionality, increase survivability under harsh gastrointestinal conditions, and enable precision-oriented probiotic interventions.

Acknowledgments The authors thank the DST-WOS-A (LS-397/2021) project for the funding to carry out probiotic antifungal research against mycotoxigenic fungal contaminants in feeds. BioRender was used to create the graphical abstract. OpenAI tools were used to assist with language editing.

Conflict of Interest The authors declare that there are no conflicts of interest regarding the publication of this chapter.

Author Contributions DBV—conceptualization, writing original draft, review, editing; KPR and PNA—review and editing; MYS—conceptualization, review, and editing.

References

Abouelela ME, Helmy YA (2024) Next-generation probiotics and novel therapeutics for improving human health: current trends and future perspectives. Microorganisms 12(3):430

Adithi G, Rakesh S, Shruthi B, Divyashree S, Deepa N, Vanitha PR, Vasundaradevi R, Deepthi BV, Sreenivasa MY (2025) Biological decontamination of myco-toxigenic *Fusarium verticillioides* using *Lactiplantibacillus plantarum* MYSN128 isolated from traditional vellappam: implications for food safety and global mycotoxin management. Curr Microbiol 82:404

Aghamohammad S, Sepehr A, Miri SY, Najafi S, Pourshafie MR, Rohani M et al (2022) Anti-inflammatory and immunomodulatory effects of *Lactobacillus* species as a preservative and therapeutic agent for IBD control. Immun Inflamm Dis 10:e635

Akkasheh G, Kashani-Poor Z, Tajabadi-Ebrahimi M, Jafari P, Akbari H, Taghizadeh M et al (2016) Clinical and metabolic response to probiotic administration in patients with major depressive disorder: a randomized, double-blind, placebo-controlled trial. Nutrition 32:315–320

Alayande KA, Aiyegoro OA, Nengwekhulu TM, Katata-Seru L, Ateba CN (2020) Integrated genome-based probiotic relevance and safety evaluation of *Lactobacillus reuteri* PNW1. PLoS One 15:e0235873

Al-Fakhrany OM, Elekhnawy E (2024) Next-generation probiotics: the upcoming biotherapeutics. Mol Biol Rep 51:505

Arjun OK, Sethi M, Parida D, Dash J, Das SK, Prakash T et al (2024) Comprehensive physiological and genomic characterization of a potential probiotic strain, *Lactiplantibacillus plantarum* ILSF15, isolated from the gut of tribes of Odisha, India. Gene 931:148882

Arora M, Baldi A (2015) Regulatory categories of probiotics across the globe: a review representing existing and recommended categorization. Indian J Med Microbiol 33:S2–S10

Ashwar BA, Gani A, Gani A, Shah A, Masoodi FA (2018) Production of RS4 from rice starch and its utilization as an encapsulating agent for targeted delivery of probiotics. Food Chem 238:287–294

Azadi H, Ho P (2010) Genetically modified and organic crops in developing countries: a review of options for food security. Biotechnol Adv 28(1):160–168

Azeem M, Saeed F, Afzaal M, Ateeq H, Ahmad A, Liaqat A et al (2023) Encapsulation of probiotics in solid lipid microparticles for improved viability and stability under stressed conditions. Int J Food Prop 26:1612–1623

Blount KF, Shannon WD, Deych E, Jones C (2019) Restoration of bacterial microbiome composition and diversity among treatment responders in a phase-2 trial of rBX2660: an investigational microbiome restoration therapeutic. Open Forum Infect Dis 6:ofz095

Bonfrate L, Di Palo DM, Celano G, Albert A, Vitellio P, De Angelis M et al (2020) Effects of *Bifidobacterium longum* BB536 and *Lactobacillus rhamnosus* HN001 in IBS patients. Eur J Clin Investig 50:e13201

Boranbayeva G, Tekebayeva Z, Temirkhanov A, Temirbekova A, Yevneyeva D, Abilkhadirov A et al (2025) Probiotic consortium from poultry strains for supporting gut immunity against pathogens. Microb Pathog 204:107584

Braat H, Rottiers P, Hommes DW, Huyghebaert N, Remaut E, Remon JP et al (2006) A phase I trial with transgenic bacteria expressing interleukin-10 in Crohn's disease. Clin Gastroenterol Hepatol 5(6):754–759

Cani PD, Depommier C, Derrien M, Everard A, De Vos WM (2022) *Akkermansia muciniphila*: paradigm for next-generation beneficial microorganisms. Nat Rev Gastroenterol Hepatol 19: 625–637

Carlino N, Blanco-Miguez A, Puncochar M, Mengoni C, Pinto F, Tatti A et al (2024) Unexplored microbial diversity from 2500 food metagenomes and links with the human microbiome. Cell 187:5775–5795

Ceylan Z, Uslu E, Ispirli H, Meral R, Gavgali M (2019) A novel perspective for *Lactobacillus reuteri*: Nanoencapsulation to obtain functional fish fillets. LWT 115:108427

Chang CJ, Lin TL, Tsai YL, Wu TR, Lai WF, Lu CC et al (2019) Next generation probiotics in disease amelioration. J Food Drug Anal 27:615–622

Charbonneau MR, Isabella VM, Li N, Kurtz CB (2020) Developing a new class of engineered live bacterial therapeutics to treat human diseases. Nat Commun 11:1738

Cheng J, Li X, Ke T, Gong J, Jiang T, Chen L (2025) Advances in polysaccharide-based probiotic delivery systems: a review. Carbohydr Polym Technol Appl 10:100804

Cordaillat-Simmons M, Rouanet A, Pot B (2020) Live biotherapeutic products: the importance of a defined regulatory framework. Exp Mol Med 52:1397–1406

Dai X, Shen L (2022) Advances and trends in omics technology development. Front Med (Lausanne) 9:911861

Danino T, Prindle A, Kwong GA, Skalak M, Li H, Allen K et al (2015) Programmable probiotics for detection of cancer in urine. Sci Transl Med 7(289):289ra84

Debnath N, Kumar A, Yadav AK (2022) Probiotics as a biotherapeutics for the management and prevention of respiratory tract diseases. Microbiol Immunol 66:227–291

Deepa N, Charith Raj AP, Sreenivasa MY (2016) Nested PCR method for the early detection of fumonisin-producing Fusarium verticillioides in pure cultures, cereal samples and plant parts. Food Biotechnol 30(1):18–29

Deepthi BV, Rao KP, Chennappa G, Naik MK, Chandrashekara KT, Sreenivasa MY (2016) Antifungal attributes of *Lactobacillus plantarum* MYS6 against fumonisin-producing *Fusarium proliferatum* associated with poultry feeds. PLoS One 11:e0155122

Deepthi BV, Rakesh S, Poornachandra Rao K, Deepa N, Dharanesha NK, Girish KS, Sreenivasa MY (2017) *Lactobacillus plantarum* MYS6 ameliorates fumonisin B1-induced hepatorenal damage in broilers. Front Microbiol 8:2317

Desmond C, Fitzgerald GF, Stanton C, Ross RP (2004) Improved stress tolerance *Lactobacillus paracasei* NFBC 338. Appl Environ Microbiol 70(10):5929–5936

Duhan P, Gupta B, Ahmed M, Bansal P (2025) Gut microbiome engineering with probiotics: current trends and future directions. Discov Appl Sci 7:1354

FAO/WHO (2006) Probiotics in food: health and nutritional properties and guidelines for evaluation. FAO Food Nutr Pap 85:1–56

Figueroa-Gonzalez I, Quijano G, Ramirez G, Cruz-Guerrero A (2011) Probiotics and prebiotics: perspectives and challenges. J Sci Food Agric 91:1341–1348

Food and Drug Administration (FDA) (2016) Early clinical trials with live biotherapeutic products: chemistry, manufacturing and control information: guidance for industry

Foligne B, Daniel C, Pot B (2013) Probiotics from research to market: the possibilities, risks and challenges. Curr Opin Microbiol 16:284–292

Fontana L, Bermudez-Brito M, Plaza-Diaz J, Muñoz-Quezada S, Gil A (2013) Sources, isolation, characterization and evaluation of probiotics. Br J Nutr 109:S35–S50

Franklin CL, Ericsson AC (2017) Microbiota and reproducibility of rodent models. Lab Anim 46: 114–122

Fuller R (2012) Probiotics: the scientific basis. Springer, Berlin/Heidelberg, pp 1–398

Gao C, Ganesh BP, Shi Z, Shah RR, Fultz R, Major A et al (2017) Gut microbe-mediated suppression of inflammation-associated colon carcinogenesis by luminal histamine production. Am J Pathol 187:2323–2336

Garai-Ibabe G, Duenas MT, Irastorza A, Sierra-Filardi E, Werning ML, Lopez P et al (2010) Naturally occurring 2-substituted (1,3)-β-D-glucan producing *Lactobacillus suebicus* and

Pediococcus parvulus strains with potential utility in the production of functional foods. Bioresour Technol 101:9254–9263

Garber K (2020) First microbiome-based drug clears phase III in clinical trial turnaround. Nat Rev Drug Discov 19:655–656

Garcia-Gamboa R, Perfecto-Avalos Y, Gonzalez-Garcia J, Alvarez-Calderon M, Gutierrez-Vilchis-A, Garcia-Gonzalez A (2024) In vitro analysis of postbiotic antimicrobial activity against *Candida* species in a minimal synthetic model simulating the gut mycobiota in obesity. Sci Rep 14:16760

Gautier T, David-Le Gall S, Sweidan A, Tamanai-Shacoori Z, Jolivet-Gougeon A, Loreal O et al (2021) Next-generation probiotics and their metabolites in COVID-19. Microorganisms 9:941

Grover S, Rashmi HM, Srivastava AK, Batish VK (2012) Probiotics for human health: new innovations and emerging trends. Gut Pathog 4:15

Gudadappanavar AM, Hombal PR, Timashetti SS, Javali S (2017) Influence of *Lactobacillus acidophilus* and *Lactobacillus plantarum* on wound healing in male Wistar rats. Int J Appl Basic Med Res 7:233

Gueniche A, Benyacoub J, Philippe D, Bastien P, Kusy N, Breton L et al (2010) *Lactobacillus paracasei* CNCM I-2116 (ST11) inhibits substance P-induced skin inflammation and accelerates skin barrier function recovery in vitro. Eur J Dermatol 20:731–737

Gullifa G, Risoluti R, Mazzoni C, Barone L, Papa E, Battistini A et al (2023) Microencapsulation by a spray drying approach to produce innovative probiotics-based products extending the shelf-life in non-refrigerated conditions. Molecules 28:860

Guo Y, Gao M, Wang L, Yuan H, Yin J, Wu J et al (2025) An engineered probiotic consortium based on quorum-sensing for colorectal cancer immunotherapy. Adv Sci 12:e12744

Heavey MK, Durmusoglu D, Crook N, Anselmo A (2022) Discovery and delivery strategies for engineered live biotherapeutic products. Trends Biotechnol 40(3):354–369

Hemmati MA, Monemi M, Asli S, Mohammadi S, Foroozanmehr B, Haghmorad D et al (2024) Using new technologies to analyze gut microbiota and predict cancer risk. Cells 13(23):1987

Hijova E (2023) Benefits of biotics for cardiovascular diseases. Int J Mol Sci 24:6292

Hizay A, Dag K, Oz N, Comak-Gocer EM, Ozbey-Unlu O, Ucak M, Keles-Celik N (2023) *Lactobacillus acidophilus* regulates abnormal serotonin availability in experimental ulcerative colitis. Anaerobe 80:102710

Hor YY, Lew LC, Lau ASY, Ong JS, Chuah LO, Lee YY et al (2018) Probiotic *Lactobacillus casei* Zhang alleviates respiratory, gastrointestinal and RBC abnormality via immune-modulatory, anti-inflammatory and anti-oxidative actions. J Funct Foods 44:235–245

Hua Z, Liu S, Yang G, Hou X, Fang Y (2024) Next generation probiotics: innovations in safety assessments. Curr Opin Food Sci 8:101238

Inc DBE, Kuang J, Wang J, Li Y, Li M, Zhao M et al (2023) Hyodeoxycholic acid alleviates non-alcoholic fatty liver disease through modulating the gut-liver axis. Cell Metab 35:1752–1766

Izuddin WL, Humam AM, Loh TC, Foo HL, Samsudin AA (2020) Dietary postbiotic *Lactobacillus plantarum* improves serum and ruminal antioxidant activity and upregulates hepatic antioxidant enzymes and ruminal barrier function in post-weaning lambs. Antioxidants 9:250

Jain S, Shukla AK, Deepika, Panwar S, Kumar A, Yadav AK et al (2024) Revolutionizing disease treatment through bioengineered probiotics and glucagon-like peptide-1 (GLP-1) based strategies: a path towards effective cures. Food Bioeng 3:280–300

Jensen GS, Benson KF, Carter SG, Endres JR (2010) GanedenBC30 cell wall and metabolites: anti-inflammatory and immune modulating effects in vitro. BMC Immunol 11:15

Jiang S, Cai L, Lv L, Li L (2021) *Pediococcus pentosaceus*, a future additive or probiotic candidate. Microb Cell Factories 20:45

Jiang T, Li Y, Li L, Liang T, Du M, Yang L et al (2022) *Bifidobacterium longum* 070103 fermented milk improves glucose and lipid metabolism disorders by regulating gut microbiota in mice. Nutrients 14:4050

Jo DM, Ko SC, Kim KW, Yang D, Kim JY, Oh GW et al (2025) Artificial intelligence-driven strategies to enhance the application of lactic acid bacteria as functional probiotics: health promotion and optimization for industrial applications. Trends Food Sci Technol 165:105309

Kasahara K, Krautkramer KA, Org E, Romano KA, Kerby RL, Vivas EI et al (2018) Interactions between *Roseburia intestinalis* and diet modulate atherogenesis in a murine model. Nat Microbiol 3(12):1461–1471

Kashyap PC, Chia N, Nelson H, Segal E, Elinav E (2017) Microbiome at the frontier of personalized medicine. Mayo Clin Proc 92(12):1855–1864

Kaya Y, Sengul M, Dertli E (2024) *Lactobacillus* spp. tarafindan uretilen post biyotiklerin gidalarda biyokoruyucu olarak kullanimi: Probiyotiklerden postbiyotiklere gecis. J Instr Sci Technol 14: 1562–1575

Khanal S, Sharma A, Pillai M, Thakur P, Tapwal A, Kumar V et al (2025) Artificial intelligence-driven innovation in *Ganoderma* spp.: potentialities of their bioactive compounds as functional foods. Sustain Food Technol 3:759

Kim JY, Kwon JH, Ahn SH, Lee SI, Han YS, Choi YO et al (2010) Effect of probiotic mix (*Bifidobacterium bifidum*, *Bifidobacterium lactis*, *Lactobacillus acidophilus*) in the primary prevention of eczema: a double-blind, randomized, placebo-controlled trial. Pediatr Allergy Immunol 21:e386–e393

Kim SW, Park KY, Kim B, Kim E, Hyun CK (2013) *Lactobacillus rhamnosus* GG improves insulin sensitivity and reduces adiposity in high-fat diet-fed mice through enhancement of adiponectin production. Biochem Biophys Res Commun 431:258–263

Kim G, Yoon Y, Park JH, Park JW, Noh MG, Kim H et al (2022) Bifidobacterial carbohydrate/nucleoside metabolism enhances oxidative phosphorylation in white adipose tissue to protect against diet-induced obesity. Microbiome 10:188

Kullisaar T, Zilmer M, Mikelsaar M, Vihalemm T, Annuk H, Kairane C et al (2002) Two antioxidative *Lactobacillus* strains as promising probiotics. Int J Food Microbiol 72(3):215–224

Kwoji ID, Aiyegoro OA, Okpeku M, Adeleke MA (2023) 'Multi-omics' data integration: applications in probiotics studies. NPJ Sci Food 7:25

Lagoumintzis G, Patrinos GP (2023) Triangulating nutrigenomics, metabolomics and microbiomics toward personalized nutrition and healthy living. Hum Genomics 17:109

Lang T, Zhu R, Zhu X, Yan W, Li Y, Zhai Y et al (2023) Combining gut microbiota modulation and chemotherapy by capecitabine-loaded prebiotic nanoparticle improves colorectal cancer therapy. Nat Commun 14:11–16

Lee AD, Spiegel BM, Hays RD, Melmed GY, Bolus R, Khanna D et al (2017) Gastrointestinal symptom severity in irritable bowel syndrome, inflammatory bowel disease and the general population. Neurogastroenterol Motil 29:e13003

Lee S, Park HO, Yoo W (2022a) Anti-melanogenic and antioxidant effects of cell-free supernatant from *Lactobacillus gasseri* BNR17. Microorganisms 10(4):788

Lee J, Kim S, Kang CH (2022b) Immunostimulatory activity of lactic acid bacteria cell-free supernatants through the activation of NF-kappaB and MAPK signalling pathways in RAW 264.7 cells. Microorganisms 10(11):2247

Lee C, Kim SW, Verma R, Noh J, Park JC, Park S et al (2024) Probiotic consortium confers synergistic anti-inflammatory effects in inflammatory disorders. Nutrients 16:790

Leylabadlo HE, Heravi FS, Soltani E, Abbasi A, Kafil HS, Parsaei M et al (2022) The role of gut microbiota in the treatment of irritable bowel syndrome. Rev Med Microbiol 33:e89–e104

Li W, Luo XG, Song R, Zhu Y, Li B, Liu SL (2016) Porous cellulose microgel particle: a fascinating host for the encapsulation, protection, and delivery of *Lactobacillus plantarum*. J Agric Food Chem 64:3430–3436

Li Y, Wang X, Ma X, Liu C, Wu J, Sun C (2021) Natural polysaccharides and their derivatives: a promising natural adjuvant for tumor immunotherapy. Front Pharmacol 12:621813

Li Y, Qiao Y, Ma Y, Xue P, Ding C (2025) AI in fungal drug development: opportunities, challenges, and future outlook. Front Cell Infect Microbiol 15:1610743

Liu Y, Ruan B, Jiang H, Le S, Liu Y, Ao X et al (2023) The weight-loss effect of GLP-1 receptor agonists in non-diabetic individuals with overweight or obesity: a systematic review with meta-analysis and trial sequential analysis. Am J Clin Nutr 118:614–626

Lv L, Ren S, Jiang H, Yan R, Chen W, Yan R et al (2024) Oral administration of *Lacticaseibacillus casei* Shirota alleviates acetaminophen-induced liver injury through accelerated acetaminophen metabolism via the liver–gut axis in mice. mSphere 9:e00672–23

Makhkamova D (2024) Current problems, prospects and modern approaches in th efield of microbiology. Sci Innovate 3(D9):182–189

Makino S, Sato A, Goto A, Nakamura M, Ogawa M, Chiba Y et al (2016) Enhanced natural killer cell activation by exopolysaccharides derived from yogurt fermented with *Lactobacillus delbrueckii* ssp. *bulgaricus* OLL1073R-1. J Dairy Sci 99(2):915–923

Manan MA (2025) The role of probiotics in personalized therapeutics: advances in gut microbe-driven interventions. Microbe 8:100497

Martin R, Langella P (2019) Emerging health concepts in the probiotics field: streamlining the definitions. Front Microbiol 10:1047

Martin R, Miquel S, Chain F, Natividad JM, Jury J, Lu J et al (2015) *Faecalibacterium prausnitzii* prevents physiological damages in a chronic low-grade inflammation murine model. BMC Microbiol 15:67

Martins FS, Nardi RM, Arantes RM, Rosa CA, Neves M, Nicoli JR (2005) Screening of yeasts as probiotic based on capacities to colonize the gastrointestinal tract and to protect against enteropathogen challenge in mice. J Gen Appl Microbiol 51(2):83–92

Martoni CJ, Srivastava S, Leyer GJ (2020) *Lactobacillus acidophilus* DDS-1 and *Bifidobacterium lactis* UABla-12 improve abdominal pain severity and symptomology in irritable bowel syndrome: randomized controlled trial. Nutrients 12:363

Mathipa MG, Thantsha MS (2017) Probiotic engineering towards development of robust probiotic strains with enhanced functional properties and for targeted control of enteric pathogens. Gut Pathog 9:28

Matsuzaki C, Shiraishi T, Chiou TY, Nakashima Y, Higashimura Y, Yokota SI et al (2022) Role of lipoteichoic acid from the genus *Apilactobacillus* in inducing a strong IgA response. Appl Environ Microbiol 88(8):e00190–22

Mazziotta C, Tognon M, Martini F, Torreggiani E, Rotondo JC (2023) Probiotics mechanism of action on immune cells and beneficial effects on human health. Cells 12:184

Mei L, He F, Zhou RQ, Wu CD, Liang R, Xie R et al (2014) Novel intestinal-targeted Ca-alginate-based carrier for pH-responsive protection and release of lactic acid bacteria. ACS Appl Mater Interfaces 6:5962–5970

Merenstein D, Pot B, Leyer G, Ouwehand AC, Preidis GA, Elkins CA et al (2023) Emerging issues in probiotic safety: 2003 perspectives. Gut Microbes 15:185034

Milward E, Shahandeh A, Heidari M, Johnstone DM, Daneshi N, Hondermarck H (2016) Transcriptomics. Encycl Cell Biol 4:160–165

Min U, Jin YJ, Jang YJ, Lim J, Kim BY (2024) Personalized probiotic strategy considering bowel habits: impacts on gut microbiota composition and alleviation of gastrointestinal symptoms via Consti-Biome and Sensi-Biome. Front Nutr 11:1302093

Mohamadzadeh M, Fazeli A, Shojaosadati SA (2024) Polysaccharides and proteins-based bionanocomposites for microencapsulation of probiotics to improve stability and viability in the gastrointestinal tract: a review. Int J Biol Macromol 259:129287

More BH, Shirke VP, Bhambure AS, Rajesh R (2023) Duocapsule and alternative shell material to gelatin: advancement in capsule formulation. World J Biol Pharm Health Sci 13:221–227

Navarro S, Cossalter G, Chiavaroli C, Kanda A, Fleury S, Lazzari A et al (2011) Oral administration of bacterial extracts prevents asthma via recruitment of regulatory T cells to the airways. Mucosal Immunol 4(1):53–65

Nie J, Liao W, Zhang Z, Zhang M, Wen Y, Capanoglu E et al (2022) A 3D co-culture intestinal organoid system for exploring glucose metabolism. Curr Res Food Sci 6:100402

Orkkatteri Krishnan A, Mudgal LN, Soni V, Prakash T (2025) ProbML: a machine learning-based genome classifier for identifying probiotic organisms. Mol Nutr Food Res 69(17):e70025

O'Toole PW, Marchesi JR, Hill C (2017) Next-generation probiotics: the spectrum from probiotics to live biotherapeutics. Nat Microbiol 2:1–6

Palade EM, Vulpoi GA, Vulpoi RA, Drug VL, Barboi OB, Ciocoiu M (2022) The biotics family: current knowledge and future perspectives in metabolic diseases. Life 12:1263

Park SE, Yoo SA, Seo SH, Lee KI, Na CS, Son HS (2016) GC-MS based metabolomics approach of kimchi for understanding *Lactobacillus plantarum* fermentation characteristics. LWT Food Sci Technol 68:313–321

Patil A, Singh N, Patwekar M, Patwekar F, Patil A, Gupta JK et al (2025) AI-driven insights into the microbiota: figuring out the mysterious world of the gut. Intell Pharm 3(1):46–52

Peixoto R, Rosado PM, Leite DC, Rosado AS, Bourne DG (2017) Beneficial microorganisms for corals (BMC): proposed mechanisms for coral health and resilience. Front Microbiol 8:341

Peleg AY, Hogan DA, Mylonakis E (2010) Medically important bacterial–fungal interactions. Nat Rev Microbiol 8:340–349

Poornachandra Rao K, Deepthi BV, Rakesh S, Ganesh T, Achar P, Sreenivasa MY (2019) Anti-aflatoxigenic potential of cell-free supernatant from *Lactobacillus plantarum* MYS44 against *Aspergillus parasiticus*. Probiotics Antimicrob Proteins 11:1238–1248

Qi Y, Huang L, Zeng Y, Li W, Zhou D, Xie J et al (2021) *Pediococcus pentosaceus*: screening and application as probiotics in food processing. Front Microbiol 12:762467

Rajoka RMS, Mehwish HM, Zhang H, Ashraf M, Fang H, Zeng X et al (2020) Antibacterial and antioxidant activity of exopolysaccharide mediated silver nanoparticles synthesized by *Lactobacillus brevis* isolated from Chinese koumiss. Colloids Surf B Biointerfaces 186:110734

Ranjan A, Arora J, Chauhan A, Basniwal RK, Kumari A, Rajput VD et al (2022) Advances in characterization of probiotics and challenges in industrial application. Biotechnol Genet Eng Rev 38:1–44

Riglar DT, Silver PA (2018) Engineering bacteria for diagnostic and therapeutic applications. Nat Rev Microbiol 16(4):214–225

Ritze Y, Bardos G, Claus A, Ehrmann V, Bergheim I, Schwiertz A, Bischoff SC (2014) *Lactobacillus rhamnosus* GG protects against non-alcoholic fatty liver disease in mice. PLoS One 9: e80169

Rodrigues A, Gonçalves A, Morais J, Araújo R, Falcão-Pires I (2023) Diet-induced microbiome's impact on heart failure: a double-edged sword. Nutrients 15(5):1223

Rohwer F, Seguritan V, Azam F, Knowlton N (2002) Diversity and distribution of coral-associated bacteria. Mar Ecol Prog Ser 243:1–10

Rokka S, Rantamäki P (2010) Protecting probiotic bacteria by microencapsulation: challenges for industrial applications. Eur Food Res Technol 231:1–12

Roselli M, Finamore A, Hynönen U, Palva A, Mengheri E (2016) Differential protection by cell wall components of *Lactobacillus amylovorus* DSM 16698(T) against membrane barrier alterations and NF-kB activation induced by enterotoxigenic F4(+) *Escherichia coli*. BMC Microbiol 16:226

Rosenberg E, Koren O, Reshef L, Efrony R, Zilber-Rosenberg I (2007) The role of microorganisms in coral health, disease and evolution. Nat Rev Microbiol 5(5):355–362

Scillato M, Spitale A, Mongelli G, Privitera GF, Mangano K, Cianci A et al (2021) Antimicrobial properties of *Lactobacillus* cell-free supernatants against multidrug-resistant urogenital pathogens. Microbiology 10(2):e1173

Singh TP, Natraj BH (2021) Next-generation probiotics: a promising approach towards designing personalized medicine. Crit Rev Microbiol 47:479–498

Sinha VR, Kumria R (2001) Polysaccharides in colon-specific drug delivery. Int J Pharm 224:193–198

Sipailiene A, Petraityte S (2018) Encapsulation of probiotics: selection of strains and influence of encapsulation technology and materials on viability. Probiotics Antimicrob Proteins 10(1):1–10

Sornsenee P, Chatatikun M, Mitsuwan W, Kongpol K, Kooltheat N, Sohbenalee S et al (2021) Lyophilized cell-free supernatants of *Lactobacillus* isolates exhibit antibiofilm and antioxidant activity and reduce nitric oxide in RAW 264.7 cells. PeerJ 9:e12586

Stanton, C., Desmond, C., Coakley, M., Collins, J. K., Fitzgerald, G., & Ross, R. P. (2003) Challenges facing development of probiotic-containing functional foods, CRC Press, 27–58

Steidler L (2003) Genetically engineered probiotics. Best Pract Res Clin Gastroenterol 17:861–876

Steidler L, Hans W, Schotte L, Neirynck S, Obermeier F, Falk W et al (2000) Treatment of murine colitis by *Lactococcus lactis* secreting interleukin-10. Science 289:1352–1355

Steidler L, Neirynck S, Huyghebaert N, Snoeck V, Vermeire A, Goddeeris B (2003) Biological containment of genetically modified *Lactococcus lactis* for intestinal delivery of human interleukin-10. Nat Biotechnol 21:785–789

Suarez N, Ferrara F, Rial A, Dee V, Chabalgoity JA (2020) Bacterial lysates as immunotherapies for respiratory infections: methods of preparation. Front Bioeng Biotechnol 8:545

Suresh V, Parida D, Minz AP, Sethi M, Sahoo BS, Senapati S (2021) Tissue distribution of ACE2 protein in Syrian golden hamster (*Mesocricetus auratus*) and implications for SARS-CoV-2 studies. Front Pharmacol 11:579330

Taube C (2022) Bacterial lysates to inhibit allergic airway disease: are they up to the challenge? J Allergy Clin Immunol 149(3):893–894

Thorakkattu P, Khanashyam AC, Shah K, Babu KS, Mundanat AS, Deliephan A et al (2022) Postbiotics: current trends in food and pharmaceutical industry. Foods 11(19):3094

Torii S, Torii A, Itoh K, Urisu A, Terada A, Fujisawa T et al (2011) Effects of oral administration of *Lactobacillus acidophilus* L-92 on symptoms and serum markers of atopic dermatitis in children. Int Arch Allergy Immunol 154:236–245

Torp AM, Bahl MI, Boisen A, Licht TR (2022) Optimizing oral delivery of next-generation probiotics. Trends Food Sci Technol 119:101–109

Vanitha PR, Somashekaraiah R, Divyashree S, Pan I, Sreenivasa MY (2023) Antifungal activity of probiotic strain *Lactiplantibacillus plantarum* MYSN7 against *Trichophyton tonsurans*. Front Microbiol 14:1192449

Vanitha PR, Divyashree S, Shruthi B, Adithi G, Deepthi BV, Sreenivasa MY (2024) *Limosilactobacillus fermentum* MYSY8 as a potential probiotic isolate from fermented rice beverage for control of *Microsporum canis*. Food Biotechnol 38(3):314–342

Vargason AM, Anselmo AC (2021) Live biotherapeutic products and probiotics for the skin. Adv NanoBiomed Res 1(12):2100118

Verma I, Banerjee B, Singh A, Kannan P, Saleena LM (2025) Exploring omics approaches in probiotics: contemporary developments and prospective pathways. J Microbiol Methods 232–234:107135

Wang J, Zhang W, Tang C, Xiao J, Xie B, Sun Z (2018) Synergistic effect of B-type oligomeric procyanidins from lotus seedpod with water-soluble *Poria cocos* polysaccharides against *Escherichia coli*. J Funct Foods 48:1341–1343

Wang LW, Janes ME, Kumbhojkar N, Kapate N, Clegg JR, Prakash S et al (2021) Cell therapies in the clinic. Bioeng Transl Med 6(2):e10214

Wang ZD, Zhang W, Liang TZ (2025) Advancements in oral delivery systems for probiotics based on polysaccharides. Polymer 17:144

Wilczewska AZ, Niemirowicz K, Markiewicz KH, Car H (2012) Nanoparticles as drug delivery systems. Pharmacol Rep 84:1020–1037

Xie T, Jin F, Jia X, Mao H, Xu Y, Zhang S (2022) High cellulose diet promotes intestinal motility through regulating intestinal immune homeostasis and serotonin biosynthesis. Biol Chem 403(3):279–292

Yang J, Li Y, Wen Z, Liu W, Meng L, Huang H (2021) *Oscillospira*: a candidate for next-generation probiotics. Gut Microbes 13:1987783

Yerlikaya F, Camlik G, Akkol EK, Degim Z, Degim IT, Sobarzo-Sanchez E (2021) Formation of quantum water in nanoparticulate systems. J Drug Deliv Sci Technol 63:102456

Yong SJ, Tong T, Chew J, Lim WL (2020) Antidepressive mechanisms of probiotics and their therapeutic potential. Front Neurosci 13:1361

Yu AQ, Li L (2016) The potential role of probiotics in cancer prevention and treatment. Nutr Cancer 68:535–544

Zaeim D, Sarabi-Jamab M, Ghorani B, Kadkhodaee R (2019) Double layer co-encapsulation of probiotics and prebiotics by electro-hydrodynamic atomization. J Food Sci Technol 56:102–109

Zhang S, Uludag H (2009) Nanoparticulate systems for growth factor delivery. Pharm Res 26: 1561–1571

Zhang H, Duan Y, Cai F, Cao D, Wang L, Qiao Z et al (2022) Next-generation probiotics: microflora intervention to human diseases. Biomed Res Int 2022:5633403

Zhang Y, Zhou CM, Tian B, Xu JL, Wang XY, Dai HX et al (2025) 3D-printed spiral tube-like cellulose scaffold for oral delivery of probiotics. Sci Adv 10:eadp3654

Zhao M, Ren Z, Zhao A, Tang Y, Kuang J, Li M et al (2024) Gut bacteria-driven homovanillic acid alleviates depression by modulating synaptic integrity. Cell Metab 36:1000–1012

Zhou HB, Chen J, Li S, Zhang J, Zhu CE, Ran H et al (2017) Preparation of acid-resistant microcapsules with shell-matrix structure to enhance stability of *Streptococcus thermophilus* IFFI 6038. J Food Sci 82:1978–1984

Zuo F, Zeng Z, Hammarstrom L, Marcotte H (2019) Inducible plasmid self-destruction assisted genome engineering in *Lactobacilli* and *Bifidobacteria*. ACS Synth Biol 8(8):1723–1729

MIX
Papier aus verantwortungsvollen Quellen
Paper from responsible sources
FSC® C105338

If you have any concerns about our products,
you can contact us on
ProductSafety@springernature.com

In case Publisher is established outside the EU,
the EU authorized representative is:
Springer Nature Customer Service Center GmbH
Europaplatz 3, 69115 Heidelberg, Germany

Printed by Libri Plureos GmbH
in Hamburg, Germany